Milkha S. Aulakh, PhD
Cynthia A. Grant, PhD
Editors

Integrated Nutrient Management for Sustainable Crop Production

Pre-publication
REVIEWS,
COMMENTARIES,
EVALUATIONS . . .

"**P**rovides a comprehensive description of efficient nutrient management practices used in diverse cropping systems throughout the world. For each major cropping region, leading international scientists thoroughly discuss nutrient use and management for the major crops and cropping systems, while concisely identifying future research needs and education priorities. Students and professionals interested in the global view of nutrient management technologies essential to world food security and protecting our natural resources will find this an invaluable resource."

Dr. John Havlin, Professor
Department of Soil Science,
North Carolina State University, Raleigh

"**T**he authors have well illustrated the critical importance of integrated nutrient management in meeting crop production and food security needs while maintaining environmental sustainability under a range of environments, agricultural systems, and societal and economic conditions. This will be a valuable desk reference and resource for students and professionals as a compendium to integrated nutrient management approaches that are uniquely applied over a range of geographic, social, and environmental conditions that influence the availability, effectiveness, and environmental consequences of fertilizers, plant and animal manures, soil organic resources, and biological fixed N for crop production."

John W. Doran
Professor Emeritus, Agronomy
& Horticulture, University of Nebraska;
Former President, Soil Science
Society of America; Co-Founder
of Renewing Earth & Its People

More pre-publication
REVIEWS, COMMENTARIES, EVALUATIONS . . .

"This book provides a comprehensive review of integrated nutrient management throughout the world. The regional perspectives allow readers to see the commonalities in nutrient management across regions as well as the uniqueness within regions because of factors such as climate, soils, and resources. I believe the regional data on agricultural production, fertilizer consumption, and nutrient balances in a single-source will be quite useful for a number of readers. As a researcher, I particularly liked the sections on future challenges for integrated nutrient management, research gaps, and future research needs. This will be a valuable reference book for years to come for all professionals interested in integrated nutrient management."

Alan Schlegel, PhD
Professor, Kansas State University

"This is a timely book. It provides a diverse fund of information on integrated nutrient use and should be on the shelf of anyone involved with international agriculture."

John Ryan, PhD, DSc
Soil Fertility Specialist,
International Center for Agricultural Research in the Dry Areas (ICARDA),
Aleppo, Syria

Integrated Nutrient Management for Sustainable Crop Production

Integrated Nutrient Management for Sustainable Crop Production

Milkha S. Aulakh, PhD
Cynthia A. Grant, PhD
Editors

CRC Press
Taylor & Francis Group
Boca Raton London New York

CRC Press is an imprint of the
Taylor & Francis Group, an **informa** business

CRC Press
Taylor & Francis Group
6000 Broken Sound Parkway NW, Suite 300
Boca Raton, FL 33487-2742

First issued in paperback 2019

ISBN-13: 978-1-56022-304-7 (hbk)
ISBN-13: 978-0-367-38773-0 (pbk)

Library of Congress Cataloging-in-Publication Data

Integrated nutrient management for sustainable crop production/Milkha S. Aulakh, Cynthia A. Grant, editors.
  p. cm.
  ISBN-13: 978-1-56022-304-7 (hard : alk. paper)
  1. Fertilizers. 2. Crops—Nutrition. 3. Cropping systems. 4. Sustainable agriculture. I. Aulakh, Milkha S. II. Grant, Cynthia A.

S633.I58 2007
631.8'1-dc22   2007000538

CONTENTS

Chapter 6. Integrated Nutrient Management:
Experience and Concepts from New Zealand   **253**
> *Antony H. C. Roberts*
> *Tony J. van der Weerden*
> *Douglas C. Edmeades*

Chapter 7. Integrated Nutrient Management:
Experience from South Asia   **285**
> *Milkha S. Aulakh*
> *Guriqbal Singh*

ABOUT THE EDITORS

Milkha S. Aulakh, PhD, is Additional Director of Research and Professor at Punjab Agricultural University. The recipient of numerous international awards and author of countless articles, book chapters, reviews, and bulletins, Dr. Aulakh's research contributions are globally acclaimed. He is the Fellow of the National Academy of Agricultural Sciences and the Indian Society of Soil Science as well as an editor of Biology & Fertility of Soils.

Cynthia A. Grant, PhD, is a Senior Research Scientist in Soil Fertility Management at Agriculture and Agri-Food Canada Research Centre in Manitoba. She has published over 85 research papers, ten reviews and book chapters, and over 500 proceeding, reports, and technology transfer articles. A popular speaker and source of information for farmers and industry agronomists, Dr. Grant is a former associate editor for the *Canadian Journal of Plant Science* and the *Canadian Journal of Soil Science* and a fellow of the Canadian Society of Agronomy.

CONTRIBUTORS

Mark M. Alley, W. G. Wysor Professor of Agriculture, Department of Crop and Soil Environmental Sciences, Virginia Tech, Blacksburg, Virginia 24061 USA; e-mail: malley@vt.edu

Denis A. Angers, Research Scientist, Agriculture and Agri-Food Canada, 2560 boulevard Hochelaga, Sainte Foy, Quebec, Canada G1V 2J3; e-mail: angersd@agr.gc.ca

Andre Bationo, Soil Scientist and African Network (AfNet) Coordinator, Tropical Soil Biology and Fertility (TSBF) Institute of the International Centre for Tropical Agriculture (CIAT) C/o World Agroforestry Centre (ICRAF), P.O. Box 30677, Nairobi, Kenya; e-mail: a.bationo@cgiar.org

Mathias Becker, Professor, Department of Plant Nutrition, Institute of Crop Science and Resource Conservation, University of Bonn, Karlrobert-Kreiten Str.13 D-53115, Bonn, Germany; e-mail: mathias.becker@uni-bonn.de

David J. Bonfil, Research Scientist, Field Crops and Natural Resources, The Institute of Plant Sciences, Agricultural Research Organization, Gilat Research Center, M.P. Negev 85280, Israel; e-mail: bonfil@volcani. agri. gov.il

Eduardo Casanova, Professor, Instituto de Edafologia, Facultad de Agronomia, Universidad Central de Venezuela, Maracay AP 4579, estado Aragua, Venezuela; e-mail: casanovaen@cantv.net

Martín Díaz-Zorita, Professor, CONICET, Cátedra de Cerealicultura, Facultad de Agronomía, Universidad de Buenos Aires, 1417 Buenos Aires, Argentina; e-mail: mdzorita@agro.uba.ar

Craig F. Drury, Research Scientist, Agriculture and Agri-Food Canada, 2585 Highway County Rd 20, Harrow, Ontario N0R 1G0, Canada; e-mail: druryc@agr.gc

Douglas C. Edmeades, Managing Director, Acknowledge, P.O. Box 9147, Hamilton, New Zealand; e-mail: doug.ed@xtra.co.nz

Roberta Farina, Researcher, Research Institute for Plant Nutrition, Agricultural Research Council, Via della Navicella, 2-4, 00184 Rome, Italy; e-mail: roberta.farina@entecra.it

Rosa Francaviglia, Senior Researcher, C.R.A.—Instituto Sperimentale per la Nutrizione delle Piante, A.R.C.—Research Institute for Plant Nutrition, Agricultural Research Council, Via della Navicella, 2-4, 00184 Rome, Italy; e-mail: rosa.francaviglia@entecra.it

Patrick Heffer, Executive Secretary, Agriculture Committee, International Fertilizer Industry Association (IFA), 28 rue Marbeuf, 75008 Paris, France; e-mail: pheffer@fertilizer.org

A. E. Johnny Johnston, Lawes Trust Senior Fellow, Department of Soil Science, Rothamsted Research, Harpenden, Herts AL5 2JQ, United Kingdom; e-mail: johnny.johnston@bbsrc.ac.uk

Adrian M. Johnston, Vice President & Coordinator, Asia Group, International Plant Nutrition Institute (IPNI), 102-411 Downey Rd., Saskatoon, Saskatchewan S7N 4L8, Canada; e-mail: ajohnston@ipni.net

Uzi Kafkafi, Professor, Emeritus, The Robert H. Smith Inst. of Plant Sciences and Genetics in Agriculture, The Hebrew University of Jerusalem, Faculty of Agriculture, P.O. Box 12, Rehovot 76100, Israel; e-mail: kafkafi@agri.huji.ac.il

Job Kihara, Assistant Scientific Officer, Tropical Soil Biology and Fertility (TSBF) Institute of the International Centre for Tropical Agriculture (CIAT) C/o World Agroforestry Centre (ICRAF), P.O. Box 30677, Nairobi, Kenya; e-mail: j.kihara@cgiar.org

Joseph Kimetu, Post Graduate Fellow, Crop and Soil Science Department, 1022 Bradfield Hall, Cornell University, Ithaca, New York 14853; e-mail: jmk229@cornell.edu

Zhihong Li, Professor, Soil and Fertilizer Institute, Chinese Academy of Agricultural Sciences (CAAS).12 zhongguancun Nandajie Beijing 100081 P. R. China; e-mail: zhhli@caas.ac.cn

Bao Lin, Professor, Soil and Fertilizer Institute, Chinese Academy of Agricultural Sciences (CAAS).12 Zhongguancun Nandajie Beijing 100081 P. R. China; e-mail: blin@caas.ac.cn

Bruce A. Linquist, Associate Project Scientist, Department of Plant Sciences, University of California, Davis, California 95616; e-mail: balinquist@ucdavis.edu

Alfredo Scheid Lopes, Emeritus Professor, Departamento de Ciência do Solo, Universidade Federal de Lavras, 37200-000 Lavras, MG, Brasil; e-mail: ascheidl@ufla.br

Luc M. Maene, Director General, International Fertilizer Association (IFA), 28, rue Marbeuf 75008 Paris, France; e-mail: lmaene@fertilizer.org

Sukhdev S. Malhi, Research Scientist, Agriculture and Agri-Food Canada, Research Farm, P.O. Box 1240, Melfort, Saskatchewan S0E 1A0, Canada; e-mail: malhis@agr.gc.ca

Gregory L. Mullins, Professor and Head, Department of Plant and Environmental Sciences, New Mexico State University, Las Cruces, New Mexico 88003; e-mail: gmullins@nmsu.edu

Dan C. Olk, Soil Scientist, USDA-ARS, National Soil Tilth Laboratory, 2150 Pammel Drive, Ames, Iowa 50011; e-mail: dan.olk@ars.usda.gov

Sushil Pandey, Agricultural Economist and Deputy Head, Social Sciences Division, International Rice Research Institute, DAPO Box 7777, Metro Manila, Philippines; e-mail: sushil.pandey@cgiar.org

Antony H. C. Roberts, Chief Scientific Officer, Ravensdown Fertiliser Co-Operative Ltd., P.O. Box 608, Pukekohe, New Zealand; e-mail: ants.roberts@ravensdown.co.nz

Kanwar L. Sahrawat, Visiting Scientist (Soil Chemist), Global Theme On Agroecosystems International Crop Research Institute for the Semi Arid Tropics (ICRISAT), Patancheru 502 324, Andhra Pradesh, India; e-mail: klsahrawat@yahoo.com

Jeff J. Schoenau, Professor, Department of Soil Science, University of Saskatchewan, 51 Campus Drive, Saskatoon, Saskatchewan S7N 5A8, Canada; e-mail: schoenau@skyway.usask.ca

Paolo Sequi, Professor of Soil Chemistry, Research Institute for Plant Nutrition, Agricultural Research Council, Via della Navicella 2-4, 00184, Rome, Italy; e-mail: paolo.sequi@entecra.it

Guriqbal Singh, Agronomist (Pulses Section), Department of Plant Breeding, Genetics and Biotechnology, Punjab Agricultural University, Ludhiana 141004, Punjab, India; e-mail: singhguriqbal@rediffmail.com

Kristen E. Sukalac, Head of Information and Communications, International Fertilizer Association (IFA), 28, rue Marbeuf 75008 Paris, France; e-mail: ksukalac@fertilizer.org

Bernardo van Raij, Scintific Researcher, Instituto Agronômico, Caixa Postal, 28 13001-970 Campinas, SP, Brazil; e-mail: bvanraij@terra.com.br

Bernard Vanlauwe, Soil Scientist, Tropical Soil Biology and Fertility (TSBF) Institute of the International Centre for Tropical Agriculture (CIAT) C/o World Agroforestry Centre (ICRAF), P.O. Box 30677, Nairobi, Kenya; e-mail: b.vanlauwe@cgiar.org

Boaz S. Waswa, Assistant Scientific Officer, Tropical Soil Biology and Fertility (TSBF) Institute of the International Centre for Tropical Agriculture (CIAT), c/o World Agroforestry Centre (ICRAF), P.O. Box 30677, Nairobi, Kenya; e-mail: b.waswa@cgiar.org

Tony J. van der Weerden, Technical Manager, Ravensdown Fertiliser Co-Operative Ltd., P.O. Box 1049, Christchurch, New Zealand; e-mail: tvw@ravensdown.co.nz

Dwayne G. Westfall, Professor, Department of Soil and Crop Sciences, Colorado State University, Fort Collins, Colorado 80523; e-mail: Dwayne.Westfall@ColoState.edu

Christian Witt, Director, Southeast Asia Program, International Plant Nutrition Institute (IPNI), a joint mission with the International Potash Institute (IPI), 126 Watten Estate Road, Singapore 287599; e-mail: cwitt@ipni.net

Ronggui Wu, Professor, Soil and Fertilizer Institute, Chinese Academy of Agricultural Science, 12 Zhongguancun Dandajie, Beijing 100081, P.R. China; e-mail: Ronggui.wu@gov.mb.ca

Jianchang Xie, Nanjing Institute of Soil Sciences, Chinese Academy of Sciences, P.O. Box 821, 71 East Beijing Road, Nanjing 210008, China; e-mail: qzfan@issas.ac.cn

Guangxi Xing, Nanjing Institute of Soil Sciences, Chinese Academy of Sciences, P.O. Box 821, 71 East Beijing Road, Nanjing 210008; China; e-mail: xinggx@issas.ac.cn

Foreword

Integrated Nutrient Management for Sustainable Crop Production is focused on the effective management of all nutrient sources to optimize crop production and environmental sustainability around the world. It is a well-balanced, region-specific review of the current and future role of integrated nutrient management in meeting the demands society places on agriculture. It will be a valuable reference to serious students of integrated nutrient management who want to consider concepts and practices employed in other parts of the world and to those seeking an overview of nutrient management issues in a particular country or set of countries.

The contributors did an excellent job of showing the need to place integrated nutrient management in context. They not only illustrate the importance of understanding the biophysical/chemical context of managing the nutrition of crops, but also the need to recognize the socioeconomic factors that may determine feasibility and override many other considerations. The context diversity in this global coverage results in a multitude of challenges and associated options for integrated nutrient management. Across the region-specific chapters, the implications of appropriate integrated nutrient management range from the economic and environmental to matters of life and death.

In this book, you clearly see an element of integrated nutrient management that is simple common sense and requires only limited understanding of the science behind the plant-soil-water-air system. It is logical to utilize the nutrient sources already available on the farm before purchasing external nutrients. However, there is also an element of integrated nutrient management that challenges the most knowledgeable scientist, the most gifted field agronomist, and the most dedicated farmer. The annual nutrient contribution of a previous legume crop, an intercropped species, applied livestock manure, composts, green manure, or crop residues to a specific growing crop, depend not only on source properties, but also on site characteristics and growing-season weather. Assessing economic impacts sometimes involves positive effects on crop growth beyond nutritional considerations, as well as collection, transportation, and application costs. One of the unique contributions of this book is that it gives the reader a new appreciation of the importance of system- or farm-level factors beyond the microeconomics of a specific enterprise.

The importance of nutrient management and the value of this book will increase with time, as shown in the following.

- Population growth, along with the anticipated continued improvement in the economic status of many regions, will increase the demand for food and fiber. Bioenergy is further increasing the demand for agricultural products and appears to be heading for considerable expansion.
- As population increases, concern over global impacts on water, air, and climate are also likely to increase. Keeping vulnerable lands out of production, by optimizing productivity on those lands in production, while minimizing nutrient losses from farm fields will be critical.
- As global nutrient ore bodies are gradually depleted and energy costs climb, the costs of commercial fertilizers common today will increase. Along with that increase will come a concomitant rise in the economic value of organic nutrient sources and nutrient-availability enhancing practices.

The demand for integrated nutrient management decision support will grow as production demand increases, environmental concerns intensify, and nutrient costs rise. Getting nutrient management decisions right will be of utmost importance: right rate, right time, right place, and right source.

Paul E. Fixen, PhD
Senior Vice President, Americas Group Coordinator
and Director of Research
International Plant Nutrition Institute

Preface

One of the great success stories for agriculture over the past century has been the increase in crop production to provide food for the expanding global population. Effective nutrient management has played a major role in this accomplishment. However, long-term security of the global food supply requires a balance between increasing crop production and environmental sustainability. Nutrient surpluses and nutrient excesses can threaten both crop productivity and environmental sustainability.

Intensification in commercial agriculture frequently leads to specialization of crop and livestock production. A concern with intensive livestock operations is excess application of manure nutrients on limited land areas. In the production of high-value crops, the relatively low cost of chemical fertilizers in proportion to the value of the agricultural products can encourage the application of excess nutrients to ensure that maximum yield is attained. Production and application of excess nutrients bears an environmental cost, including greenhouse gas and ammonia emission into the atmosphere, nitrate and phosphate movement into water, and land degradation from heavy metals and acidification.

At the other extreme, nutrient mining has occurred in many production systems due to a lack of affordable nutrient sources. Nutrient depletion becomes a self-accelerating process, as the restricted nutrient supply decreases crop production and fewer organic residues are returned to the system. Declining soil organic matter further reduces productivity and increases the risk of erosion, creating a cycling of land degradation. Rising population pressure intensifies the impact of nutrient depletion on the sustainability of crop production.

Integrated nutrient management (INM) is an essential step to address the twin concerns of nutrient excess and nutrient depletion. The basic principle of INM is to optimize all available nutrient sources for economic and environmental sustainability. Nutrient resources available to crops include: the native soil reserves augmented by nutrients added through atmospheric deposition, released by soil biological activity and recycled from crop residues, organic manures, and other urban and industrial sources; and nutrients applied in chemical fertilizers. Effective use of all nutrient sources combined with effective agronomic management is needed to ensure both production efficiency and agroecosystem health.

Challenges to INM differ with different environments, agricultural systems, and societal and economic pressures. This book has attempted to take a global view of the challenges in nutrient management faced in various regions of the world. Of the twelve chapters, ten focus on INM in different regions including Africa, North and South America, Europe, South and Southeast Asia, China, the Middle East, and New Zealand. The differences in issues from region to region provide an opportunity to learn from the experiences of other regions and to anticipate emerging issues for future planning. The problems of nutrient depletion and excess are explored and the challenges in effective adoption of INM to face production and environmental challenges in agricultural production systems around the world are discussed. The authors of the chapters have a wealth of knowledge in addressing nutrient management impacts on production and the environment. Their experiences and their thoughtful evaluation of the challenges and potential of INM provide an interesting and informative look at the issue. INM is one of the great long-term issues in global agriculture. We hope that this book helps to outline the range of options available, and identify future directions to improve nutrient management in agricultural systems around the world.

Chapter 1

Global Food Production and Plant Nutrient Demand: Present Status and Future Prospects

Luc M. Maene
Kristen E. Sukalac
Patrick Heffer

INTRODUCTION

During the past forty years, world population has doubled from 3.08 billion in 1961 to 6.15 billion in 2001. During the same period, the combined use of high-yielding varieties, fertilizers, and improved crop management practices has allowed food production to keep pace with, and even exceed, the fast-growing food, feed, fiber, and biofuel demand. Nevertheless, many farmers still do not have access to modern agricultural technology and largely depend on the state of development of their domestic agriculture. In addition, not all consumers have access to adequate supplies of nutritious and safe food.

On the occasion of the World Food Summit (1996), the Millennium Summit (2000), the World Food Summit *five years later* (2002), and the World Summit on Sustainable Development (2002), policymakers adopted and reaffirmed the ambitious goal of halving the number of the hungry by 2015. The World Summit on Sustainable Development in Johannesburg explicitly highlighted the importance of increasing food production in a sustainable manner, especially in sub-Saharan Africa where declining soil fertility is a major problem. Governments there acknowledged that agriculture is the key to poverty alleviation. Producing more nutritious and safe food and facilitating access to agricultural technology were also emphasized.

Unfortunately, current trends regarding the evolution of people suffering from undernourishment indicate that the goals of the 1996 World Food Summit will not be reached, notwithstanding the possibility of significant progressive reduction of hunger.

Within this global context, this chapter provides an overview of the progress that has been made in meeting global demands for food, feed, and fiber over the past forty years. Manufactured fertilizers have played a key role in the achievements so far, which have prevented millions of people from starving while protecting uncultivated land from the plow. This chapter examines that progress and looks to the future contribution that fertilizers will make to ending hunger and malnutrition. Unwanted impacts and constraints are included in the overview.

In order to reflect a diverse range of situations, three case studies have been chosen: France, India, and sub-Saharan Africa (SSA). These can be considered as broadly representative of "developed," "developing," and "stagnating" agricultures, respectively. South Africa is generally not included in the data quoted for SSA in this paper, as the Food and Agriculture Organization (FAO) of the United Nations considers South Africa to be "developed" from an agricultural perspective.

This chapter opens with a review of global production and consumption trends for food, feed and fiber during the past four decades. It then provides information on the evolution of plant nutrient demand over that same period and discusses the medium to long-term prospects for fertilizer demand.

TRENDS IN GLOBAL FOOD, FEED, AND FIBER PRODUCTION AND CONSUMPTION

Evolution over the Past Four Decades

World Population

World population has doubled over the past forty years (see Table 1.1). These aggregate figures hide a dramatic shift: urbanization. Between 1961 and 2001, urban population rose by 179 percent to 2.9 billion people, while the rural population grew only 59 percent to 3.2 billion. The number of people active in agricultural production grew by an even smaller margin, reaching only 2.6 billion in 2001, a 44 percent growth over the past four decades. The urban population represented 48 percent of world population in 2001 against 34 percent in 1961. The percentage of the population involved in agriculture declined from 58 to 42 percent (FAO, 2004b).

TABLE 1.1. Evolution of world rural, agricultural, and urban population (billion people)

	1961	1971	1981	1991	2001	2001 vs. 1961 (%)
Rural	2.03	2.38	2.71	3.00	3.22	+59
Agricultural	1.79	2.02	2.24	2.46	2.58	+44
Urban	1.05	1.39	1.80	2.35	2.93	+179
Total	3.08	3.77	4.51	5.35	6.15	+100

Source: Adapted from FAO (2004b), FAO (1992).

Very soon, more than half of the world population will live in cities, and this trend is not expected to reverse itself (FAO, 2004b). This will have a huge impact on agriculture, which will continue its progressive shift from subsistence to commercial objectives, in part driven by a declining workforce dedicated to meeting growing demand. The expansion of cities outward will also lead to a reduction in the arable land area, especially as dense population centers tend to be located in the most fertile areas. Urban expansion will also lead to even more competition than there already is for water supplies. At the same time, urban populations, with higher average incomes, will continue to diversify their diets with more meat, fruits, and vegetables.

World Cereal Production and Yields

Over the past four decades, world cereal output has increased by 140 percent, from 877 million metric tonnes (Mt) in 1961 to 2,107 Mt in 2001 (Table 1.2). Most regions have made significant gains in cereal output during that period. However, a given country's agricultural development status determines whether these increases have come primarily from technological innovation leading to higher yields or from a jump in the cultivated area.

At the global level, most of the increase in cereal production comes from yield gains (Table 1.3). From 1961 to 2001, average global cereal yields rose ten times more than the total cultivated area (Table 1.4). In France, total cereal production tripled over the past four decades, from 20.8 to 60.3 Mt. This increase has been obtained solely through higher yields, as the cultivated area in France actually declined over that period. The average cereal yield grew from 2.3 metric tonnes (t) per hectare (t ha^{-1}) in 1961 to 6.7 t ha^{-1} in 2001.

TABLE 1.2. Evolution of cereal production (Mt)

	1961	1971	1981	1991	2001	2001 vs. 1961 (%)
World	877.0	1,299.9	1,632.8	1,889.4	2,106.9	+140
France	20.8	37.0	45.9	60.3	60.3	+190
India	87.4	113.2	147.6	193.1	243.4	+178
SSAᵃ	31.4	39.1	45.9	64.1	77.4	+146

Source: Adapted from FAO (2004b).

[a]Sub-Saharan Africa excluding South Africa.

TABLE 1.3. Evolution of cereal yields (kg ha^{-1})

	1961	1971	1981	1991	2001	2001 vs. 1961 (%)
World	1,353	1,892	2,246	2,684	3,118	+130
France	2,276	3,876	4,729	6,538	6,740	+196
India	947	1,136	1,399	1,926	2,431	+157
SSAᵃ	760	807	1,018	972	992	+31

Source: Adapted from FAO (2004b).

[a]Sub-Saharan Africa excluding South Africa.

TABLE 1.4. Evolution of cultivated (arable + permanent) land (Mha)

	1961	1971	1981	1991	2001	2001 vs. 1961 (%)
World	1,356.7	1,405.1	1,442.4	1,504.1	1,532.1	+13
France	21.4	18.7	18.9	19.2	19.6	−8
India	161.0	164.4	168.4	169.3	169.9	+5.5
SSA*	119.5	130.0	140.7	149.8	164.5	+38

Source: Adapted from FAO (2004b).

[a]Sub-Saharan Africa excluding South Africa.

In India, the Green Revolution boosted agriculture through the successful introduction of high-yielding varieties (semi-dwarf and dwarf rice and wheat varieties), the use of crop protection products and manufactured fertilizers, irrigation, and mechanization. As a result, cereal production increased 2.8 times, from 87 to 243 Mt, between 1961 and 2001. Cereal yields rose almost 2.6 times, from 0.9 to 2.4 t ha^{-1}. Current yield levels in India remain, however, far below their potential, estimated at 4.7 t ha^{-1} for rice and 4.8 t ha^{-1} for wheat.

Unfortunately, SSA was bypassed by the Green Revolution and the regional situation is characterized by persistently small harvests. Average cereal yields in the region were only 1.0 t ha^{-1} in 2001, compared with 3.1 t ha^{-1} at the world level. It is worrying to note that cereal yields in SSA are stagnating, with a yield growth as low as 31 percent over the past forty years. In some places, production is falling, a trend that is worsened by the devastating effects of AIDS on rural populations.

Another notable trend is the current slowdown in cereal yield growth across the globe. While the average annual world growth rate was 3.4 percent in the 1960s, it declined regularly over the next three decades to only 1.2 percent in the 1990s (Table 1.5).

As Figure 1.1 shows, world cereal stocks have decreased over the past four consecutive years. Dwindling stocks combined with the slowdown in yield growth rates raise concerns in some policy circles regarding the ability to maintain or improve world food security. This situation is driven by a number of factors. Soil fertility degradation in some parts of the world is an important agronomic reason: 75 percent of the agricultural land in Central America, 20 percent in Africa, and 11 percent in Asia is estimated to be seriously degraded (Scherr, 1999). At the same time, there have been policy decisions aimed at reducing cereal stocks. Market forces that respond to

TABLE 1.5. Average annual cereal yield growth rate (% year^{-1})

	1961-1971	1971-1981	1981-1991	1991-2001
World	3.4	2.3	2.1	1.2
France	5.8	3.5	3.0	1.0
India	2.1	2.2	3.4	2.4
SSA[a]	0.7	2.6	0.6	0.5

Source: Calculated using FAO (2004b).

[a]Sub-Saharan Africa excluding South Africa.

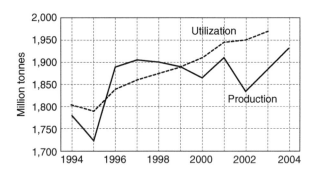

FIGURE 1.1. World cereal production and utilization. *Source:* Adapted from FAO (2004a).

depressed prices from abundant stocks also discourage production. Furthermore, environmentally oriented policies in the industrialized countries, which give priority to environment preservation over increased productivity, are of growing importance. Although there is a higher demand for feed and bioenergy, this does not fully compensate for the slowdown in cereal demand expansion for food purposes that results from a declining population growth rate. As demand increases at a more moderate pace, the argument for extensive stocks has weakened.

Land Use: Cultivation and Irrigation

Globally, the total cultivated land area increased only marginally (+13 percent) over the past four decades (Table 1.4). In France, the cultivated area dropped from 21.4 million hectares (Mha) to 19.6 Mha (−8 percent), essentially to the benefit of forests. In India, arable and permanent cropland covered 9 Mha more in 2001 than in 1961, a marginal increase of 5.5 percent. This contrasts sharply with that country's 160 percent growth in cereal production over the same period (Table 1.3).

Figure 1.2 shows the cultivated area needed to produce today's total volume of cereals if yields had remained the same as in 1961. Much of the land spared by yield increases is marginal, thus preserving fragile soils from degradation and safeguarding habitats that harbor biological diversity. Without the spectacular yield gains that have been achieved, today's agricultural output would require nearly three times as much land as half a century ago. Given competing demands for land and the relatively small proportion of suitable land not already in production, this additional area is simply not

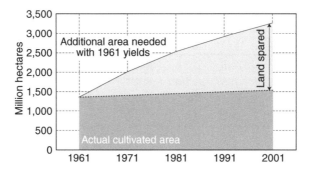

FIGURE 1.2. Evolution of cultivated land at the global level. *Source:* Calculated using FAO (2004b).

available, leaving no choice but to continue to increase yields in the most productive areas.

In SSA, contrary to France or India, a large portion of the production gains in recent decades came from an increase in the cultivated land area, from 120 Mha in 1961 to 165 Mha in 2001. This represents a 38 percent growth of the cultivated area, compared with the low 31 percent gain in cereal yield over the same period. Future expansion of agricultural production in SSA in this way would be to the further detriment of wilderness areas with fragile soils. The environmental cost (e.g., desertification) is excessive and unsustainable. Therefore, agricultural production in SSA must come from yield increases, which would be fairly simple to achieve if access to the necessary inputs and output markets could be ensured.

Between 1961 and 2001, the development of irrigation contributed largely to the agricultural productivity growth at the global level. The total area of irrigated land doubled from 139 to 276 Mha (Table 1.6). Irrigation was instrumental in bringing the Green Revolution to Asia. In 2001, 36 percent of the cultivated land in India was irrigated, against only 15 percent in 1961 (Table 1.7). In 2001, 18 percent of cultivated land at the global level was irrigated, but only 3 percent in SSA.

Further expansion of irrigation, essentially through small-scale, low-investment projects, will play an increasingly important role in world food production in future, but water use efficiency will also have to improve (e.g., through expansion of drip irrigation) since agriculture is increasingly in competition with other sectors for water. Industrial activities and human needs for sanitation and drinking water are just two of these sectors.

TABLE 1.6. Evolution of irrigated land (Mha)

	1961	1971	1981	1991	2001	2001 vs. 1961 (%)
World	139.1	171.8	213.6	248.8	275.9	+98
France	0.5	0.8	1.4	2.1	2.6	+420
India	24.7	31.1	38.8	47.4	57.2	+132
SSA[a]	2.7	3.2	4.1	4.9	5.2	+93

Source: Adapted from FAO (2004b).

[a]Sub-Saharan Africa excluding South Africa.

TABLE 1.7. Share of irrigated versus cultivated land (%)

	1961	1971	1981	1991	2001
World	10.3	12.2	14.8	16.5	18.0
France	2.3	4.5	7.4	10.9	13.3
India	15.3	18.9	23.0	28.0	35.5
SSA[a]	2.3	2.5	2.9	3.3	3.2

Source: Adapted from FAO (2004b).

[a]Sub-Saharan Africa excluding South Africa.

Climate change could reduce rainfall in some regions, which would drive better water use efficiency in agriculture in those parts of the world.

As soil organic matter plays a key role in moisture retention, the need for increased water use efficiency provides a strong argument in favor of integrated plant nutrient management.

What Is to Come for World Agriculture?

Priorities for agricultural and food policies have evolved significantly in recent years, with clear differentiation between developed and developing countries. Quality, safety, and environmental aspects of food production and consumption increasingly top the policy agenda in developed countries, while the main focus in most developing countries remains on securing adequate quantities of food, although quality aspects have gained in importance.

Priorities in the Developing Countries

According to FAO's global estimates for 1999-2001, 842 million people were undernourished (consuming too few calories per day). Ten million of these were in industrialized countries, 34 million in countries in transition, and the balance (798 million) in developing countries (Table 1.8). The undernourished represent 14 percent of the world population and 17 percent of the people living in the developing nations. The most affected regions are SSA and South Asia with 33 and 22 percent of their respective populations lacking food. Hunger still dogs 145 million people in East Asia.

Unfortunately, this situation has improved only slightly since the 1996 World Food Summit, when the heads of state and government agreed to halve the number of undernourished people by 2015. Unless there is a dramatic shift, this target is unlikely to be met within the agreed timeframe. Although the focus remains on food availability in many developing countries, there is a growing awareness among the scientific community and the policymakers in these countries regarding the urgent need to improve overall nutrition. Particular attention is paid to proteins, essential amino acids, vitamins, and micronutrients.

Incredibly, there are more than 3 billion people, that is, half the world population, suffering from one or more micronutrient deficiencies globally

TABLE 1.8. Prevalence of undernourishment in developing countries in 1999-2001 (million people)

Region	Total population (million people)	Number of undernourished people (million people)	Proportion of undernourished people (%)
Developing countries	4,712	798	17
Asia	3,205	505	16
East Asia	1,353	145	11
Southeast Asia	517	66	13
South Asia	1,330	293	22
Latin America/Caribbean	512	53	10
Near East/North Africa	250	35	14
Sub-Saharan Africa[a]	603	198	33

Source: Adapted from FAO (2003).

[a]Excluding South Africa.

(Welch, 2003). According to the United Nations, more than two billion people today suffer from severe iron deficiency, making anemia the most widespread disease in the world. One third of the world population is also at risk from inadequate zinc intake (Alloway, 2004). Although considerable progress has been made in the past decade, deficiencies in vitamin A, iodine, and many other essential elements are also widespread and have dramatic health consequences. Some 100 to 250 million pre-school age children are affected by severe vitamin A deficiency, and 740 million people suffer from goiter, which indicates a lack of iodine (Fritschel, 2001). Such deficiencies can also severely affect crop yields and livestock health. Today, this "hidden hunger" affects some five billion people worldwide, about four-fifths of the total population. Nutrition security can be achieved only through a blend of policy and technical tools, which are aimed at promoting, among others, food diversification and the production of more nutritious food.

In some ways, nutritional balance has been a victim of the success in increasing food availability. During the Green Revolution, priority was put on cereals, to the detriment of other crops, in particular of pulses, an important source of proteins and other essential nutrients (Welch, 2003). There has been a recent trend to rediversify food consumption, essentially as a result of urbanization and of the fast-rising income of certain population segments. For instance, in Asia, the proportion of rice in the average daily diet is declining, while there is a trend of increasing consumption of meat, vegetable oil, fruits, and vegetables (FAO, 2004b).

On a global scale, this movement has catalyzed greater meat production, which more than tripled over the past four decades (Table 1.9). In comparison, milk production has increased less. Taking industrialized and developing countries together, cereal and milk supply per capita increased only slightly between 1961 and 2001, by 14 and 4 percent (Figure 1.3). Over the same period, the average individual intake of fruits, vegetables, meat, and eggs grew by 60 to 80 percent, and the consumption of vegetable oils more than doubled. In contrast, the consumption of pulses per capita dropped by 40 percent.

TABLE 1.9. Evolution of world meat and milk production (Mt)

	1961	1971	1981	1991	2001	2001 vs. 1961 (%)
Meat	71.3	104.8	139.2	185.0	239.3	+236
Milk	344.2	394.8	469.6	533.3	585.8	+70

Source: Adapted from FAO (2004b).

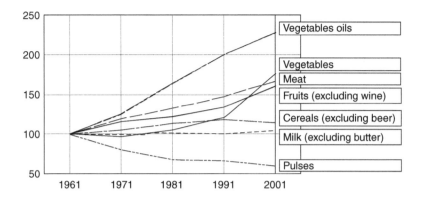

FIGURE 1.3. Evolution of world food supply per capita (1961 level = 100). *Source:* Calculated using FAO (2004b).

Figure 1.3 shows a generally improving average human diet, although tempered by undernourishment in some regions and obesity in others. The overall trend is a progressive increase of animal proteins against plant proteins and a shift from animal fats to vegetable oils. Vegetables and fruits are gaining importance as a source of vitamins and micronutrients. The trend of food diversification is likely to continue, particularly in countries where the GDP evolves positively. Meat demand is anticipated to almost double in the developing countries between 1997 and 2020 (Rosegrant et al., 2001).

Evolution of Priorities in the Industrialized Countries

Food availability is no longer a problem in the industrialized countries, except for small segments of the population that are affected by unemployment or social exclusion. Despite the widespread availability of affordable high-quality foodstuffs, price remains the major criteria for most consumers' food-purchasing decisions. This is in contrast to what consumers often state in surveys, but behavior patterns clearly show price as the primary factor. This implies that increasing productivity in order to reduce food prices further is likely to remain a major objective, even if new priorities have emerged in recent years.

Indeed, consumers are now increasingly concerned about food quality. As interest has grown in the protein, vitamin, and micronutrient contents, such information has been added to the labels of many prepacked foods. Consumers are particularly keen to know the content of items that should be ingested within reasonable limits or that are undesirable, such as fats, sugar, and salt. Such preoccupations have grown alongside our knowledge about the links between diet and certain diseases: cardiovascular disorders, cancers, diabetes, and not least of all, obesity. Formerly a scourge of rich countries, obesity is also becoming a major public health issue in many developing countries, which are seeing a divide between the underweight and overweight.

Following a number of high-profile food safety crises, mostly in Europe but also in North America to a lesser extent, consumer demands on food safety have grown. This has pushed regulators to develop and implement identity preservation or traceability schemes.

Concerns about the effect of agriculture on the environment and guarantees that growers in developing countries earn a minimum living wage ("fair trade" products) are also taken into account by some consumers.

The increasingly vibrant consumer culture influences how farmers produce food. Many face a multitude of high standards for quality and safety as well as environmental and social issues. It is difficult enough for farmers with broad access to information and agricultural techniques to keep up. Farmers in developing countries are challenged to keep track of all of these requirements, to adopt them, and to document that "acceptable" methods have indeed been used.

In parallel to consumer-driven changes, there is a move toward producing special molecules for the food, pharmaceutical, and petrochemical industries through agriculture (molecule farming). This is characterized by bioenergy and biomaterial production, which has been growing by more than 10 percent a year and is expected to further develop as world oil stocks shrink and oil prices remain high. Some plant varieties have been specifically bred for their high content in compounds such as starch or other molecules needed by the food and chemical industries. With the coming of age of biotechnology, the next decade should see the rapid development of "functional foods" or "nutraceuticals." Such crops will, for example, have their fatty acid or amino acid content altered to make them healthier or more readily usable by the food and pharmaceutical industries.

The debate that is currently raging about biotechnology and other cutting-edge technologies will be one of the major issues likely to shape the future of food and agriculture in the industrialized countries. The flames have been fanned by a series of recent food crises that have undermined the faith that consumers, particularly in Europe, have in regulators, scientists, and industry. The rapid growth of the organic food market in Europe and the long-lasting European Union moratorium on the cultivation of genetically modified (GM) crops stem from this lack of confidence. Many European consumers perceive organic food as safe and GM food as presenting potential risks for health, despite empirical evidence. There is no end in sight for this conflict, which is likely to have a significant impact on the evolution of food production in Europe and in those countries from which it imports food. Analysts speculate that European consumers may be more open to GM foods that provide benefits directly to consumers (so-called output traits) and not only to farmers ("input" traits). The first such plant varieties are expected to be released by 2010.

World Population and Agricultural Trends

In the medium to long-term perspective, world population will continue to increase, but at a lower pace than in recent decades. FAO projects world population to reach 8.9 billion in 2050 (Figure 1.4). This corresponds to a 47 percent rise from 2000, much lower than the surge in world population during the second half of the twentieth century when a 141 percent change was registered. Most of the growth is projected to occur in urban areas. During the next three decades, rural population is expected to flatten and even to start declining in absolute terms between 2020 and 2030. These demographic trends have two major implications for world agriculture:

- Growing world population will increase the demand for food, feed, fiber, and bioenergy. Agricultural output will need to outstrip the climbing population in order to meet the triple challenge of feeding extra people, providing enough calories to the 800 million people who currently suffer from hunger, and producing bioenergy, especially given limits on fossil fuel sources. Rising average income per capita is likely to drive meat consumption further, and therefore, result in higher grain requirements for feed production.
- From 2010 onward, more than half of the world population will be living in cities and will rely on commercial farming for food. Unless a significant number of current subsistence farmers moves toward commercial activities, a major part of the new agricultural supplies for cities will have to come from existing commercial farms. Since little additional arable land is available for cultivation, this will require further intensification of the land currently under the plow through the adoption and more efficient use of modern technologies and inputs, such as manufactured sources of crop nutrients, water, crop protection products, and genetic resources.

According to the International Food Policy Research Institute (IFPRI, 2002), world cereal demand will rise from 1,843 Mt in 1997 to 2,497 Mt in 2020, more than-four fifths of the difference coming from developing

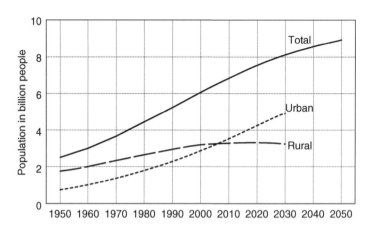

FIGURE 1.4. Evolution of world population. *Source:* Adapted from FAO (2004b).

countries. Over the same period, cereal yield growth would continue its slowdown, averaging 1.2 percent per year in the developing countries and only 0.7 percent in the industrialized countries. At the same time, the demand for meat is expected to rise from 209 to 327 Mt, almost entirely due to increased meat consumption in the developing nations. This situation would trigger stronger cereal imports by developing countries.

TRENDS IN GLOBAL NUTRIENT DEMAND

Evolution over the Past Four Decades

Fertilizer Consumption

There was a sustained increase in world fertilizer consumption from 31 Mt nutrients ($N + P_2O_5 + K_2O$) in 1961 to 143 Mt in 1989, almost 6 percent annually. From 1989/1990 to 1993/1994, world fertilizer consumption fell from 143 to 120 Mt nutrients, due to a sharp decline of fertilizer use in the countries of central Europe and of the Former Soviet Union. This fall was partly offset by increases in Asia. From 1993/1994 to 1999/2001, total world fertilizer consumption started to climb again from 120 to 139 Mt (Table 1.10).

In 2001, China accounted for a quarter of world fertilizer consumption, the United States for 14 percent, and India for 12 percent. Together these

TABLE 1.10. Evolution of fertilizer consumption (1,000 tonnes nutrients)

	1961	1971	1981	1991	2001	2001 vs. 1961 (%)
World	31,151	69,497	110,020	128,473	138,787	+346
France	2,603	4,939	5,570	5,565	4,171	+60
India	418	2,383	5,724	12,728	17,359	+4,054
SSA[a]	161	517	1,057	1,266	1,260	+681

Source: Adapted from IFA (2003).

[a]Sub-Saharan Africa excluding South Africa.

three countries accounted for more than half of total world consumption. France represented 3 percent of global demand, and SSA, excluding South Africa, only 1 percent.

From 1961 to 2001, world fertilizer consumption increased by almost 350 percent. Moderate growth rates were registered in the industrialized countries, which had a relatively high base for fertilizer use in the early 1960s. Furthermore, fertilizer use in these countries peaked in the 1980s and the early 1990s and started to decline as a result of stringent environmental regulations on the use of nutrients, coupled with increased fertilizer use efficiency arising from better practices. This trend is reflected in the figures for France. In contrast, demand grew very quickly in developing countries such as India, which multiplied domestic consumption forty times. In SSA, fertilizer consumption grew in the 1960s and 1970s, but then stagnated, largely due to the repercussions of the removal of fertilizer subsidies and the decline of state-supported distribution.

Fertilizer application rates per hectare also increased, rising from a world average of 23 kg nutrients per hectare of cultivated land (kg ha^{-1}) in 1961 to 91 kg ha^{-1} in 2001 (Table 1.11). During the same period, fertilizer application rates rose from 122 to 213 kg ha^{-1} in France (with a peak around 300 kg ha^{-1} in the 1980s) and from less than 3 to 102 kg ha^{-1} in India. This is in sharp contrast to the negligible fertilizer use in SSA, where average rates are as low as 7.7 kg ha^{-1}. This is a major contributor to the stagnating yields on that continent.

Plant Nutrient Requirements

Yield increases result in a proportional growth in plant nutrient requirements. Each harvest effectively exports nutrients from the local system.

TABLE 1.11. Evolution of fertilizer application rates (kg nutrient ha^{-1} cultivated land)

	1961	1971	1981	1991	2001	2001 vs. 1961 (%)
World	23.0	46.8	76.3	86.6	90.6	+294
France	121.6	264.1	294.7	289.8	212.8	+75
India	2.6	14.5	34.0	75.2	102.2	+3,031
SSA[a]	1.3	4.0	7.5	8.5	7.7	+492

Source: Calculated using IFA (2003) and FAO (2004b).

[a]Sub-Saharan Africa excluding South Africa.

In intensive farming systems, indigenous nutrient sources such as soil nutrients, atmospheric deposition, crop residues, animal manure, and biological nitrogen fixation (BNF) no longer suffice to replenish the nutrients extracted by high-yielding varieties. External supplies need to be applied in order to maintain the balance between nutrient inputs and outputs, and to prevent soil fertility degradation (Mosier et al., 2004).

India and many other Asian countries have experienced a positive agricultural development in recent decades, yet it is estimated that the Indian subcontinent will have to increase its average fertilizer use by some 10 percent per hectare by the end of the decade to keep up with fast growing domestic requirements for food, feed, fiber, and bio energy. Moreover, native sources of secondary and micronutrients have been exhausted by higher yields, creating a need to replenish these often overlooked elements in order to maintain production levels.

In the case of sulfur, deficiencies have also emerged in many countries because of the reduction in atmospheric deposition as a direct result of stricter regulations regarding sulfur emissions from industrial sites. The increased use of high-analysis fertilizers, which contain less sulfur, has also played a role.

Fertilizer Use by Crop Type

The evolution of fertilizer use by crop type is difficult to assess because accurate time series are scarce. Available figures show an increase of the fertilizer application rates for the main cereal crops in developing countries. For instance, between 1990 and 1997, fertilizer application rates in China rose by 50, 14, and 3 percent for maize, rice, and wheat respectively

(Table 1.12). In India, application rates for wheat increased by 5 percent between 1989 and 1997/1998. However, application rates in the United States dropped for maize (–3 percent) and wheat (–2 percent) between 1991 and 1998.

Fertilizer Use Efficiency

As mentioned previously, global food production has increased drastically over the past forty years, thanks in part to higher fertilizer use and, over the past two decades, improved plant nutrition management practices. In particular, the steady increase of cereal yields at a global level correlates with the level of nitrogen (N) used in agriculture (Figure 1.5).

TABLE 1.12. Evolution of fertilizer use by crop type (kg ha^{-1})

	Crop	1990	1997	1997 vs. 1990 (%)
China	Maize	140	210	+50
	Rice	215	245	+14
	Wheat	230	237	+3
		1989	**1997/1998**	**1989 vs. 1997/1998**
India	Wheat	116	122	+5
		1991	**1998**	**1991 vs. 1998**
United States	Maize	321	310	–3
	Wheat	146	143	–2

Source: Adapted from FAO (2002).

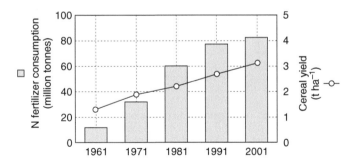

FIGURE 1.5. Evolution of world N fertilizer consumption and of global average cereal yield. *Source:* Adapted from IFA (2003) and FAO (2004b).

Taking France as an example, Figure 1.6 shows the evolution of fertilizer use efficiency as indicated by the amount of fertilizer N needed to produce one unit of cereal. According to this measure, fertilizer use efficiency dropped significantly in the 1960s and 1970s. This trend reached its lowest point in the early 1980s, when approximately 35 kg of cereal were produced per kg of fertilizer N applied. Since then, both yields and fertilizer use efficiency in France have increased, with the latter reaching values at the end of the century close to those observed in the early 1970s (about 45 kg of cereal per kg of fertilizer N applied). Thanks to modern technologies and improved agricultural practices, including fertilization, lower nutrient consumption has not depressed yields. However, the recent slowing of fertilizer use efficiency gains across most of west Europe might signal the early stages of soil nutrient mining (not necessarily of N), which is unsustainable from both an agronomic and an environmental perspective.

At the global level, fertilizer use efficiency also dropped rapidly during the Green Revolution (in the 1960s and 1970s), although grain production grew in the developing world, in particular through the use of fertilizers. The 1980s and 1990s were not only characterized by continuously growing world food and feed demands, but also by greater attention to environmental concerns. As a consequence, fertilizer use efficiency stabilized and then started a slight but steady recovery, as major agricultural regions reached "mature" status. The efficiency of N fertilizer use can be categorized on a national basis, depending on the level of agricultural development, as summarized in Table 1.13.

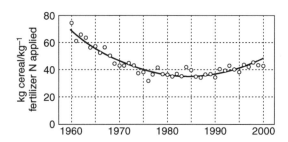

FIGURE 1.6. Evolution of N fertilizer use efficiency in France. *Source:* Adapted from UNIFA (2004). *Note:* In the absence of more precise data on fertilizer use by crop, Figure 1.6 is an estimation based on the ratio between cereal production and N fertilizer consumption. This measures "partial factor productivity" from applied N (kg of product per kg N applied). "Agronomic efficiency" (amount of product increase per kg N applied) would provide a more precise idea of improvements in fertilizer use efficiency, but national-level data are not available for this calculation.

Balanced Fertilization

The ratio between nitrogen (N), phosphate (P_2O_5), and potash (K_2O) in world fertilizer consumption evolved greatly since 1961, favoring N as a result of the rapid development of N fertilizer production (Table 1.14) and because the immediate effect of N on yields is highly appealing to farmers.

The nutrient ratios in many parts of the world show serious imbalances, with far too little phosphorus (P) and potassium (K) used relative to the quantity of N (Aulakh and Malhi, 2004). A recent survey of North American soils indicates that 45 percent test at medium or low for P and K (PPI-PPIC-FAR, 2001). However, a growing awareness of the benefits of balanced

TABLE 1.13. Phases of N fertilizer use in agricultural development

Type of agriculture	Example	Cultivated area	Yields	N fertilizer use	N fertilizer use efficiency
Stagnating	Sub-Saharan Africa	Increasing	Low and stagnating, or slowly increasing	Low and stable on average. Significant variation between fields	High at the country level
Growing	South Asia	Stable in most cases	Rapidly increasing	Growing rapidly	Decreasing rapidly
Mature	West Europe	Stable or decreasing	High, steadily increasing at a moderate pace	Stable or slowly decreasing	Moderate and increasing slowly

Source: Adapted from IFA (2003).

TABLE 1.14. Evolution of the NPK ratio at world level

	1961	1971	1981	1991	2001
Fertilizer consumption ('000 t)					
N	11,824.5	31,579.5	60,309.0	77,131.9	82,444.5
P_2O_5	10,258.2	17,874.9	25,836.8	28,533.3	33,372.3
K_2O	9,068.4	16,253.0	23,874.8	24,596.1	22,970.5
NPK ratio					
$N:P_2O_5:K_2O$	1:0.87: 0.77	1:0.57: 0.51	1:0.43: 0.40	1:0.37: 0.32	1:0.40: 0.28

Source: Adapted from IFA (2003).

plant nutrition has started to reverse the situation. Throughout the 1990s, phosphate fertilizer consumption rose faster than nitrogen consumption (Figure 1.7). The most recent figures for potassium fertilizer demand indicate the beginning of a similar turnaround during the past decade. The ratio was 0.253 in 1996/1997 and rose to 0.279 in 2001/2002. Sulfur and micronutrient (particularly zinc, boron, and iron) fertilizer consumption are also increasing in response to growing deficiencies of these nutrients and knowledge of their roles in plant nutrition.

Use of Organic and Inorganic Nutrient Sources

Inorganic fertilizers represent about half of the total N currently supplied to agricultural land. In 1996/1997, world fertilizer N consumption was assessed at 82.6 Mt N (IFA, 2003). Smil (1999) estimated that the other N inputs into crop production in 1996 reached 25-41 Mt N for biological N fixation, 12-20 Mt N for crop residues and 12-22 Mt N for animal manure. In 1996, total organic sources of N to agricultural land amounted, on average, to 66 Mt N, that is, almost 40 percent of total N supplies. The balance comes from atmospheric deposition and irrigation water, estimated to contribute 21-27 Mt N annually (Smil, 1999). The ratio between the use of organic and inorganic sources of N may vary in the future. In a short- to medium-term perspective, it is anticipated that the ratio will evolve toward greater use of manufactured fertilizers in most developing countries, and more recycling of organic sources of nutrients in developed countries. Similar trends are expected for sources of P and K.

Agronomic objectives and environmental imperatives both favor the use of complementary and readily available nutrient sources. This approach, called Integrated Plant Nutrient Management (IPNM), aims to provide the

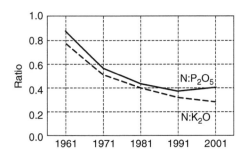

FIGURE 1.7. Evolution of the global NPK ration (N level = 1). *Source:* Calculated using IFA (2003).

necessary nutrients to the crop while also improving the physical and biological soil properties. Furthermore, this practice tends to minimize nutrient losses to the environment.

The basic principles of IPNM are to:

- reduce losses of nutrients from the ecosystem;
- take into account nutrients supplied by soil and atmospheric deposition;
- optimize the use of other available indigenous nutrient sources;
- enhance soil biological activity;
- calculate nutrient budgets in line with yield objectives;
- improve the efficiency of nutrient uptake;
- add the required plant nutrients in appropriate form(s); and
- combine IPNM with integrated water, pest, and other crop management practices.

This approach takes into account which nutrient sources are readily available and then weighs the relative advantages and disadvantages of using each in a site-specific context. In combination with products and management practices that improve nutrient use efficiency, IPNM is the best way to achieve sustainable fertilization that will increase yields while reducing nutrient losses to the environment.

Environmental Issues

There are a number of environmental issues related to fertilizer use, including soil acidification and the accumulation of naturally occurring impurities in soils to which fertilizers are applied. These are covered in another chapter of this book (Grant et al., 2008). This section focuses on two of the most relevant questions: the application of excess nitrogen to some agricultural lands and the inadequate supplies of crop nutrients experienced in other world regions.

Nitrogen Losses. Regardless of the source, the application of excess N can result in a number of undesirable effects such as the emission of greenhouse gases (NO, NO_2, and N_2O) through denitrification and, to some extent, nitrification, the volatilization of ammonia (NH_3), the leaching of nitrate (NO_3^2), and the loss of particulate N through erosion. As a consequence, there is a trend in many countries to reduce N applications to an absolute minimum. This has been particularly true in areas with intensive livestock production, in order to optimize the recycling of animal wastes. In Denmark, for example, nitrogen fertilizer consumption dropped considerably in the 1990s, mainly because of environmental regulations aimed at controlling the N load on agricultural land. The consumption of manufactured N fertilizers in that country decreased steadily from about 0.4 Mt in 1990 to approximately 0.2 Mt in 2002. Over the same period, the use of organic sources of N remained fairly stable (Figure 1.8). In 2001, for the first time in several decades, consumption of N from manures was higher than that from manufactured fertilizers.

In many other European countries and in North America, regulators are increasingly concerned about the recycling of animal manure, which constitutes the main source of N losses to the environment today. In countries with already high fertilizer application, this development has contributed to declining rates as reflected by the United States and west Europe in Tables 1.12 and 1.13. Farmers are increasingly adopting nutrient budgets, both to improve income and to limit nutrient losses to the environment. Site-specific nutrient management, either with highly sophisticated spatial techniques (e.g., global positioning system) or with simple tools (e.g., leaf color chart), further contributes to achieving these goals.

The development and use of better performing fertilizers, such as slow- and controlled-release products and urease or nitrification inhibitors, as well as performance enhancing application techniques, like deep placement of urea super-granules, also offer great promises for preserving the environment.

On a cautionary note, the current downward trend in fertilizer use initiated in industrialized countries can have negative impacts, if not compensated by other nutrient sources or increased use efficiency. For instance, in Denmark again, the reduction in N fertilizer use has led to lower cereal protein content (Figure 1.9). If this trend continues, grain quality could

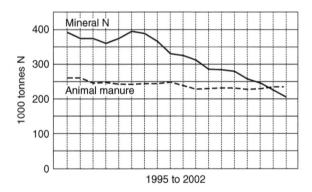

FIGURE 1.8. Use of N manufactured fertilizer and animal manure in Denmark. *Source:* Grant et al., cited in Knudsen (2003). Reprinted with permission.

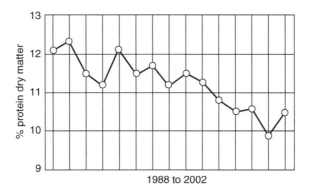

FIGURE 1.9. Protein content of winter wheat used for pig feeding in Denmark. *Source:* Vils et al., cited in Knudsen (2003). Reprinted with permission.

deteriorate appreciably. Therefore, the right balance has to be found between food production, food quality, and protection of the environment.

Soil Fertility Degradation. In stark contrast to the introduction of excess N into an ecosystem, insufficient nutrient application is detrimental to soil fertility. This situation is particularly critical in SSA, where transport and other distribution costs prevent most farmers from accessing manufactured fertilizers and other nutrient sources at affordable prices. Indeed, fertilizers in Africa may cost several times more than in industrialized countries, essentially because of transportation costs. In the context of rapidly growing food demand leading to the shortening of fallow periods, the result is depletion of soil nutrient reserves. Depressed yields will follow (Table 1.3), because nutrients such as N and P are often the limiting factors. Extreme conditions can trigger desertification. In such situations, urgent measures are needed to provide the nutrients required to sustain agricultural production. However, a number of constraints must first be alleviated, requiring strong policy commitments (see "Fertilizer Use Constraints").

Future Prospects for Fertilizer Demand

Agronomic Context

Medium to long-term fertilizer demand will be strongly influenced by the need to feed, clothe, and provide energy to an increasing population. One option is to expand the cultivated area, as is currently the case in Brazil where a large part of the savanna is being converted to soybean and maize farming. However, only a few countries still have such reserves of potential arable land. Moreover, converting pristine areas to cultivated systems can have a significant environmental cost, especially if not implemented carefully. Therefore, most of the increased production is likely to come from higher yields, even if the growth rate over the next two decades is less dramatic than has been the case over the past half-century.

At the same time, there is a pressing need to restore soil health in numerous countries where the soil base has been mined for decades, leading to soil fertility degradation. Rehabilitating these systems, provided desertification is not too advanced, requires substantial amounts of fertilizers in order to rebuild the soil nutrient pool. To the extent possible, fertilizer use should be coupled with incorporation of organic matter into the soil. This is a priority in many parts of SSA and of Central America and the Caribbean, where damage may soon become irreversible for soils that are currently salvageable.

Greater attention should also be paid to sulfur and micronutrients, deficiencies of which are on the rise. If these nutrients are not taken into account, they will strongly affect the use efficiency of the macronutrients (N, P, and K).

Water availability under both irrigated and rainfed conditions will be a critical factor, especially in the context of climate change and rising demand from competing users such as industries and cities. Farm management practices should strive to improve water use efficiency. In this connection, fertilizers have a key role to play, since appropriate fertilizer use improves the efficiency of agricultural water use (Aulakh and Malhi, 2005).

Other Driving Forces

Environmental policies and regulations will certainly increase the pressure on fertilizer use. They will require maximal recycling of organic matter (animal manures and urban wastes in particular) in order to mitigate the negative impact of losses to the environment. They will also call for the optimization of biological N fixation, with the inclusion of leguminous crops in crop rotations and improvement of the legume-*Rhizobium* symbiosis. In many locations, they are also likely to impose nutrient budgeting, at least for N and possibly for P, to improve the nutrient input/output balance, while using all available inputs in an integrated manner. In some sensitive areas or ecosystems, regulators might consider nutrient quotas.

The development of new technologies will also strongly influence future fertilizer demand, and more specifically, the efficiency with which fertilizers are used. Improved nutrient use efficiency can be achieved through better performing fertilizer products, through better application techniques, or through new management practices. In this connection, slow- and controlled-release fertilizers, urease and nitrification inhibitors, deep placement of urea super-granules, fertigation (the delivery of crop nutrients through irrigation water), and site-specific nutrient management offer great promises. However, some of these products or practices remain restricted to limited market segments because of significant price differentials with conventional products and methods. At this time, it is difficult to predict how the market for such options will evolve since incentive measures enacted by the policymakers could be highly influential.

Biotechnology is another sector that could impact medium to long-term fertilizer demand, but for the next twenty years, no significant changes to fertilizer use patterns are expected from this quarter. In the long run, the main development that might significantly alter fertilizer demand would be an improvement of the plant's ability to use nutrients more efficiently, through improved uptake, translocation, metabolism, or storage. Research on this subject is ongoing, but no application is foreseen in the medium term.

More efficient symbiosis between legumes and rhizobia could make leguminous crops more competitive compared to cereals, and boost their cultivation. However, genetic control of symbiosis is not well understood, so any application is still distant. Transferring the genes responsible for the symbiosis or for BNF to cereals is likely to remain theoretical for the next few decades, due to the complexity of achieving the feat and the probable tradeoffs, which are expected to be high.

Fertilizer Use Constraints

Another critical issue that will drive changes in fertilizer demand over the coming decades is the evolution of current barriers to fertilizer use in developing countries and in countries with economies in transition. These constraints are manifold. In some cases, small markets are unattractive for international suppliers. Weak infrastructures and distribution networks prevent timely and affordable access to fertilizers. The lack of adequate credit facilities at reasonable rates is also problematic for many farmers. Entrepreneurs in many locations lack incentives to take risks. Inefficient extension services in many countries are an obstacle to the transfer of improved technologies and agricultural practices. In some cases, the absence of other key inputs such as improved seeds and crop protection products makes fertilizer use a moot point. If fertilizers are not profitable because of low prices for harvested produce, farmers will not bother with their use. Alleviating these constraints, which are particularly troublesome in SSA, requires strong policy commitments.

Fertilizer Demand Projections

With the exception of the persistence of constraints on fertilizer use, the above-mentioned trends should have a relatively positive impact on fertilizer demand in a five-year timeframe (from 2003/2004 to 2008/2009). During that period, world fertilizer demand is forecast to grow on average by 2.1 percent annually, to reach 163 Mt nutrients (IFA, 2004). Looking at the nutrients individually, projections indicate higher growth rates for phosphate and potash fertilizers (+2.7 percent annually) than for nitrogen (+1.7 percent annually).

Annual fertilizer consumption growth rates above 3 percent are forecast in Eastern Europe and central Asia, South America, South Asia, and Oceania. Three of these regions (Eastern Europe and central Asia, South Asia, and Oceania) are recovering from a depressed year in 2003/2004, thus artificially inflating their year-on-year growth rate. In contrast, fertilizer demand is expected to steadily decline in western Europe and northeast Asia and to grow only marginally in North America.

The Food and Agriculture Organization of the United Nations provides longer-term forecasts in *Fertilizer Requirements in 2015 and 2030* (FAO, 2000). In order to attain projected yields, FAO foresees fertilizer consumption increasing from 138 Mt nutrients in 1995-1997 to between 167 and 199 Mt per year by 2030. This represents an annual growth rate of between 0.7 and 1.3 percent, compared with an average annual increase of 2.3 percent between 1970 and 2000. According to FAO, most of the increase would be in South Asia, East Asia, and South America.

The recent IFA forecasts cited indicate a more sustained growth in world fertilizer demand than the lower FAO projection. According to IFA, the lower FAO projection for 2030 would be already reached at the end of the current decade. IFA believes that the adoption of technologies improving fertilizer use efficiency will be more moderate than FAO's "increase efficiency scenario" assumes.

CONCLUSIONS

Projections up to 2050 show that there will be a continuous growth of the world population, especially in urban centers in the developing countries. Population growth will be more moderate than during the second half of the twentieth century, but it will nonetheless lead to a considerable increase in the world demand for food, feed, fiber, and bioenergy. Increasing agricultural production to cope with demand should be carried out in a way that preserves soils, water, the air, and biological diversity.

The production of adequate and diverse food (of which more will be meat) to ensure world food and nutrition security will require greater use of crop nutrients over the coming decades. However, in order to limit the negative impacts on human health and the environment that can be associated with the excess use of nutrients, an integrated approach is needed. This approach should promote improved nutrient use efficiency, the complementary use of all available nutrient sources and the adoption of site-specific management practices.

Integrated Plant Nutrient Management is an important component of sustainable agriculture, together with other elements of integrated farming such as integrated pest and water management. The adoption of IPNM by farmers requires supportive agricultural, environmental, and trade policies. The use of manufactured fertilizers is an important component of IPNM. However, in many countries, fertilizer use remains constrained by insufficient access to input and output markets, poor transportation infrastructures, lack of affordable rural credit or limited access to information. In the absence of appropriate policy measures, such constraints prevent agriculture in regions affected by soil fertility degradation, such as SSA, from emerging out of stagnation.

REFERENCES

Alloway, B.J. (2004). *Zinc in Soils and Crop Nutrition.* Brussels, Belgium: International Zinc Association.

Aulakh, M.S. and S.S. Malhi. (2004). Fertilizer nitrogen use efficiency as influenced by interactions of N with other nutrients. In Mosier, A., K. Syers, and J.R. Freney (eds.) *Agriculture and the Nitrogen Cycle: Assessing the Impacts of Fertilizer Use on Food Production and the Environment.* pp. 181-191. Covelo, CA: Island Press.

Aulakh, M.S. and S.S. Malhi. (2005). Interactions of nitrogen with other nutrients and water: Effect on crop yield and quality, nutrient use efficiency, carbon sequestration and environmental pollution. *Advances in Agronomy.* 92 (In press).

FAO. (1992). *Fertilizer Use by Crop,* 1st Edition. Rome, Italy: Food and Agriculture Organization of the United Nations.

FAO. (2000). *Fertilizer Requirements in 2015 and 2030.* Rome, Italy: Food and Agriculture Organization of the United Nations.

FAO. (2002). *Fertilizer Use by Crop,* 5th Edition. Rome, Italy: Food and Agriculture Organization of the United Nations.

FAO. (2003). *The State of Food Insecurity in the World, 2003.* Rome, Italy: Food and Agriculture Organization of the United Nations.

FAO. (2004a). *Food Outlook, April 2004.* Rome, Italy: Food and Agriculture Organization of the United Nations.

FAO. (2004b). FAOSTAT (http://apps.fao.org) (accessed September 10, 2004).

Fritschel, H. (2001). Fighting hidden hunger. In Pinstrup-Andersen, P. and R. Pandya-Lorch (eds.) *The Unfinished Agenda: Perspectives of Overcoming Hunger, Poverty and Environmental Degradation.* Washington, DC: International Food Policy Research Institute. pp. 31-36.

Grant, C.A., M.S. Aulakh, and A.E. Johnston. (2008) Integrated nutrient management: Present status and future prospects. In Aulakh, M.S. and C.A. Grant (eds.) *Integrated Nutrient Management for Sustainable Crop Production.* Binghamton, NY: The Haworth Press.

IFA. (2003). *IFADATA Statistics from 1973/74 to 2001/02.* Paris, France: International Fertilizer Industry Association.

IFA. (2004). *World Agricultural Situation and Fertilizer Demand, Global Fertilizer Supply and Trade, 2003/04-2008/09—Summary Report.* International Fertilizer Industry Association, Paris, France.

IFPRI. (2002). Sustainable food security for all by 2020. Proceedings of an International Conference, September 4-6, 2001, Bonn, Germany. International Food Policy Research Institute. Washington, DC.

Knudsen, L. (2003). Nitrogen input controls on Danish farms: Agronomic, economic and environmental effects. *Proceedings No. 520, International Fertiliser Society,* November 2003.

Mosier, A.R., J.K. Syers, and J.R. Freney. (2004). Nitrogen fertilizer: An essential component of increased food, feed, and fiber production. In Mosier, A.R., J.K. Syers, and J.R. Freney (eds.) *SCOPE Series 65: Agriculture and the Nitrogen Cycle: Assessing the Impacts of Fertilizer Use on Food Production and the Environment.* Washington/Covelo/London: Island Press. pp. 3-15.

PPI-PPIC-FAR. (2001). Technical Bulletin 2001-1: Soil Test Levels in North America. Norcross, GA: Potash and Phosphate Institute.

Rosegrant, M.W., M.S. Paisner, S. Meijer, and J. Witcover. (2001). *Global Food Projections to 2020.* Washington, DC: International Food Policy Research Institute.

Scherr, S.J. (1999). *Soil Degradation: A Threat to Developing-Country Food Security by 2020?* Washington, DC: International Food Policy Research Institute.

Smil, V. (1999). Nitrogen in crop production: An account of global flows. *Global Biogeochemical Cycles.* 13: 647-662.

UNIFA. (2004). La lettre de l'UNIFA No. 12. Union des industries de la fertilisation, Paris, France.

Welch, R.M. (2003). Farming for nutritious foods: Agricultural technologies for improved human health. *Proceedings of the IFA/FAO Agriculture Conference,* March 26-28, 2003, Rome, Italy. Paris, France: International Fertilizer Industry Association.

Chapter 2

Integrated Nutrient Management: Present Status and Future Prospects

Cynthia A. Grant
Milkha S. Aulakh
A. E. Johnny Johnston

INTRODUCTION

One of the most important challenges facing humanity today is that of increasing food production while avoiding environmental degradation. As the world's population grows, the yields of crops on land currently under production must be increased to maintain food security if marginal land or natural ecosystems are not to be converted to agriculture. This increase in production will require an adequate supply of plant-available nutrients to support both crop yield potential and nutritional quality. Sustainable agricultural production requires that the soil contains an adequate supply of plant-available nutrients to support crop growth, and that the nutrients removed in the harvested material or in the exported product of livestock systems are replaced so that soil fertility is not depleted over time. At the same time, excess nutrient accumulation must be avoided to reduce the risk of nutrients moving out of the rooting zone in the soil to the air or water, where they may pose an environmental risk. Integrated nutrient management (INM) requires an understanding of nutrient dynamics throughout the soil-microbe-plant system to effectively manage organic and inorganic nutrient inputs and the nutrient pools within the soil to address long- and short-term production and environmental goals. Ideally, nutrient management synchronizes nutrient demand by the crop with nutrient supply from the soil, considering the amount and availability of nutrients within the soil that can be accessed by actively growing roots.

Historically, crop production relied on the input of nutrients from sources such as animal manures, bone meal, and plant ash, growing leguminous crops to supply nitrogen (N), together with fallowing to allow accumulation of nutrients from the weathering of soil minerals. Integration of livestock and crop production aided nutrient management as animal wastes could be recycled as a source of crop nutrients. For this reason the Norfolk four-course rotation, developed in England in the mid-eighteenth century and practiced in many temperate countries, was particularly successful. Animal manures and any other available organic inputs were applied for the root crop, usually turnips or swedes grown as feed for sheep or cattle. The root crop was followed by spring barley, with the grain used for human consumption and the straw for animal feed. The third crop was a legume, either peas or beans for human food or a legume used as hay for animal feed. The legume left a residue of N for the succeeding winter wheat crop, with the grain used for bread and the straw for animal bedding to make farmyard manure (FYM) during the winter. As only animal products and cereal grain were sold off, the farm nutrient export was minimal at a time when external nutrient inputs were not generally available. In the long-term this system would not have been sustainable because although the export of nutrients was small, it was larger than the amount recycled. It is interesting that long-term paddy rice production in countries such as China was more sustainable because nutrients were added to the paddy fields in soil eroded from land higher up the mountainside.

Development of synthetic chemical fertilizers allowed replacement of nutrients removed in the harvested crop. The production of superphosphate from the mid-1840s was followed by the first exploitation of the German potash deposits in the mid-1860s. Nitrogen in the latter half of the nineteenth century was only available as ammonium sulphate (a by-product of town gas production from coal) or as sodium nitrate (from the naturally occurring deposits in South America). Increasing use of these N sources was restricted because many soils were deficient in P and until this was corrected crops did not respond to added N, demonstrating the importance of balanced fertility in integrated nutrient management. Nevertheless because these supplies were limited, in 1898, Sir William Crookes in his Presidential Address to the British Association for the Advancement of Science warned that there was a danger that food supplies could be restricted because of lack of N. Fortunately, mastery of the industrial fixation of atmospheric N came in the early years of the twentieth century although the major use of synthetic N fertilizers in agriculture did not start until the 1950s.

In the past fifty years, following the introduction of crop cultivars with a large yield potential and the ability to protect that potential from the adverse effects of weeds, pests, and diseases, there has been a great increase in both crop yields and synthetic fertilizer use. This is clearly seen in the close relationship between the global increase in grain production and N fertilizer production and use (Galloway, 1998). Experiments and experience have shown that productivity could be maintained or increased in the absence of return of nutrients in organic manures, by those in inorganic fertilizers, applied using appropriate management practices to efficiently meet crop requirements. A good example of this has been the changing yields of winter wheat between 1850 and 2003 in the Broadbalk experiment at Rothamsted (Figure 2.1). The data in Figure 2.1, as other data from long-term experiments at Rothamsted and other places throughout the world, is a good example of long-term sustainability of agricultural production provided that the soil, the crop, the climate, and the management, including the maintenance of soil fertility and the input of nutrients, are all appropriate.

In a review of long-term nutrient management studies, Edmeades (2003) reported that nutrient inputs increased crop production by 300-400 percent

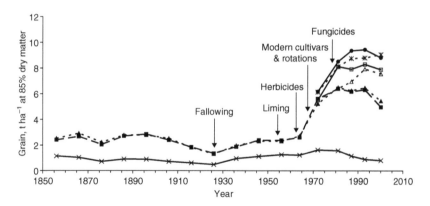

FIGURE 2.1. Yields of winter wheat grown on Broadbalk, Rothamsted, UK, 1852 - 2005 and year in which major cultural changes were introduced. *Source:* Johnston (1994) and LAT (2006). *Note:* Treatments are wheat grown continuously and unmanured χ, with FYM ■, with N3PK ▲, with FYM + N2 □, with best NPK △. Wheat grown in rotation with FYM + N2 ●, with best NPK ∗. Annual applications per ha: FYM, 35 t; N2 and N3, 96, and 144 kg; Best N, amount larger than 144 kg that gave the best yield each year; P, 35 kg P, K 96 kg K. Before the introduction of modern cultivars in 1968, yields are ten-year means, thereafter the yield shown is the mean yield for the period in which a cultivar was grown. Reprinted with permission.

and that chemical fertilizers generally had similar effects on crop yields as manure applications. In the 1840s and 1850s, the Classical experiments at Rothamsted, showed that FYM and fertilizers used correctly, could both give the same yield (Table 2.1). It is interesting to note that Lawes and Gilbert, Rothamsted's founding fathers, never claimed that fertilizers were better than FYM, rather they realized even in the mid-nineteenth century that there would never be sufficient organic manure of any description to produce enough food to feed the rapidly expanding population. Today, we have to look ever more to the use of chemical fertilizers but they must be used wisely and in ways that complement other available nutrient sources, namely INM.

The cost of chemical fertilizers is often lower than the cost of an application of an organic nutrient source, especially in developed regions where labor costs are high. Therefore, intensification of agriculture has commonly led to specialization of animal and crop production and, as a result, de-linkage of nutrient cycles. In the absence of a return of nutrients in the organic wastes from livestock production, nutrient off-takes through crop removal must be replenished by external inputs to avoid nutrient depletion and the loss of soil fertility. In contrast, nutrients may accumulate to excess in intensive

TABLE 2.1. Yields, t ha^{-1}, of winter wheat and spring barley grain, at 85 percent dry matter, and roots of mangolds and sugar beet at Rothamsted, England

Experiment	Crop	Period	Yield with FYM	NPK fertilizers[a]
Broadbalk	Winter wheat	1852-1861	2.41	2.52
		1902-1911	2.62	2.76
		1970-1975	5.80	5.47
Hoosfield	Spring barley	1856-1861	2.85	2.91
		1902-1911	2.96	2.52
		1964-1967	4.60	3.36
		1964-1967	5.00	5.00[b]
Barnfield	Mangolds	1876-1894	42.20	46.00
	Mangolds	1941-1959	22.30	36.20
	Sugar beet	1946-1959	15.60	20.10

Source: Johnston (1986).

[a]Application per ha: FYM, 35 t; N to winter wheat, 144 kg; to spring barley, 48 kg.

[b]With the exception of 96 kg (because new variety had a larger yield potential); mangolds and sugar beet, 96 kg.

livestock systems, especially where nutrients are brought onto the farm in purchased feedstuffs, and the cost of transport and application of manures, especially off the farm, is greater than the value of the nutrients they contain. Excessive nutrient accumulation may also occur where commodity prices are high in relation to the cost of fertilizers, as nutrients may be applied in excess to ensure that yield is not lost due to nutrient deficiency. Such excessive nutrient applications are not sustainable in the long term. The production and transport of fertilizers require inputs of nonrenewable resources, and the transfer of nutrients, mainly N and phosphorus (P) from over-enriched soil to water can lead to the adverse effects of eutrophication, while nitrous oxide (N_2O) emissions to the atmosphere may lead to increased global warming. In addition, phosphatic fertilizers produced from some phosphate rocks can add heavy metals, such as cadmium (Cd), to the soil.

The challenge for both environmental stewardship and economic efficiency is to develop systems that manage nutrients effectively to avoid either excess or depletion. The most pressing challenge or constraint will vary with the location. Population density, environment, stage of agricultural development, soil biological, chemical, and physical conditions, and agricultural policies all influence nutrient requirement and potential losses. Although the principles of INM apply to all nutrients, it must be remembered when examining these challenges that there are some important differences among nutrients. For example, although nitrate from N fertilizer and manure applications in excess of that required to produce a crop may remain in the soil at harvest, under wet conditions it is either denitrified or lost by leaching. Therefore, except in relatively arid environments, very little N accumulates in soil other than in increased soil organic matter (SOM), and the amount of SOM produced from any one application of organic matter is usually extremely small. Thus, for high-yielding crops grown under favorable conditions, N has to be added each year and soil analysis for plant-available soil N may be costly and difficult to interpret. The situation is quite different for residual P and potassium (K), both of which can accumulate in readily and less-readily available forms which can, if an appropriate method of analysis is used, be estimated with some reliability. Critical levels of both P and K can be determined experimentally and fertilizer amendments applied only when their availability is below the critical level.

An example of this important difference comes from the long-term experiments at Rothamsted. In the Broadbalk experiment, where winter wheat has been grown each year since 1843, one plot receives 96 kg N ha^{-1} annually. In 1987 after 135 years, a total of 19,440 kg N had been applied. Assuming that over the whole period N was used with 50 percent efficiency then some 10,000 kg N has to be accounted for. The increase in total soil

N however, amounts to only 570 kg N ha^{-1} in the 0-23 cm surface soil (Glendining et al., 1996). This increase in soil N as SOM occurred in the first fifty years and has remained unchanged at this equilibrium level since. Although this SOM releases some mineral N each year, because it is in equilibrium the same amount of N is retained in "new" SOM. In contrast, P and K both accumulate in the silty clay loam soil at Rothamsted when there is a positive balance. For example, in the Exhaustion Land Experiment where single superphosphate was applied between 1876 and 1901, the positive P balance was 1,217 kg P ha^{-1} and of this 1,085 kg P ha^{-1} was found as an increase in total P in the top 23 cm of the soil in 1903 (Johnston and Poulton, 1977). Similarly on the sandy loam soil at Woburn, P was applied as single superphosphate between 1876 and 1927, the positive P balance was 1,197 kg P ha^{-1} of which 1,154 kg P was found as an increase in total soil P. Of this total soil P, about 13 percent was readily plant-available (Olsen P) and this made a major contribution to the P taken up each year by the crop.

THE CHALLENGE OF NUTRIENT DEPLETION

Nutrient depletion is a major threat to long-term agricultural sustainability in many areas of the world. Negative nutrient balances (nutrient applied *minus* nutrient removed in the harvested crop) have been identified in tropical agricultural systems of Africa and Latin America (Stoorvogel and Smaaling, 1998), in areas of North America (Fixen and Johnston, 2002), Australia (Holford et al., 1992; Holford and Crocker, 1997), China (Lin et al., 2007), and India (Aulakh and Bahl, 2001). In England and Wales, the British Survey of Fertiliser Practice shows that nationally, on average, there has been for the last ten years a negative P balance for potatoes, sugar beet, cereals and oilseed rape (Johnston and Dawson, 2005). Also in England and Wales, current analysis of soil samples taken as part of the Representative Soil Sampling Scheme does not show an appreciable decline in Olsen P values because P from the less-readily available pool of soil P is being released to the readily available pool as measured by the Olsen method, but this release will not go on indefinitely.

Nutrient depletion becomes a self-accelerating process where nutrient deficiency leads to reduction in biomass production. Soil fertility depends to a greater or lesser extent on the soil type, linked to the content of SOM. Soil tilth and rootability (the ease with which roots can grow throughout the soil mass to take up nutrients and water), the water-holding capacity, and the ability of the soil to store and release nutrients are influenced by the

SOM content. The organic matter content of a soil moves towards an equilibrium value, related to soil type and climate, to the rate of decomposition of existing SOM and the amount and rate of decomposition of organic matter returned to the soil, including organic amendments and crop residues (Campbell, Zentner, Gregorich et al., 2000; Campbell, Zentner, Selles et al., 2000; Johnston, 1986; Peterson et al., 1998). Rates of increase or decrease in organic C vary with many factors such as initial organic matter in the soil, the quality of the organic matter added, fertilizer application rate and the length of time that management is imposed (Paustian et al., 1992; Peterson et al., 1998; Aulakh et al., 2001a; Carlgren and Mattsson, 2001). On the silty clay loam soil at Rothamsted, when old arable soils are put down to permanent grass, it takes 100 years to increase the SOM content of an old arable soil (1.7 percent SOM) to that of a permanent grassland soil (5.0 percent) and about twenty-five years to reach the halfway stage (Johnston, 1986). The buildup of SOM on the sandy loam soil at Woburn was related to the amount and type of organic matter added, FYM, sewage sludge or compost, but the rate of decline was independent of the form in which the organic matter was added. Presumably this was because SOM is largely the end product of microbial decomposition of any organic matter added to the soil and the decline in SOM would then be independent of the type of organic matter added (Johnston et al., 1989). Decreasing crop productivity associated with declining soil fertility will reduce the amount of residue returning to the soil, further decreasing SOM content and soil fertility (Bationo et al., 1998; Aulakh et al., 2003).

Increasing population pressure increases the impact of nutrient depletion on the sustainability of crop production. In many tropical regions, traditional agriculture relied on shifting cultivation (Sommer et al., 2004). Land was prepared by clearing existing vegetation by slash-and-burn, after which crops were cultivated for several years. Nitrogen in the standing crop would be lost on burning but the ash would contain mineral nutrients, and mineralization of SOM would provide N. Following cultivation, which continued until nutrient depletion meant that the size of the crop no longer justified the effort to grow it, the site was abandoned and a new site taken into cultivation. On the abandoned site the reestablishment and growth of native vegetation for a number of years allowed the accumulation of nutrients in preparation for the next crop. When population density was low, farmers could allow sufficient years after abandoning the land for it to accumulate enough nutrients in the vegetation and SOM to again grow food crops for a number of years. However, in many parts of the world where such methods of cultivation are practiced, demographic pressure from an increasing population has restricted the opportunity for farmers to access new land when nutrient depletion limits fertility (Batiano et al., 1998; Sommer et al., 2004). This has led to shorter fallow periods, less nutrients in the natural vegetation and smaller yields of the cultivated crops. In some cases, it has led to marginal soils being brought into cultivation and this has led to loss of important habitats or increased risk of soil erosion. In these situations, soil fertility decline in small farm holdings is a fundamental reason for declining food production (see Chapter 11 in this book). Organic matter declines of 3-5 percent per year were reported in northern Nigeria after eighteen years of continuous cropping in the absence of nutrient inputs (Jones, 1971). As SOM declines, the soil is less able to store water, and in the absence of a permanent vegetation cover, soil erosion and loss of nutrients become an increasing problem. Declining fertility leads to reduced productivity, placing even greater stress on the supply of food. Development of INM approaches that couple efficient use of organic and inorganic nutrient sources with cropping practices that reduce erosion and organic matter depletion are critical in these regions. For example, in areas where regrowth is substantial during the fallow periods, adoption of slash-and-mulch rather than slash-and-burn could reduce nutrient loss by volatilization and loss of ash by wind erosion. This could allow intensified cultivation with less deterioration of soil fertility (Sommer et al., 2004).

Fallowing is still used in many areas of the Great Plains of North America to store water and nutrients through one year to support a crop in the following growing season. It is estimated that losses of organic matter in Canada since converting uncultivated land to agriculture were in the range of 15-50 percent, with most of the loss occurring in the first ten years after conversion (Gregorich et al., 1995). Similarly, long-term field studies in central Sweden illustrate that continuous fallow or cereal cropping without straw addition on a sandy clay loam decreased SOM in the 0-20 cm layer from 3.8 kg C m^{-2} to 2.7 kg C m^{-2} in the fallow treatment, whereas cropping mainly with cereals (fertilized with 80 kg N ha^{-1} as $Ca(NO_3)_2$) and incorporation of straw residues increased SOM to 4.3 kg C m^{-2} (Paustian et al., 1992; Kirchmann et al., 1994). Incorporation of green manure plant material in continuous cereal cropping also increased SOM to about the same level as straw plus fertilizer N inputs. It should be noted, however, that the organic amendment increased SOM by only the same amount as the fallowing decreased it. Even with the organic amendment, SOM would not go on increasing indefinitely but would reach a new equilibrium level.

Soil organic matter depletion will lead to an overall decline in the ability of the soil to supply mineralized N (Campbell, Lafond, and Zentner, 1993; Campbell and Zentner, 1993). In studies at Indian Head, Saskatchewan, Canada, average cereal yield, organic N and initial potential rate of N mineralization were all reduced after thirty-four years by fallow systems when compared to continuous cropping where N and P fertilizers were applied (Table 2.2) (Campbell, Lafond, and Zentner, 1993). Nitrogen mineralization rates after 10 years at Mandan, North Dakota, USA, were decreased by a crop-fallow system as compared to annual cropping with reduced tillage intensity (Weinhold and Halvorson, 1999). In eastern Colorado, USA, potential N mineralization capacity and N mineralization in the surface 5 cm of soil were less after three and a half years of a wheat-fallow rotation compared to a wheat-corn-millet-fallow rotation under no-till (Wood et al., 1990). The differences were related primarily to larger SOM concentrations in the surface soil caused by larger plant residue additions with more intensive cropping, although there may also have been an impact of differential erosion under the varying cropping systems. The extended rotation also received larger amounts of fertilizer N. Similarly, in a nine-year study of continuous no-till spring wheat in Saskatchewan, the N-supplying capacity

TABLE 2.2. Average annual rate of change in yield trends over thirty-four years and organic N, available-P, and potential N-supplying capacity of soil in a thin Black Chernozem at Indian Head, Saskatchewan, Canada

Rotation[a] and fertilizer treatment	Average yield increase[b] (kg ha^{-1} year^{-1})	Organic N (t ha^{-1} year^{-1})	Potential N-supplying capacity[c] (kg ha^{-1}wk^1)	Available P[d] in 0-120 cm (kg ha^{-1})
F-W-(W)[e]	−28	3.00	28	37
F-W-(W) (N + P)	30	3.24	38	55
Cont. (W)	−11	3.17	36	31
Cont. (W) (N + P)	33	3.46	55	42
GM-W (W)	−26	3.37	44	26
F-W-(W)-H-H-H	−5	3.58	48	24
LSD (p < 0.10)	ND	0.31	8	13

Source: Campbell, Meyers, and Curtin, 1995. Managing nitrogen for sustainable crop production. *Fertilizer Research,* 42:277-296. Reprinted with kind permission of Springer Science and Business Media.

[a]( ) denotes rotation phase sampled.

[b]Average slope of yield trend lines over time.

[c]The initial potential rate of N mineralization (i.e., $N_0 \times k$) where N_0 is the potentially mineralizable N, and k is the rate constant.

[d]Bicarbonate extractable P (Olsen-P).

[e]W = wheat; F = fallow; GM = green manure; H = hay.

of the soil was improved by a combination of fertilizing, reducing tillage, and cropping more frequently (Campbell, Zentner, Selles, et al., 1993). Therefore, although continuous cropping will reduce the amount of residual mineral N in the soil, in the long-term it may increase the potential ability of a soil to supply N because of the increased amount of crop residues returned and their subsequent mineralization.

Phosphorus depletion is also a concern with crop production in the absence of replenishment. McKenzie et al. (1992a,b) evaluated the effect of cropping system and fertilizer management on P in two long-term rotation studies in Alberta, Canada. They found that without fertilizer addition, continuous cropping resulted in the greatest reduction of almost all soil organic and inorganic P pools. By increasing the frequency of fallowing, the drain on the soil P pools was generally decreased because less P was removed by crops. Similarly, in a twenty-four-year study in the Brown soil zone at Swift Current, Saskatchewan, Canada, annual P removal in the grain in fertilized systems decreased with an increase in fallow frequency, being 4.8 to 5.3 kg P ha^{-1} for continuous wheat, 3.5 to 3.8 kg P ha^{-1} for fallow-wheat-wheat and 3.1 kg P ha^{-1} for fallow-wheat (Selles et al., 1995). However, if continuous cropping was coupled with the addition of N and P fertilizers there was a positive P balance and an increase in plant-available soil P (Table 2.3). The residual P fertilizer enriched the inorganic labile pools (resin Pi and bicarbonate Pi), the P held in the microbial biomass, and the moderately labile NaOH Pi (Table 2.4). Bowman and Halvorson (1997) also reported increases in P availability under continuous cropping compared to a wheat-fallow system, even though P inputs were generally greater in the latter system. The authors attributed the increased P availability to redistribution of soil P from lower depths through bio-cycling in residue and litter production.

Soil P depletion may be of particular concern in organic farming systems, if soluble P fertilizers are not allowed and manure is not available. In a case study of five organic dairy farms in Norway, soil concentrations of P decreased over time because the P balance was negative (Løes and Øgaard, 2001). In the absence of applied P, soil organic matter can supply P to growing plants, leading to depletion over time, as was shown by a 50 percent reduction in organic P in a no P treatment in a twenty-five-year study on a subtropical soil (Aulakh et al., 2003). P deficiency can have a major influence on crop yield particularly for legume crops, which would in turn restrict the N fixation. A major study in the United Kingdom at the time of writing this chapter seeks to find ways that would be acceptable to organic farmers, of making the P in phosphate rock readily available for plants grown on neutral and alkaline soils (Johnston, personal communication).

TABLE 2.3. Phosphorus balance for rotations during a twenty-four-year period (1967-1990) at Swift Current, Saskatchewan, Canada

Rotation[a]	Grain Production (Mg ha^{-1})	Mean P[b] concentration (g kg^{-1})	P removed with grain	P applied	Balance[c]
			kg ha^{-1}		
F-W (N + P)	22.6	3.30	74	114	40
F-W-W (N + P)	26.3	3.54	92	152	60
F-W-W (N)	24.4	3.44	84	0	-84
F-Flx-W (N + P)	16.9	3.80	64	134	69
F-Rye-W/ CF-WW-WW[d] (N + P)	26.2	3.36	85	137	53
Cont-W (N + P)	33.0	3.88	126	228	102
Cont-W (P)	28.9	4.09	115	228	113
W-Lent (N + P)[e]	30.0	3.86	116	225	107
LSD (0.05)[f]	0.9	0.07	4	1	4

Source: Selles et al. (1995). Effect of cropping and fertilization on plant and soil phosphorus. *Soil Science Society of America* 59:140-144. Reprinted with permission.

[a]F = fallow, CF = chemical fallow, W = spring wheat, WW = winter wheat, Flx = flax, Rye = fall rye; Lent = grain lentil, Cont = continuous, fertilization in parentheses.

[b]Means weighted by grain yields.

[c]Calculated as inputs-outputs.

[d]Rotation was F-Rye-W until 1984 and CF-WW-WW thereafter.

[e]Rotation was Cont-W (N + P) until 1978 and W-Lent thereafter.

[f]Based on mean squares of the error for ANOVA of crop data with years, rotation, and blocks as class variables.

Using fertilizers in an unbalanced manner may lead to depletion of some nutrients and excess of others. The nutrients N, P, and K are most commonly deficient for crop production, although others may also become limiting. Balanced fertilization ensures that the supply of these nutrients in the soil is maintained at a level such that crop demand at any stage of growth can be satisfied and that long-term soil fertility is not jeopardized. The potential imbalance of P and K fertilization poses a long-term concern for crop production, particularly in areas with degraded soils, or where crop residues are not returned to the soil (Mosier, 2002). For example, in China

TABLE 2.4. Phosphorus fractions (0-15 cm depth) in three selected cropping systems at Swift Current, Saskatchewan, Canada based on the Hedley fractionation procedures.

Fractions[a]	Cont-W (N + P)	F-W-W (N + P)	F-W-W (N)	LSD (0.05)
	kg ha^{-1}			
Resin Pi[b]	81a[c]	84a	44b	5
Microbial P	19a	21a	16b	2
Bicarbonate Po	55	74	69	NS[d]
Bicarbonate Pi	40a	43	24b	4
NaOH Po	196	210	197	NS
NaOH Pi	101a	112a	86b	11
HCl Pi	279	289	269	NS
Sum of all Po fractions	270	305	282	NS
Sum of all Pi fractions	501a	528a	423b	52
Sum of all fractions	771ab	834a	703b	87
Total P	868ab	943a	810b	119

Source: Selles et al. (1995). Effect of cropping and fertilization on plant and soil phosphorus. *Soil Science Society of America* 59:140-144. Reprinted with permission.

[a]Cont-W = continuous wheat, F-W-W = fallow-wheat-wheat.

[b]Pi = inorganic P, Po = organic P.

[c]Means followed by different letters are significantly different ($P < 0.05$).

[d]NS = not significantly different ($P > 0.05$).

a great deal of emphasis has been placed since the 1970s on N fertilization to enhance crop yields to support the large population (see Chapter 8 in this book). Currently China is the world's largest consumer of chemical fertilizers and China and India together account for almost 50 percent of global N fertilizer use (Mosier, 2002). However, the balance of nutrient application is poorly matched with removal. While the situation varies considerably with location, nationwide there is a positive N balance and a negative K balance in India (Aulakh and Malhi, 2004). The availability of native soil K may be adequate on many soils, where the soil mineralogy results in release of K as the soils weather. However, in many regions an adjustment of N, P, and K application to match crop demand more closely would potentially increase production, and economic, and environmental sustainability.

THE CHALLENGE OF NUTRIENT EXCESS

Application of nutrients in excess or out of synchrony with crop requirements can increase both the cost of production and the risk of adverse environmental effects. Nitrous oxide emissions to the atmosphere (Lemke et al., 1999), nitrate leaching to groundwater (Izaurralde et al., 1995; Aulakh et al., 2000; Malhi et al., 2002), hypoxia in coastal waters (Mitsch et al., 1999), P movement to surface waters (Malhi et al., 2003), soil acidification (Malhi et al., 1998), and heavy metal accumulation in soils (Johnston and Jones, 1995 and papers therein) can all be associated with inefficient or excess applications of fertilizers and manures. In addition, the manufacture and transport of fertilizers involves the utilization of nonrenewable resources, providing further environmental impact.

Nutrient accumulation may occur in systems where the cost of nutrients is low in proportion to the value of the crop being produced. In this case, the economic risk associated with lost yield due to underfertilization outweighs the direct cost of excess nutrient inputs. Although this was certainly the case in the past and can occur today, in many situations the excessive use of nutrients can have an adverse environmental impact. For example, nitrate accumulation in groundwater is associated with vegetable and potato production in many areas of the world (Aulakh, 1994; Ramos et al., 2002; Bélanger et al., 2003; Kraft and Stiles, 2003), particularly when crops are grown on sandy soils where excess irrigation rapidly leads to nutrient leaching in water draining through the soil. In Japan, large inputs of N in tea plantations, orchards, and intensive vegetable production have been associated with steady increases in nitrate accumulation in the groundwater over the past two decades (Kumazawa, 2002). With P, enriching a soil so that the plant-available P exceeds the level needed for optimum crop production can cause eutrophication if the soil is eroded into water courses. An inability to accurately predict precise nutrient requirements for optimal crop yield will increase the likelihood of over-application of nutrients.

Overapplication can also occur where nutrients are present in animal manures, biosolids, or other production by-products. The cost associated with the transport and application of these by-products is large relative to the value of the nutrients they contain, particularly when the nutrient concentration in the product is low. They are, therefore, frequently viewed as waste products for disposal rather than as valuable sources of nutrients. Because the costs of disposal increase with increasing transport distance, the greatest direct economic gain is achieved by applying large amounts of the product near the point of origin. In the absence of regulation, this encourages excess nutrient application and accumulation of nutrients in the soil near production sites. This problem is exemplified by intensive livestock operations with imported feed and minimal land area available for manure application (Zebarth et al., 1999; Sharpley et al., 2001). Besides the downside of excess nutrient applications, there can be another serious adverse effect. Frequently, nutrients removed in products intended for animal feed produced in one location and transported to another, often across continents, are not replaced in the soils on which the feeds were produced. This nutrient relocation leads to a loss of soil fertility in the area of production, coupled with accumulation of nutrients in the area of deposition. Excess application of animal slurry, especially to fields in hilly terrain and high-rainfall areas, can lead to large losses of nutrients when heavy rainfall leads to excess water flowing over the soil surface and transporting the entrained nutrients to water courses.

Phosphorus accumulation is of particular concern with livestock production, as the N:P ratio of livestock manure tends to be lower than crop requirements. So, for example, in Europe, the amount of animal manure that can be applied in designated "Nitrate Vulnerable Zones" must not supply more than 170 kg N ha^{-1} but this amount of manure will always supply more P than a crop will remove in one growing season. Although applying slurry periodically would seem the best solution, intensive animal production units rarely have enough land available to do this. In an assessment of fourteen long-term field trials comparing the effects of fertilizers and manures on soil properties, Edmeades (2003) reported manured soils tended to be enriched with P and K in the topsoil, particularly if the application rates were balanced for N requirements. Similarly, solid dairy cattle manure applied at 20 Mg ha^{-1} for many years significantly increased the labile P fractions in a soil cropped to silage corn (Tran and N'dayegamiye, 1995), while larger applications (>100 Mg ha^{-1} every two years) led to excessively large soil P concentrations (Tran et al., 1996). In a Québec catchment, soils under dairy production had significant enrichment of labile P as compared to forest soils (Simard et al., 1995). While the risk of P movement to water is normally through erosive transport of P-enriched soil, overland transport or leaching of dissolved P may occur in specific situations. For example, in sandy soils amended with large amounts of pig and cattle slurry/manure, leaching of P, both as organic and inorganic P, can occur in areas with high groundwater levels (Gerritse, 1981; Chardon et al., 1997; Sims et al., 1998; Koopmans et al., 2003). Application of slurry or compost to no-till systems can lead to a risk of overland movement of dissolved P (Eghball and Gilley, 1999). On the other hand, applications of adequate amounts of organic manures along with N, P, and K fertilizers may reduce the risk of P loss to groundwater. In Rothamsted experiments, where approximately equal amounts of P have been applied in superphosphate and FYM over many years and there has been a positive P balance, the increase in total soil P and Olsen P is about the same but a larger proportion of the P in the FYM-treated soils is soluble in 0.01 M $CaCl_2$. The difference is presumably because the extra SOM provides more low energy bonding sites for P (Johnston and Poulton, 1992). Similarly, the extra SOM provides more cation exchange sites on which K can be held.

CHALLENGES FOR INTEGRATED NUTRIENT MANAGEMENT

A historical perspective shows that throughout its long existence mankind has until comparatively recently relied on each family or small group of families producing just sufficient food for themselves, with crops produced by relying on biological fixation of N and weathering of soil minerals to release P, K, and other nutrients. Such a system supported very few people. In England, in the eighteenth and nineteenth centuries, as in other parts of Europe, self sufficiency was lost as land was amalgamated into larger farms and the displaced country dwellers migrated to the towns. Even so in an ideal situation the land might have been expected to feed more people, if nutrients taken from the land in crops to feed people in the towns could have been recycled with perfect efficiency. Practically such a situation cannot be achieved because nutrients removed from the farm in food for an urban population are rarely returned efficiently. However, there are a number of areas that should be addressed to improve nutrient management, bearing in mind that N must be made available annually while P and K reserves in the soil may provide for the annual needs of a crop.

Prediction of Nutrient Availability

Optimum crop production requires that nutrients be available in the correct amount and proportion, in a position where they can be accessed by the roots when needed. The management of plant nutrients either as fertilizers or manures aims to supplement the soil supply so that the total available quantity is sufficient for optimum crop growth. Thus, effective nutrient management requires accurate prediction of the amount and timing of the nutrient supply from the soil so that the quantity added can be adjusted to meet crop demand. In the absence of accurate information on nutrient supply and requirements, producers will either over- or underapply nutrients, compromising economic and environmental sustainability.

Crop nutrient supply includes inorganic nutrients present in the soil, nutrients released to the plant through the mineralization of organic matter or weathering of soil minerals over the growing season, atmospheric deposition, addition in rainfall or irrigation water, and biological fixation to supply N. Nutrient losses include removal in the crop, retention on soil particles in forms not immediately available to plants, reaction to insoluble compounds, leaching, erosion, and surface runoff, and immobilization of N, P, and S by microbial action. For N there are additional losses by volatilization and denitrification. An effective soil test would measure both the current pool of inorganic nutrients present in the system and the potential supply from soil reserves that will become available to the crop over the growing season.

Accurate prediction of availability is of particular importance for N, as this nutrient is the most commonly limiting for crop production and is the nutrient generally applied in the largest amount in most crop production systems. Growing larger crops by applying N increases the amount of crop residues returned to the soil and this can help maintain existing SOM or slightly increase it, although as noted previously, in the temperate climate at Rothamsted the long continued application of fertilizer N to cereal crops has only increased SOM by about 10 percent. Accurate prediction of the soil N supply to adjust the amount added in fertilizers and manures has the potential to minimize nitrate loss to the groundwater and N_2O emissions to the atmosphere. Results from experiments at Rothamsted show that when N fertilizer has been applied at the optimum amount and time, most of the nitrate in the soil in the autumn following the harvest comes from the mineralization of SOM (McDonald et al., 1989). Soil N supply can be large and is usually highly variable among fields and years (Zebarth et al., 2001). In western Canada, soil testing laboratories generally base N recommendations on a single preplant soil nitrate test and ignore or estimate the contribution from organic N (Walley and Yates, 2002). In England and Wales, soil testing for mineral N to 90 cm is usually only done where there is likely to be significant accumulation of residual N following the ploughing of clover-rich grass swards or vegetable crops given much N (MAFF, 2000). In India and several other Asian countries, the current practice of soil testing for mineral N involves sampling only to 15 or 30 cm depths, whereas applied fertilizer N could rapidly move to a 60 cm depth, especially in coarse-textured porous soils under irrigated cropping systems (Aulakh et al., 2000). In humid environments, although there can be significant amounts of mineral N in the root zone in some fields and years (Zebarth and Milburn, 2003), much of the soil N supply to the growing crop will come from the mineralization of SOM particularly in soils with a large content of SOM.

Nitrogen mineralization from SOM and crop residues is a microbial process, influenced greatly by environmental factors such as moisture and temperature (Goncalves and Carlyle, 1994). Soil characteristics such as aeration, the amounts of soil N and organic C, pH, soil texture, and microbial biomass are also important (Herlihy, 1979; Walley and Yates, 2002). In addition, manure management, cropping history (Zebarth et al., 2001), and other management factors are important. Therefore, the rate and pattern of N mineralization may vary substantially from location to location. Nutrient release patterns are likely to change with shifts in management, such as long- and short-term cropping sequences, residue return, nutrient application practices, and tillage systems (Grant et al., 2002). For example in a nine-year study of continuous no-till spring wheat, soil N supply was improved by a combination of fertilizing, reducing tillage, and intensified cropping (Campbell et al., 1993b). Changing tillage practices has a substantial influence on nutrient dynamics (Mahli et al., 2001; Grant et al., 2002), therefore N management should be changed in response to changes in tillage management to achieve agronomic and environmental objectives. Improved understanding and prediction of N release patterns from SOM and organic residues would allow better synchronization of additional N applications with crop uptake. This prediction is challenging, in part because both the crop and the soil microbial population are in competition for the N released by mineralization and that applied as fertilizers.

The greater the potential for season to season variability in crop yield potential, the more difficult it is to predict the N application to achieve optimum yield. On the sandy loam soil at Woburn, the effect of previous cropping sequences, including all arable crops and short- and long-term leys, on the yields of winter wheat has been studied over a twenty-year period. In each year, the response curve to N fertilizer was measured for each cropping sequence and the optimum economic yield and its associated N application were determined and compared with the fertilizer recommendation given in RB 209, the book of fertilizer recommendations published for England and Wales (MAFF, 2000). Averaged over twenty years, the economic N maximum (N_{emax}) for each cropping sequence differed from RB 209's N recommendation (N_{rec}) by 2 to 6 kg N ha^{-1}. However, the number of individual years when N_{emax} varied by less than 10 kg N ha^{-1} from N_{rec} ranged from four to seven depending on crop sequence. On average, in nine years N_{emax} was smaller than N_{rec} so that too much N would have been applied by following the recommendations, leading to the risk of more than normal N losses. Also, on average, there were only five years when N_{emax} was larger than N_{rec} and farmers could have lost yield and hence income by applying too little N (Johnston and Poulton, personal communication).

So far these differences appear to be related more to seasonal weather than to soil N supply.

The use of in-crop assessments of nutrient sufficiency, especially for N, has the potential to improve nutrient management and reduce environmental risk. Tissue testing, petiole sap testing, and a number of reflectance methods are available to estimate N sufficiency in the growing crop. In-crop N sufficiency estimates used in conjunction with in-crop N application could allow for reduction of N application prior to or at seeding and delay an application of N until N stress in the crop indicates that the application is required.

Return of Nutrients to Agriculture in Residues, Manures, Effluents, and Biosolids

Export of agricultural products from the farm leads to a removal of nutrients from the site of production and their accumulation near urban areas. Return of these exported nutrients to agricultural systems to balance crop demand is a key component of integrated nutrient management. Even on farms there are problems with nutrient return after harvesting crops. When crop residues are returned to the fields on which they were grown the nutrients they contain are returned to the soil. Most of the K will be available to the following crop and the P will not be lost from the system. For N, the decomposition of straw will invariably lead to an initial net immobilization of N in SOM which can be mineralized later. Where home grown produce is fed to animals on the farm the fate of the N, P, and K depends on how and where the manure is used. Even under ideal conditions where animals are fed indoors and every effort is made to conserve all the excreta in farmyard manure, about 15 percent of the N leaving the animal will have been lost during the making of the FYM and a further 35 percent during storage and application to the soil (Voelcker, 1923). If the manure is distributed evenly over the farm then the nutrients it contains should replace part of those removed from different fields, but this rarely happens. Generally less accessible fields do not get applications of slurry and this leads to nutrient depletion whilst fields on which slurry is being disposed have considerable nutrient buildup. An additional challenge to the effective use of nutrients in organic manures is the difference in nutrient ratio between that needed for optimal crop production and that provided in crop residues and manures. As residues and manures provide both rapidly and gradually available nutrients, both the total nutrient input and the pattern of release for crop uptake become important. Effective utilization of organic resources will require balancing the ratio of nutrients in organic and inorganic sources through INM.

Nutrients lost to the farm in exported food for human consumption can have serious implications for soil fertility if they are not replaced. Unfortunately, recycling of nutrients is not easy even in countries with a well-developed infrastructure for the collection and treatment of urban sewage. In treating sewage, almost all the K is lost in the effluent water and much of the N in the readily decomposable organic matter N will be converted to nitrate during microbial decomposition. This nitrate will also be lost in the effluent water if it is not converted to N_2 gas or N_2O under anaerobic conditions. The effluent water should be a source of K and nitrate for irrigation but often it is cheaper to discharge it to the nearest river. There is also the argument that the return of effluent water to the river is essential to maintain the flow of water in the river. Microbial decomposition of organic P during treatment leads to losses of inorganic P in the effluent water. This effluent P was identified as the principle source of P causing the adverse effects of eutrophication in the 1960s and 1970s. Such losses of P must now be controlled from larger treatment works within the EU. One way of removing P would be to precipitate it as struvites, usually magnesium ammonium phosphate, in which the P is readily plant-available (Johnston and Richards, 2003). Organic matter resistant to decomposition during treatment will, when added to soil, be decomposed microbially to produce SOM that then decomposes like any other SOM (Johnston et al., 1989). The organic P in sewage sludge applied to land will increase Olsen P over time (Johnston, 1975) to the benefit of crop production. In some communities, especially in developing countries, human waste is composted with other organic material and although some of the N will be lost, much of the other nutrients should be returned to the soil.

In most developing countries, including those in South Asia, industries still discharge untreated effluents into the sewage system, and being free or cheap, farmers often apply such waters to irrigate crops. The composition of such waters varies depending upon the nature of industry. Studies have shown that potentially toxic elements (Pb, As, Cr, Cd, Ni, and Co) could accumulate in soil and plants due to the application of such sewage waters emanating from industrial towns (Azad et al., 1984; Singh and Kansal, 1985; Brar et al., 2000, 2002; Khurana et al., 2003). However, sewage water from non-industrial towns can be used directly and safely for irrigation when salinity and pH are within safe limits (Brar et al., 2000, 2002; Khurana et al., 2003). This utilization of biosolids and effluents for crop production helps decrease the costs of fertilizer P and improves soil fertility, which is especially beneficial for poor and resource-limited farmers.

Many industrial and municipal waste products can be applied to agricultural land to recycle nutrients. Pressmud cake, a biosolid by-product from the processing of sugarcane in South Asia, is an excellent source of N and P for field crops, especially when used in conjunction with chemical fertilizers. Several studies have shown that pressmud cake could supply significant amounts of N and P to crops and increase the efficiency of applied P fertilizers (Narval et al., 1990; Singh et al., 1997; Singh et al., 2005), increase the solubility of native P (Khanna and Roy, 1956, 2005), and improve soil productivity by improving SOM and available P (Singh et al., 2005). Incinerator waste, ash, and lime-mud from pulp mills, food-processing wastes, and stillage from ethanol production can also be effectively applied to the land, recycling nutrients and potentially improving soil tilth and quality, if managed properly. Integrated management of organic and inorganic nutrient sources can be used to enhance the overall nutrient use efficiency. For example, research is currently being conducted in the UK to improve the availability of phosphate rock by fermenting and/or composting it with organic waste to produce a P source acceptable in organic farming.

Enhanced Understanding and Harnessing of Soil Microbial Activity to Improve Nutrient Utilization by Crops

Soils often contain large amounts of total nutrients held in forms that cannot be directly accessed by crops. Microbiological agents increase the availability of these nutrients for plant uptake, potentially reducing the need to apply fertilizers and manures. Microbes which increase the solubilization of P and elemental sulfur (S) could improve the availability of nutrient sources suited for organic production systems, such as phosphate rock or S. Sulfur-oxidizing bacteria have been investigated to enhance conversion of elemental to plant-available sulfate (Germida and Janzen, 1993; Sholeh et al., 1997; Aulakh, 2003). Similarly, soil and environmental factors have been identified for optimizing the oxidation of elemental S (Table 2.5). The actual period required for the oxidation of S to a plant-available form is determined by soil pH, moisture, and temperature (Jaggi et al., 1999; Aulakh et al., 2002; Aulakh, 2003). Jaggi et al. (1999) observed that in semiarid subtropical soils the oxidation rate of applied S was greatest at 36°C, irrespective of soil pH, suggesting that the application of S would be more beneficial to summer crops than to winter crops, since the application to winter crops occurs when the temperature remains quite low. Where the temperature remains low, S would need to be applied well before seeding to synchronize the availability of S with plant need. Aulakh et al. (2002) demonstrated a strong influence of moisture and soil pH on the rate of S oxidation. Sulfur oxidation was soil pH and moisture dependant, being more rapid in alkaline than in acidic soils and at a higher rather than lower moisture content.

TABLE 2.5. Oxidation of elemental S in acidic, neutral, and alkaline soils incubated for forty-two days at different temperature and moisture regimes

Treatment	Oxidation of applied S (%)		
	Acidic soil	Neutral soil	Alkaline soil
Temperature (°C)			
12	17	23	25
24	24	30	38
36	39	45	52
LSD(0.05) for Temp = 2.0, Soil = 2.0 and Temp × Soil = 4.1			
Moisture regime (%)			
Upland-40 WFPS[a]	18	21	27
Upland-60 WFPS	24	30	38
Flooded-120 WFPS	0	0	0
LSD(0.05) for Moisture = 2.6, Soil = 2.6, and Moisture × Soil = 5.3			

Source: Adapted from Jaggi et al. (1999) and Aulakh et al. (2002).

[a]Water-filled pore space.

Microbial agents could also aid in bioremediation of sites containing excess toxic elements.

Phosphorus solubilizing microorganisms may increase the availability of sparingly soluble P in soils, by secreting organic acids and lowering rhizosphere pH. *Penicillium bilaii* can increase available P in soil and improve the availability of phosphate rock (Kucey, 1983, 1987, 1988; Asea et al., 1988; Kucey and Leggett, 1989). In fifty-five field experiments in Alberta, Saskatchewan, and Manitoba, Gleddie et al. (1991) found that seed inoculation with *P. bilaii* generally increased the yield of wheat grain compared to the yield produced by phosphate fertilizer at 4.4 kg P ha^{-1}. Other studies have shown benefits of *P. bilaii* on field pea (Vessey and Heisinger, 2001) and alfalfa (Beckie et al., 1998), although some studies have shown no advantage in the use of *P. bilaii* in flax (Grant, Dribnenki et al., 1999) and durum wheat (Grant et al., 2002).

Mycorrhizal fungi occur in soils in close association with plant roots and are divided into two groups, ectotrophic and endotrophic mycorrhizae. Ectotrophic mycorrhizae (ECM) cover roots and rootlets with a thick mantle of hyphae that spread between the cortical cells of the roots ensuring close contact with the root. These fungi are mainly found on the roots of trees and shrubs and are of economic importance to the growth of forest trees mainly on nutrient poor soils. Many plant species, including most crops, will form associations with endotrophic mycorrhizae, predominately arbuscular mycorrhizae (AM). The fungus has mycelia that penetrate cells in the root cortex and are connected to external mycelia in the root rhizosphere and soil. Both in the cortex and rhizosphere the hyphae branch extensively and some hyphae may extend more than ten centimeters from root surfaces (Jakobsen et al., 1992). This is a hundred times further than most root hairs, and the small diameter of the hyphae (20-50 μ) allows access to soil pores that cannot be entered by roots. Thus, the mycelial network can access nutrients that would be otherwise unavailable to the plant. The fungus only infects living roots and the infection does not damage the root. The host provides the fungus with life-sustaining organic compounds while the fungus assists the roots in exploiting the soil for water and nutrients. These beneficial effects have been extensively studied in relation to the P nutrition of crops (Tinker, 1980). Moreover, mycorrhizal colonization may induce formation of lateral roots or increase root branching (Citernesi et al., 1998; Schellenbaum et al., 1991), further increasing the volume of soil explored. Mycorrhizal plants can absorb more P at a lower concentration in the soil solution than nonmycorrhizal plants (Plenchette and Morel, 1996). In addition, AM inoculation can increase N, K, Mg, Ca, S, Fe, Zn, and Cu in plants (Kothari et al., 1990; Thompson, 1996; Liu et al., 2000; Ortas and Sari, 2003), especially where little P is applied (Table 2.6). The enhanced nutrient availability may

TABLE 2.6. Effect of P application and mycorrhizal inoculation on the nutrient concentration in the shoots of sweet corn at the silking stage when grown in Turkey

	N	P	K	Zn	Cu	Fe	Mn
Applied P (kg P_2O_5 ha^{-1})	**g kg^{-1} dry weight**			**mg kg^{-1} dry weight**			
Without AM inoculation							
0	23	1.6	8	10.5	3.0	92	90
50	24	2.0	9	14.5	3.8	103	93
100	24	2.2	6	15.0	3.5	160	103
With AM inoculation							
0	26	2.2	12	16.2	3.1	98	110
50	28	2.3	13	15.8	4.0	110	100
100	26	2.4	10	15.4	4.7	130	117

Source: Adapted from Ortas and Sari (2003).

increase crop growth, particularly if P is limiting. For example, on a sandy clay loam soil in England where potatoes, spring barley, and sugar beet were grown in rotation on P-deficient soils, the yield of barley after potatoes was larger than that of barley after sugar beet. This was considered to be so because sugar beet is non-mycorrhizal and, in consequence, there were fewer spores to inoculate the following barley than there were after potatoes (Johnston et al., 1986).

Silverbush and Barber (1983) observed that there is a very small P concentration gradient around hyphae because the radius of hyphae is much smaller than that of roots plus root hairs. Hence, the P concentration in the soil solution around hyphae is higher than in the P depletion zone around roots and, in consequence, hyphae may absorb more P in low P soil even without having a higher affinity for P than do roots and root hairs. Mycorrhizae also have biochemical and physiological characteristics that can enhance P availability. They can acidify the rhizosphere through increased proton efflux or pCO_2 enhancement (Rigou and Mignard, 1994), which can mobilize P (Bago and Azcon-Aguilar, 1997), particularly in neutral or calcareous soils. In acidic soils, where P is mainly bound with Fe or Al, excretion of chelating agents (citric acid and siderophores) by mycorrhizae can increase the bioavailable P supply of the soil (Cress et al., 1986; Hazelwandter, 1995). Mycorrhizae also produce phosphatases which can mobilize P from organic sources (Tarafdar and Marschner, 1994a,b). Importantly, mycorrhizal association does not allow the use of highly insoluble mineral P sources (Tinker, 1980). In a recent review of 51 published papers, data for different combinations of soil type, plant species and P fertilizers showed that the specific activities of P for mycorrhizal and non-mycorrhizal plants did not differ, and that plant-available P, as determined by the L value, did not differ after mycorrhizal inoculation (Morel, 2002). Therefore it appears that the primary benefit of mycorrhizae is in the extension of the zone of P uptake.

It is important to note that, while mycorrhizal associations may be beneficial, they do not necessarily enhance P uptake sufficiently to maximize crop yields on P deficient soils compared to yields on P sufficient soils. Ryan and Ash (1999, 2000) reported that in spite of enhanced mycorrhizal association in biodynamic pastures, the level of P in the forage was below that of conventionally fertilized pastures and that P deficiency may have restricted yields in the biodynamic system. In studies in southeast Australia, neither field pea nor autumn-sown wheat showed a benefit in yield or P uptake from enhanced mycorrhizal colonization, even under low soil-P conditions. This was possibly due to the temperature or moisture during the growing season not being optimum for crop growth and P availability or to the type of AM fungal community present (Ryan and Angus, 2003). It must also be remembered that the carbon compounds that are used by the mycorrhizae for them to function are the same compounds that could go to increase the yield of saleable product. This may well explain in part why yields of crops with mycorrhizal associations on P deficient soils are not as large as those of crops without mycorrhizal infection on P sufficient soils. While improving the availability of nutrients using microbial agents may be beneficial, it leads to mining of nutrient reserves in the soil. Whilst this is acceptable to "kick start" agriculture, especially in developing countries, the extra income derived from the additional yields should be used, at least in part, to replace the nutrients removed in the crops to maintain and improve soil fertility. Enhancement of nutrient availability using microbial agents may be of particular value in situations where it is desirable to avoid increases in already large levels of nutrients in soil, such as locations where nutrient movement to surface waters are of concern. Microbial agents may improve plant uptake of nutrients at a lower bulk soil solution concentration, thus improving crop yield potential while reducing environmental risk.

A number of N-fixing organisms, both symbiotic and non-symbiotic, are known and incorporation of legume crops is a widely-used method of providing fixed N into cropping systems. Symbiotic N fixation will reduce the fertilizer N requirement for optimum yield of the legume crop (Izaurralde et al., 1992). If incorporated into the soil as a green manure, a significant amount of available N can be mineralized if the C:N ratio, is less than 40:1 (Kabba and Aulakh, 2004). Sawatsky and Soper (1991) reported that up to 44 percent of N fixed by legumes remained in the soil after roots were physically removed from the soil, and this N was presumably present in irrecoverable root material, or in organic material lost from the plant root by sloughing and exudation. Some of the N remaining in the soil would become available for subsequent crops. Thus production of legume crops can be an important component in integrated nutrient management, provided that they contribute fully to the economic viability of the farm. It is important to account for the increased N supply to crops grown after legumes or green manures, otherwise over-fertilization could reduce economic efficiency and increase losses of N to the air and water.

There can be other beneficial effects from growing legumes in rotation with cereals. In the Rothamsted experiments with winter wheat, the fertilizer N equivalent for wheat following a crop of field beans (*Vicia faba*) was only about 20 kg N ha^{-1} but the break from wheat diminished the carryover of *Gaeumannomyces graminis,* a soil-borne fungal pathogen, and the yield of wheat responded to more fertilizer N than did wheat grown continuously.

While production of legume crops is widely practiced, utilization of other N-fixing systems may be able to play a greater role in integrated nutrient management. Nitrogen fixation from a wide range of free-living and associative microorganisms has been assessed in rice systems and the N input from fixation may be substantial (Rao et al., 1998). *Azolla* (a water fern) is used as a green manure organic fertilizer in rice production systems in many Asian countries (Singh, 1981). Further work to determine their potential contribution in a wider range of production systems could reduce requirements for N additions in fertilizers and manures.

Microbial agents have great potential benefits in INM but more research is needed both to develop production packages which use them effectively and to determine their impact on nutrient dynamics in the crop-soil system and on nutrient movement in agroecosystems.

Development of Cost-Effective Fertilizer Technologies for Broad-Acre Agriculture to Match the Time of Nutrient Availability to Crop Demand

Where the use of nutrient inputs are required for optimizing crop productivity, it is important that they are utilized as efficiently as possible to improve financial returns, maintain soil quality and reduce the likelihood of adverse effects on the environment. Traditional methods for improving fertilizer use efficiency include selection of an amount of nutrient matched to the requirement of a crop and utilizing a combination of fertilizer source, timing, and placement that reduces nutrient loss and optimizes nutrient recovery by the crop. Many recent papers have reviewed techniques for improving fertilizer use efficiency in various production systems (Dobermann and White, 1998; Mohanty et al., 1998; Malhi et al., 2001; Schlegel et al., 2005) and therefore this topic will not be covered in detail here. While management practices for improving fertilizer use efficiency are widely known and to some extent practiced, there is still considerable room for further adoption. Current trends toward increased costs of fertilizer inputs and increased regulation for the reduction of off-site nutrient enrichment may encourage continued improvement in nutrient use efficiency.

In principle, nutrient use efficiency will be optimized when the nutrient supply to the crop is closely synchronized to crop demand, both in terms of amount and timing. Synchrony between plant-available nutrients in the rooting zone and the crop nutrient uptake will minimize the risk of nutrient losses prior to plant uptake. Applications of N in small increments frequently during the growing season rather as than a single large application at the beginning of the season, can be used to more closely match N supply with the period of maximum N demand (Power et al., 2000). In rice production, split applications of N can improve the synchrony between N supply and crop N demand (Cassman et al., 2002) and decrease the concentration of ammonia in the water, thus reducing volatilization. Where drip irrigation is available, N can be added in the irrigation water to respond to crop requirements (Neilsen and Neilsen, 2002). Split applications can also be applied with trickle and center pivot systems. However, multiple applications of fertilizer may be impractical for low value crops in non-irrigated broad acre agriculture, due to the cost associated with the extra applications. In drier regions where rainfall is erratic, in-crop applications may remain on the soil surface and not be taken up by the crop. Foliar fertilization may be effective in applying relatively small amounts of N, for example for protein enhancement for bread-making wheat, but may not be able to supply the bulk of the N to support yield. Compared to in-soil banded application of nutrients, surface applied nutrients may be subject to greater loss from the soil-plant system because banding reduces the potential for volatilization to the air, leaching or surface runoff in water, and immobilization by soil microorganisms. Volatilization losses from surface applications can be minimized by selecting nitrate-based fertilizers rather than urea or ammonium-based fertilizers for use on calcareous soils. Methods for combining the efficiencies of banding with application of nutrients during crop growth could further improve nutrient use efficiency and reduce the risk of negative environmental impacts.

Spatial variability in soil nutrient supply influences the effects of nutrient management practices on crop yield and nutrient movement to both air and water. For example, in-field yield maps identify areas of good and poor yield and soil chemical analysis and in-field physical examination can often identify the cause. If the effects are due to nutrient deficiency or excess then remedial measures can be taken. In-crop detection of crop N deficiency can allow for improved prediction of N requirements and better timing of in-crop applications (Power et al., 2000; Cassman et al., 2002). Chlorophyll meters are available to detect N status of growing crops (Schlegel et al., 2005). Remote-sensing techniques, based on the use of visible and near infrared spectral analysis can be used to determine crop nutrient stress (Ma et al., 1996). Ground-based sensors utilize the normalized difference vegetation index (NDVI) to estimate crop stress and N requirements (Stone et al., 1996; Raun et al., 2002). Site-specific assessment of nutrient deficiencies in conjunction with surface or in-soil application can improve nutrient use efficiency. Precision guidance systems and the modification of fertilization application equipment would also prevent overlapping applications in fields due to obstacles such as trees and ponds.

Product formulations to control nutrient release, improve the synchrony between nutrient supply and crop demand, enhance availability, and restrict losses are either available or becoming available. Urease and nitrification inhibitors are chemical treatments that may help to reduce N losses from fall, spring, or in-crop N applications, by inhibiting microbial N conversions in the soil (Aulakh et al., 2001b; Watson, 2000). Recent reviews have discussed urease inhibitors (Watson, 2000) and nitrification inhibitors (Zerulla et al., 2000). Large urea granules, co-formulations of finely ground elemental S with P, N, or sulfate, and seed coatings containing P or micronutrients have also been developed to improve nutrient availability. Various coatings for N and P fertilizers have been developed to slowly release nutrients over time, providing a continuous, low concentration of nutrients designed to match the demand of the crop while reducing the risk of seedling damage and of loss of nutrients to the environment (Shaviv, 2001). While the cost of these coatings has been too great for many crops, lower cost formulations are being developed and may be applicable to a broader range of production systems.

Food Quality and Crop Nutrient Management

Crop nutrient management generally focuses on increasing crop yield. However, management practices can also be used to enhance the safety and quality of crops. Management for crop quality requires recognition of the value of improved quality, a mechanism for capturing the value and a method of rewarding the producer for delivering a higher than normal benefit to the customer. While this is often critical for high-value fruit and vegetable crops and for characteristics such as the protein content in wheat for breadmaking and of barley for malting, management for other quality parameters is often not considered. By utilizing an integrated management program including well-planned crop rotations, targeted inputs of inorganic and organic nutrients, genetic selection and consideration of soil and environmental potentials, food and feed can be produced with greater nutritional and functional quality.

Functional Quality

In crops produced for a targeted final function, such as oil production, baking, or malting, quality characteristics such as protein and oil content are critical. Because of the importance of these characteristics for end use, considerable effort is invested in developing cultivars and management practices to optimize functional quality. For example, breeding programs and management practices aim to produce optimum protein content and quality in wheat grown for bread making, pasta, pastry, and noodles. To encourage quality, premiums are paid for optimum protein content and penalties imposed when protein content falls outside the desired range. Attempting to satisfy these criteria normally implies a higher protein level and this can lead to excessive inputs of N, which if not fully used can lead to environmental problems. Similarly, the importance of protein level in malting barley is reflected in the net return to the producer, as samples will be accepted or rejected for the higher price malting market, based in part on protein content. In this case, the issue is to decrease the N input without jeopardizing yield. Therefore, producers will adopt cultivars and management practices that may reduce final yield, but help to maintain the final protein content within required limits.

Food Safety

Nutrient management can influence the presence of both heavy metals and other contaminants in crops. Because B, Cu, Mn, Mo, Ni, and Zn are phytotoxic at concentrations that pose little risk to human health, they are unlikely to be a risk to food safety, except on soils contaminated from industrial activity. In contrast, Cd, Hg, Pb, As, Co, Mo, and Se pose human health risks at plant tissue concentrations that are not generally phytotoxic, leading to greater potential for entry into the food chain (McLaughlin et al., 1999; Dhillon and Dhillon, 1997, 2003).

Cadmium (Cd) is of particular interest because it is a contaminant in P fertilizers (Williams and David, 1976; Grant, Bailey et al., 1999). Codex Alimentarius suggests an upper limit of 0.2 mg kg^{-1} for grains traded internationally. Crops and some animal products with Cd above specified levels are currently excluded from some markets. Flaxseed and sunflower with low Cd concentrations are being grown under contract to meet market demands. Cadmium accumulation in crops is influenced by crop genetic factors and Cd activity in the soil solution. Plant breeding and management practices can be used to reduce the level of Cd in the edible portion of crops (Grant, Bailey et al., 1999). Macronutrient fertilizers, such as N, P, and K, can decrease or increase crop Cd concentration, the latter if the increase in Cd accumulation is proportionally larger than the increase in crop yield (Grant et al., 1996; Mitchell, 1997; Grant and Bailey, 1998). The effect can be minimized by using application rates based on accurate soil testing and reasonable target yields (Grant, Bailey et al., 1999). Specific sources of N, P, or K may have different effects on Cd, but field confirmation is limited (McLaughlin et al., 1995). Zinc fertilization may reduce Cd concentration in crops, particularly where Zn levels are marginal to deficient for crop growth (McLaughlin et al., 1995; Grant and Bailey, 1997). On acid soils, lime application may reduce crop Cd, but results have not been consistent (Grant, Bailey et al., 1999). On high-pH soils, liming is unlikely to reduce Cd concentration in crops and may even increase it due to Cd in the lime. Long-term accumulation of Cd in soils from Cd in fertilizers and soil amendments can be limited by reducing the amount of Cd added through the use of low Cd fertilizers and appropriate rates and placement of added fertilizer. Crops with a high potential for transfer of Cd into the food chain should be grown on soils with low-phyto available Cd. Soils with Cl-based salinity or use of high-Cl irrigation water should be avoided because Cl enhances Cd uptake (McLaughlin et al., 1997).

Nutritional Quality and Nutraceutical Benefits

The first priority in accessing food is to attain calories, mainly through the staple food of the geographical region. When caloric needs are met, extra income goes to purchase non-staple foods, largely animal products, vegetables, and fruits. The Green Revolution helped to increase global food production, but the nutritional quality of the diet has not kept pace with the caloric increase (Welch and Graham, 1999). The role of micronutrients in human health has been discussed recently by Arthur (2004 and references therein). Deficiencies of Fe, Zn, I, and vitamin A affect more than half of the world's population, with impact primarily on low-income women, infants, and children (Welch and Graham, 1999). Improving the nutritional quality of staple foods such as wheat and rice can directly improve human nutrition.

Nutritional deficiencies also occur in more affluent countries. Anemia and Zn deficiency are widespread in North America, particularly among poor pregnant females. Micronutrient deficiencies are more commonly associated with low family income and vegetarian diets. Reducing meat in the diet may increase the incidence of Fe and Zn deficiencies. Increasing cereals in the diet can also increase micronutrient deficiencies, as they contain anti-nutritional factors that lower micronutrient absorption. Even where incomes are such that meat forms a larger part of the diet, there may be a need to add supplements to some types of food. Many basic foods such as flour and milk are fortified with vitamins and/or minerals and vitamins and mineral supplements are widely used, but not always appropriately.

There is also great interest in functional foods that contain nutraceutical phytochemicals that may aid in the prevention and treatment of cancer, hypertension, heart disease, osteoporosis, Alzheimer's disease, arthritis, and a range of other ailments. Compounds such as beta-carotene, alpha-carotene, lycopene, selenium (Se), and vitamin C are associated with a reduction in several cancers, although effects may be due to an increased consumption of fruits and vegetables (Temple, 2000). In spite of questions on the effectiveness of nutraceuticals in isolation, it is projected that the U.S. nutraceutical market will become a $500 billion a year industry by 2010 (Duxbury and Welch, 1999).

As the micronutrient element content of crops varies considerably with cultivar, there is potential to enhance micronutrient content through plant breeding. Plant breeding for improved concentrations of micronutrients is costly but once a seed source is in place, it can be self-generating so the benefit persists after the funding ceases. Plant breeding for greater micronutrient uptake may increase crop yield, by improving the efficiency of micronutrient uptake from the soil, provided that the soil contains a sufficient supply in plant-available forms. A consortium of several research centers, including the Universities of Adelaide and Copenhagen and the USDA-ARS Plant, Soils and Nutrition Laboratory, Ithaca NY, is attempting to provide more Fe and Zn to the world's poorest people by improving the micronutrient content of rice, wheat, maize, cassava, and dry beans.

Apart from cultivar selection, the micronutrient concentration in crops may be affected by management practices such as crop sequence, soil amendment, site selection, manuring, and tillage systems. Direct application of micronutrients, through foliar, soil, or seed-dressings can increase the micronutrient concentration of the grain. Soil applications of Zn can increase grain Zn concentrations in cereal crops by factors of two to three, depending on the species, genotype, and soil characteristics (Rengel et al., 1999). Micronutrient availability is related to soil pH (Johnston, 2004 and references therein). Selenium content can also be enhanced by changes in management, with Se concentration of durum wheat increasing from 200 to more than 1,000 $\mu g\ kg^{-1}$ by application of Se seed coating (Grant et al., 2007). Although high levels of Se are toxic (Dhillon and Dhillon, 1997, 2003), Se deficiency is also a serious problem. In North America, Se deficiencies can cause white muscle disease in cattle, while deficiencies in Finland and China have been associated with increased risk of human cardiovascular diseases and cancer (Rengel et al., 1999). In Finland, Se is frequently added to fertilizers to increase its availability and hence crop content. Low Se also may be a concern in vegetarian diets. There is a reasonable amount of information indicating that elevated levels of Se may reduce cancer incidence and tumor growth. Organic forms of Se such as selenomethionine may be more bioavailable than inorganic selenium salts, so enhanced Se in the grain used to produce whole wheat pasta or bread could be effective in increasing dietary Se.

Macronutrient application can influence micronutrient content of grain, especially if the increase in yield is not matched by an increase in micronutrient uptake. For example, phosphate application reduces the Zn concentration of a variety of crops including durum wheat. Nitrogen management can also increase or decrease the Zn and Cu concentration of durum wheat, through its impact on crop growth, ionic strength of the soil solution, or dilution-concentration effects (Mitchell, 1997). Phytochemicals can also be influenced by macronutrient management, with the lycopene content of tomatoes increasing to 67 percent by application of K (Bruulsema, 2000). The beta-glucan content of whole grains, the lignin and omega-3 fatty acid content of flax, and the isoflavone and saponin content of soybeans all may be of benefit as nutraceuticals and may be influenced by cultivar and management. For instance, optimized supply of N and P produced a larger proportion of polyunsaturated fatty acids in soybean (Aulakh et al., 1995) and sunflower (Bahl et al., 1997), thereby improving the dietary quality of their edible oils. Further information is needed on the effects of nutrient balance, tillage systems, crop rotation, microbial activity, and other cropping practices on both nutrient concentration and nutrient availability.

FUTURE CHALLENGES

The basic principle of INM is to optimize the use of all available nutrient sources, for enhanced economic and environmental sustainability and for the security and quality of the food supply. Nutrient resources available to crops include those in chemical fertilizers, nutrients supplied by soil and by atmospheric deposition, nutrients released by soil biological activity and nutrients recycled from crop residues, organic manures, and other urban and industrial sources. Effective management involves calculation of nutrient requirements to support yield objectives, improvement in the efficiency of nutrient uptake and production efficiency, and reduction of nutrient loss from the agroecosystem. Nutrient management must also be integrated with optimized management of the overall cropping system. Tillage management, crop genetics, pest control, water management, and soil tilth must all be managed effectively so that the crop is able to convert the nutrients supplied into usable yield and crop quality with the greatest efficiency. Yield variability due to controllable environmental stress must be reduced, as the greatest risk to nutrient use efficiency is crop failure. As the yield potential of the crop increases due to factors other than nutrient input, such as better use of water, overall soil fertility, higher yielding cultivars, disease, pest and weed control, timeliness of operations, or improvements in soil structure, nutrient use efficiency will be improved. Hence, all resources will be used more effectively if yield potential is increased.

In spite of dire predictions for imminent famine, the genetic improvement developed during the "Green Revolution" and improved agronomic practices have allowed food production to keep pace with the growing population until the present (Khush, 2001). Population growth continues and it is estimated, based on demographic projections that world food production will need to increase by a further 50 percent (Khush, 2001). The increase must occur in spite of long-term resource depletion, climate change, land degradation, and scarcity of land and water. Stability of food supply is based on the availability of an adequate nutrient supply to nourish the crops and animal products on which the diet of the human population depends. Therefore, throughout the world, INM will play a critical role in economically and environmentally sustainable agricultural production.

REFERENCES

Arthur, J.R. (2004). Micronutrients in the diet for human health and welfare. *Proc. 542. The International Fertiliser Society.* 20p. York, UK.

Asea, P.E.A., R.M.N. Kucey, and J.W.B. Stewart. (1988). Inorganic phosphate solubilization by two *Penicillium* species in solution culture and soil. *Soil Biology and Biochemistry.* 20: 459-464.

Aulakh, M.S. (1994). Integrated nitrogen management and leaching of nitrates to groundwater under cropping systems followed in tropical soils of India. *Transactions of 15th World Congress of Soil Science.* 5(a): 205-221. Acapulco, Mexico.

Aulakh, M.S. (2003). Crop responses to sulphur nutrition. In Y.P. Abrol and A. Ahmad (eds.) *Sulphur in Plants.* pp. 341-358. Boston/London: Kluwer Academic Publishers.

Aulakh, M.S. and G.S. Bahl. (2001). Nutrient mining in agro-climatic zones of Punjab, India. *Fertilizer News.* 46(4): 47-61.

Aulakh M.S., R.C. Jaggi, and R. Sharma. (2002). Mineralization-immobilization of soil organic S and oxidation of elemental S in subtropical soils under flooded and nonflooded conditions. *Biology and Fertility of Soils.* 35(3): 197-203.

Aulakh, M.S., B.S. Kabba, H.S. Baddesha, G.S. Bahl, and M.P.S. Gill. (2003). Crop yields and phosphorus fertilizer transformations after twenty-five years of applications to a subtropical soil under groundnut–based cropping systems. *Field Crops Research.* 83: 283-296.

Aulakh, M.S., T.S. Khera, J.W. Doran, and K.F. Bronson. (2001a). Managing crop residue with green manure, urea, and tillage in a rice—wheat rotation. *Soil Science Society of America Journal.* 65: 820-827.

Aulakh, M.S., T.S. Khera, J.W. Doran, Kuldip-Singh, and Bijay-Singh. (2000). Yields and nitrogen dynamics in a rice—wheat system using green manure and inorganic fertilizer. *Soil Science Society of America Journal.* 64: 1867-1876.

Aulakh, M.S., Kuldip-Singh, and J. Doran. (2001b). Effects of 4-amino 1,2,4-triazole, dicyandiamide and encapsulated calcium carbide on nitrification inhibition in a subtropical soil under upland and flooded conditions. *Biology and Fertility of Soils.* 33: 258-263.

Aulakh, M.S. and S.S. Malhi. (2004). Fertilizer nitrogen use efficiency as influenced by interactions of N with other nutrients. In A. Mosier, K. Syers, and J.R. Freney (eds.) *Agriculture and the Nitrogen Cycle: Assessing the Impacts of Fertilizer Use on Food Production and the Environment.* pp. 189-191. Covelo, CA: Island Press.

Aulakh, M.S., N.S. Pasricha, A.S. Azad, and K.L. Ahuja. (1995). Influence of fertilizer P application on N, P, S and oil content, and fatty acid profile of soybean (*Glycine max* L.). *Journal of Research Punjab Agricultural University.* 32(2): 140-142.

Azad, A.S., B.R. Arora, B. Singh, and G.S. Sekhon. (1984). Nature and extent of heavy metal pollution from industrial units in Ludhiana. *Indian Journal of Ecology.* 2: 1-5.

Badaruddin, M. and D.W. Meyer. (1990). Green-manure legume effects on soil nitrogen, grain yield, and nitrogen nutrition of wheat. *Crop Science.* 30: 819-825.

Badaruddin, M. and D.W. Meyer. (1994). Grain legume effects on soil nitrogen, grain yield, and nitrogen nutrition of wheat. *Crop Science.* 34: 1304-1309.

Bago, B. and C. Azcon-Aguilar. (1997). Changes in the rhizospheric pH induced by arbuscular mycorrhiza formation in onion (*Allium cepa* L). *Zeitschrift Fur Pflanzenernahrung Und Bodenkunde.* 160: 333-339.

Bahl, G.S., N.S. Pasricha, and K.L. Ahuja. (1997). Effect of fertilizer nitrogen and phosphorus on the grain yield, nutrient uptake and oil quality of sunflower. *Journal of the Indian Society of Soil Science.* 45: 292-296.

Bationo, A., F. Lompo, and S. Koala. (1998). Research on nutrient flows and balances in west Africa: State-of-the-art. *Agriculture, Ecosystem and Environment.* 71: 19-35.

Beckie, H., J.D. Schlechte, A.P. Moulin, S.C. Gleddie, and D.A. Pulkinen. (1998). Response of alfalfa to inoculation with *Penicillium bilaii* (Provide). *Canadian Journal of Plant Science.* 78: 91-102.

Bélanger, Gilles, Noura Ziadi, John R. Walsh, John E. Richards, and Paul H. Milburn. (2003). Residual soil nitrate after potato harvest. *Journal of Environmental Quality.* 32: 607-612.

Bowman, R.A. and A.D. Halvorson. (1997). Crop rotation and tillage effects on phosphorus distribution in the central Great Plains. *Soil Science Society of America Journal.* 61: 1418-1422.

Brar, M.S., M.P.S. Khurana, and B.D. Kansal. (2002). Effect of irrigation by untreated effluents on the micro and potentially toxic elements in soils and plants. *Proceedings of the 17th World Congress of Soil Science.* 198: 1-10. Bangkok, Thailand.

Brar, M.S., S.S. Malhi, A.P. Singh, C.L. Arora, and K.S. Gill. (2000). Sewage water irrigation effects on some potentially toxic trace elements in soil and potato plants in northwestern India. *Canadian Journal of Soil Science.* 80: 465-471.

Bruulsema, T.W. (2000). Functional food components: A role for mineral nutrients. *Better Crops with Plant Food.* 84: 4-5.

Campbell, C.A., R.J.K. Meyers, and D. Curtin. (1995). Managing nitrogen for sustainable crop produc tion. *Fertilizer Research.* 42:277-296.

Campbell, C.A. and R.P. Zentner. (1993). Soil organic matter as influenced by crop rotations and fertilization. *Soil Science Society of America Journal.* 57: 1034-1040.

Campbell, C.A., R.P. Zentner, E.G. Gregorich, G. Roloff, B.C. Liang, and B. Blomert. (2000). Organic C accumulation in soil over thirty years in semiarid southwestern Saskatchewan—Effect of crop rotations and fertilizers. *Canadian Journal of Soil Science.* 80: 179-192.

Campbell, C.A., R.P. Zentner, F. Selles, V.O. Biederbeck, B.G. McConkey, B. Blomert, and F.G. Jefferson. (2000). Quantifying short-term effects of crop rotations on soil organic carbon in southwestern Saskatchewan. *Canadian Journal of Soil Science.* 80: 193-202.

Campbell, C.A., R.P. Zentner, F. Selles, B.G. McConkey, and F.B. Dyck. (1993). Nitrogen management for spring wheat grown annually on zero-tillage: Yields and nitrogen use efficiency. *Agronomy Journal.* 85: 107-114.

Carlgren, K. and L. Mattsson. (2001). Swedish soil fertility experiments. *Acta Agriculturae Scandinavica.* 51: 49-76.

Cassman, K.G., A. Dobermann, and D.T. Walters. (2002). Agroecosystems, nitrogen-use efficiency, and nitrogen management. *Ambio.* 31(2): 132-140.

Chardon, W.J., O. Oenema, P. del Castilho, R. Vriesema, J. Japenga, and D. Blaauw. (1997). Organic phosphorus in solutions and leachates from soils treated with animal slurries. *Journal of Environmental Quality.* 26: 372-378.

Citernesi, A.S., C. Vitagliano, and M. Giovannetti. (1998). Plant growth and root system morphology of *Olea europa* L. rooted cuttings as influenced by arbuscular mycorrhizae. *Journal of Horticulture Science and Biotechnology.* 73: 647-654.

Cress, W.A., G.V. Johnson, and L.L. Barton. (1986). The role of endomycorrhizal fungi in iron uptake by *Hilaria jamesii. Journal of Plant Nutrition.* 9: 547-556.

Dhillon, K.S. and S.K. Dhillon. (1997). Distribution of seleniferous soils in northwest India and associated toxicity problems in the soil–plant–animal–human continuum. *Land Contamination and Reclamation.* 5: 313-322.

Dhillon, K.S. and S.K. Dhillon. (2003). Distribution and management of seleniferous soils. *Advances in Agronomy.* 79: 119-184.

Dobermann, A. and P.F. White. (1998). Strategies for nutrient management in irrigated and rainfed lowland rice systems. *Nutrient Cycling in Agroecosystems.* 53: 1-18.

Duxbury, J.M. and R.M. Welch. (1999). Agriculture and dietary guidelines. *Food Policy* 24: 197-209.

Edmeades, D.C. (2003). The long-term effects of manures and fertilisers on soil productivity and quality: A review. *Nutrient Cycling in Agroecosystems.* 66: 165-180.

Eghball, B. and J.M. Gilley. (1999). Phosphorus and nitrogen in runoff following beef cattle manure or compost application. *Journal of Environmental Quality.* 28: 1201-1210.

Fixen, P.E. and A.M. Johnston. (2002). Nutrient budgets in North America. In *Plant Nutrient Use in North American Agriculture.* (pp. 79-85). PPI/PPIC/FAR Technical Bulletin 2002-1. Norcross, GA: Potash & Phosphate Institute.

Galloway, J.N. (1998). The global nitrogen cycle: Changes and consequences. *Environmental Pollution.* 102: 15-24.

Germida, J.J. and H.H. Janzen. (1993). Factors affecting the oxidation of elemental sulfur in soils. *Fertilizer Research.* 35: 101-114.

Gerritse, R.G. (1981). Mobility of phosphorus from pig slurry in soils. In T.W.G. Hucker and G. Catroux (eds.). *Phosphorus in Sewage Sludge and Animal Waste Slurries. Proceedings of the EEC Seminar.* pp. 347-369. Groningen, The Netherlands, June 12-13, 1980. Dordrecht, The Netherlands: D. Reidel Publ.

Gleddie, S.C., G.L. Hnatowich, and D.R. Polonenko. (1991). A summary of wheat response to Provide™ (*Penicillium bilaii*) in western Canada. *Proceedings of Alberta Soil Science Workshop.* pp. 306-313. February 19-21, 1991, Lethbridge, Alberta, Canada.

Glendining, M.J., D.S. Powlson, P.R. Poulton, N.J. Bradbury, D. Palazzo, and X. Li. (1996). The effect of long-term applications of inorganic nitrogen fertilizer on soil nitrogen in the Broadbalk Wheat Experiment. *Journal of Agricultural Science Cambridge.* 127: 347-363.

Goncalves, J.L.M. and J.C. Carlyle. (1994). Modelling the influence of moisture and temperature on net nitrogen mineralization in a forested sandy soil. *Soil Biology and Biochemistry.* 26: 1557-1564.

Grant, C.A. and L.D. Bailey. (1997). Effects of phosphorus and zinc fertilization management on cadmium accumulation in flaxseed. *Journal of the Science of Food and Agriculture.* 73: 307-314.

Grant, C.A. and L.D. Bailey. (1998). Nitrogen, phosphorus and zinc management effects on grain yield and Cd concentration in two cultivars of durum wheat. *Canadian Journal of Plant Science.* 78: 63-70.

Grant, C.A., L.D. Bailey, J.T. Harapiak, and N.A. Flore. (2002). Effect of phosphate source, rate and cadmium content and use of *Penicillium bilaii* on phosphorus, zinc and cadmium concentration in durum wheat grain. *Journal of the Science of Food and Agriculture.* 82: 301-308.

Grant, C.A., L.D. Bailey, M.J. McLaughlin, and B.R. Singh. (1999). Management factors which influence cadmium concentrations in crops. In M.J. McLaughlin, and B.R. Singh (eds.) *Cadmium in Soils and Plants.* pp. 151-198. Dordrecht, The Netherlands: Kluwer Academic Publishers.

Grant, C.A., L.D. Bailey, and M.C. Therrien. (1996). The effect of N, P and KCl fertilizers on grain yield and Cd concentration of malting barley. *Fertilizer Research.* 45: 153-161.

Grant, C.A., W.T. Buckley, and R. Wu. (2007). Effect of selenium fertilizer source and rate on seed yield and selenium and cadmium concentration of durum wheat. *Canadian Journal of Plant Science.*

Grant, C.A., J.C.P. Dribneneki, and L.D. Bailey. (1999). A comparison of the yield response of solin (cv. Linola 947) and flax (cvs. McGregor and Vimy) to application of nitrogen, phosphorus and provide (*Penicillium bilaii*). *Canadian Journal of Plant Science.* 79: 527-533.

Grant, C.A., G.A. Peterson, and C.A. Campbell. (2002). Nutrient considerations for diversified cropping systems in the Northern Great Plains. *Agronomy Journal.* 94: 186-198.

Gregorich, E.G., D.A. Angers, C.A. Campbell, M.R. Carter, C.F. Drury, B.H. Ellert, P.H. Groenevelt, et al. (1995). Changes in soil organic matter. In D.F. Acton, and L.J. Gregorich (eds.) *The Health of Our Soils. Toward Sustainable Agriculture in Canada.* pp. 41-50 Agriculture and Agri-Food Canada Publication 1906/E/. Ottawa, ON, Canada.

Herlihy, M. (1979). Nitrogen mineralisation in soils of varying texture, moisture and organic matter. *Plant Soil.* 53: 255-267.

Holford, I.C.R. and G.J. Crocker. (1997). A comparison of chickpeas and pasture legumes for sustaining yields and nitrogen status of subsequent wheat. *Australian Journal of Agricultural Research.* 48: 305-315.

Holford, I.C.R., A.D. Doyle, and C.C. Leckie. (1992). Nitrogen response characteristics of wheat protein in relation to yield responses and their interactions with phosphorus. *Australian Journal of Agricultural Research.* 43: 969-986.

Izaurralde, R.C., Y. Feng, J.A. Robertson, W.B. McGill, N.G. Juma, and B.M. Olson. (1995). Long-term influence of cropping systems, tillage methods, and N sources on nitrate leaching. *Canadian Journal of Soil Science.* 75: 497-505.

Izaurralde, R.C., W.B. McGill, and N.G. Juma. (1992). Nitrogen fixation efficiency, interspecies N transfer, and root growth in barley-field pea intercrop on a Black Chernozemic soil. *Biology and Fertility of Soils.* 13: 11-16.

Jaggi, R.C., M.S. Aulakh, and R. Sharma. (1999). Temperature effects on soil organic sulphur mineralization and elemental sulphur oxidation in subtropical soils of varying pH. *Nutrient Cycling in Agroecosystems.* 54: 175-182.

Jakobsen, I.L.K. Abbott, and A.D. Robson. (1992). External hyphae of vesicular arbuscular mycorrhizal fungi associated with *Trifolium subterraneum* L. I. Spread of hyphae and phosphorus inflow in roots. *New Phytologist.* 120: 509-516.

Johnston, A.E. (1975). The Woburn Market Garden experiment, 1942-69. II. Effects of treatments on soil pH, soil carbon, nitrogen, phosphorus and potassium. *Rothamsted Experimental Station Report for 1974,* Part 2, 102-131.

Johnston, A.E. (1986). Soil organic matter, effects on soils and crops. *Soil Use and Management.* 2: 97-105.

Johnston, A.E. (1994). The Rothamsted classical experiments. In: R.A. Leigh and A.E. Johnston (eds). *Long-term Experiments in Agricultural and Ecological Sciences.* pp. 9-37. Wallingford, UK: CAB International.

Johnston, A.E. (2004). Micronutrients in soil and agrosystems: Occurrence and availability. *Proceedings No. 544. The International Fertiliser Society.* 32p. York, UK.

Johnston, A.E. and C.J. Dawson. (2005). *Phosphorus in Agriculture and the Aquatic Environment.* Peterborough, UK: Agricultural Industries Confederation.

Johnston, A.E. and K.C. Jones. (1995). The origin and fate of cadmium in soil. *Proceedings No. 366, The International Fertiliser Society.* 39p.York, UK.

Johnston, A.E., P.W. Lane, G.E.G. Mattingly, P.R. Poulton, and M.V. Hewitt. (1986). Effects of soil and fertilizer P on yields of potatoes, sugar beet, barley and winter wheat on a sandy clay loam soil at Saxmundham, Suffolk. *Journal of Agricultural Science, Cambridge.* 106: 155-167.

Johnston, A.E., S.P. McGrath, P.R. Poulton, and P.W. Lane. (1989). Accumulation and loss of nitrogen from manure, sludge and compost: Long-term experiments at Rothamsted and Woburn. In J.A.A. Hansen and K. Henriksen (eds.) *Nitrogen in Organic Wastes Applied to Soils.* pp. 126-139. London, New York: Academic Press.

Johnston, A.E. and P.R. Poulton. (1977). Yields on the Exhaustion Land and changes in the NPK contents of the soils due to cropping and manuring, 1852-1975. *Rothamsted Experimental Station Report for 1976,* Part 2, 53-85.

Johnston, A.E. and P.R. Poulton. (1992). The role of phosphorus in crop production and soil fertility: 150 years of field experiments at Rothamsted, United Kingdom. In J.J. Schultz (ed.) *Phosphate Fertilizers and the Environment.* pp. 45-64. Muscle Shoals, USA: International Fertilizer Development Centre.

Johnston, A.E. and I.R. Richards. (2003). Effectiveness of different precipitated phosphates as phosphorus sources for plants. *Soil Use and Management.* 19: 45-49.

Jones, M.J. (1971). The maintenance of organic matter under continuous cultivation at Samaru, Nigeria. *Journal of Agricultural Science Cambridge.* 77: 473-482.

Kabba, B.S. and M.S. Aulakh. (2004). Climatic conditions and crop residue quality differentially affect N, P and S mineralization in soils with contrasting P status. *Journal of Plant Nutrition and Soil Science.* 167: 596-601.

Khanna, K.L. and P.K. Roy. (1956). Studies on factors affecting soil fertility in sugar belt of Bihar. VI. Influence of organic matter on phosphate availability under calcareous soil conditions. *Journal of the Indian Society of Soil Science.* 4: 189-192.

Khurana, M.P.S., V.K. Nayyar, R.L. Bansal, and M.V. Singh. (2003). Heavy metal pollution in soils and plants through untreated sewage water. In V.P. Singh, and R.N.Yadava (eds.) *Groundwater Pollution.* pp. 487-495. New Delhi, India: Allied Publishers Pvt Limited.

Khush, G.S. (2001). Green evolution: The way forward. *Nature Reviews Genetics.* 2: 815-822.

Kirchmann, H., J. Persson, and K. Carlgren. (1994). *The Ultuna Long-term Soil Organic Matter Experiment (1956-1991).* Reports and Dissertations No. 17. Department of Soil Sciences, Swedish University of Agricultural Sciences Upsalla.

Koopmans, G.F., W.J. Chardon, J. Dolfing, O. Oenema, P. van der Meer, and W.H. van Riemsdijk. (2003). Wet chemical and phosphorus-31 nuclear magnetic resonance analysis of phosphorus speciation in a sandy soil receiving long-term fertilizer or animal manure applications. *Journal of Environmental Quality.* 32: 287-295.

Kothari, S.K., H. Marchner, and V. Romfield. (1990). Effect of VA mycorrhizal fungi and rhizosphere microorganisms on root and shoot morphology, growth and water relationship in maize. *New Phytologist.* 116: 303-311.

Kraft, G.J. and W. Stites. (2003). Nitrate impacts on groundwater from irrigated-vegetable systems in a humid north-central US sand plain. *Agriculture, Ecosystem and Environment.* 100: 63-74.

Kucey, R.M.N. (1983). Phosphate-solubilizing bacteria and fungi in various cultivated and virgin Alberta soils. *Canadian Journal of Soil Science.* 63: 671-678.

Kucey, R.M.N. (1987). Increased phosphorus uptake by wheat and field beans inoculated with a phosphorus-solubilizing *Penicillium bilaii* strain and with vesicular-arbuscular mycorrhizal fungi. *Applied Environmental Microbiology.* 53: 2699-2703.

Kucey, R.M.N. (1988). Effect of *Penicillium bilaii* on the solubility and uptake of P and micronutrients from soil by wheat. *Canadian Journal of Soil Science.* 68: 261-270.

Kucey, R.M.N. and M.E. Leggett. (1989). Increased yield and phosphorus uptake by Westar canola (*Brassica napus* L.) inoculated with a phosphorus-solubilizing isolate of *Penicillium bilaii* strain and with vesicular-arbuscular mycorrhizal fungi. *Canadian Journal of Soil Science.* 69: 425-432.

Kumazawa, K. (2002). Nitrogen fertilization and nitrate pollution in groundwater in Japan: Present status and measures for sustainable agriculture. *Nutrient Cycling in Agroecosystem.* 63: 129-137.

LAT. (2006). *Rothamsted Long-term Experiments.* Lawes Agricultural Trust Co. Ltd. Harpenden, Herts, AL5 2JQ, UK.

Lemke, R.L., R.C. Izaurralde, M. Nyborg, and E.D. Solberg. (1999). Tillage and N source influence soil-emitted nitrous oxide in the Alberta Parkland region. *Canadian Journal of Soil Science.* 79: 15-24.

Lin, B., J. Xie, R. Wu, G. Xing, and Z. Li. (2007). Integrated nutrient management—experience from China. In M.S. Aulakh, and C.A. Grant (eds.) *Integrated Nutrient Management for Sustainable Crop Production.* New York/London/Oxford: Haworth Press, Inc. (in this volume)

Liu, A., C. Hamel, R.I. Hamilton, B.L. Ma, and D.L. Smith. (2000). Acquisition of Cu, Zn, Mn and Fe by mycorrhizal maize (*Zea mays* L.) grown in soil at different P and micronutrient levels. *Mycorrhiza.* 9: 331-336.

Løes Anne-Kristin and Anne Falk Øgaard. (2001). Long-term changes in extractable soil phosphorus (P) in organic dairy farming systems. *Plant and Soil.* 237: 321-332.

Ma, B.L., M.J. Morrison, and L. Dwyer. (1996). Canopy light reflectance and field greenness to assess nitrogen fertilization and yield of maize. *Agronomy Journal.* 88: 915-920.

MacDonald, A.J., D.S. Powlson, P.R. Poulton, and D.S. Jenkinson. (1989). Unused nitrogen fertilizer in arable soils—Its contribution to nitrate leaching. *Journal of the Science of Food and Agriculture.* 46: 407-419.

MAFF. (2000). *Fertiliser Recommendations for Agricultural and Horticultural Crops (RB 209)*. The Stationery Office, London: Ministry of Agriculture, Fisheries and Food. 177p.

Malhi, S.S., S.A. Brandt, D. Ulrich, R. Lemke, and K.S. Gill. (2002). Accumulation and distribution of nitrate-nitrogen and extractable phosphorus in the soil profile under various alternative cropping systems. *Journal of Plant Nutrition*. 25: 2499-2520.

Malhi, S.S., C.A. Grant, A.M. Johnston, and K.S. Gill. (2001). Nitrogen fertilization management for no-till cereal production in the Canadian Great Plains: A review. *Soil and Tillage Research*. 60: 101-122.

Malhi, S.S., J.T. Harapiak, R. Karamanos, K.S. Gill, and N. Flore. (2003). Distribution of acid extractable P and exchangeable K in a grassland soil as affected by long-term surface application of N, P and K fertilizers. *Nutrient Cycling in Agroecosystems*. 67: 265-272.

Malhi, S.S., M. Nyborg, and J.T. Harapiak. (1998). Effects of long-term N fertilizer-induced acidification and liming on micronutrients in soil and in bromegrass hay. *Soil Tillage and Research*. 48: 91-101.

McKenzie, R.H., J.W.B. Stewart, J.F. Dormaar, and G.B. Schaalje. (1992a). Long-term crop rotation and fertilizer effects on phosphorus transformations: I. In a Chernozemic soil. *Canadian Journal of Soil Science*. 72: 569-579.

McKenzie, R.H., J.W.B. Stewart, J.F. Dormaar, and G.B. Schaalje. (1992b). Long-term crop rotation and fertilizer effects on phosphorus transformations: I. In a Luvisolic soil. *Canadian Journal of Soil Science*. 72: 581-589.

McLaughlin, M.J., N.A. Maier, K. Freeman, K.G. Tiller, C.M.J. Williams, and M.K. Smart. (1995). Effect of potassic and phosphatic fertilizer type, fertilizer Cd concentration and zinc rate on cadmium uptake by potatoes. *Fertilizer Research*. 40: 63-70.

McLaughlin, M.J., D.R. Parker, and J.M. Clarke. (1999). Metals and micronutrients—food safety issues. *Field Crops Research*. 60: 143-163.

McLaughlin, M.J., K.G. Tiller, and M.K. Smart. (1997). Speciation of cadmium in soil solution of saline/sodic soils and relationship with cadmium concentrations in potato tubers. *Australian Journal of Soil Research*. 35: 1-17.

Mitchell, L.G. (1997). *Solubility and Phytoavailability of Cadmium in Soils Treated with Nitrogen Fertilizers*. MSc Thesis. University of Manitoba. p. 90.

Mitsch, William J., John W. Day, Jr. J. Wendell Gilliam, Peter M. Groffman, Donald L. Hey, Gyles W. Randall, and Naiming Wang. (1999). *Reducing Nutrient Loads, Especially Nitrate–Nitrogen, to Surface Water, Ground Water, and the Gulf of Mexico: Topic 5 Report for the Integrated Assessment on Hypoxia in the Gulf of Mexico*. NOAA Coastal Ocean Program Decision Analysis Series No. 19. NOAA Coastal Ocean Program, Silver Spring, MD. p. 111.

Mohanty, S.K., U. Singh, V. Balasubramanian, and K.P. Jha. (1998). Nitrogen deep-placement technologies for productivity, profitability, and environmental quality of rainfed lowland rice systems. *Nutrient Cycling in Agroecosystems*. 53: 43-57.

Morel, C. (2002). Caractérisation de la phytodisponiblté du P du sol par la modélisation du transfert des ions phosphate entre le sol et la solution. Habilitation à Diriger des Recherches. INPL-ENSAIA Nancy. 80p. Téléchargeable à l'adresse http://www.bordeaux.inra.fr/tcem/rubrique: Actualités. Accessed May 4, 2004.

Mosier, A.R. (2002). Environmental challenges associated with needed increases in global nitrogen fixation. *Nutrient Cycling in Agroecosystems.* 63: 101-116.

Narval, P., A.P. Gupta, and R.S. Antil. (1990). Efficiency of triple super phosphate and Mussoorie rock phosphate mixture incubated with sulphitation process pressmud. *Journal of the Indian Society of Soil Science.* 38: 51-55.

Neilsen, D. and G.H. Neilsen. (2002). Efficient use of nitrogen and water in high density apple orchards. *HortTechnology.* 12: 19-25.

Ortas, I and N. Sari. (2003). Enhanced yield and nutrient content of sweet corn with mycorrhizal inoculation under field conditions. *Agricoltura Mediterranea.* 133: 188-195.

Paustian, K., W.J. Parton, and J. Persson. (1992). Modeling soil organic matter in organic amended and nitrogen-fertilized long-term plots. *Soil Science Society of America Journal.* 56: 476-488.

Peterson, G.A., A.D. Halvorson, J.L. Havlin, O.R. Jones, and D. Tanaka. (1998). Reduced tillage and increasing cropping intensity in the Great Plains conserves soil C. *Soil and Tillage Research.* 47: 207-218.

Plenchette, C. and C. Morel. (1996). External phosphorus requirement of mycorrhizal and non-mycorrhizal barley and soybean plants. *Biology and Fertility of Soils.* 21: 303-308.

Power, J.F., R. Wiese, and D. Flowerday. (2000). Managing nitrogen for water quality—Lessons from management systems evaluation area. *Journal of Environmental Quality.* 29(2): 355-366.

Ramos, C., A. Agut, and A.L. Lido. (2002). Nitrate leaching in important crops of the Valencian Community region (Spain). *Environmental Pollution.* 118: 21-223.

Rao, V. R., B. Ramakrishnan, T.K. Adhya, P.K. Kanungo, and D.N. Nayak. (1998). Review: Current status and future prospects of associative nitrogen fixation in rice. *World Journal of Microbiology and Biotechnology.* 14: 621-633.

Raun, W.R., J.B. Solie, G.V. Johnson, M.L. Stone, R.W. Mullen, K.W. Freeman, W.E. Thomason, and E.V. Lukima. (2002). Improving nitrogen use efficiency in cereal grain production with optical sensing and variable rate application. *Agronomy Journal.* 94: 815-820.

Rengel, Z., G.D. Batten, and D.E. Crowley. (1999). Agronomic approaches for improving the micronutrient density in edible portions of field crops. *Field Crops Research.* 60: 27-40.

Rigou, L. and E. Mignard. (1994). Factors of acidification of the rhizosphere of mycorrhizal plants. Measurement of pCO_2 in the rhizosphere. *Acta Botanica Gallica.* 141: 533-539.

Ryan, M.H. and J.F. Angus. (2003). Arbuscular mycorrhizae in wheat and field pea crops on a low P soil: Increased Zn-uptake but no increase in P-uptake or yield. *Plant Soil.* 250: 225-239.

Ryan, M. and J. Ash. (1999). Effects of phosphorus and nitrogen on growth of pasture plants and VAM fungi in SE Australian soils with contrasting fertiliser histories (conventional and biodynamic). *Agriculture Ecosystems and Environment.* 73: 51-62.

Ryan, M. and J. Ash. (2000). Phosphorus controls the level of colonisation by arbuscular mycorrhizal fungi in conventional and biodynamic irrigated dairy pastures. *Australian Journal of Experimental Agriculture.* 40: 663-670.

Sawatsky, N. and R.J. Soper. (1991). A quantitative measurement of the nitrogen loss form the root system of field peas (*Pisum avense* L.) grown in the field. *Soil Biology and Biochemistry.* 23: 255-259.

Schellenbaum, L., G. Berta, F. Ravolanirina, B. Tisserant, S. Gianinazzi, and A.H. Fitter. (1991). Influence of mycorrhizal infection on root morphology in a micropropagated woody plant species (*Vitis vinifera* L.). *Annals of Botany.* 68: 135-141.

Schlegel, A.J., C.A. Grant, and J.L. Havlin. (2005). Challenging approaches to nitrogen fertilizer recommendations in continuous cropping systems in the Great Plains. *Agronomy Journal.* 97: 391-398.

Selles, F., C.A. Campbell, and R.P. Zentner. (1995). Effect of cropping and fertilization on plant and soil phosphorus. *Soil Science Society of America Journal.* 59: 140-144.

Sharpley, A.N., R.W. McDowell, and P.J.A. Kleinman. (2001). Phosphorus loss from land to water: Integrating agricultural and environmental management. *Plant and Soil.* 237: 287-307.

Shaviv, A. (2001). Advances in controlled release fertilizers. *Advances in Agronomy.* 71: 1-49.

Sholeh, R., D.B. Lefroy, and G.J. Blair. (1997). Effect of nutrients and elemental sulfur particle size on elemental sulfur oxidation and the growth of *Thiobacillus thiooxidans*. *Australian Journal of Agricultural Research.* 48: 497-501.

Silverbush, M. and S.A Barber. (1983). Sensitivity of simulated phosphorus uptake to parameters used by a mechanistic-mathematical model. *Plant Soil.* 74: 93-100.

Simard, R.R., D. Cluis, G. Gangbazo, and S. Beauchemin. (1995). Phosphorus status of forest and agricultural soils from a watershed of high animal density. *Journal of Environmental Quality.* 24: 1010-1017.

Sims, J.T., R.R. Simard, and B.C. Joern. (1998). Phosphorus loss in agricultural drainage: Historical perspective and current research. *Journal of Environmental Quality.* 27: 277-293.

Singh, C.P., N. Singh, N.S. Dangi, and B. Singh. (1997). Effect of rock phosphate enriched pressmud on dry matter yield, phosphorus and sulphur nutrition of mung bean (*Vigna radiata* L. Wilczek). *Indian Journal of Plant Physiology.* 2: 262-266.

Singh, H., Y. Singh, and K.K. Vashist. (2005). Evaluation of pressmud cake as source of phosphorus for rice-wheat rotation. *Journal of Sustainable Agriculture.* 26: 5-21.

Singh, J. and B.D. Kansal. (1985). Amount of nutrients and heavy metals in the sewage water of different towns of Punjab and its evaluation for irrigation. *Journal of Research Punjab Agricultural University.* 22: 17-24.

Singh, P.K. (1981). Use of *Azolla* and blue-green algae in rice cultivation in India. In Bose P.V. and A.P. Ruschel (eds.) *Associative N₂-Fixation.* (2: 236-242). Boca Raton, FL: CRC. Press.

Sommer, R., R.L.G. Vlek, T.D.A. Sá, K. Vielhauer, R.F.R. Coelho, and H. Fölster. (2004). Nutrient balance of shifting cultivation by burning or mulching in the eastern Amazon—Evidence for subsoil nutrient accumulation. *Nutrient Cycling in Agroecosystems.* 68: 257-271.

Stone, M. L., J.B. Solie, W.R. Raun, R.W. Whitney, S.L. Taylor, and J.D. Ringer. (1996). Use of spectral radiance for correcting in-season fertilizer nitrogen deficiencies in winter wheat. *Transactions of the ASAE.* 39: 1623-1631.

Stoorvogel, J.J. and E.M.A. Smaling. (1998). Research on soil fertility decline in tropical environments: Integration of spatial scales. *Nutrient Cycling in Agroecosystems.* 50: 153-160.

Tarafdar, J.C. and H. Marschner. (1994a). Phosphatase activity in the rhizosphere and hyphosphere of VA mycorrhizal wheat supplied with inorganic and organic phosphorus. *Soil Biology and Biochemistry.* 26: 387-395.

Tarafdar, J.C. and H. Marschner. (1994b). Efficiency of VAM hyphae in utilisation of organic phosphorus by wheat plants. *Soil Science and Plant Nutrition.* 40: 593-600.

Temple, N.J. (2000). Antioxidants and disease: More questions than answers. *Nutrition Research.* 20: 449-459.

Thompson, J.P. (1996). Soil sterilization methods to show VA-Mycorrhizae aid P and Zn nutrition of wheat in vertisols. *Soil Biology and Biochemistry.* 22: 229-240.

Tinker, P.B. (1980). Role of rhizosphere microorganisms. In Khasawneh F.E., E.C. Sample, and E.J. Sample. (eds.) *The Role of Phosphorus in Agriculture.* pp. 617-654. Madison WI: ASA CSSA SSSA Publication.

Tran, T.S. and A.N'dayegamiye. (1995). Long-term effects of fertilizers and manure application on the forms and availability of soil phosphorus. *Canadian Journal of Soil Science.* 75: 281-285.

Tran, T.S., D. Côté, and A. N'dayegamiye. (1996). Effets ds apports prolongés de fumier et de lisier sur l'évolution des teneurs du sol en éléments nutritifs majeurs et mineurs. *Agrosol.* 9: 21-30.

Vessey, J.K. and K.G. Heisinger. (2001). Effect of *Penicillium bilaii* inoculation and phosphorus fertilization on root and shoot parameters of field-grown pea. *Canadian Journal of Plant Science.* 81: 361-366.

Voelcker, J.A. (1923). The Woburn experimental farm and its work, 1876-1921. *Journal of the Royal Agricultural Society of England.* 63: 76-114.

Walley, F. and T. Yates. (2002). Estimating soil nitrogen release-what factors need to be considered? *Proceedings of the Manitoba Agronomists Conference 2002.* University of Manitoba. http://www.umanitoba.ca/faculties/afs/agronomists _conf/2002/pdf/walley.pdf (February 4, 2004).

Watson, C.J. (2000). Urease activity and inhibition—Principles and practice. *Proceedings No. 454. The International Fertilizer Society.* York, UK. 39p.

Weinhold, B.J. and A.D. Halvorson. (1999). Nitrogen mineralization responses to cropping, tillage, and nitrogen rate in the northern Great Plains. *Soil Science Society of America Journal.* 63: 192-196.

Welch, R.M. and R.D. Graham. (1999). A new paradigm for world agriculture: Meeting human needs: Productive, sustainable, nutritious. *Field Crops Research.* 60: 1-10.

Welty, L.E., L.S. Prestbye, R.E. Engel, R.A. Larson, R.H. Lockerman, R.S. Speilman, J.R. Sims, L.I. Hart, G.D. Kushnak, and A.L. Dubbs. (1988). Nitrogen contribution of annual legumes to subsequent barley production. *Applied Agricultural Research.* 3: 98-104.

Williams, C.H. and D.J. David. (1976). The accumulation in soil of cadmium residues from phosphate fertilizers and their effect on the cadmium content of plants. *Soil Science.* 121: 86-93.

Wood, C.W., D.G. Westfall, G.A. Peterson, and I.C. Burke. (1990). Impacts of cropping intensity on carbon and nitrogen mineralization under no-till dryland agroecosystems. *Agronomy Journal.* 82: 1115-1120.

Zebarth, B.J. and P.H. Milburn. (2003). Spatial and temporal distribution of soil inorganic nitrogen concentration in potato hills. *Canadian Journal of Soil Science.* 83: 183-195.

Zebarth, B.J., J.W. Paul, and R. Van Kleeck. (1999). The effect of nitrogen management in agricultural production on water and air quality: evaluation on a regional scale. *Agriculture, Ecosystems and Environment.* 72: 35-52.

Zebarth, B.J., J.W. Paul, M. Younie, and S. Bittman. (2001). Fertilizer N recommendations for silage corn in high-fertility environment based on the pre-side dress soil nitrate test. *Communications in Soil Science and Plant Analysis.* 32: 2721-2739.

Zerrulla, W., K-F. Kummer, A. Wissemeier, and M. Roidle. (2000). The development and testing of a new nitrification inhibitor. *Proceedings No. 455, The International Fertiliser Society,* York, UK. 23p.

Chapter 3

Integrated Nutrient Management: Experience and Concepts from the United States

Mark M. Alley
Dwayne G. Westfall
Gregory L. Mullins

INTRODUCTION

Nutrient management for agronomic crop production from the late 1800s and through the 1970s in the United States focused on using nutrients efficiently. Nutrients were relatively expensive, and many areas were greatly deficient in nutrients, especially nitrogen (N) and phosphorus (P). Classic textbooks such as *Soil Fertility and Fertilizers* (Tisdale and Nelson, 1956), *The Nature and Properties of Soils* (Buckman and Brady, 1960), and *Economics of Agricultural Production and Resource Use* (Heady, 1952) discuss nutrient management in terms of fertilizer use and efficiency of production, as well as sustainability of soil productivity. It was not until the 1980s and 1990s that "nutrient management" in the United States became synonymous with the management of nutrients to reduce potential enrichment of surface and groundwaters with N and P. This change is clearly reflected in the most recent editions of *Soil Fertility and Fertilizers* (Havlin et al., 2005) and the *Nature and Properties of Soils* (Brady and Weil, 2002). In addition,

The authors wish to acknowledge Rhonda Shrader for assistance with manuscript preparation, and Steve Nagle and Catherine Byers for assistance in data summarization and table preparation.

many current nutrient management efforts focus on managing manure nutrients that have resulted from concentrated animal feeding operations (USDA-NRCS, 2004a,b).

The current U.S. situation in regard to nutrient management programs mainly referring to efforts to reduce nonpoint source pollution from organic wastes has resulted from the trends of increasing farm specialization and size. Animal production farms are concentrated for efficiency of production and processing and are not, in many cases associated with grain crop production. This system has also resulted in the concentration of manure nutrients in localized regions.

Yields of maize (*Zea mays* L.), soybean (*Glycine max.* L.), wheat (*Triticum aestivum* L., *T. durum* Desf.), and cotton (*Gossypium* spp.) have increased at a moderate rate from 1960 to 2000 as shown in Figure 3.1. Although the increase from year to year fluctuates, the general trends do result in significant increased nutrient needs when one considers a ten- to twenty-year time period. For example, maize yields averaged approximately 6 tonnes (t) ha^{-1} in 1980 and were slightly above 8.0 t ha^{-1} in 2000.

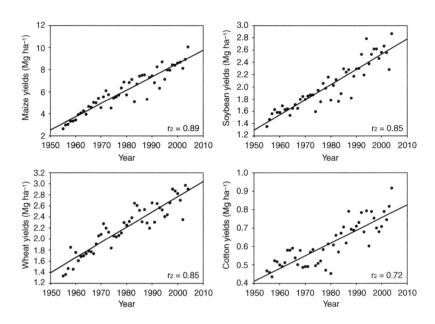

FIGURE 3.1. Maize, soybean, wheat, and cotton yields, 1960-2000 for the United States. *Source:* USDA-National Agricultural Statistics Service (2004).

Soybean and wheat yield increases were much smaller, but represent significant increases that result in higher nutrient requirements and removals. Dobermann and Cassman (2002) suggest that average farm yields are 40 to 65 percent of the attainable yield potential and thus with improved management, significant yield increases are possible in many instances. Increased nutrient removal will occur as crop yields increase.

The objectives of this chapter are to (1) discuss the general nature of the major agronomic crop production regions in the United States; (2) present general nutrient balances for nutrient removal versus applications; (3) outline approaches to nutrient management in these regions; and (4) provide ideas on nutrient management research needs.

MAJOR CLIMATIC REGIONS AND CROP PRODUCTION

The continental United States excluding Alaska has diverse soil and climate regimes. Climatic conditions range from subtropic wet in southeastern United States to "arid steppe" in the western regions (Figure 3.2). Average temperature and precipitation ranges for the continental United States are

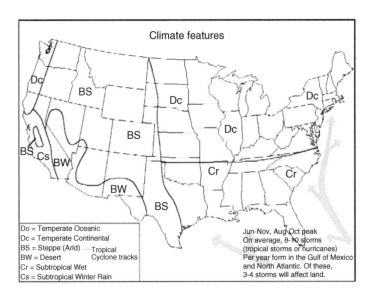

FIGURE 3.2. Climatic zones in the continental United States. *Source:* USDA-Office of the Chief Economist (2007).

shown in Figure 3.3, and illustrate the wide array of conditions that exist for agronomic crop production (see also color gallery).

Soils vary widely, as would be expected, depending on climatic conditions and parent material. The highly productive north central region known as the "corn belt" has a temperate continental climate and is predominated by mollisols (Figure 3.4, see also color gallery). The southeastern United States is predominated by highly weathered ultisols associated with acidic parent materials, and/or coarse-textured sandy materials in the coastal regions. The Great Plains extend from Canada in the north to Mexico in the south, and are east of the Rocky Mountains. The climate in this region is defined as being an "arid steppe" and many soils are developed from loess.

Major and minor growing areas for maize (corn), soybean, wheat, cotton, barley (*Hordeum vulgare* L.), and rice (*Oryza sativa* L.) are shown in Figure 3.5 (see also color gallery). These regions reflect the soil and climatic conditions that are most suitable for the production of various crops, and the locations in which the crops have a competitive advantage. Total areas devoted to various crops in 2003 are shown in Table 3.1.

The agronomic crops maize, wheat, barley, soybean, cotton, rice, sorghum (*Sorghum bicolor* L. Moench), millet (*Pennisetum americanum* L.), oats and sunflower (*Helianthus annuus* L.) are harvested from approximately 89,030.945 ha annually in the United States, and alfalfa (*Medicago sativa* L.) is harvested from approximately 10,119,102 ha annually (Table 3.1). These crops provide the basis for food, feed, fiber and fuel production in the United States, as well as contributing significant amounts of products for world

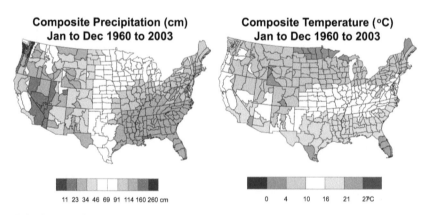

FIGURE 3.3. Composite average precipitation and temperature data for the continental United States from 1960 to 2003. *Source:* www.cdc.noaa.gov.

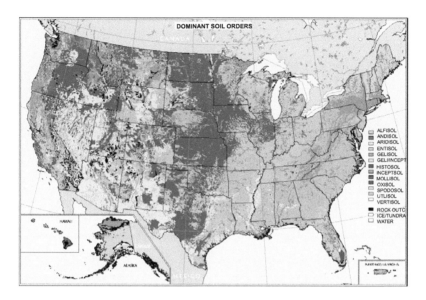

FIGURE 3.4. Dominant soil orders in the United States. *Source:* http://www.nrcs
.usda.gov/technical/land/meta/m4025.html.

trade. Agricultural exports from the United States were valued at slightly over US$59 billion in 2003 (http://www.ers.usda.gov/Data/ FATUS/).

CROP PRODUCTION AND NUTRIENT REMOVALS IN HARVESTS

Nutrient removals by agronomic crops in the United States are best discussed by region due to diverse climatic conditions and thus highly different yield levels. Nutrient removals were calculated by multiplying harvested area in hectares by average yield per ha times the percent of the nutrient in the harvested portion of the crop. Values for the nutrient content in the harvested portion of the various crops are shown in Table 3.2, and are from data summaries compiled by the Potash and Phosphate Institute (Anon., 2001). Nutrient contents of harvested crops vary greatly depending on fertility level, yield level, and specific growing conditions. However, the values utilized are reasonable estimates and are the best available for the general purpose of this discussion.

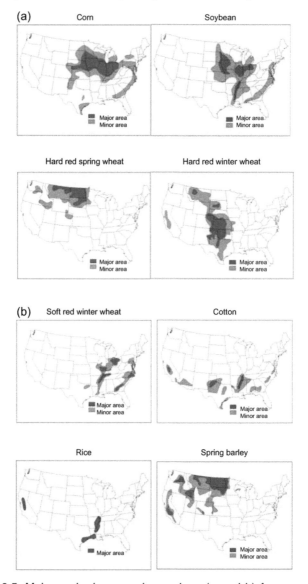

FIGURE 3.5. Major and minor growing regions (a and b) for corn, soybean, wheat, cotton, barley, and rice in the continental United States. *Source:* USDA-Office of the Chief Economist (2007). Major world crop areas and climatic profiles, North America. Available online: http://www.usda.gov/oce/weather/pubs/Other/MWCACP/world_crop_country.htm#northamerica (accessed April 2007).

TABLE 3.1. Crop, area harvested, and average yields for 2003 in the United States

Crop	Area (ha)	Average yield (Mg ha^{-1})
Maize (grain)	28,737,023	9.17
Maize (silage)	1,967,371	36.80
Soybean	25,229,150	2.33
Wheat	21,486,429	3.29
Cotton	2,603,239	0.96
Barley	1,914,775	3.67
Oats	911,089	2.35
Rice	935,223	7.22
Alfalfa	10,119,102	8.26
Sorghum	4,130,061	3.47
Millet	263,250	1.86
Sunflower	853,335	1.24
Total	99,150,047	–

Source: USDA-National Agricultural Statistics Service (2004). http://www.nass.usda.gov/index.asp.

TABLE 3.2. Nutrient removal in harvested portions of crops

Crop	N	P	K
	(kg mg^{-1})		
Maize (grain)	13.4	3.4	4.3
Maize (silage)	4.1	0.7	3.4
Soybean	66.7	5.8	19.4
Cotton (lint)	64.0	12.0	31.5
Wheat (10 percent protein)	18.3	3.6	4.8
Barley	22.9	3.6	6.1
Rice	12.7	2.9	3.5
Alfalfa	28.0	3.2	24.9

Source: Anonymous (2001), Potash and Phosphate Institute.

Western Region

The area of the United States we refer to as the western United States includes all of the states west of, and including, North and South Dakota, Nebraska, Kansas, Oklahoma, and Texas, or specifically the following: - Arizona, Colorado, California, Idaho, Kansas, Montana, Nebraska, Nevada, New Mexico, North Dakota, Oklahoma, Oregon, South Dakota, Texas, Utah, Washington, and Wyoming.

The major agronomic crops produced in this region are wheat, maize, sorghum, alfalfa, barley, sunflower, and millet. Wheat is the major crop and is planted on over 21 Mha in the region, essentially all under dryland (nonirrigated) conditions (USDA-National Agricultural Statistics Service (NASS), 2004). However, due to crop abandonment (failure) not all of this acreage is harvested, as explained below. Three types of wheat make up this production, winter wheat, spring wheat, and durum. Winter wheat is typically planted to about 15 Mha while spring and durum occupy smaller acreages, 4.8 and 1.2 Mha, respectively. Winter wheat is grown in every state in this region except Arizona and Nevada. The major spring wheat producing states are Colorado, Idaho, Montana, North and South Dakota, Oregon, Utah, Washington, and Wyoming. Durum wheat production is concentrated in Arizona, California, Montana, and North and South Dakota.

Wheat

The majority of the winter wheat is grown in a summer fallow cropping system that comprises a winter wheat-fallow (WF) system. Limited rainfall is the major growth-limiting factor in this semiarid environment (Peterson and Westfall, 2004) and impacts fertilizer management practices. One crop is produced every two years with the fallow year being used to store soil water in this semiarid climate. The WF system consists of a ten-month crop growth period and a fourteen-month fallow period. Summer fallow is usually practiced in areas receiving 255 mm of annual precipitation or less (Haas et al., 1974). About 30 Mha are devoted to winter wheat production using this WF system in the western United States. No-till intensive cropping systems that maximize the capture of natural precipitation are slowly being adopted. These systems decrease the length of the fallow period by integrating summer crops such as maize, millet, and sunflower into the cropping system. This is driven by the need for residue retention on the soil surface to prevent wind erosion and improved economics (Peterson et al., 1993; Kaan et al., 2002). When producers adopt no-till intensive cropping systems, 83 percent more N fertilizer is required in comparison to the traditional WF system (Westfall et al., 1996). These authors also reported N use efficiencies for WF of 25-28 percent, however, the N use efficiency of winter wheat-maize-fallow systems ranged from 40 to 47 percent.

Crop failure due to drought is not uncommon in the winter wheat production areas. As stated above, about 15 Mha are planted to winter wheat, however, averaged over the years from 1991 to 2000, 18 percent of the acreage planted was not harvested (crop failure). The western United States came under a severe drought which began in 2000-2001 and did not abate till 2004. This drought resulted in a larger percentage of the winter wheat not being harvested; averaged over the years from 2001 to 2004, 25 percent of the acreage planted was not harvested. In addition, severe weather, mainly hail storms, impacts production, but generally in limited geographic areas. Much of the failed winter wheat acreage is replanted in the spring to summer crops such as millet, maize, and sunflower. However, farmers are hesitant to put large fertilizer expenditures into winter wheat production because of the high percentage of crop failure.

In general, farmers take a very conservative approach to fertilization, applying minimal amounts of N and P for crop growth. Other nutrients are rarely applied. Yields of winter wheat are relatively low, averaging 3.1 Mg ha^{-1} in 2003 (USDA-NASS, 2004). In years with adequate rainfall, lack of adequate N fertilizer inputs limit production. The inability of producers to match the N inputs with yields, as driven by variable precipitation, is indicated by years of lower and years of higher than average yields. Characteristically, wheat will have a low grain protein content in years of higher yields and high grain protein content in years of lower yields. Goos et al. (1982) reported that if the grain protein content was less than 11.5 percent, wheat yields were limited by inadequate N fertility; however, grain yield was not limited by N fertility if the protein content was greater than 12.0 percent.

The historic yield of all wheat types (spring, durum, and winter) in the western region has increased over the past fifty years (Figure 3.6). In 1955, the average yield was 1.28 Mg ha^{-1}, increasing to 3.17 Mg ha^{-1} by 2003. In 1999, before the onset of drought, the average yield was 3.30 Mg ha^{-1} (USDA-NASS, 2004). Continued advancements in variety development, genetic engineering, and improved management practices will result in a continued increase in wheat yields into the future. If more accurate long-term weather predictions were possible, it could have a major impact on fertilizer use through the improved ability of producers to match the potential yield with nutrient needs, as well as decreased crop abandonment (failure).

The area of all types of wheat harvested increased from 1950 into the 1980s; however, from 1982 to 2003 the area of wheat harvested in the western United States decreased from a maximum of 23.9 Mha to 17.9 Mha in 2003 (Figure 3.7). This decrease was mainly due to competition from other

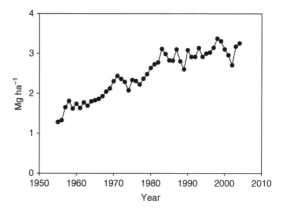

FIGURE 3.6. Average wheat yield from 1955 to 2003 in the western United States. *Source:* Data adapted from USDA-National Agricultural Statistics Service (2004).

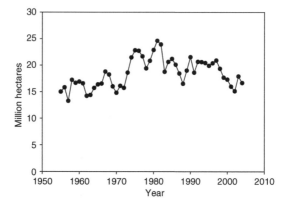

FIGURE 3.7. Total wheat area harvested from 1965 to 2003. *Source:* Data adapted from USDA-National Agricultural Statistics Service (2004).

crops that are more profitable and federal programs such as the Conservation Reserve Program that encouraged conversion of highly erosive land to permanent grass for soil conservation purposes.

Nutrient removal by wheat constitutes the greatest amount of removal for all nonlegume crops in the western United States. Nutrient removal increased from 1954 to 1982 (Figures 3.8 through 3.10). In 1954, nutrient removal

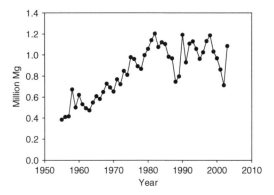

FIGURE 3.8. Total N removed in wheat in the western United States, from 1965 to 2003. *Source:* Calculated from USDA-National Agricultural Statistics Service (2004), and Anonymous (2001b). Nutrients removed in harvested portion of crop.

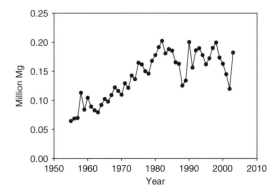

FIGURE 3.9. Total P Removed in wheat in the western United States, from 1965 to 2003. *Source:* Calculated from USDA-National Agricultural Statistics Service (2004), and Anonymous (2001b). Nutrients removed in harvested portion of crop.

levels were 383,600 Mg N, 64,500 Mg P, and 85,700 Mg K. By 1982, the nutrient removal had increased to 1,202,900 Mg N, 202,200 Mg P, and 268,800 Mg K. After 1982 there was a decrease in nutrient removal, as driven by the decrease in acreage of winter wheat and winter wheat grain production. Spring and durum wheat production continued to increase slightly after 1982. In 2003, all types of wheat removed 1,083,400 Mg of N, 182,100 Mg P, and 242,100 Mg K, respectively (Table 3.3). Spring wheat

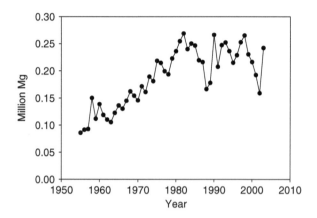

FIGURE 3.10. Total K removed in wheat in the western United States, from 1965 to 2003. *Source:* Calculated from USDA-National Agricultural Statistics Service (2004), and Anonymous (2001b). Nutrients removed in harvested portion of crop.

TABLE 3.3. Crop, area, average yield, and nutrient removal in major agronomic crops in the western region of the United States in 2003

Crop	Area (ha)	Avg. yield (mg ha^{-1})	Nutrient removal per mg		
			N	P	K
Maize (grain)	7,460,100	9.12	805,566	209,317	262,026
Maize (silage)	126,481	43.72	137,848	59,790	137,848
Sorghum	3,761,640	3.32	314,100	68,631	66,045
Wheat (all types)	17,910,315	3.17	1,083,389	182,093	242,098
Oats	451,170	2.19	24,398	3,332	5,062
Barley	1,728,135	3.73	123,984	19,702	32,743
Sunflower	853,335	1.24	29,656	5,091	5,275
Millet[a]	263,250	1.86	9,170	1,499	2,116
Alfalfa	6,000,075	9.03	1,246,233	145,987	1,108,257
Total	38,554,501		3,774,344	695,442	1,861,470

Source: USDA-NRCS (2005). Nutrient Content of Crops: Nutrients Removed by Harvest. http://npk.nrcs.usda.gov (accessed June 22, 2007).

[a]2003 data, Millet NPK removal data.

grain resulted in 23 percent, 24 percent, and 24 percent of the total N, P, and K removal, respectively. Among all wheat types, durum production only removed 4 percent, 5 percent, and 5 percent of the total N, P, and K removed, respectively.

Millet and Sunflower

Dryland millet and sunflower are produced in smaller areas in this region, generally in rotation with wheat. Only approximately 263,000 ha of millet and 853,000 ha of sunflower are grown in the western United States. However, this represents essentially 100 percent of these crops produced in the United States. Yields are low since they are grown under nonirrigated conditions. Crop abandonment of millet was 10 percent in 2001 and 51 percent in 2002. From 1991 to 2000 the average acreage abandonment for sunflower was 18 percent; in 2001 and 2002 acreage abandonment was 26 percent, and 32 percent, respectively (USDA-NASS, 2004). The driving factor was the severe drought in 2002. Long-term data are not available for millet. The average yields of millet and sunflower were 1.86 Mg ha^{-1} and 1.24 Mg ha^{-1}, respectively, in 2003 (Table 3.3). Nutrient removal is also relatively small (Table 3.3). Millet and sunflower are often planted following winter wheat abandonment with little, if any, additional fertilizer applied. When these crops are grown in a planned cropping sequence, low fertilizer rates are used because yields are limited by inadequate rainfall. Farmers are hesitant to put high production costs into these crops due to the high level of risk.

Barley

The production of both spring and winter barley occupies about 1.7 Mha in the western region. Barley is produced in all the western states except New Mexico, Nevada, Oklahoma, and Texas. However, 89 percent of the barley production occurs in the states of Idaho, Montana, North Dakota, and Washington (USDA-NASS, 2004). In the Great Plains region, it is mainly grown under dryland conditions while west of the Rocky Mountains it is mainly grown under irrigation. Crop abandonment is about 10 percent, mainly occurring under dryland production. The average yield of barley is 3.72 Mg ha^{-1}. Obviously, the higher yields are achieved under irrigation and lower yields, similar to dryland wheat, are achieved under dryland conditions.

The total nutrient removal by barley was about 124,000 Mg N, 19,700 Mg P, and 32,700 Mg K, respectively in 2003 (Table 3.3). Under dryland conditions, fertilizer rates are low while under irrigated conditions higher rates of fertilizer are used. Malting barley is included in these figures. Nitrogen management is very important in malting barley production since high N fertility rates can cause the protein content to be too high, resulting in the barley grain not being acceptable for malting. Some of the first research conducted to develop deep soil nitrate soil testing procedures, and correlations to make N fertilizer recommendations, were conducted on malting barley (Geist et al., 1972). The model developed by Reuss and Geist (1972) is still the basis for deep soil nitrate soil testing and N recommendations in this region. This procedure has proved to be successful in developing N fertilizer recommendations and is still used today, with some regional modifications.

Maize

Maize production occupies about 7.4 Mha for grain production and 126,400 ha for silage; the second largest nonlegume acreage crop produced in the western United States (Table 3.3). It is grown under both dryland and irrigated conditions. Crop abandonment is not significant under irrigation. As producers have adopted more intensive no-till cropping systems in rainfall areas less than 45 mm annual precipitation, crop abandonment occurs in some regions each year. Unlike winter wheat, the acreage of maize for grain production has increased steadily from 1955 to 2003. In 1955, 4.9 Mha of maize were grown in this region, while in 2003, 7.4 Mha were produced (USDA-NASS, 2004). Yields during this time period increased from 2.28 to 9.12 Mg ha^{-1}. Nutrient removal has followed the same trend, increasing from 95,300 Mg N, 27,800 Mg P, and 31,000 Mg K in 1955 to 805,600 Mg N, 209,300 Mg P, and 262,000 Mg K in 2003 (Table 3.3).

Sorghum

In the western United States, sorghum is produced in all states east of the Rocky Mountains, with the exception of North Dakota. However, Texas and Kansas produce 80 percent of the total sorghum in this region. The acreage devoted to sorghum production has decreased in this region over the past fifty years. From 1955 to the mid-1970s, there was about 5 Mha of sorghum grown in this region (USDA-NASS, 2004). Starting in the mid-1970s, the acreage started to decline and in 2003, about 3.7 Mha were produced. However, yields have increased steadily from 1954 to the present time. In 1954, the average yield was 1.19 Mg ha^{-1} while in 2003 the average yield was 3.3 Mg ha^{-1}. Sorghum is grown under both irrigated and dryland conditions.

Nutrient removal is about 314,800 Mg N, 68,631 Mg P, and 66,045 Mg K (Table 3.3).

Alfalfa

Alfalfa is produced in every state in the western United States. Over 6.0 Mha are devoted to alfalfa production with an average yield of 9.03 Mg ha^{-1}. The area planted with alfalfa in the western United States has remained relatively constant over the past fifty years. In 1954, there were 5.6 Mha of alfalfa produced and in 2003, 6.0 Mha. No upward trend in acreage production has occurred. However, the average yield has increased substantially from 5.11 Mg ha^{-1} in 1954 to 9.0 Mg ha^{-1} in 2003 (USDA-NASS, 2004). The highest yields are achieved in California and Arizona where the growing season is long, alfalfa is irrigated, and numerous cuttings are harvested each year. Yields in these states average from 15.5 to 19 Mg ha^{-1}.

Alfalfa stands are left for three to five years in most environments. It is grown in varied combinations of rotations with other crops. The N-fixing characteristic of alfalfa has a significant N input into the following crops N management program, and is credited to the following crops' N management calculations. Alfalfa removes large quantities of P and K. About 146,000 Mg P and 1.1 million Mg K are removed annually by alfalfa in this region (Table 3.3). Most producers apply P fertilizers, but since most of the soils in the western United States have large native K contents, K fertilization is limited. However, under high yields, K fertilization is practiced.

Midwestern Region

The "Midwestern" region refers to the states of Minnesota, Iowa, Wisconsin, Illinois, Missouri, Indiana, Ohio, and Michigan which are many times considered to be the "corn belt" of the United States. The soils and climate of this region are very suitable for maize production and this is reflected in the large area and high yields of maize grown in this region (Table 3.4). Most of the maize in this region is rotated with soybean as shown by the almost similar area devoted to soybean production (Table 3.4). Alfalfa and winter wheat are the next most widely grown crops in the region with smaller amounts of sorghum, barley, rice, and cotton.

Maize

Maize yields have increased steadily from just over 2.0 Mg ha^{-1} in 1955 to over 9.0 Mg ha^{-1} in 2003 (Figure 3.1). This large yield increase is associated with improved genetics and production practices, and interestingly,

TABLE 3.4. Crop, area, average yield and nutrient removal in major agronomic crops in the midwestern region of the United States in 2003

Crop	Area (ha)	Avg. yield (t ha^{-1})	Nutrient removal/mg		
			N	P	K
Maize (grain)	17,275,304	9.36	2,166,065	546,426	695,162
Maize (silage)	973,684	36.70	147,781	28,011	122,659
Soybean	18,834,008	2.28	2,861,215	250,070	831,183
Sorghum	127,530	4.93	9,440	2,048	1,985
Wheat (all types)	2,346,559	4.26	216,568	36,400	48,395
Oats	310,931	2.60	20,202	2,759	4,192
Barley	91,093	3.14	7,886	1,253	2,083
Alfalfa	2,801,619	7.03	551,136	64,512	490,118
Rice	69,231	6.87	6,038	1,392	1,381
Cotton	157,895	0.97	10,159	1,942	5,006
Total	42,987,854	–	5,996,490	934,813	2,202,164

Source: USDA-NASS (2004), Anonymous (2001b), Potash and Phosphate Institute.

occurred as planted area which increased from approximately 16.0 Mha in 1955 to almost 19.0 Mha in 2004 (USDA-NASS, 2004). The crop rotation in the region changed from maize, small grain (wheat or oats), and alfalfa-grass, to maize-soybean systems.

Nutrient removals by maize grown in this region for both grain and silage were approximately 2.4 million Mg N, 0.62 million Mg P, and 0.86 million Mg K, respectively (Table 3.4). Almost all maize in this region is grown in rotation with soybean, and thus some of the maize N requirement is met with residual N from the previous soybean crop, and is considered in fertilizer recommendations. Maize N fertilizer is generally reduced by 50 kg N ha^{-1} in Iowa recommendations (Blackmer et al., 1997) while in Indiana, Ohio, and Michigan, soybean is expected to contribute approximately 35 kg N ha^{-1} to the following maize crop (Vitosh et al., 2004). Nutrient needs for maize, besides the N supplied by legumes in the rotation, are met with fertilizer and/or manures.

Soybean was produced on approximately 18.8 Mha in 2003 in this region (Table 3.4) with an average yield of 2.28 Mg ha^{-1}. The average yield in 2003 was below the trend-line (Figure 3.1) average of approximately 2.6 Mg ha^{-1} due to unfavorable growing conditions. Yields rebounded in

2004, but the 2003 data are used for consistency of presentation. Almost all soybean is rotated with maize. Yields are increasing gradually, as shown in Figure 3.1 and can be expected to continue to increase with improved genetics and management practices. Nutrient removals were 2.8 million Mg N, 0.24 million Mg P, and 0.81 million Mg of K, respectively. Soybean, being a legume, does not receive N fertilization, but all P and K are usually supplied with fertilizers. Generally P and K applications for both the maize and soybean crop will be applied prior to planting maize. Manure applications are generally reserved for maize in order to utilize more efficiently the N content of the manures.

Winter Wheat

Most winter wheat in the region is soft red winter wheat and is utilized for pastry and cracker flour, or livestock feed, depending on price and quality. Winter wheat area in 2003 was 2.35 Mha. The major production areas are in Indiana, Ohio, Michigan, southern Illinois, and eastern Missouri (Figure 3.5). Winter wheat is generally rotated with soybean in most of the region, but in southern Indiana, Ohio, and Illinois, the crop rotation would involve maize, winter wheat, and double crop soybean (three crops in two years). Nutrient removal by winter wheat in 2003 is estimated to be 0.16, 0.04, and 0.05 million Mg of N, P, and K respectively. Essentially all nutrients removed by winter wheat in this region are supplied with fertilizers, either as direct application, or at some time in the crop rotation. Nitrogen fertilization for winter wheat production is generally split between a small at-planting application followed by a late winter, early-spring application of the remainder of the crop N requirement. Phosphorus and K fertilizers are applied preplant according to soil test recommendations.

Alfalfa

Alfalfa is the other major broad-acre crop planted in this region, with an average yield of 7.4 Mg ha^{-1} harvested from approximately 2.8 Mha in 2003. Alfalfa is mainly utilized for the dairy industry and is usually grown in a minimum three-four year rotation with maize. Nutrients removed by alfalfa include approximately 0.49 million Mg N, 0.058 million Mg P, and 0.44 Mg K, respectively (Table 3.4). *Rhizobium* growing in symbiosis with alfalfa provides needed N, and alfalfa contributes significant N to maize crops following in the rotation. For example, Iowa recommends specific N applications of 0-35 kg N ha^{-1} to maize following alfalfa, and rates of 0-70 kg N ha^{-1} for second crop maize following alfalfa (Blackmer et al.,

1997). Normal fertilizer N applications to maize would be 175-200 kg N ha^{-1} in this region. Alfalfa has a high K requirement and moderate P requirement, all of which must be satisfied with fertilizer applications in order to sustain production.

Other Crops

Other crops grown in the region include sorghum, barley, rice, and cotton, but occupy small areas (Table 3.4). While some barley is grown for livestock feed in almost all states, cotton and rice are confined to the southeastern part of Missouri along the Mississippi River (Figure 3.5). Sorghum production is small and is mainly in western Missouri. Nutrients for these crops must be supplied with fertilization. Nitrogen rates are directly related to expected yield, and P and K should be applied according to soil test recommendations.

Southern Region

The southern region of the United States is comprised of an area extending south of the states of Kentucky and Virginia on the north, and east of Arkansas and Louisiana on the west. The climate in the northern part of the region is defined as "temperate continental" but the majority of the region is classified as "subtropical wet" (Figure 3.2). Rainfall in this region is relatively high, averaging 1,140-1,600 mm per year, and the region is subject to intense hurricanes from July through November, with several making landfall somewhere in the region each year. Soils are inherently low in fertility in this region due to soil forming factors such as high annual rainfall and parent materials, such as marine sediments, with low nutrient content.

Soybean and Winter Wheat

Soybean and winter wheat are major crops in this region (Table 3.5). Soybean was planted on 4.1 Mha in 2003 and 1.0 Mha of winter wheat. Soybean is grown as a full-season crop in many areas, but most winter wheat would have a second or "double" crop of soybean following the winter wheat because of the long growing season in this region. Nutrient removal by soybean was 0.64, 0.055, and 0.19 million Mg of N, P, and K, respectively, in 2003. Nitrogen for soybean is supplied via fixation, but P and K must be supplied with fertilizers.

Winter wheat area has declined from a high of 3.42 Mha in 1982 to 1.65 Mha in 1996 and finally to 1.0 Mha in 2003 (USDA-NASS, 2004). This

TABLE 3.5. Crop, area, average yield, and nutrient removal in major agronomic crops in the southern region of the United States in 2003

Crop	Area (ha)	Avg. yield (t ha^{-1})	Nutrient removal/mg		
			N	P	K
Maize (grain)	1,953,036	7.97	208,545	52,609	66,929
Maize (silage)	185,425	38.10	29,266	5,547	24,291
Soybean	4,077,733	2.50	680,843	59,506	197,785
Sorghum	237,247	5.07	17,875	3,879	3,759
Wheat (all types)	1,023,077	3.17	70,231	11,804	15,694
Oats	29,150	2.04	1,487	203	309
Barley	27,126	3.35	2,082	331	550
Alfalfa	181,376	7.90	40,127	4,697	35,684
Rice	865,992	7.25	79,769	18,390	18,246
Cotton	2,445,344	0.96	150,500	28,774	74,168
Total	11,025,506	–	1,280,725	185,740	437,415

Source: USDA-NASS (2004), Anonymous (2001b), Potash and Phosphate Institute.

decline is associated with economics of production costs and wheat price. Area planted to wheat would be expected to increase rapidly if economic returns improved relative to full season soybean. Nutrient removals by wheat in 2003 were 0.053, 0.012, and 0.016 million Mg of N, P, and K, respectively, all of which must be supplied by fertilizers. While most P and K can be applied preplant and include the P and K required by the following double-crop soybean, N must be applied as a 2 or 3 split application. Split N applications are required in order to obtain maximum N use efficiency and optimum economic yields (Scharf and Alley, 1993, 1994; Scharf et al., 1993).

Cotton

Cotton area increased from 870,000 ha in 1983 to a high of 3.1 Mha in 2001 (USDA-NASS, 2004). This increase in area planted to cotton has been associated with boll weevil eradication programs that reduced pesticide costs and production risks greatly. In addition, transgenic cotton varieties that include glyphosate resistance and resistance to lepidopteran insects have reduced the complexity of cotton production. Cotton N requirements are moderate, 55-100 kg N ha^{-1} depending on residual soil N from organic matter decomposition or legumes in the rotation, and must be managed carefully. Excessive available N results in greater vegetative growth relative to reproductive growth. N fertilizer applications are generally split between a small application (25-30 kg N ha^{-1}) at-planting and a second application near the time the first flower buds (squares) begin to appear (approximately forty-five days after emergence). This is a very important practice since rainfall can be high, subjecting N to leaching and/or denitrification losses depending on soil texture and drainage class.

Potassium requirement for cotton is much greater than the P requirement. Potassium availability is especially critical at fruit set, and split applications of K may be required on low cation exchange capacity soils that have experienced leaching rainfall (Crozier, 2004). The use of manure for supplying nutrients to cotton is complicated by the potentially detrimental effects of excessive levels of plant-available N. The use of manure is generally recommended to build soil P and K levels through application to other crops in the rotation, but not be applied to the cotton crop.

Maize

Maize for grain and silage was harvested from approximately 2.2 Mha in 2003 (Table 3.5). The area planted to maize fluctuates with economics of maize versus cotton on many farms. Maize yields are lower in this region than in the midwestern region. Yield variation is also much greater for maize grown in the southern region compared to the midwestern region due to low water-holding capacity soils in many areas. Maize is grown in rotation with soybean, cotton, and peanut, and is grown continuously for silage on many dairy farms. Nutrient removal by maize is significant and must be supplied via fertilization.

Other Crops

Other crops in the region include rice and small areas of alfalfa, barley, and sorghum. These are important crops in localized areas and do require significant amounts of fertilizer nutrients. As with the other crops, fertilizers are utilized and split application of N to match the time of crop N need is the standard practice for crops requiring N.

North Atlantic Region

The north Atlantic region of the United States is comprised of states north of Virginia and east of Ohio. The total planted area in this region is relatively low compared to other regions of the United States. This is due to the fact that many of the states in this region are small relative to other regions, overall soil and climatic conditions are not as conducive to agronomic crop production, and the region has a high degree of urbanization. However, there are important crop production regions, especially in Pennsylvania, New York, and Maryland. From an agronomic crop production standpoint, this region is grain deficit due to the concentration of dairy and poultry production in certain areas within the region. The problem of managing manure nutrients that are brought into the region through imported grain, mainly from the midwestern region, is significant.

Maize

Maize for grain and silage was planted on approximately 2.7 Mha in the region in 2003 (Table 3.6). Yields and associated nutrient removals are similar to those from other regions. Area planted to maize has remained relatively stable in recent years reflecting the need for silage and grain for the dairy industry, but this area is significantly less than the 3.3 Mha planted in 1981 (USDA-NASS, 2004). The reduction in area planted to maize reflects the loss of farms in the region and the specialization of livestock farms that rely on feed grains transported from other regions.

TABLE 3.6. Crop, area, average yield, and nutrient removal for major agronomic crops in the North Atlantic region of the United States in 2003

Crop	Area (ha)	Avg. yield (mg ha^{-1})	Nutrient removal/mg		
			N	P	K
Maize (grain)	2,048,583	8.86	243,151	61,339	78,035
Maize (silage)	681,781	35.30	99,528	18,865	82,609
Soybean	2,317,409	2.68	414,295	36,209	120,353
Sorghum	3,644	4.88	267	58	56
Wheat (all types)	206,478	2.92	13,044	2,193	2,915
Oats	119,838	2.36	7,074	966	1,468
Barley	68,421	3.18	4,987	792	1,317
Alfalfa	1,136,032	7.28	231,441	27,091	205,817
Total	6,582,186		1,013,787	147,513	492,570

Source: USDA-NASS (2004), Anonymous (2001b), Potash and Phosphate Institute.

Alfalfa

Alfalfa was harvested from 1.1 Mha in this region in 2003 and is used mainly for the dairy industry (Table 3.6). Alfalfa is generally planted in rotation with maize with three to four years of alfalfa followed by three to four years of corn. Maize area for silage and alfalfa area were approximately equal in 2003 reflecting this rotation. Alfalfa removes the largest amount of K of any crop in the region, being slightly greater than K removal by maize silage. Phosphorus removal is greatest for maize for grain followed by alfalfa.

Manure nutrients supply a large amount of N, P, and K in the region, although exact estimates are difficult to make. Manure application to alfalfa is generally avoided because alfalfa is a legume. However, manure where available would be used in the rotation to build P and K levels prior to planting alfalfa, and N from the alfalfa would be credited toward maize N requirements in maize following alfalfa. Nutrient management plans in this region specifically recommend efficient use of manure nutrients in conjunction with supplemental fertilizer applications.

Other Crops

Other crops grown in the region include soybean, wheat, and barley (Table 3.6). Planted areas and associated nutrient removals are generally low relative to maize and alfalfa.

Total Nutrient Removal in Harvested Major Agronomic Crops

Total nutrient removal for the United States in 2003 by major agronomic crops is shown in Table 3.7. Nitrogen removed by maize is almost three times greater than for wheat or alfalfa. However, if we are considering fertilizer N needs, then wheat has the second largest requirement for N, and all other crops are much less. Phosphorus removal by maize for grain is also much greater than any other crop, followed by soybean, and again, over three times that removed by wheat or alfalfa (Table 3.7). Potassium removal by soybean is greatest, followed by maize for grain and silage, and alfalfa, and K removal by these crops is much greater than K removal by any other crops (Table 3.7). Total N, P, and K removed by agronomic crops in 2003 in the United States was approximately 19.4 million Mg N, 3.8 million Mg P, and 5.3 million Mg K. These removal data illustrate the need for the large amounts of nutrients that must be provided through residual soil supplies or supplemental fertilization with commercial fertilizers, manures, or crop residues.

TABLE 3.7. Crop, area, average yield, and nutrient removal in major agronomic crops in the United States in 2003

Crop	Area (ha)	Avg. yield (t ha^{-1})	Nutrient removal		
			N	P	K
Maize (grain)	28,737,023	9.17	3,423,327	869,691	1,102,152
Maize (silage)	1,967,371	36.80	414,432	71,288	367,407
Soybean	28,229,150	2.33	3,956,353	345,605	1,149,321
Sorghum	4,130,061	3.47	341,682	74,616	71,845
Wheat (all types)	21,486,429	3.29	1,383,232	232,490	309,102
Oats	911,089	2.35	53,161	7,260	7,331
Barley	1,914,775	3.67	138,939	22,078	36,693
Alfalfa	10,119,102	8.26	2,068,937	242,287	1,839,876
Rice (all types)	935,223	7.22	85,807	19,782	19,627
Cotton	2,603,239	0.96	160,659	28,774	79,174
Millet	263,250		9,170	1,499	2,116
Sunflower	853,335		29,656	5,091	5,275
Total	102,150,047		12,065,355	1,920,461	4,989,919

Source: USDA-NASS (2004), Anonymous (2001b), Potash and Phosphate Institute.

MONITORING SOIL FERTILITY LEVELS IN THE UNITED STATES

Soil Testing and Soil Test Methods

Soil testing is the most widely utilized soil fertility evaluation method in the United States. Plank (1998) estimated that approximately 5 million soil samples are analyzed annually in the United States for nutrient recommendations for crop production, but more recent estimates indicate 8 to 9 million soil samples are analyzed annually in North America in approximately 190 laboratories (Robert Miller, 2004, personal communication). While many states still have soil testing laboratories operated by universities, or state agencies, Dr. Robert Miller, head of the North American Proficiency Testing (NAPT) program for soil testing labs, estimates that 95 percent of soil samples are now analyzed by private laboratories (Robert Miller, 2004, personal communication).

Soil testing services are available throughout the United States. In addition, with the almost universal availability of courier services and use of the internet to transmit data, laboratories receive samples from throughout the country and report data and recommendations to growers in a timely manner. Costs for these services range from approximately US$7.00 to US$25.00 depending on the analyses conducted.

Extensive soil test methodology work began in the United States in the 1940s and some work continues to this day. Black (1993) presents an excellent, detailed discussion of basic principles associated with "Soil Testing and Fertilizer Requirement." Topics covered in Black's book include chemical indexes of nutrient availability, theory of nutrient supply, multinutrient extractants, and calibration and interpretation of various soil test methods. In addition, the Soil Science Society of America published a summary of soil test methods, principles associated with the methods, and a review of the scientific literature associated with the various soil test methods (Westerman, 1990) as well as where the various methods are applicable.

Jones (1998) published a short summary of extractants widely used in the United States and points to the movement of labs toward "universal" extractants. The term "universal" extractant was first proposed by Van Raij (1994) and refers to those extractants capable of evaluating the soil status for a number of elements or ions. Universal extractants reduce the time and cost of conducting soil analyses. Examples of such extractants are the Mehlich 3 method (Mehlich, 1984) for acid soils, and the AB-DTPA extractant (Soltanpour and Schwab, 1977) for alkaline soils. Finally, currently used methods are described in detail in the *Soil Analysis Handbook of Reference Methods* (Anon., 2000) along with brief discussions of soils that are most suitable for testing by the various procedures.

A typical soil analysis usually includes soil-water pH, and plant-available P, K, Ca, and Mg, regardless of location in the United States. However, testing for N availability is most widely conducted in semiarid regions where nitrate does not move greatly in the soil profile. Soils tested for plant-available N (nitrate and exchangeable ammonium extracted with 1M KCl) (Mulvaney, 1996) in semiarid regions are preferably sampled to depths of 0.6 m or 0.9 m. If the entire rooting depth is not sampled, the residual nitrate level of the rooting zone is estimated based upon the concentrations in the sampled depth(s). Algorithms have been developed for all agronomic crops in the western region with local modifications made by various states. An example of this N testing and recommendation approach is presented by Mortvedt et al. (2004).

The only widely used N testing procedure in humid regions of the United States is the "Pre-sidedress soil nitrate test" (PSNT) (Magdoff et al., 1984;

Magdoff, 1991). This procedure is used for adjusting N fertilizer applications in maize and utilizes a soil sample collected to a 30 cm depth when corn is 30 to 40 cm in height. The procedure has been widely calibrated in humid regions and found to be useful where manure applications or a previous legume crop, such as alfalfa, are a major N source (Roth et al., 1991; Meisinger et al., 1992; Evanylo and Alley, 1997).

Soil pH is generally measured on 1:1 soil: water slurries (Jones, 2001) or saturated paste extracts that are being analyzed for soluble salts. The saturated paste pH method is used mainly in the western region (75 percent of laboratories) of the United States where alkaline soils predominate and soluble salts accumulate. Lime requirement tests are available in most laboratories with the Shoemaker, McLean, and Pratt (SMP) buffer (Shoemaker et al., 1962), the Adams-Evans buffer (Adams and Evans, 1962), and the Mehlich buffer (Mehlich, 1976) being most widely used. The SMP buffer was developed for use on soils with pH of ≤ 5.8, lime requirements of 2.0 Mg ha^{-1} or more, and organic matter levels of <10 percent (Shoemaker et al., 1962). The Adams-Evans buffer was developed for use on low cation exchange capacity (<8 cmol charge/kg) with low organic matter content, that is, <3 percent. Finally, the Mehlich buffer method was developed to be more universally applicable, but this procedure has not been widely adopted (Jones, 2001).

The most common soil test method for P in the western region is the sodium bicarbonate extract, commonly referred to as the Olson procedure (Kuo, 1996). This procedure was developed for soils with pH levels greater than 7.0. For acid soils, the Bray P1 (Bray and Kurtz, 1945), and the Mehlich 1 (Mehlich, 1953), and Mehlich 3 (Mehlich, 1984) procedures are utilized. While the Bray P1 test is specific for P, the Mehlich extractants also evaluate available Ca, Mg, and K, and data from the Mehlich 3 procedure has been shown to be well correlated with Bray P1 values (Jones, 2001). Because of this correlation, and the reduction in costs associated with using a single extractant, the Mehlich 3 procedure is gaining in popularity.

Potassium availability in soil is estimated by the 1N ammonium acetate extractable K test (Haby et al., 1990; Helmke and Sparks, 1996). This procedure is widely used throughout the United States because the extraction procedure also enables the estimation of available Ca and Mg along with K. However, the Mehlich 3 extractant has been shown to be well correlated with the values from the ammonium acetate extraction procedure (Beegle and Oravec, 1990), and is thus gaining favor in several soil testing programs as a universal extractant.

Soil Test Levels

The Potash and Phosphate Institute (reorganized in 2006 into the International Plant Nutrition Institute) summarizes soil test data on a national basis to examine trends in plant-available P and K levels as well as soil pH. A summary in 2001 included results from approximately 2.5 million soil samples collected and analyzed during the autumn of 2000 and spring of 2001. The data revealed that for all of North America, 47 percent of the samples tested medium or lower in plant-available P (Figure 3.11a), 43 percent tested medium or lower in plant-available K (Figure 3.11b), and 34 percent of samples had a pH in water (1:1 soil: water slurry) of 6.0 or less

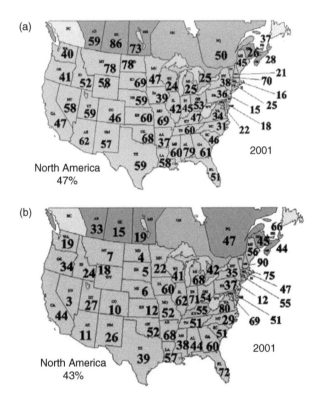

FIGURE 3.11. Soil samples testing medium or lower for P(a) and K(b) in North America. *Source:* Fixen, P.E. (2002). Soil test levels in North America. *Better Crops with Plant Food,* 86:12-15. Reprinted by permission of the International Plant Nutrition Institute.

(Figure 3.12). Results varied by region as shown in Figures 3.11 and 3.12. The lowest pH levels occurred in the southeastern United States while the highest potassium levels are associated with the soils in the Great Plains region and western United States. Analysis of the data showed that over 50 percent of the samples had Bray P1 equivalent values of 5 to 30 ppm, while approximately 50 percent of the samples had ammonium acetate extractable K levels of 40 to 160 ppm. A more recent update of these data is available from the International Plant Nutrition Institute (http://www.ipni.net/ bettercrops).

Future surveys of this type will be useful to determine if available plant nutrient levels are decreasing, increasing, or remaining relatively the same. However, as pointed out by Fixen (2002), it is difficult to separate non-agronomic crop samples, that is, turf or construction site reclamation, etc., from agronomic samples in these surveys, and such samples could bias the results of the survey, especially in states with large urban centers. In addition, there is no way of estimating the amount of agronomic crop production land that is not tested, which adds to the uncertainty surrounding the actual soil fertility levels in crop production areas in the United States.

These data illustrate that even with a relatively long history of commercial fertilizer use (over sixty years), in conjunction with significant education and research efforts from land grant universities, the federal government, and companies, there still appear to be areas where increased nutrient use could be profitable. However, with farm consolidation in the United States much of

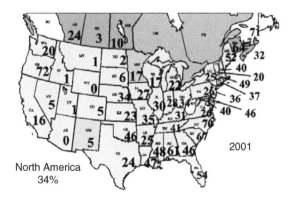

FIGURE 3.12. Soil samples testing pH 6.0 or below (water pH values) in North America. *Source:* Fixen, P.E. (2002). Soil test levels in North America. *Better Crops with Plant Food,* 86:12-15. Reprinted by permission of the International Plant Nutrition Institute.

the land is managed by farmers with management skills that optimize economic returns. These growers realize that maintaining medium to high soil test levels with respect to P and K are generally the most profitable soil fertility levels, and that yields and nutrient use efficiency is limited with low soil pH.

MANURE AND BIOSOLIDS NUTRIENT SOURCES

Animal manures supply essential plant nutrients and as a source of organic matter can improve soil physical conditions which can be beneficial to crop production. Using data from the 1997 USDA National Agricultural Statistics Service (USDA-NASS) Animal Inventories, Powers and Van Horn (2001) estimated that confined animal feeding operations in the United States produce 23,910 million kg of dry matter annually (Table 3.8). This corresponds to an annual production of approximately 1,470 million kg N, 726 million kg P and 1,505 million kg K (Table 3.8).

In a separate evaluation, data from the U.S. Census of Agriculture were used by Lander et al. (1998) and Kellogg et al. (2000) to estimate livestock populations, quantities of manure produced, and land available for manure

TABLE 3.8. Estimated annual production of recoverable dry matter and nutrients from confined animal feeding operations in the United States

Species[a]	Dry matter[b]	Nitrogen[c]	Phosphorus	Potassium
			$\times 10^6$	
Dairy Cattle	23,807	623.7	277.0	752.8
Feedlot Beef Cattle	9,190	309.4	136.2	354.0
Swine	5,066	192.0	80.3	160.4
Layers	1,182	72.8	55.5	55.4
Broilers	4,991	208.0	138.3	145.1
Turkeys	1,348	66.4	39.1	37.4
Total	45,584	1,472.0	726.0	1,505.0

Source: Adapted from Powers and Van Horn (2001).

[a]Estimated based on 1997 USDA National Agricultural Statistics Service animal inventories.

[b]Assumes that 20 percent of original dry matter is lost through anaerobic or aerobic decomposition.

[c]Assumes that 60 percent of N excreted is lost through volatilization.

application throughout the United States for 1982, 1987, 1992, and 1997. Their study showed that small- and medium-sized livestock operations in the United States have been replaced by larger operations over time and that the number of confined animals per operation has increased, thus livestock populations have become more spatially concentrated in high-producing areas (Figure 3.13, see also color gallery). This situation for all livestock types has resulted in increased challenges for the utilization of manure and manure nutrients. As livestock production has become more spatially concentrated, the amount of manure nutrients relative to the assimilative capacity of land available on farms for application has grown, especially in high production areas.

The density of livestock operations based on the 1997 census data on a county basis in the United States are presented in Figure 3.13. Estimated manure N and P produced by county and amounts of manure N and P estimated to be in excess of crop needs by county are presented in Figure 3.14 (see also color gallery). These data illustrate clearly that areas of high livestock density are producing more nutrients as manure than can be utilized by available cropland. In the future greater emphasis will be needed on off-farm utilization of manure nutrients especially as P-based nutrient management planning is implemented on permitted livestock operations.

In the United States, an estimated 95 billion liters of wastewater are processed by municipal, private, and federally owned treatment works each day, generating more than 7 million dry Mg of biosolids each year, and is pro-

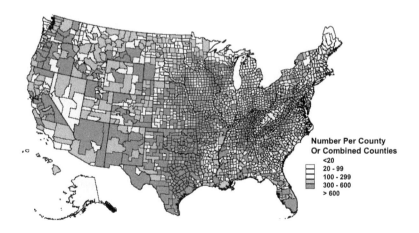

FIGURE 3.13. Confined animal feed operations in the United States in 1997.

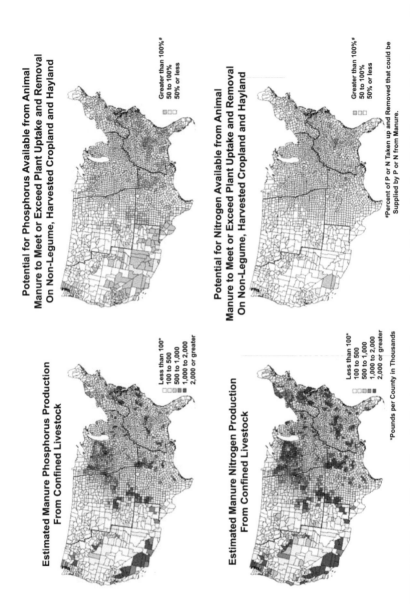

FIGURE 3.14. Estimated N and P production in animal manures and potential crop land available to assimilate these nutrients in the United States.

jected to be 8.2 million dry Mg per year by 2010 (U.S. EPA, 1999). Using typical biosolids total nutrient concentrations of 39 kg N Mg per year, 25 kg P Mg per year, 4 kg K Mg per year (Sommers, 1977) (similar values were observed for forty-five biosolids samples collected in Virginia and North Carolina in 2004; G. E. Evanylo, 2005, personal communication), biosolids generated in the United States could potentially supply approximately 273,000 Mg N, 173,600 Mg P, and 28,000 Mg K per year. Based on EPA estimates, approximately 40 percent of these biosolids are beneficially re-used via land application (U.S. EPA, 1999), so approximately 109,000 Mg N, 69,440 Mg P, and 11,200 Mg K per year are directly land applied as biosolids in the United States.

NUTRIENT USE

Nutrient (N-P-K) use in the United States from 1961 to 2002 is shown in Figure 3.15. Nitrogen use increased steadily from 1960 to about 1980 as crop yields increased, especially maize. The sharp decline in 1982 for N is associated with a government program to reduce maize and wheat plantings in an attempt to manage production. Nitrogen consumption increased grad-ually through the late 1980s and early 1990s to approximately 10.5 to 11.0 million Mg per year.

Consumption of P and K increased from 1960 to about 1980 with maxi-mum consumption reaching a level of 5.1 million Mg for P in 1978 and K

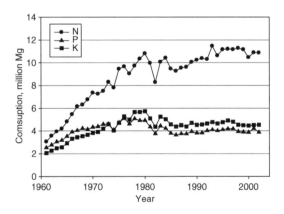

FIGURE 3.15. Nutrient (N-P-K) use in the United States from 1961 to 2002.
Source: FAOSTAT data (2004).

use peaking at 5.7 million Mg in 1980. From these maximum use levels, P and K consumption has declined to approximately 4.0 million Mg P and 4.5 million Mg K per year.

A detailed discussion of inorganic nutrient use in North America from the 1800s to the early 2000s can be found in the article by Stewart and Roberts (2002). These authors relate the development of the commercial fertilizer industry in North America along with more specific reasons affecting fertilizer consumption. In addition, they discuss crop production to fertilizer use ratios for selected crops.

Although the use levels of N, P, and K have been rather stable for the past ten years, recent cost increases for natural gas used to manufacture N fertilizers, increased transportation costs, and increased world demand for fertilizers has resulted in significant fertilizer price increases to U.S. growers. Consumption in the short-term can be expected to decline, especially if grain prices do not increase proportionally.

Nutrients applied, or currently available for utilization in the United States, are shown in Table 3.9. Estimates include recoverable manure nutrients, biosolids that are land-applied, and commercial fertilizer. On a national basis, recoverable nutrients in manure will never totally replace nutrients applied as commercial fertilizer. However, manure nutrients are a significant source, especially on a local basis, and as pointed out in an earlier section, these nutrients can be an environmental problem in areas of high livestock concentration due to over application of manure. This excessive application of manure nutrients has been the driving force behind the development of and requirement for nutrient management plans, most of which

TABLE 3.9. Recoverable manure nutrients, biosolids-nutrients, and commercial fertilizer nutrients utilized, or available for utilization in the United States

Nutrient source	N	P	K
	mg $\times$ 10^6		
Manure (recoverable)[a]	1.17	0.65	1.73
Biosolids (land-applied)[b]	0.11	0.07	0.01
Commercial Fertilizer[c]	10.90	3.90	4.50
Total	12.18	4.57	6.24

[a]Kellogg et al., 2000. Estimated from 1997 data on livestock farms.

[b]USEPA, 1999. Estimated that 40 percent of biosolids are land-applied for nutrient use.

[c]FAOSTAT data, 2004. Fertilizer consumption of 2002.

focus on the utilization of manure nutrients in conjunction with supplemental commercial fertilizer nutrients.

Biosolids are a relatively insignificant nutrient source with respect to the total nutrients applied in the United States. However, these materials are important nutrient sources in areas near large urban population centers. Land application of biosolids was estimated to be only 40 percent of the total in 1999 (U.S. EPA, 1999) but is expected to increase as disposal costs by other means, that is land-filling, continue to increase.

Commercial fertilizers comprise the vast majority of nutrients applied in the United States, and it is important to consider that a significant portion of the nutrients in manures and biosolids originated in commercial fertilizers that were applied to crops utilized for either livestock or human food. Smil (2001), in discussing fertilizer N production and use states: "Its [U.S.] applications of synthetic nitrogen fertilizers . . . have made it possible to provide diets unusually high in animal foods supply and to expand food exports to the extent unmatched by any other country" (p. 166). While this is a success story from a food production standpoint, as discussed in the previous paragraph, the localized concentration of animal feeding operations has resulted in the development of potential water-quality problems associated with excessive manure applications. This issue of excessive nutrient applications to land has been the driving force for the development of nutrient management programs in the United States which is well illustrated by the "Comprehensive Nutrient Management Planning" program of the USDA-NRCS (2004).

Overall nutrient removal by agronomic crops shown in Table 3.7 is 19.4 million Mg N, 3.8 million Mg P, and 5.3 million Mg K. The sum of recoverable manure nutrients, land-applied biosolids, and commercial fertilizer in the United States (Table 3.9) is 12.2 million Mg N, 4.6 million Mg P, and 6.2 million Mg K. However, these values cannot be compared directly because much of the manure and commercial fertilizers are applied to other crops such as pastures and hay land, vegetables, and managed turf. In addition, legumes considered in the work (soybean and alfalfa) fix their N requirement, and thus it is to be expected that much larger amounts of N are removed in the harvested crops than is applied as fertilizer N.

The most detailed study on nutrient budgets for North America is presented by Fixen and Johnston (2002). In this comprehensive study, they conclude that the amount of N removed in harvested crops equals approximately 77 percent of inputs, and that P removal exceeds P applied as fertilizer by 29 percent but when recoverable manure is included in the evaluation, removal represents 95 percent of inputs. Finally, K removal exceeds input by 44 percent when both fertilizer and recoverable manure nutrients are considered. Supporting data for their analysis is extensive and included in their publication.

NUTRIENT SOURCES
AND SOIL FERTILITY MAINTENANCE

Organic Nutrient Sources

An exhaustive study on manure nutrients production and utilization in the United States was published by Kellogg et al. in 2000. Data from this study indicates that livestock production facilities are generally increasing in size with a resulting concentration of manure nutrients. However, because the nutrient content of manures is relatively low (0.5-3.0 percent N, 0.1-1.4 percent P, 0.2-1.2 percent K by weight), the transport cost for application of these materials is a significant economic issue (Mullins, 2000). In addition, the major environmental issue associated with long-term manure application has been the buildup of soil P levels where manure has been applied to supply crop N needs which are greater than P needs for most crops.

Most states in the United States are now moving toward "P-based" nutrient management plans. Where soil test P levels have reached a level considered to represent a risk to water-quality, manure applications may be based on P as mandated by state law. For example, most of the states in the tributaries of the Chesapeake Bay Watershed in the eastern United States, including Delaware, Maryland, and Virginia, have passed legislation that requires affected producers to implement P-based nutrient management plans (Sims, 2000). In P-based management, P application levels are determined based on soil test P levels, expected crop removal and/or site-specific conditions and P source management factors (Simpson, 1998; Sims, 1999). Manure applications will not be allowed in some cases as USEPA's new regulations for Concentrated Animal Feeding Operations are adopted (U.S. EPA, 2003). These regulations mandate P-based manure and waste water applications (Sharpley et al., 2003).

Nutrient management planning programs have focused on the proper application of manure nutrients with supplementation of crop nutrient needs with commercial fertilizers. This focus has developed because of the concentration of livestock production and the specific water quality impairments that can be traced to the concentration of livestock production. However, as Ludwick and Johnston (2002) point out, only 17 percent of corn production and 6 percent of soybean production in the United States receive manure. Biosolids nutrient applications are much smaller than manure applications

(Table 3.9), are highly regulated, and are generally made at no-cost to growers as land-application is a "least-cost" method for disposal of biosolids by municipalities. These applications are generally made to supply crop N needs or to satisfy liming requirements, and are usually made only once every three to five years. Thus, while organic nutrient sources are important in many areas, fertilizers, which also are probably the major source of nutrients in manures and biosolids, are the major source for sustaining plant nutrient availability in the United States.

Commercial Fertilizers

The United States has a highly developed distribution system for fertilizer nutrients. Most fertilizers for agronomic crops are "bulk-blended" near the application site and are made to specifications supplied by growers. Dry bulk blending of nutrient sources such as urea, diammonium phosphate (DAP), and potassium chloride (KCl) probably supplies the majority of plant nutrient elements for agronomic crops. However, blending of clear and suspension liquid fertilizers is important in many regions.

Use of various fertilizers in the United States during recent years is shown in Figure 3.16. Anhydrous ammonia utilized for direct application to maize, wheat, and cotton has leveled and perhaps declined somewhat in recent years due to cost, while the amount of N solutions (urea-ammonium nitrate, 28, 30, 32 percent N) is increasing. This increase for N solutions is probably the result of the flexibility of using N solutions for split application programs, herbicide carrier, and/or more ease of placement for increased N use efficiency. DAP and other N-P fertilizers, that is, monoammonium phosphate, and ammonium polyphosphate, have also increased while the use of superphosphate fertilizers has declined dramatically. KCl continues to be the leading K source in the United States (Taylor, 1993; Terry and Kirby, 1994-2004).

Integrated Nutrient Management

Nutrient sources for agronomic plant production include the following: (1) residual soil nutrient supplies; (2) nutrients available from previous crops, that is, legumes; (3) manures; (4) biosolids; and (5) commercial fertilizers. Integrated nutrient management involves accounting for all nutrient sources and developing a program to optimize, economically and environmentally, plant nutrient use for crop production.

Residual soil nutrient supplies are evaluated by soil sampling and testing for plant-available nutrients. Soil testing services are available throughout

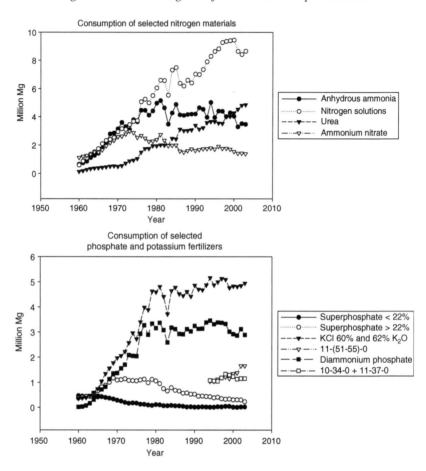

FIGURE 3.16. Tonnage of various commercial fertilizers in the United States, from 1960 to 2003. *Source:* Terry, D. L. and B. J. Kirby (1994-2004). Commercial Fertilizers (published yearly). Association of American Plant Food Control Officials and The Fertilizer Institute. Washington, DC: The Fertilizer Institute.

the United States and the methodology utilized in the various regions has been addressed in a previous section. While no valid estimate can be made for the percentage of agronomic crop production land that is sampled on a regular basis, agronomists generally comment on the lack of regular soil sampling programs by many growers, and/or the lack of a current soil test result when diagnosing a problem. However, agronomists continue to promote the effectiveness of soil testing for P, K, and soil acidity for the humid regions of the United States, while testing for N, P, K, and soil alkalinity is consistently recommended in the semiarid regions.

Nutrients from preceding crops in rotations are considered in making plant nutrient recommendations. The most common example of such recommendations is the adjustment of N fertilizer applications for grass crops following a legume. In Iowa, N fertilizer recommendations for maize following alfalfa (three to four years of alfalfa) would be 0 to 35 kg N ha^{-1}, while maize following maize is expected to need 170 to 220 kg N ha^{-1} depending on the soil productivity level (Blackmer et al., 1997). In a more recent work, Nafziger et al. (2004) analyzed seventy-two site-years of data in Illinois and eighty-one site-years of data in Iowa to develop economic optimum N fertilization rates for maize, and found that high yields of maize usually required N fertilizer rates of less than 170 kg N ha^{-1}, and most times, the economic rates were much lower. In the southeastern United States, N fertilizer recommendations for maize are recommended to be reduced by 22 to 34 kg N ha^{-1} when maize follows peanut (*Arachis hypogaea* L.) or soybean, and by 90 to 110 kg N ha^{-1} when following alfalfa (Kissel, 2003; Hardy et al., 2003; Larson and Oldham, 2004). Similar credits for legumes in crop rotations are found in most fertilizer recommendation programs.

Laboratory testing is used to determine nutrient content in organic wastes. Sampling to obtain representative samples for analysis is a significant problem due to the large volumes and variability of manure in most livestock operations. However, specific sampling procedures are readily available and an example of such procedures can be found in the *Nutrient Management Handbook* for Virginia (Kindig, 2002) as well as in the article by Shaffer and Sheffield (2004) from Purdue University. Manures are generally analyzed for total Kjeldahl N (TKN), ammonium N, total P, and total K. Organic N is the difference between TKN and ammonium N. Availability of the organic N fraction in manures is estimated by the use of mineralization coefficients that are specific for various climatic regions and manure types. For example, the availability of organic N from spring applied poultry manure is estimated to be 60 percent during the summer growing season in Virginia, while that of dairy manure is estimated to be 35 percent. The degree of inaccuracy in estimating N availability from manures is a recognized problem. Though most agronomists feel that more research is needed to improve the accuracy of organic N availability, it is generally recognized that current programs have greatly improved the utilization of manure N in the United States.

The availability of P from manures is a subject of current interest. Many nutrient recommendation programs have indicated that all P in manures and biosolids is plant-available. However, research by Sharpley and Moyer (2000) indicated that the organic P content was 5 to 25 percent of the total P in a range of manures and biosolids and that availability for soil reaction, movement, and plant uptake was much greater for the inorganic forms of P. While these differences in availability are reasonable, routine manure analyses do not measure the organic and inorganic forms of P. Maguire et al. (2001) found that treatment processes also affected P availability from biosolids. The general trend in their research was for higher levels of extractable P in soils amended with biosolids produced without the use of Fe and Al salts, followed by biosolids treated with metal salts and lime, and the lowest availability being from biosolids treated with only metal salts. These differences in P availability are not generally accounted for in nutrient recommendations because P has been generally applied in excess of crop needs as the organic nutrient application rates have been calculated to supply crop N requirements. Phosphorus-based nutrient management will require more precise determination of P availability from organic sources.

Potassium in most organic wastes occurs as soluble inorganic K^+, and thus the total K content of manures and biosolids can be used as reasonable estimates of K available to plants (Havlin et al., 2005). Amounts of K in organic wastes range from 0.2 to 2 percent depending on the organic source and the handling method for the waste.

The commercial fertilizer requirement for a particular site is determined after consideration is given to residual soil nutrient supplies (soil testing), the crop to be grown, and organic nutrient sources that might be available. The fertilizer recommendation can be viewed as supplementing the other available sources. Although certain areas have high levels of plant-available nutrients, that is K availability in many western United States soils due to an abundance of K-bearing minerals, most soils in the United States will require supplemental fertilizer applications to optimize productivity and to sustain productivity with continued crop removal.

Integrated nutrient management programs in the United States have been driven by environmental concerns associated mainly with manure nutrient applications. However, concern for efficient nutrient use, especially N, in crop production is increasing for both economic and environmental reasons. It is reasonable to expect that integrated nutrient management based on soil testing and other techniques such as real time remote sensing will increase in importance in the United States in the future.

NUTRIENT MANAGEMENT PLANS

Nutrient management plans are documents that outline detailed nutrient management strategies for a specific tract of land. The development of nutrient management plans in the United States is directly related to concerns regarding enrichment of surface and ground waters with N and P. Some of the earliest nutrient management efforts occurred in the Chesapeake Bay Watershed on the eastern U.S. coast. The Chesapeake Bay program is a comprehensive effort to restore water quality in the largest estuary in the United States (Anon., 2004b) (http://www.chesapeakebay.net/overview.htm). A major focus has been the reduction of N and P loading from agricultural sources to surface waters, and one of the most widely used "best management practices" for reducing nutrient pollution in the Bay is nutrient management planning (Anon., 2004a) (http://www.chesapeakebay.net/index_cbp.cfm).

All states (Virginia, Maryland, Pennsylvania, and New York) in the Chesapeake Bay Watershed have nutrient management programs that focus on soil testing for residual nutrient sources, utilizing organic nutrient sources for supplying needed N and/or P, and supplementing with commercial fertilizers. These plans have the following elements:

1. field boundary maps that clearly identify the location of each field, and identify environmental concerns such as streams and wells;
2. soil series maps for each field along with information outlining expected yield levels for each crop in the rotation for the specific soil series;
3. current soil test levels for each field and/or major soil series within the field;
4. manure nutrients available on the farm;
5. crop rotation for each field for a three to five year period;
6. manure, lime, and supplemental commercial fertilizer recommendations for each field for the entire crop rotation;
7. recommendations for timing and methods of manure and fertilizer applications; and
8. a general discussion of the overall objectives of the plan and particular concerns for the specific farm.

Virginia's *Nutrient Management Handbook,* (Kindig, 2002) provides complete details on the development of nutrient management plans, while details on Pennsylvania's program can be found at http://panutrientmgmt. cas.psu.edu/. While most states have state-specific nutrient management programs, the federal government has established guidelines for "Comprehensive Nutrient Management Plans" (USDA-NRCS, 2004a). The focus for all of these programs is to better utilize nutrients from organic wastes because excessive application of manure has been determined to be the major source of nutrient pollution in areas of concentrated animal production. Finally, when animal numbers reach certain levels in many states, nutrient management plans become a legal requirement to obtain a permit to operate the livestock farm.

Besides using manure nutrients more efficiently, nutrient management plans focus on establishing realistic yield expectations for specific fields and utilizing the most efficient methods and timing of application to improve nutrient use efficiency. For example, split N applications will be recommended for maize and wheat production in order to increase applied N uptake, with a resulting reduction in potential loss to the environment. Such recommendations have improved nutrient use efficiency on some farms using only commercial fertilizer nutrients. In addition, plans will call for injection of manure in certain situations, or specify that no additional P should be applied to a certain field until the soil-test P level declines below a certain level. The plans are all derived to reduce the potential for nonpoint source nutrient pollution of water. However, improved economic returns are sometimes generated through better use of on-farm nutrients, but the cost of moving low nutrient content manures off-farm can result in added costs to the farm operation.

Nutrient Management Plan Implementation

Formal nutrient management plans that are required for operation of livestock farms are implemented to varying degrees, and have generally improved the use of manure nutrients by reducing excessive manure nutrient applications. Analysis of manures has increased so that growers are more aware of manure nutrient content. Also, these plans have created an awareness of efficiencies to be gained with timing of nutrient applications to better coincide with periods of maximum crop nutrient need. This is especially true for N applications. Finally, the plans have highlighted the potential for buildup of excessive levels of P from continuous manure application at rates that supply all crop N needs. More P-based manure applications will be made in future years.

Formal nutrient management plans (current U.S. view of managing organic nutrients) for farms using only commercial fertilizers are not widespread in the United States. Only in some areas, such as the Chesapeake Bay region, irrigated areas of states such as Nebraska, Kansas, and Colorado, and other sensitive watersheds, where selected conservation practices require nutrient management plans for receiving governmental monetary support, are formal nutrient management plans common. However, in most areas of agronomic crop production, nutrient management takes the form of "developing a fertilization program that is economically efficient."

Fertilization programs hopefully rely on soil testing to identify residual soil nutrient supplies and the potential for crop response to applied nutrients. In the humid eastern United States, soil acidity must be neutralized with appropriate lime applications in order to efficiently utilize plant nutrients, especially P. Fertilizer recommendations consider legumes in the rotation and soil productivity levels. The higher the yield levels that can be obtained (more productive soils), the greater the amounts of nutrients that will be removed over time, and thus the greater the potential for response to applied fertilizer nutrients. However, the most difficult challenge for nutrient management is determining the appropriate N fertilizer application rate for a specific field in a specific year.

The most widespread practice for determining N rates for agronomic crops is to select a suitable "yield goal" for the crop to be grown in a specific field and then calculate the necessary N required to grow the predetermined yield. For example, applied N recommended to produce a Mg ha^{-1} of maize grain is 17.9 kg N in Virginia (Donohue and Heckendorn, 1994) while levels of 23.2 kg N Mg^{-1} grain are recommended in Mississippi (Larson and Oldham, 2004). The approach is similar for N recommendations in Michigan, Ohio, and Indiana (Vitosh et al., 2004) except that N recommended for yield potentials of 6.3 Mg ha^{-1} or less is 17.9 kg N Mg^{-1} grain, while for yield potentials of 11.3 Mg grain ha^{-1}, N recommendations increase to 21.8 kg N Mg^{-1} grain. The required N fertilizer application rate is adjusted to reflect previous contributions from legumes in the rotation, or residual soil nitrate levels. In the humid region, where residual N levels in soils can vary greatly from year to year, this approach has been problematic at best. Sawyer (2004) and Nafziger et al. (2004) have recently examined numerous N rate studies for maize in Illinois and Iowa and conclude that economics, that is, cost of fertilizer N and value of grain produced, must be included in formulating N fertilizer recommendations for maize. In the semiarid western region, residual soil N is included in the N recommendation algorithm, but yield goals and N required per unit of production are still the major starting points for all recommendations (Mortvedt et al., 2004; Hergert and Penas, 1992).

Given the degree of response, or lack of response, to applied N, much work has been done in the past ten years in the United States to develop sensors that adjust N fertilizer rates as the applicator moves across the crop. Solie et al. (1996), working in Oklahoma, proposed using an optical sensor to vary N rates within small areas, that is, 1.0 m^2, for improving N fertilizer applications. The system utilizes optical sensors to determine NDVI values for an algorithm that predicts yields for the crop (Stone et al., 1996). The system has been commercialized and is marketed under the "Greenseeker" brand (http://www.ntechindustries.com/greenseeker-home.html). Scharf et al. (2002, 2005), working in Missouri, are also addressing the problem of optimizing N fertilization rates with the use of sensors.

The other major approach for improving the efficiency of fertilizer applications is defining management zones within fields that respond to different rates of nutrient applications. Fiez et al. (1994), working with winter wheat in Washington state, determined that wheat yield potential varied by as much as 63 percent depending on landscape position, and that variable N fertilizer applications could improve N use efficiency and profitability. Kitchen et al. (1999) determined that soil conductivity measurements could be used to determine differences in soils that might lead to differences in crop productivity, and associated N fertilizer needs. However, numerous other management factors such as plant population and date of planting had to be considered along with soil properties. Recent research in Colorado found that variable rate N applications associated with management zones increased profitability as compared to uniform N rates (Koch et al., 2004). Similar research by Anderson-Cook et al. (2002) determined that electromagnetic conductivity and crop yield maps could be used to differentiate soil types with widely varying productivity levels (and associated nutrient requirements) in the mid-Atlantic Coastal Plain. The development of management zones has been shown in almost all cases to be equal to or superior to grid soil sampling to identify areas with varying nutrient requirements. In addition, the use of the management zone approach reduces the cost of soil sampling and analysis compared with grid soil sampling.

Variable nutrient rate application technologies can be expected to increase as a part of nutrient management in the United States in coming years. For the relatively nonmobile nutrients P and K, soil sampling in zones associated with crop yields and removals will provide the basis for monitoring nutrient levels, and applying appropriate rates. This approach also appears to be valid for adjusting N fertilizer rates in the semiarid crop production regions where N leaching is not a major issue. However, in the humid regions where in-season N losses can be great, the use of real-time sensors for making in-season adjustments with split N applications appears to offer important opportunities for improving N use efficiency and overall crop production efficiency.

RESEARCH NEEDS

A major concern with current fertilizer recommendation programs is that most of the soil test calibration data are old, dating at least prior to the 1980s, or even the 1950s, and 1960s. The exception to this concern is the P and K calibration work recently conducted in Iowa (Mallarino et al., 2004) and the research cited in the previous section to optimize N fertilization of maize and wheat. Calibration of soil test levels and associated crop response to fertilizer nutrients is needed with modern cultivars, reduced tillage systems, and other practices that continue to increase yields. The lack of modern calibration research is also needed to meet increased environmental concerns that relate to application of N and P fertilizers. Funding to support soil test calibration research is currently lacking in most regions of the United States.

Prediction and application of appropriate N fertilizer rates for wheat, maize, and cotton continues to be an area of need. As Raun and Johnson (1999) point out, an increase of 1 percent in the efficiency of N use worldwide could lead to a savings of over $2.3 million dollars per year in fertilizer costs. In addition, the recent Scientific Committee on Problems of the Environment (SCOPE) publication entitled *Agriculture and the Nitrogen Cycle* (Mosier et al., 2004) clearly stresses the environmental needs to improve N use efficiency to address water quality and other ecosystem issues. Research with real-time sensors and variable rate N applications, as well as development of new N fertilizer sources that more readily match N release with crop nutrient requirements are needed to meet economic and environmental challenges.

Nutrient release from organic sources is variable and unpredictable in many cases. The degree of nonuniformity in organic wastes must be reduced, and nutrients concentrated into forms that can re-enter our fertilizer nutrient distribution system. Low nutrient content wastes will not be transported great distances due to costs. Thus, new processing methods must be developed to cost-effectively concentrate nutrients in organic wastes such as poultry litter, dairy and swine slurry, and feedlot manure.

Finally, continued work with placement and timing of nutrient applications is needed to improve nutrient use efficiency in modern crop production systems that are continuing to reduce tillage. Nutrient placement programs that enhance fertilizer nutrient uptake in production systems that minimize disturbance of surface residues will be critical in coming years as sediment and nutrient losses are reduced in agronomic crop production programs.

Perhaps the greatest challenge for this research area in the United States is funding. Although significant funds have been focused on dealing with excessive manure applications that affect localized areas, the larger issue of improved nutrient use in agriculture receives small support in federal, state, and industry research funding. The area of fertilizer use is not currently perceived to be an area of high return for research expenditures, especially as food production in the United States is quite adequate, and environmental concerns are being addressed to a lesser degree than in the 1990s. However, agricultural production in the United States is technology driven and highly efficient in terms of labor use per unit of production. There is no reason to believe that yield and efficiency increases will not continue in the foreseeable future.

REFERENCES

Adams, F. and C.E. Evans. (1962). A rapid method for measuring lime requirement of red-yellow podzolic soils. *Soil Science Society of America Proceedings*. 26: 355-357.

Anderson-Cook, C.M., M.M. Alley, J.K.F. Roygard, R. Khosla, R.B. Noble, and J.A. Doolittle. (2002). Differentiating soil types using electromagnetic conductivity and crop yield maps. *Soil Science Society of America Journal*. 66: 1562-1570.

Anonymous. (2000). *Soil Analysis Handbook of Reference Methods*. Soil and Plant Analysis Council, Inc., Boca Raton, FL: CRC Press LLC.

Anonymous. (2001a). Soil test levels in North America, Summary Update. PPI/PPIC/FAR Technical Bulletin 2001-1, Potash and Phosphate Institute, Norcross, GA.

Anonymous. (2001b). Nutrients removed in harvested portion of crop. Available On-line: http://www.ppi-ppic.org/ppiweb/ppibase.nsf/$webindex/article=FC 18933385256A00006BF1AD5F8663ED. (accessed December 28, 2004). Potash and Phosphate Institute. Norcross, GA.

Anonymous. (2004a). Overview of the Bay Program, America's premier watershed restoration project. Chesapeake Bay Program Office, Annapolis, MD. http://www.chesapeakebay.net/overview.htm (accessed December 2004).

Anonymous. (2004b). Reducing nutrient pollution. Chesapeake Bay Program Office, Annapolis, MD. http://www.chesapeakebay.net/overview.htm (accessed December 2004).

Beegle, D.B. and T.C. Oravec. (1990). Comparison of field calibrations for Mehlich-III P and K with Bray P1 and ammonium acetate K for corn. *Communications in Soil Science and Plant Analysis*. 21: 1025-1036.

Black, C.A. (1993). Soil testing and fertilizer requirement. In *Soil Fertility Evaluation and Control*. pp. 271-443. Boca Raton, FL: Lewis Publishers.

Blackmer, A.M., R.D. Voss, and A.P. Mallarino. (1997). Nitrogen fertilizer recommendations for corn in Iowa. Pub. Pm-1714. Iowa State University Extension, Ames, IA.

Brady, N.C. and R.R. Weil. (2002). *The Nature and Properties of Soils.* 12th edition. Upper Saddle River, NJ: Prentice-Hall, Inc.

Bray, R.H. and L.T. Kurtz. (1945). Determination of total, organic and available forms of phosphorus in soils. *Soil Science.* 59: 39-45.

Buckman, H.O. and N.C. Brady. (1960). *The Nature and Properties of Soils.* 6th edition. New York: Macmillan Publishing Co.

Crozier, Carl R. (2004). Fertilization. In Edmisten, K.L., A.C. York, F.H. Yelverton, J.F. Spears, and D.T. Bowman (eds.) *Cotton Information,* North Carolina State University pub. AG-417 (revised), Raleigh, NC. Available online: http://ipm .ncsu.edu/Production_Guides/Cotton/contents.html (accessed January 2005).

Dobermann, A. and K.G. Cassman. (2002). Plant nutrient management for enhanced productivity in intensive grain production systems of the United States and Asia. *Plant Soil.* 247: 153-175.

Donohue, S.J. and S.E. Heckendorn. (1994). *Soil Test Recommendations for Virginia.* Blacksburg, VA: Virginia Cooperative Extension.

Evanylo, G.K. and M.M. Alley. (1997). Pre-sidedress soil nitrogen test for corn in Virginia. *Communications in Soil Science and Plant Analysis.* 28: 1285-1301.

FAOSTAT data (2004). Food and Agriculture Organization of the United Nations. World Agriculture Information. FAOSTAT. Fertilizer Consumption. http://faostat .fao.org/site/422/default.aspx.

Fiez, T.E., B.C. Miller, and W.L. Pan. (1994). Assessment of spatially variable nitrogen fertilizer management in winter wheat. *Journal of Production Agriculture.* 7: 86-93.

Fixen, P.E. (2002). Soil test levels in North America. *Better Crops.* 86:12-15.

Fixen, P.E. and A.M. Johnston. (2002). Nutrient budgets in North America. In *Plant Nutrient Use in North American Agriculture.* pp. 79-85. PPI/PPC/FAR Technical Bulleting 2002-1. Norcross, GA: Potash & Phosphate Institute.

Geist, J.M., J.O. Reuss, and D.D. Johnson. (1972). Prediction or the nitrogen requirement of field crops. Part II. Application of theoretical models to malting barley. *Agronomy Journal.* 62: 385-389.

Goos, R.J., D.G. Westfall, A.E. Ludwick, and J.E. Goris. (1982). Grain protein as an indicator of nitrogen sufficiency. *Agronomy Journal.* 74: 130-133.

Haas, H.J., W.O. Willis, and J.J. Bond. (1974). Summer fallow in the Western United States. USDA Conservation Res. Rpt. 17. 160 p.

Haby, V.A., M.P. Russell, and Earl O. Skogley. (1990). Testing soils for Potassium, Calcium, and Magnesium. In Westerman R.L. (ed.) *Soil Testing and Plant Analysis.* pp. 181-227. Madison, WI: Soil Science Society of America, Inc.

Hardy, D.H., M.R. Tucker, and C.E. Stokes. (2003). *Crop fertilization based on North Carolina soil tests.* Circular No. 1. North Carolina Dept. of Agriculture and Consumer Services, Agronomic Division, Raleigh, NC.

Havlin, J.L., J.D. Beaton, S.L. Tisdale, and W.L. Nelson. (2005). *Soil Fertility and Fertilizers.* 7th edition. Upper Saddle River, NJ: Pearson Prentice Hall.

Heady, Earl O. (1952). *Economics of Agricultural Production and Resource Use.* pp. 21, 125-126, 253. New York: Prentice-Hall, Inc.

Helmke, P.A. and D.L. Sparks. (1996). Lithium, sodium, potassium, rubidium and cesium. In Sparks D.L. (ed.) *Methods of Soil Analysis, Part 3. Chemical Methods.* pp. 551-601. Madison, WI: Soil Science Society of America, Inc.

Hergert, G.W. and E.J. Penas. (1992). New Nitrogen Recommendation for Corn. University Neb. NebFacts NF. pp. 91-111. Univ. of Nebraska, Cooperative Extension, Lincoln, NE.

Jones, J. Benton, Jr. (1998). Soil test methods: Past, present, and future use of soil extractants. *Communications in Soil Science and Plant Analysis.* 29: 1543-1552.

Jones, J. Benton, Jr. (2001). *Laboratory Guide for Conducting Soil Tests and Plant Analysis.* Boca Raton, FL: CRC Press.

Kaan, D.A., D.M. O'Brein, P.A. Burgener, G.A. Peterson, and D.G. Westfall. (2002). An economic evaluation of alternative crop rotations compared to wheat-fallow in Northeastern Colorado. Technical Report TR 02-03. Agric. Exp. Stn. Colo. State University, Fort Collins, CO.

Kellogg, R.L., C.H. Lander, D.C. Moffitt, and N. Gollehon. (2000). Manure nutrients relative to the capacity of cropland and pastureland to assimilate nutrients: Spatial and temporal trends for the United States. USDA-NRCS. www.nhq.nrcs .usda.gov/land/index/publication.html (accessed December 2004).

Kindig, D., ed. (2002). *Nutrient Management Handbook,* 3rd edition. Richmond, VA: Virginia Dept. of Conservation and Recreation, Division of Soil and Water Conservation.

Kissel, D. (2003). *Soil Test Handbook for Georgia.* Athens, GA: Georgia Coop. Extension Service, University of Georgia.

Kitchen, N.R., K.A. Sudduth, and S.T. Drummond. (1999). Soil electrical conductivity as a crop productivity measure for clay pan soils. *Journal of Production Agriculture.* 12: 607-617.

Koch, B., R. Khosla, W.M. Frasier, D.G. Westfall, and D. Inman. (2004). Economic feasibility of variable-rate nitrogen application utilizing site-specific management zones. *Agronomy Journal.* 96: 1572-1580.

Kuo, S. (1996). Phosphorus. In Sparks D.L. (ed.) *Methods of Soil Analysis, Part 3. Chemical Methods.* pp. 869-919. Madison, WI: Soil Science Society of America., Inc.

Lander, C.H., D. Moffitt, and K. Alt. (1998). Nutrients available from livestock manure relative to crop growth requirements. Resource Assessment and Strategy Planning Working Paper 98-1, USDA-NRCS. Available online: http://www .nhq.nrcs.usda.gov/land/index/publication.html (accessed December 2004).

Larson, E. and L. Oldham. (2004). Corn fertilization. Information Sheet 864. Mississippi State University Extension Service. Available online: http://msucares .com/pubs/infosheets/is0864.htm (accessed December 2004).

Ludwick, A.E. and A.M. Johnston. (2002). Organic nutrients. In *Plant Nutrient Use in North American Agriculture.* pp. 33-38. PPI/PPIC/FAR Technical Bulletin 2002-1. Norcross, GA: Potash and Phosphate Institute.

Magdoff, F.R. (1991). Understanding the Magdoff pre-sidedress nitrate test for corn. *Journal of Production Agriculture.* 4: 297-305.

Magdoff, F.R., D. Ross, and J. Amadon. (1984). A soil test for nitrogen availability to corn. *Soil Science Society of America Journal.* 48: 1301-1304.

Maguire, R.O., J.T. Sims, S.K. Dentel, F.J. Coale, and J.T. Mah. (2001). Relationship between biosolids treatment process and soil phosphorus availability. *Journal of Environmental Quality.* 30: 1023-1033.

Mallarino, A.P., D.J. Wittry, and P.A. Barbagelata. (2004). Iowa soil-test field calibration research update: potassium and the Mehlich-3 ICP phosphorus test. Available online: http://extension.agron.iastate.edu/soilfertility/info/recmallpk02.pdf (accessed December 2004).

Mehlich, A. (1953). *Determination of P, Ca, Mg, K, Na, and NH_4 (mimeo).* Raleigh, NC: North Carolina Soil Test Division.

Mehlich, A. (1976). New buffer pH method for rapid estimation of exchangeable acidity and lime requirement of soils. *Communications in Soil Science and Plant Analysis.* 7: 637-652.

Mehlich, A. (1984). Mehlich 3 soil test extractant: A modification of Mehlich 2 extractant. *Communications in Soil Science and Plant Analysis.* 15: 1409-1416.

Meisinger, J.J., V.A. Bandel, J.S. Angle, B.E. O'Keefe, and C.M. Reynolds. (1992). Pre-sidedress soil nitrate test evaluation in Maryland. *Soil Science Society of America Journal.* 56: 1527-1532.

Mortvedt, J.J., D.G. Westfall, and R.L. Croissant. (2004). Fertilizing corn. Colorado State University Coop. Ext. Pub. No. 0.538. Available online: http://www.ext.colostate.edu/pubs/crops/00538.html (accessed January, 2005).

Mosier, A.R., J.K. Syers, and J.R. Freney, eds. (2004). *Agriculture and the Nitrogen Cycle, Assessing the Impacts of Fertilizer Use on Food Production and the Environment.* Washington, DC: Island Press.

Mullins, G.L. (2000). *Phosphorus, Agriculture & the Environment.* Blacksburg, VA: VA Coop. Ext. Pub. 424-029.

Mulvaney, R.L. (1996). Nitrogen-inorganic forms. In Sparks D.L. (ed.) *Methods of Soil Analysis, Part 3. Chemical Methods.* pp. 1123-1184. Soil Science Society Of America Book Series 5. Madison, WI: Soil Science Society Of America, Inc.

Nafziger, E.D., J.E. Sawyer, and R.G. Hoeft. (2004). Formulating N recommendations for corn in the corn belt using recent data. *Proceedings of the North Central Extension-Industry Soil Fertility Conference, Des Moines, IA,* Nov. 17-18.

Peterson, G.A. and D.G. Westfall. (2004). Managing precipitation use in sustainable dryland agroecosystems. *The Annals of Applied Biology.* 144: 127-138.

Peterson, G.A., D.G. Westfall, and C.V. Cole. (1993). Agrecosystem approach to soil and crop management research. *Soil Science Society of America Journal.* 57: 1354-1360.

Plank, C.O. (1998). *Soil, Plant, and Animal Waste Analysis: Status Report for the United States, 1992-1996.* Soil and Plant Analysis Council. Lincoln, NE.

Powers, W.J. and H.H. Van Horn. (2001). Nutritional implications for manure nutrient management planning. *Applied Engineering in Agriculture.* 17(1): 27-39.

Raun, W.R. and G.V. Johnson. (1999). Improving nitrogen use efficiency for cereal production. *Agronomy Journal.* 91: 357-363.

Reuss, J.O. and J.M. Geist. (1972). Prediction or the nitrogen requirement of field crops. Part I. Theoretical models for nitrogen release. *Agronomy Journal.* 62: 381-384.

Roth, G.W., D.B. Beegle, and R.H. Fox. (1991). Soil nitrate quick-test kit for corn evaluation project. Test kit operating instructions. Dept. of Agronomy, Penn State University. Prepared for Hawk Creek Laboratory Inc., Glen Rock, PA.

Sawyer, J.E. (2004). With high fertilizer prices is it business as usual or should fertilization practices change? 2004 Integrated Crop Management Conference. Dept. of Agronomy, Iowa State University, Ames, IA.

Scharf, P.C. and M.M. Alley. (1993). Spring nitrogen on winter wheat: II. A flexible multicomponent rate recommendation system. *Agronomy Journal.* 85: 1186-1192.

Scharf, P.C. and M.M. Alley. (1994). Residual soil nitrogen in humid region wheat production. *Journal of Production Agriculture.* 7: 81-85.

Scharf, P.C., M.M. Alley, and Y.Z. Lei. (1993). Spring nitrogen on winter wheat: I. Farmer-field validation of tissue test-based rate recommendations. *Agronomy Journal.* 85: 1181-1186.

Scharf, P.C., N.R. Kitchen, K.A. Sudduth, J.G. Davis, V.C. Hubbard, and J.A. Lory. (2005). Field-scale variability in optimal N fertilizer rate for corn. *Agronomy Journal.* (In press).

Scharf, P.C., J.P. Schmidt, N.R. Kitchen, K.A. Sudduth, S.Y. Hong, J.A. Lory, and J.G. Davis. (2002). Remote sensing for N management. *Journal of Soil and Water Conservation.* 57: 518-524.

Shaffer, K. and R. Sheffield. (2004). Land application records and sampling. Lesson 35. In Livestock and Poultry Environmental Stewardship (LPES) Curriculum. Purdue University. Available on-line: http://www.lpes.org/Lessons/Lesson35/35_Application_Records.html (accessed December 2004).

Sharpley, A.N. and B. Moyer. (2000). Phosphorus forms in manure and compost and their release during simulated rainfall. *Journal of Environmental Quality.* 29: 1462-1469.

Sharpley, A.N., J.L. Weld, D.B. Beegle, P.J.A. Kleinman, W.J. Gburek, P.A. Moore, and G. Mullins. (2003). Development of phosphorus indices for nutrient management planning strategies in the United States. *Journal of Soil and Water Conservation.* 58(3): 137-152.

Shoemaker, H.E., E.O. McLean, and P.F. Pratt. (1962). Buffer methods for determination of lime requirement of soils with appreciable amount of exchangeable aluminum. *Soil Science Society of America Proceedings.* 25: 274-277.

Simpson, T.W. (1998). A Citizen's Guide to Maryland's Water Quality Improvement Act. University of MD Coop. Extension, College Park, MD.

Sims, J.T. (1999). Delaware's state nutrient management program: Overview of the 1999 Delaware Nutrient Management Act. College of Agric. and Nat. Resources., University of Delaware, Newark.

Sims, J.T. (2000). The Role of Soil Testing in Environmental Risk Assessment for Phosphorus. In Sharpley A.N. (ed.) *Agriculture and Phosphorus Management.* pp. 57-81. The Chesapeake Bay. Lewis Publishers, NY: CRC Press LLC.

Smil, V. (2001). *Enriching the Earth, Fritz Haber, Carl Bosch, and the Transformation of World Food Production.* Cambridge, MA: The MIT Press.

Solie, J.B., W.R. Raun, R.W. Whitney, M.L. Stone, and J.D. Ringer. (1996). Optical sensor based field element size and sensing strategy for nitrogen application. *Transactions of the ASAE.* 39(6): 1983-1992.

Soltanpour, P.N. and A.P. Schwab. (1977). A new soil test for simultaneous extraction of macro- and micro-nutrients in alkaline soils. *Communications in Soil Science and Plant Analysis.* 8: 195-207.

Sommers, L.E. (1977). Chemical composition of sewage sludges and analysis of their potential use as fertilizers. *Journal of Environmental Quality.* 6: 225-232.

Stewart, W.M. and T.L. Roberts. (2002). Inorganic nutrient use. In *Plant Nutrient Use in North American Agriculture.* pp. 23-31. PPI/PPIC/FAR Technical Bulletin 2002-1. Potash & Phosphate Institute, Norcross, GA.

Stone, M.L., J.B. Solie, W.R. Raun, R.W. Whitney, S.L. Taylor, and J.D. Ringer. (1996). Use of spectral radiance for correcting in-season fertilizer nitrogen deficiencies in winter wheat. *Transactions of the ASAE.* 39: 1623-1631.

Taylor, Harold H. (1993). Fertilizer Use and Price Statistics, 1960-93. USDA-Economic Research Service. Statistical Bulletin Number 893. Washington, DC.

Terry, D.L. and B.J. Kirby. (1994-2004). *Commercial Fertilizers.* (published yearly). Association of American Plant Food Control Officials and The Fertilizer Institute. Washington, DC: Publication available through The Fertilizer Institute. Washington, D.C., U.S.

Tisdale, S.L. and W.L. Nelson. (1956). *Soil Fertility and Fertilizers.* New York: Macmillan Publishing Co.

USDA—National Agricultural Statistics Service. (2004). Historical Data. http://www.nass.usda.gov/index.asp.

USDA—NRCS. Nutrient Content of Crops: Nutrients Removed By Harvest. http://npk.nrcs.usda.gov (accessed 22 June 2007).

USDA—Office of the Chief Economist. (2007). Major world crop areas and climatic profiles. http://www.usda.gov/oce/weather/pubs/Other/MWCACP/index.htm.

U.S. Dept. of Commerce, National Oceanic and Atmospheric Administration, Earth System Research Laboratory. (2007) www.cdc.noaa.gov (accessed May 2007).

U.S. Environmental Protection Agency. (1999). Biosolids Generation, Use and Disposal in the United States. EPA530-R-99-009. http://www.epa.gov/epaoswer/non-hw/compost/biosolid.pdf.

U.S. Environmental Protection Agency. (2003). Concentrated Animal Feeding Operations (CAFO)—Final Rule. http://cfpub.epa.gov/npdes/afo/cafofinalrule.cfm (accessed December 2004).

USDA—NRCS. (2004a). Comprehensive Nutrient Management Planning Technical Guidance. National Planning Procedures Handbook. Available online: http://www.nrcs.usda.gov/programs/afo/cnmp_guide_600.54.html (accessed December 2004).

USDA—NRCS. (2004b). Southern Extension Research Activity. Minimizing Phosphorus Losses from Agriculture National Research Project. http://www.sera17.ext.vt.edu/background.htm (accessed December 2004).

USDA—NRCS. (2007). Dominant Soil Orders.

USDA—Office of the Chief Economist. (2007). Major world crop areas and climatic profiles, North America. Available on-line: http://www.usda.gov/oce/weather/pubs/Other/MWCACP/world_crop_country.htm#northamerica (accessed April 2007).

U.S. Department of Commerce. National Oceanic and Atmospheric Administration. Earth System Research Laboratory. Boulder, Colorado, U.S. http://www.cdc.noaa.gov/USclimate/USclimdivs.html.

Van Raij, B. (1994). New diagnostic techniques, universal soil extractants. *Communications in Soil Science and Plant Analysis.* 25: 799-816.

Vitosh, M.L., J.W. Johnson, and D.B. Mengel. (2004). Tri-State Fertilizer Recommendations for Corn, Soybeans, Wheat and Alfalfa. Bulletin E-2567.

Westerman, R.L. (1990). *Soil Testing and Plant Analysis.* 3rd edition. Madison, WI: Soil Science Society of America.

Westfall, D.G., J.L. Havlin, G.W. Hergert, and W.R. Raun. (1996). Nitrogen management in dryland cropping systems. *Journal of Production Agriculture.* 9: 192-199.

Chapter 4

Integrated Nutrient Management: Experience and Concepts from Canada

Sukhdev S. Malhi
Cynthia A. Grant
Denis A. Angers
Adrian M. Johnston
Jeff J. Schoenau
Craig F. Drury

INTRODUCTION

Agriculture is a major industry in Canada and the agri-food sector accounted for 8.3 percent of Canada's GDP in 2000 (http://www.agr.gc.ca/cb/apf/pdf/bg_con_overvu_e.pdf, accessed June 3, 2004). Nearly half the agricultural production is exported, resulting in 10 percent of Canada's total trade surplus. The type of agricultural production varies greatly across the country. Red meat, grains, oilseeds, dairy, poultry, and eggs are the major contributors to farm cash receipts. Production and export of these products involves the utilization and removal of large amounts of nutrients that must be replaced to ensure sustainability of production into the future.

The authors thank Dr. K. S. Gill for the internal review and revision of the manuscript, and Joan Martin for technical help in collecting information (statistics) on crop production, and nutrients in inorganic and organic sources in Canada. Thanks are also extended to David Fuller and Jennifer Thomas who contributed to the collection and presentation of information on nutrient management in the Maritime and British Columbia agricultural systems.

In Canada, as in many other countries around the world, specialization of cropping and concentrated livestock production has led to a decoupling of nutrient cycling. Nutrients are removed from farms involved in crop production and concentrated in urban areas and on farms involved in intensive livestock production. Thus, both depletion and excess accumulation of nutrients are of concern in Canadian agriculture.

Nutrient depletion can be avoided by replacing nutrients removed in crop production with inputs of organic or inorganic nutrients. Inorganic fertilizers have played an important role in increasing crop production in Canada. Between 1965 and 2000, grain production in Canada increased by approximately 70 percent, while the consumption of fertilizer N, P, and K increased by 576 percent (PPI, 2002; CFI, 2003). Organic sources of nutrients, including animal manure, composts, crop residues, and N fixation, play an important role in increasing crop production, maintaining yield stability and improving soil quality through their beneficial effects on the physical, chemical, and biological properties of soil (Beauchamp, 1983; N'dayegamyie and Angers, 1990; Schoenau, 1997; Schoenau et al., 1999, 2000). Unfortunately, the supply and distribution of organic sources of nutrients in many parts of Canada cannot meet the total crop nutrient demand and thus cannot completely replace the use of inorganic fertilizers. In fact, estimates in 2001 indicate that only 32 percent of N, 21 percent of P, and 19 percent of K removed by crops grown are replaced by organic sources in Canada.

Nutrient accumulation is also of major concern in Canada. More nutrients are supplied in livestock feed than are exported in animal products from a farm, with the excess being excreted in manure and urine. Application of manure on a restricted land area can lead to excessive accumulation of nutrients over time, and lost opportunity for soil improvement on unmanured land.

Integrated nutrient management (INM) refers to the efficient use of available nutrient sources to maintain a viable nutrient balance on the farm. This is essential to ensure environmental, social, and economic sustainability of agricultural systems. INM may involve the use of synthetic fertilizers and amendments, and organic sources (e.g., cattle, hog and poultry manure, compost, legume crops in rotations and green manure, crop residues, microbial inoculants, and biofertilizers), as nutrient sources, either alone or in combination. Use of INM should reduce the cost of production, maintain high yields, improve soil quality, maximize water and nutrient use efficiency, and minimize nutrient loss. Long-term INM would serve to improve both on-farm nutrient balance and crop production. This chapter focuses on INM considerations for the dominant cropping systems in the major agroecological regions of Canada.

MAJOR SOIL/CLIMATIC REGIONS
AND CROPPING SYSTEMS

Canada has a land area of 998 Mha, encompassing many agroecological zones (Soil Agroecological Zone Map; Figure 4.1, see also color gallery). Farmland comprises 67.2 Mha (6.7 percent of total area), and about 46 Mha of this farmland is improved and used for crops and pasture (http://members .shaw.ca/bethcandlish/world.htm, accessed June 4, 2004). The major portion of the arable land (about 55 Mha) is located on the Canadian prairies (Alberta, Saskatchewan, and Manitoba). Central Canada (Ontario and Quebec) has about 9 Mha of arable land, while British Columbia and the Atlantic Provinces have <4 Mha.

Semiarid Prairie

The semiarid prairie region of western Canada is part of the northern Great Plains of North America. This area is dominated by the Brown and

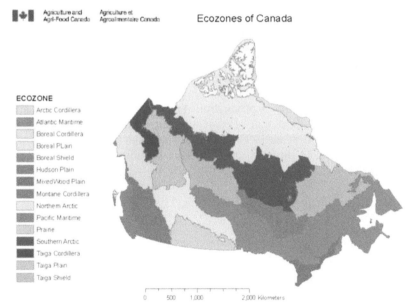

FIGURE 4.1. Agroecozones of Canada (map generated by A. Waddell from data on http://sis.agr.gc.ca/cansis/nsdb/ecostrat (accessed April 27, 2007). *Source:* Adapted from AAFC (2005).

Dark Brown Chernozemic soils (Aridic and Typic Borolls), representing a total of 15.2 Mha of crop land. The climate of this region is continental, with a long cold winter, short warm summer, and potential for large daily changes in temperature, winds, and precipitation (Padbury et al., 2002). Annual precipitation across this region ranges from 334 to 413 mm, while potential evapotranspiration ranges from 635 to 729 mm, resulting in large annual moisture deficits (Campbell et al., 1990). As a result, the success of crop production in this region centers on water conserving practices, such as summer fallow, snow management, and conservation tillage. The main nutrient requirement for optimum production in this region is N, and to a lesser degree, P.

The dominant cropping systems in the semiarid region focus on cereal crops, with oilseed and grain legumes (pulses) often included in the rotation (Campbell et al., 2002). Beef cattle production is concentrated in a part of this region, and large pig operations both present opportunities, and pose some challenges for INM and sustainable crop production. The area seeded to cereals has remained fairly static during the 1976 to 1998 period (Campbell et al., 2002), and is dominated by hard red spring wheat (*Triticum aestivum* L.), durum wheat (*Triticum turgidum* L.), and malt barley (*Hordeum vulgare* L.). Bare summer fallow, a practice that leaves the land uncropped for a year to mineralize nutrients from organic matter and conserve water while using tillage and/or herbicides to control weeds, is still practiced by many producers. However, the area under summer fallow has declined in the past two decades, at a rate of 1.25 percent year^{-1} in the Brown, and 7.5 percent year^{-1} in the Dark Brown soil zone (Campbell et al., 2002). Summer fallow is being replaced mainly by production of canola (*Brassica napus* L.), lentil (*Lens culinaris*), field pea (*Pisum sativum* L.), and chickpea (*Cicer arietinum* L.). In the late 1990s, there was 8.0 Mha in cereals, 4.1 Mha in fallow, 2.2 Mha in oilseeds, and 0.9 Mha in pulse crops in this region (Campbell et al., 2002).

Subhumid Prairie

The subhumid prairie region comprises the Parkland region of western Canada, which is the transitional area between grassland and forest land. The majority of the soils in this region are Gray Luvisols (Alfisols), Dark Gray Chernozem, and Black Chernozem (Udic Borolls) (Canadian Soil Classification System, 1978). Mean annual precipitation in the subhumid prairie varies from 400 to 550 mm and about 60-70 percent occurs during the growing season from May to early September. The temperature is about

16°C in July, −20°C in January, with an average of 5°C in the growing season. The average annual frost-free period ranges between 75 and 110 days. The short frost-free growing season and low temperatures are the main factors limiting crop growth and seed yield. Soil moisture is usually adequate for high crop yields, so rotations are more intensive and summer fallow is less common than in the semiarid prairies. In this region, nutrient supply, especially N, P, and S, is a primary concern.

Major agricultural crops are cereals (spring wheat, barley, and oat-*Avena sativa* L.), oilseeds (canola-*Brassica napus* L. or *Brassica rapa* L., flax-*Linum usitatissimum* L., and mustard-*Brassica juncea* L.), and pulses (field pea, fababean-*Vicia faba* L., lentil, chick pea, and soybean-*Glycine max* (L.) Merrill.). Other annual crops (sunflower-*Helianthus annus* L., potato-*Solanum tuberosum* L., and sugar beet-*Beta vulgaris* L.) are also grown. This region has a large amount of beef production and grows perennial forages (alfalfa-*Medicago sativa* L., bromegrass-*Bromus inermis* Leyss or *Bromus bibesteini* Roem and Shultz, clover-*Trifolium* spp., and Timothy-*Phleum pratense* L.). Major cropping systems include various rotations of annual-annual and annual-perennial crops. As in other areas, large beef cattle and pig operations present opportunities and pose some challenges for INM and sustainable soil productivity.

Central Canada

In the context of this chapter, central Canada includes Ontario and Quebec. The agricultural area is located largely in the southern parts of these two provinces (lower Great Lakes and the St. Lawrence River valley), with climate broadly defined as humid continental. Mean annual temperature ranges from 1 to 8°C and the annual precipitation ranges from 800 to 1,100 mm. Soils vary greatly, but generally belong to the Podzolic (Spodosols), Gleysolic (Aquolls), Brunisolic (Inceptisols) and Luvisolic (Alfisols) soil orders (Canadian Soil Classification System, 1978).

In Quebec, dairy-based cropping systems occupy a large proportion of the agricultural land and involve a range of annual and perennial crops grown in rotation. Annual crops are usually cereals (barley, oat, wheat, and rye-*Secale cereale* L.) and corn (*Zea mays* L.). Perennial crops can be grasses (largely timothy, but also bromegrass, and orchardgrass-*Dactylis glomerata* L.) or legumes (red clover—*Trifolium pratens* L., and alfalfa), and are used to produce hay or are grazed in the field. The length of the rotation is also quite variable but usually involves one year of cereals underseeded with perennial forages to be in place for three to seven years. Manure from the dairy cattle, supplemented by symbiotic biological N_2 fixation (BNF) by legumes, is a major source of nutrients for the crops. Mineral fertilizers are also used to augment the supply of N, P, and K. In many ways, these cropping systems involve an INM approach.

In southern Ontario, annual cropping systems involve intensive crop production. In 2004, the dominant cash crops (i.e., greater than 50,000 ha) included soybean (930,800 ha), grain corn (647,500 ha), winter wheat (303,500 ha), fodder corn (117,400 ha), and barley (95,100 ha) (Ontario Ministry of Food and Rural Affairs, http://www.omafra.gov.on.ca/english/ stats/crops/estimate_metric.html, accessed September 13, 2005). In comparison, there were 920,700 ha of hay production in Ontario in 2004. In 2001, intensively managed crops were grown on 2,268,000 ha in Ontario and on 894,600 ha in Quebec, comprising 65.4 and 50.9 percent of the total arable land in the respective provinces.

Atlantic Maritime

The Atlantic Maritime region is located on the eastern shore of Canada, in the provinces of New Brunswick, Prince Edward Island, Nova Scotia, and Newfoundland, collectively referred to as the Atlantic Provinces. The climate is classified as a modified continental type. The frost-free period averages 120-130 days (www.gov.nf.ca/agric/pubfact/Fertility/soilclimate .htm#4, accessed June 4, 2004). Annual precipitation averages 1,000 mm, with approximately 500 mm received in the growing season between May and September. Excessive moisture, particularly in spring and autumn, is usually a more serious problem than drought. The Appalachian uplands are covered by glacial till, and Humo-Ferric Podzols are the dominant soils (www.ec .gc.ca/soer-ree/English/Framework/NarDesc/atlmar_e.cfm, accessed June 4, 2004). The Northumberland coastal plain coastal lowland areas have deep, more fertile Luvisolic soils. Most of the agricultural production occurs in Nova Scotia, Prince Edward Island, and New Brunswick.

Potato (*Solanum tuberosum* L.) is the most important crop in the Atlantic region (Sanderson, 2003), grown on nearly 70,000 ha in New Brunswick and Prince Edward Island in 1998 (Holmstrom et al., 2000). Crops grown in rotation with potatoes include barley, forage grasses such as Italian ryegrass, and legumes such as red clover (Carter et al., 2003). Production in this region also includes a variety of fruit (apples-*Malus pumila,* blueberries-*Vaccinium corymbosum* L., and raspberries-*Rubus idaeus* L.) and vegetable (carrots-*Daucus carota* L., corn-*Zea mays* L., rutabagas-*Brassica napobrassica* L., cabbage-*Brassica oleracea* L., cauliflower-*Brassica oleracea*

L. var. botrytis, broccoli-*Brassica oleracea* L. (Plenck) var. italica, beans-*Phaseolus vulgaris* L., and brussel sprouts-*Brassica oleracea* L D. C. var. gemmifera) crops. Apple production is important in regions such as the Annapolis valley of Nova Scotia. Blueberries are grown on over 12,000 hectares in Nova Scotia, where there is considerable interest in organic berry production using composts as nutrient sources (Warman, 2003). Forage grasses (timothy and ryegrass-*Lolium perenne* L.) are used as a primary crop for livestock and as a rotational crop placed in between crops such as potatoes (Belanger and Richards, 1998). Alfalfa is also grown as a forage legume, but production difficulties may be encountered on some soils due to acidic subsoil (Carter and Richards, 2000).

Pacific Maritime and Moran Cordillera

The Pacific Maritime region, in coastal British Columbia, is characterized by a temperate rainforest climate, with wet winters. The mean summer temperatures range from 10°C in the north to 15.5°C in the south, while winters are mild with temperatures ranging from –0.5°C to 3.5°C (http://www.ec.gc.ca/soer-ree/English/Framework/Nardesc/pacmar_e.cfm, accessed June 11, 2004). Typically, annual precipitation is between 1,500 and 3,000 mm, but may range from a low of 600 mm to over 4,000 mm. Soils are largely Podzolic and Brunisolic.

The Moran Cordillera Ecozone, in the interior of British Columbia, includes arctic tundra, dense conifer forests, sagebrush, and grassland areas (http://www.ec.gc.ca/soer-ree/English/Framework/Nardesc/moncor_e.cfm, accessed June 11, 2004). Much of the region is mountainous, with valleys that are used for arable agriculture. Mean temperature ranges from 11°C to 16.5°C in summer and from –11°C to –1°C in winter. Some portions of the interior in intermountain regions have a semiarid or desert-like climate with precipitation below 300 mm in the southern valleys and plateaus. Annual precipitation is approximately 500-800 mm in the north and interior, and rises to 1,200 mm in the mountain ranges along the British Columbia-Alberta border (http://www.ec.gc.ca/soer-ree/English/Framework/Nardesc/moncor_e.cfm, accessed June 11, 2004). Luvisols and Brunisols are the most common soils. Podzol soils occur in the mountain ranges in the wetter eastern portion of the ecozone, while the grassland soils of the lower valley floors to the south are often Chernozems.

Predominant production systems in British Columbia include agroforestry, horticultural crops, dairy, and poultry industries. Apple production is estimated to occupy about 5,300 ha (Government of British Columbia,

2004). Blueberry production is concentrated in the Fraser valley region on acidic soils. Sweet corn is grown in the Okanagan valley as well as the lower Mainland and Vancouver Island regions and is the second most commonly grown field vegetable in British Columbia. The dairy and poultry industries are concentrated in the Fraser valley region. Forage grasses are used extensively for feeding the ruminant livestock.

AGRICULTURAL PRODUCTION AND NUTRIENT BALANCES

Sustainable agriculture requires that nutrients removed by the production system be balanced with inputs of plant-available nutrients. These can be in the form of synthetic fertilizers or from various organic materials.

The estimated annual organic nutrient supply in Canada generated from manures, crop residues, sewage sludge and food/pulp industry waste, and biofertilizers is about 1.01 M Mg of NPK (Chambers et al., 2001; PPI, 2002). Recoverable nutrients are 127,914 Mg N from a total of 384,000 Mg N in animal manures and 3,000 Mg N in composts (including hog and poultry), and 79,220 Mg P from a total of 139,000 Mg P in animal manures and 5,500 Mg P in composts. The crop residues produced in Canada can potentially supply 775,700 Mg N, 94,430 Mg P, and 1,268,810 Mg K. The nutrient value of industrial wastes (biosolids) is estimated to be equivalent to 8,400 Mg N and 5,300 Mg P. Leguminous green manures and forage or food legumes are estimated to fix 596,930 Mg N from the atmosphere. In addition, 43,000 Mg of N is provided to cropland through wet deposition of nitrate and ammonium.

In most parts of Canada, the majority of N, P, and K in crop production are supplied by synthetic fertilizers. Consumption of synthetic N, P, and K fertilizers has increased from 170,000 Mg N, 113,537 Mg P, and 107,884 Mg K in 1965, to 1,682,072 Mg N, 291,652 Mg P and 281,555 Mg K in 2000 (PPI, 2002; CFI, 2003; Table 4.1). This represents an increase of nearly 10-fold for N and 2.6-fold for P and K, with a decrease in the P/N ratio from 0.67 to 0.17, and the K/N ratio from 0.65 to 0.17. From 1965 to 2000, the ratio of P fertilizer consumed relative to N fertilizer declined substantially because of a large increase in the amount of N fertilizer used. Historically, many Canadian producers, in the prairies in particular, relied on summer fallow to provide N by mineralization from organic matter. Reduced reliance on fallow led to a reduction in N supplied through SOM mineralization and a resulting increased dependence on N fertilization during this time period. The increase in N consumption reflects the combined

TABLE 4.1. Annual production, and N, P, K fertilizer input, removal, fertilizer input: Removal ratio and balance in Canada

Parameters	1965	1970	1975	1980	1985	1990	1995	2000
Production								
Grain production (Mg)	34,219,685	31,809,921	39,143,900	43,757,300	52,986,500	62,156,100	57,072,000	58,956,400
Hay production (Mg)	19,140,700	25,642,500	23,526,200	23,749,200	23,787,800	33,112,600	27,064,400	23,145,100
Fertilizer use								
N (Mg)	170,000	267,686	531,243	806,277	1,254,411	1,196,292	1,448,356	1,682,072
P (Mg)	113,537	122,536	219,095	255,178	317,152	267,938	274,563	291,652
K (Mg)	107,884	144,883	171,602	285,754	333,447	298,600	257,156	281,555
P:N ratio	0.67	0.47	0.41	0.326	0.25	0.22	0.19	0.17
K:N ratio	0.65	0.54	0.32	0.35	0.278	0.25	0.18	0.17
Crop removal of nutrients								
N (Mg)	1,230,000	1,3000,000	1,500,000	1,610,000	1,840,000	2,270,000	2,210,000	2,300,000
P (Mg)	174,672	192,397	218,341	240,175	275,109	3,362,444	336,244	358,079
K (Mg)	605,809	730,290	771,784	788,382	838,174	1,087,137	979,253	937,759
Ratio of fertilizer use: Crop removal								
N	0.14	0.20	0.36	0.50	0.68	0.53	0.66	0.73
P	0.66	0.65	1.00	1.05	1.14	0.80	0.81	0.81
K	0.17	0.20	0.22	0.36	0.40	0.28	0.26	0.30

TABLE 4.1 (continued)

Parameters	1965	1970	1975	1980	1985	1990	1995	2000
Nutrient balance[a]								
N (Mg)	−1,060,000	−1,032,314	−968,757	−803,723	−585,589	−1,073,708	−761,644	−617,928
P (Mg)	−61,135	−69,861	+754	+15,003	+42,043	−68,306	−61,681	−66,427
K (Mg)	−497,925	−585,407	−600,184	−502,628	−504,727	−788,537	−722,097	−656,204

Source: Adapted from PPI (2002); CFI (2003).

[a]Nutrient balance = Nutrient consumption as fertilizer − nutrient removal by crop (i.e., uptake in grain and hay).

effect of declining N mineralization potential of the soils, and increasing cropping intensity and crop yields.

The data on nutrient removal in crops and consumption of N, P, and K from synthetic fertilizers show that there is a substantial net negative balance for these nutrients, especially for N and K (Table 4.1). The K imbalance reflects the high native K supply of many Canadian soils. Depending on the mineralogy and weathering of the soil parent material, this K release may be sustainable in the long-run, or may lead to depletion of the native soil fertility. Part of the N imbalance may be offset by biological N fixation in legume crops. It is important that the supply of all macro- and micronutrients be managed in a balanced fashion to avoid long-term depletion or excess that would in turn reduce crop utilization efficiency of nutrients and water, degrade soil, water and air quality, and endanger long-term sustainability of crop production.

FERTILIZERS AND SUSTAINABLE DEVELOPMENT

The amount of available N in most soils, P in many soils, and S and K in some soils, is insufficient for optimum crop yield; therefore application of these nutrients is required to optimize crop production. Prior to the introduction of synthetic fertilizers, Canadian producers relied primarily on organic nutrient sources, including use of legume crops and green manures to provide additional N from symbiotic N fixation to nonleguminous crops such as corn and wheat. Management practices including tillage and summer fallow were also employed to mine N and other nutrients from soil organic matter. Nutrient removal was frequently in excess of nutrient input, particularly in cropping systems with limited access to manure (Table 4.1). However, reliance solely on summer fallow in the Prairies resulted in depletion of the soil organic matter (Campbell et al., 1986; McGill, Dormaar, and Reinl-Dwyer, 1988), degradation of native soil fertility, and reduction in the nutrient-supplying power of soils (Campbell, Lafond et al., 1991; Campbell and Zentner, 1993). Currently, the supply of organic nutrients is inadequate to replace crop removal and is not spatially linked to areas of intensive crop production. Therefore, synthetic fertilizers are often required to optimize crop yield and avoid nutrient depletion. Intensification of crop production and increased annualized crop yield (Campbell et al., 2004, 2005) will further increase nutrient removal and response to nutrient application (Campbell et al., 1990; Campbell, Bowren et al., 1991; Peterson, 1996). If nutrient inputs do not reflect removal, nutrient depletion over time may lead to the requirement for application of additional nutrients to ensure balanced crop nutrition, or to a long-term degradation of the organic matter content of the soil.

Increased crop productivity associated with the judicious use of synthetic fertilizers can benefit both crop productivity and soil quality. Soil organic matter is linked to desirable physical, chemical, and biological properties of soil (Campbell, 1978), and is closely associated with sustainable high crop production. It plays an important role in nutrient cycling and increases availability of nutrients to crops (SAF, 1999). Therefore, enhancement of SOM is a key factor in soil quality and long-term sustainability of the soil resource. Effective fertilizer management can increase return of organic residues to the soil, which in turn can enhance soil organic matter levels (Gregorich et al., 1995; Gregorich and Drury, 1996; Drury et al., 1998; Nyborg et al., 1999; Table 4.2).

Both under- and overapplication of nutrients are uneconomical, but application in excess of crop needs also has potential for environmental damage through N_2O emissions to the atmosphere (Lemke et al., 1999), nitrate leaching to groundwater (Izaurralde, Feng et al., 1995; Gangbazo et al., 1999; Malhi et al., 1991, 2002), and movement of P to surface and groundwater (Malhi, Harapiak et al., 2003). Trace elements such as Cd are present in phosphate fertilizers and can accumulate in the soil over time (Williams and David, 1973). Therefore, care must be taken to avoid excess fertilizer applications and to use application technologies to optimize nutrient use efficiency.

TABLE 4.2. Effect of balanced N and S fertilization (1981-1991) to grassland on forage dry matter yield, and total organic and light fraction organic C and N in 0-37.5 cm soil at Canwood, Saskatchewan, Canada

Fertilizer treatments			Annual hay yield (Mg ha^{-1})	TOC (Mg C ha^{-1})	LFOC (Mg C ha^{-1})	TN (Mg N ha^{-1})	LFN (kg N ha^{-1})
Nutrient	kg N ha^{-1}	kg S ha^{-1}					
Nil	0	0	1.48	114	12.0	10.2	575
N alone	112	0	1.51	104	9.7	10.4	529
S alone	0	11	1.30	114	11.4	10.6	550
N + S	112	11	4.50	120	20.4	10.6	1,098
LSD$_{(0.05)}$				18	3.3	1.8	171

Source: Adapted from Nyborg et al. (1999).

TECHNICAL REQUIREMENTS FOR INM

For a cropping system to be sustainable, every effort should be made to balance the nutrients removed with nutrient additions and ensure that plant-available nutrients be present in the correct amount and proportion so as to satisfy crop needs throughout the growing season (Grant et al., 2002). Improved synchrony between crop nutrient needs and nutrient supply in terms of amount and timing can enhance both the economic and environmental benefits of nutrient management. Matching nutrient requirements with nutrient supply requires accurate assessment of crop nutrient requirements, and of the amount and pattern of nutrient supply to the crop throughout the growing season.

Crop nutrient requirements are generally predicted based on crop type and yield potential, with quality goals, such as protein or oil enhancement sometimes taken into account. Establishment of yield goals is commonly based on past experience of the producer. In areas where stored soil water is of great importance, and yield is highly variable, yields may be forecast based on soil water reserves. However, growing season rainfall frequently plays a large role in determining crop yields. Therefore, improvements in seasonal weather forecasting would improve prediction of crop yield, and hence nutrient requirements.

An accurate assessment of crop nutrient supply is necessary to allow application rates to be matched to crop needs, to avoid the risk of environmental damage or economic loss. Prediction of nutrient supply becomes more challenging as a greater proportion of crop nutrients are derived from nutrient mineralization during the growing season rather than from available nutrients stored in the soil prior to seeding. Changes in production practices, such as more continuous cropping, use of high-yielding cultivars, reduced tillage, manure application, and inclusion of forage and pulse crops in the rotation, have changed both crop yield potential and the potential for mineralization of N from the soil, possibly influencing the reliability of the traditional nitrate-N soil test. For example, it is recognized that inclusion of legume crops in a rotation will result in increased N availability for following crops (Grant et al., 2002). However, quantification of the amount and timing of increased N is more difficult. For a crop following a legume, the estimated N credits were 32 kg N ha^{-1} (Beckie and Brandt, 1997; Beckie et al., 1997) and 19 kg N ha^{-1} (Izaurralde, Choudhary et al., 1995), while Beckie (1997) and Beckie and Brandt (1997) suggested a credit of 4 kg N Mg^{-1} pea seed yield in the Dark Brown soil zone.

Campbell et al. (1994) and Zentner et al. (2004) showed that legumes increased the N supplying power of soils, which could enhance the risk of nitrate leaching if the contribution was not considered when calculating N applications. Zentner et al. (2004) also observed that nitrate-N test based on 0-60 cm soil depth is not appropriate for systems that include legumes and suggested that a full rooting depth (e.g., 0-120 cm in semiarid prairie) would be more appropriate.

Tillage system may also influence the N mineralization rate. If conversion to a reduced tillage system results in an increase in soil organic matter, N availability may decrease as the N is sequestered in the accumulating soil organic matter. McConkey et al. (2002) concluded that a reduced rate of net N mineralization from organic matter, with no-till on medium and fine-textured soils, limited the N supply, leading to lower grain protein concentrations. They concluded that use of fertilizer N recommendations derived from conventional tillage management may be inadequate for no-till systems. These types of uncertainties with soil testing have contributed, in part, to underutilization of soil testing by Canadian producers. New soil testing methods including various exchange membranes and assays for assessment of potentially mineralizable N may improve the reliability of soil testing. However, research is needed to calibrate both traditional and novel soil testing methods under conditions representative of modern farming practices. Further development of "virtual" soil testing techniques, which make use of models of crop N uptake, and prediction of N supply from surrogate parameters (e.g., soil organic matter), to estimate crop available N without traditional soil-test calibration is also needed.

Research to improve estimates of N mineralization to characterize the contribution from various organic N sources (manures, composts, and biosolid) is also required to allow application rates to be more accurately matched with crop requirements. The residual effect of organic sources needs to be accounted for to formulate fertilizer recommendations for the entire cropping system. The changes in soil organic matter due to application of organic sources of nutrients are slow, so the effect of manure or compost addition and straw removal or retention on soil organic matter may not be measured effectively in short-term experiments. Also, the net release of nutrients from organic sources is a function of their chemical composition and physical form, of soil and climatic conditions, and of cultural practices. For example, nutrient concentration in animal manure varies depending on the source of manure (Schoenau et al., 2000). As a result, the concentrations of N, P, and K vary from 1.5 to 5.0, 0.1 to 2.0, and 0.8 to 2.0 kg Mg^{-1}, respectively, in effluent hog manure, and from 5 to 15, 5 to 15, and 8 to 15 kg Mg^{-1} in fresh cattle manure. From the total N in effluent hog manure and cattle manure, about 30-90 percent and 10-20 percent, respectively, of N is present in ammonium form (Schoenau et al., 2000). Biosolids, hog manure, cattle manure, and fertilizer also influence the dynamics of different P forms in soil (Abul-Kashem et al., 2004). More research is necessary to develop soil nutrient level and plant response correlation techniques for INM of more complex nutrient input combinations from organic, inorganic, readily plant-available, and slow release sources.

The use of in-crop assessments of nutrient sufficiency has the potential to circumvent some of the problems associated with soil testing, and thus improve nutrient management, and reduce environmental risk. Tissue testing, petiole sap testing, and a number of reflectance methods are available to estimate nutrient sufficiency in the growing crop (Lukima et al., 2001; Raun et al., 2002). In-crop nutrient sufficiency estimates, used in conjunction with in-crop nutrient application, could allow for reduction of nutrient application prior to or at seeding, and delay of application until nutrient stress in the crop indicates that the application is required. Research is required to develop correlation measurements between sufficiency indices, and frequency of yield response to applied nutrients, and to assess the impact of related management systems on nutrient use efficiency, crop quality, and nutrient loss to the environment.

Improved fertilizer formulations and application technology may also play an important part in INM. In-soil banding is widely accepted as an efficient method for application of N and P (Grant et al., 2001, 2002; Malhi et al., 2002). Coulter, spoke-wheel or pressure-injection equipment may be used to combine the efficiencies of banding with application of nutrients during crop growth to further improve nutrient use efficiency. However, there is not always a clear benefit of in-crop application relative to application of all N prior to or at the time of seeding. Use of controlled release products, nitrification inhibitors (Bailey, 1990), and urease inhibitors (Grant et al., 1996) may also be used to slow nutrient release and loss, thus potentially improving nutrient use efficiency (NUE). Further research on these products is required under Canadian conditions to determine if there is significant economic or environmental benefit from their use. "Green-detector" systems and site-specific nutrient application technology in conjunction with surface or in-soil application also requires assessment. Precision guidance systems and the modification of fertilization application equipment could prevent overapplication of N from equipment overlap in fields with obstacles such as trees and ponds. Careful assessment of the environmental and economic benefits of new technologies is required before their widespread adoption.

Biofertilizer technology also shows promise for INM through biological nitrogen fixation (BNF) and enhanced soil P availability to crops. Use of

Rhizobium inoculants for legume crops is commonly practiced in legume crops across Canada. Kutcher et al. (2002) indicated that there was generally a beneficial effect of inoculation on yield and protein concentration of field pea. McKenzie et al. (2001) reported that field pea yield was significantly increased by *Rhizobium* inoculation in 41 percent of the trials with an average increase of 14 percent. The yield response to inoculation was dependent on soil nitrate-N at seeding (29 percent increase in yield for <20 kg N ha^{-1} and 10 percent for >20 kg N ha^{-1}). An average yield increase of 56 percent independent of nitrate-N level in soil was observed by Clayton et al. (2004a,b).

Inoculant failures have been observed to result from poor quality inoculants and from environmental stresses such as dry seedbed conditions or soil acidity (Rice et al., 2000). In response to inoculation failures, farmers have applied higher N fertilizer rates. Fertilizer N has at times proven beneficial to plant development, and subsequent nodulation, and N fixation (Mohan and Child, 1979), leading to increased biomass, but not yield (Jensen, 1986).

Because of the ease of use, common practice for delivery of *Rhizobium* has been seed application of liquid or peat-based inoculants (Hynes et al., 1995). Soil inoculation was demonstrated to be better than conventional seed inoculation in alfalfa (Rice and Olsen, 1992), fababean (*Vicia faba* L.) (Dean and Clark, 1977), and soybean (Muldoon et al., 1980). Kutcher et al. (2002) observed greater increases in yield and protein concentration of pea with granular than liquid inoculants. Relative ranking as granular $>$ peat powder $>$ liquid inoculants was observed for field pea nodulation and N fixation (Clayton et al., 2004a), and granular inoculants resulted in 17 and 50 percent more seed yield than peat and liquid inoculants (Clayton et al., 2004b).

Biological fertilizers may also be used to improve P availability for crop uptake. A mycorrhizal association is a symbiotic relationship between mycorrhizal fungi and a plant. The plant supplies the fungi with carbohydrate and the fungal hyphae improve the ability of the crop to access P from the soil (Grant et al., 2005). Mycorrhizal hyphae may extend more than 10 cm from root surfaces (Jakobsen et al., 1992), and the small diameter of hyphae (20-50 µ) allows access to soil pores that cannot be explored by roots. Therefore, a root system that has formed a mycorrhizal network will have a greater effective surface area to absorb nutrients, and explore a greater volume of soil than non-mycorrhizal roots. Several arbuscular mycorrhizal (AM) inoculants for broad crop production have been developed and are being evaluated in Canada. However, because of problems in isolation and multiplication of pure strains of AM fungi, as of the time of publication, it has not been possible to use this as a biofertilizer on a broad commercial scale.

Phosphate-solubilizing microorganisms (PSM) increase availability of sparingly soluble P in soil through secretion of organic acids, which lower soil pH. *Penicillium bilaii* has been known to increase the amount of available P in soil (Kucey, 1983, 1987; Asea et al., 1988). In treatments with or without rock phosphate, *P. bilaii* inoculation increased dry matter, seed yield, and P uptake over uninoculated wheat (Kucey, 1987, 1988; Kucey and Leggett, 1989; Table 4.3). In the above studies, increases in dry matter/seed production, P concentration, and P uptake in plant from inoculation with *P. bilaii*, together with rock phosphate were equivalent to those obtained with application of triple superphosphate or monoammonium phosphate (MAP) at 20 mg P kg^{-1} soil. From fifty-five field experiments in Alberta, Saskatchewan, and Manitoba, Gleddie et al. (1991) found that seed inoculation with *P. bilaii* in general increased wheat grain yield over the lower check rates of phosphate fertilizer at 4.4 kg P ha^{-1}.

In a field study on pea in Manitoba, *P. bilaii* inoculation increased growth and P uptake in soils with low levels of available P (Vessey and Heisinger, 2001), most likely due to increase in root-hair production (Gludden and Vessey, 2000). Beckie et al. (1998) reported that inoculation of alfalfa with *P. bilaii* increased forage yield and P uptake by 3 to 18 percent, usually in the establishment year.

CASE STUDIES

Semiarid Prairie (Diverse Annual Crop Rotations of Cereals, Oilseeds, and Grain Legumes)

The semiarid prairie is located in the Brown and Dark Brown soil zones of Alberta and Saskatchewan. Success of crop production in the semiarid

TABLE 4.3. Effect of seed row inoculation of soil with *P. bilaii* on wheat yield and P uptake under field

	Yield (g plot^{-1})		Uptake of P
Treatment	Seed	Straw	(mg P plot^{-1})
Control	181	379	742
Rock phosphate (RP)	187	388	771
P. biaii	198	402	850
P. bilaii + RP	208	403	834
Monoammonium phosphate	209	420	975

Source: Adapted from Kucey (1987, 1988).

prairie centers on water availability and use of conservation practices, with N and, to a lesser degree, P being the most critical nutrients. For example, about one-third to one-half of the year to year variation in wheat yield in long-term crop rotation experiments was explained by water availability (Zentner and Campbell, 1988; Zentner et al., 2003; Campbell et al., 2004). Summer fallow represents approximately 27 percent of the cultivated land and plays a major role in the cropping systems used in dryland semiarid regions (Campbell et al., 2002). Much of the wheat grown in this region is durum wheat (*Triticum durum* L.). Conservation tillage systems are also being adopted at an increasing rate, for moisture conservation, erosion control, and to improve economics of production (Zentner et al., 2002). Therefore, the interactions between fallow, crop rotation, and tillage management play an important role in INM in this area.

Tillage System

Conservation tillage has gained a dominant position in semiarid regions for its ability to improve moisture conservation and increase cropping intensity (Lafond et al., 1992; Larney and Lindwall, 1995; Halvorson et al., 1999a,b). The extra water retained by adopting reduced tillage is the buffer needed to take full advantage of the often low and erratic growing season precipitation of the Great Plains (Campbell et al., 1988). Combinations of conservation tillage seeding practices, snow management, increased diversity in the cropping systems, and proper nutrient management have the potential to improve whole farm economics (Zentner et al., 2002).

With reduced tillage, residue from previous crops is left on the soil surface rather than incorporated into the soil, reflecting light and insulating the soil (Doran and Smith, 1987). As a result, soil moisture content is generally higher under reduced tillage as compared to conventional tillage (Carefoot et al., 1990; Lafond et al., 1992), while temperatures are generally slightly cooler during the spring and summer (Gauer et al., 1982; Carefoot et al., 1990; Cox et al., 1990) and warmer during the autumn and winter (Gauer et al., 1982). Retention of residues on the soil surface, combined with the slightly cooler soil temperatures during the spring and summer, may slow microbial activity and reduce organic matter loss under reduced tillage. In addition, organic matter in the soil is frequently occluded within macroaggregates where it is protected from decomposition. Tillage exposes this protected organic matter to microbial action, enhancing its rate of decomposition (Doran and Smith, 1987). Therefore, with conversion to reduced tillage, organic matter tends to accumulate in the surface soil horizons and the rate of release of nutrients from organic matter decomposition may be initially decreased. Thus, although soil organic matter is increased by reduced tillage, the nutrient supply to plants in situ may be decreased at least temporarily. For example, in the Brown soil zone in Saskatchewan, McConkey et al. (2002) reported that grain yield and protein content of wheat on silt loam and clay soils were lower when no-till was used compared to conventional and minimum tillage. However, on a sandy loam soil, no difference in grain protein was recorded between tillage systems, and grain yields were higher with no-till. The authors concluded that reduced rate of net N mineralization from organic matter with no-till on medium and fine-textured soils limited the N supply.

Nitrogen mineralization and soil nitrate-N has been shown to increase with the number and intensity of tillage operations (Malhi and Nyborg, 1987; Nyborg and Malhi, 1989), while intensive tillage to control weed growth during the fallow period has been shown to reduce soil organic C levels (Larney et al., 1997). In studies conducted in Alberta, tilled fallow plots contained 24 kg ha^{-1} more nitrate-N than untilled plots in the 0-30 cm soil depth in November, due to greater rate of N mineralization in tilled soil than in untilled soil (Nyborg and Malhi, 1989). This accumulated nitrate-N in fallow soil can be partially lost through denitrification and/or leaching, if sufficient moisture is present (Campbell et al., 1984, 1994; Malhi and Nyborg, 1986). After sixteen years of fallow management at Lethbridge, Alberta, intensive tillage treatments (discs and cultivators) had 2.2 Mg ha^{-1} less total organic C, and 13-18 percent less mineralizable N, than the low intensity tillage treatments (blade cultivator and use of herbicides in place of tillage) (Johnston et al., 1995). These low intensity tillage treatments also had higher levels of plant-available nitrate-N at seeding, leading to higher grain yields. In contrast, yield, grain N concentration, and total N uptake of wheat and canola were highest for residue incorporated with autumn plowing, and lowest for residue incorporated with spring plowing or direct seeding. Net returns were highest when the residue was removed and plots were autumn plowed (Smith et al., 2004). The difference in productivity reflected the difference in N availability, and these differences in productivity between low and high intensity tillage treatments were reduced when N rates were increased. The maximum productivity occurred when N rates were between 100 and 150 kg N ha^{-1}.

Summer Fallow and Cropping Intensity

Summer fallow-based cropping systems have historically been used in the semiarid prairie region to conserve water and provide most of the crop N needs through mineralization. Adoption of no-till cropping systems has reduced the reliance on summer fallow in many parts of the semiarid prairies. The extra moisture stored with good snow management and no-till has allowed producers in the semiarid zone to shift away from rotations that include high proportions of monoculture cereals and summer fallow to more intensified and diversified cropping rotations. For example, in a twelve-year study in Swift Current, Saskatchewan, higher annualized yields were obtained compared to a F-W-W, F-W-W-W or a continuous wheat system when wheat was planted in the years when reduced tillage and snow management practices resulted in more than 7.5 cm of available water (Zentner et al., 2003). The annualized wheat yield in these rotations increased with an increase in cropping intensity (less frequent summer fallow) despite lower (24-31 percent) yield on wheat stubble than after fallow. The smallest difference in wheat yield between summer fallow and wheat stubble was when minimum tillage and tall stubble were used to trap snow and increase stored soil moisture (Zentner et al., 2003; Campbell et al., 2004).

Continuous wheat and wheat-pea sequences also increased organic matter and microbial biomass in the soil, and reduced leaching of nitrate-N, as compared to rotations with more fallow (Zentner et al., 2001), indicating improved environmental sustainability. Continuous cropping systems, augmented with recommended rates of fertilizer N and P show a significant improvement in N availability over fallow-based systems (Campbell et al., 1993; Weinhold and Halvorson, 1999). When combined with reduced tillage, the benefits of continuous cropping, and soil test-based fertilizer use further increase soil N mineralization potential (Campbell et al., 1993). However, increased cropping intensity can increase crop removal of nutrients, potentially increasing the need for addition of organic or synthetic nutrients to avoid excess nutrient depletion (Grant et al., 2002).

Cropping intensity will also influence P dynamics. Many soils in the Canadian prairies initially had high levels of total available P and supported crop production for fifty to seventy years with no P additions. Even over the past thirty years of cropping, fertilizer P inputs have rarely exceeded grain P removal. Long-term crop rotation studies have shown that continuous cropping without fertilizer P additions resulted in a significant reduction in both the organic and inorganic soil P pools (McKenzie et al., 1992a,b). Increasing the frequency of fallow in the rotation reduced P removal and the drain on the soil P pool, whether the cropping systems were fertilized with P or not. When continuously cropped wheat systems were fertilized with soil test recommended rates of N and P, a positive P balance resulted in the enrichment of the labile, moderately labile, and microbial biomass P pools in soil (Selles et al., 1995).

Effects of Crop Residues

Crop residues from previous years of cropping may have an important impact on nutrient availability, as they serve to replenish the organic nutrient pool in the soil. The role that crop residues play in soil fertility is affected by the crops grown and tillage practices (Schoenau and Campbell, 1996). As discussed previously, the changes in moisture content, temperature, and aggregation of surface soil under reduced tillage influence nutrient cycling. Crop type will influence both the amount and the C:N ratio of residues returned to the system. Fertilizer addition also increases the initial N content of crop residues left in the field, and their rate of decomposition and nutrient release to subsequent crops. In studies conducted by Janzen and Kucey (1988), decomposition of residues was found to be directly related to their N concentration. Well-fertilized wheat and canola residues, with relatively high N concentrations, were found to decompose at a similar rate as grain legume residue, and release similar amounts of N into the soil. Nitrogen deficiencies during the early years of conversion from conventional to no-till systems can result from crop residues being retained on the soil surface rather than incorporated by tillage. However, any increased demand for N with adoption of no-till should decrease over time because of reduced losses by erosion and the buildup of the mineralizable organic N pool under no-till systems (Schoenau and Campbell, 1996).

Crop residue amount and type also influence soil P levels. Decomposition conditions mediate the release of the residue P, with the original concentration in the different crop residues influencing the amount released (Li et al., 1990). Release of P from crop residue has been shown to occur quickly on contact with soil, and not to differ between tilled and no-till management (Lupwayi et al., 2003). While the decomposition rate of some crop residues may be slower than others based on N content, the total amount of P applied in residue will ultimately influence the availability of P to subsequent crops in rotation (Grant et al., 2002).

Legumes in Rotation

Grain legumes have recently become a major component of crop diversification in the semiarid regions of western Canada (Campbell et al., 2002). The N fixation capabilities of these crops help to reduce N requirements and fertilizer costs when grown on soils with low levels of plant-available N. In a long-term rotation study using a wheat-lentil rotation, soil N was reported to be on average 11 kg N ha^{-1} more following lentil than following wheat (Zentner et al., 2001). Miller et al. (2003b) reported that postharvest soil mineral N was 36, 28, and 25 kg ha^{-1} on field pea, lentil, and chickpea stubbles, respectively, which were in turn 103, 60, and 43 percent higher than the postharvest N recorded under wheat stubble. In the spring following grain legume production, loam and clay-textured soils had 12 and 28 kg N ha^{-1}, respectively, more plant-available N in the surface 61 cm than wheat stubble (Miller et al., 2003b). While these pulse crops remove most of the N accumulated during the growing season, they have been shown to leave the soil with higher residual N than cereal crops (Beckie and Brandt, 1997; Zentner et al., 2001; Miller et al., 2003a). Beckie and Brandt (1997) reported that an N credit of 4 kg N ha^{-1} could be expected from every 1,000 kg ha^{-1} pea seed yield. Nitrogen-fixing forage crops, such as alfalfa (*Medicago sativa* L.), also provide an N benefit to the following grain crops (Mohr et al., 1999), but in the semiarid regions crop yields of the following grain crops may be depressed due to dry soil conditions resulting from forage crop water use relative to summer fallow (Zentner et al., 1990).

Soil testing laboratories often try to develop techniques to assign N credit for the effects of pulse crops on N supply to subsequent crops. While the goal is to use this type of N credit to help the grower to assess better crop N requirements, the impact of soil-climatic conditions on the amount and timing of N release often makes the benefit of such credits difficult to predict.

Legume Green Manures

Legume crops used for green manure, rather than harvested for seed or forage, can be an important N source in an INM system. Slinkard et al. (1987) and Biederbeck et al. (1993) suggested that legume green manure, combined with snow trapping to replenish water used by the legume, could be used to replace summer fallow in the semiarid prairies. Other researchers have also suggested that snow management was essential for legume green manure-cereal crop rotations in the semiarid prairies (Brandt, 1990; Townley-Smith et al., 1993). However, most of the earlier studies in the semiarid prairies tended to show lower or similar grain yield and protein, with legume green manure than summer fallow, due to a reduction in available water at seeding of the subsequent crop (Brandt, 1996, 1999; Zentner et al., 1996; Vigil and Nielsen, 1998). In a twelve-year study conducted at Swift Current, where Zentner et al. (1996) reported reduced grain yield with legume green manure in the first six years, changing to earlier seeding (from early or mid-May to mid- to late April) and earlier cessation of legume growth (from late July or early August to early July) in the final six years of the study led to equivalent wheat yield, better water use efficiency, higher grain protein, and lower fertilizer N requirements, leading to higher net returns than with the fallow rotations (Zentner et al., 2004). Zentner et al. (2004) concluded that savings in N fertilizer and tillage costs and higher net returns with green manure relative to summer fallow more than offset the added seed and management costs for green manure, and thus green manure is a viable alternative to summer fallow, if managed properly.

Cover crops in the fallow period may also be of benefit in cycling N and reducing N losses in some regions. In field studies at Lethbridge, Alberta, Moyer et al. (2000) reported that fall rye cover crop treatment had more soil moisture, and less soil available N and weeds than no cover treatment (Table 4.4). Blackshaw et al. (2001) reported that underseeded sweet clover residues provided excellent ground cover to reduce the risk of soil erosion throughout the twenty-month fallow period. They also obtained 16-56 kg ha^{-1} more available soil N in the following spring and 47-75 percent more wheat grain yield in sweet clover green manure than in summer fallow treatments. While the management of green manure and cover crops has been well established for the semiarid region, their use is not yet a common practice by producers.

Organic Nutrient Resources

Manure. Use of livestock manure as a nutrient source for annual crop production is less common in the semiarid prairies than in the subhumid prairies, so this aspect will be discussed in more detail in that section. Manures are particularly beneficial in improving the quality of degraded soils. The findings of a field study at Lethbridge, Alberta, comparing organic amendments and synthetic fertilizers demonstrated that productivity on eroded soils can be restored with the addition of livestock manure and crop

TABLE 4.4. Effect of cover crop treatment on available N (NH_4-N + NO_3-N) in the 0-60 cm soil in April after the fallow periods at Lethbridge, Alberta, and Canada

Treatment	Amount of available N (kg N ha^{-1})	
	1994	1996
Fall rye	43	35
Spring rye	35	44
No cover crop	52	70
SEM	3	3

Source: Adapted from Moyer et al. (2000).

residues (Larney and Janzen, 1996). Application of organic manure at proper rates can also reduce nitrate-N accumulation in the soil profile compared to when synthetic fertilizers are applied (Yanan et al., 1997), but repeated applications of manure at excessively high rates can adversely affect soil quality and crop yield (Chang et al., 1990) and lead to large ammonia volatilization (Beauchamp et al., 1982) and other N losses (Meek et al., 1982; Chang and Entz, 1996). Often the rate of manure application has been based on crop N requirements and as the P:N ratio in some manure sources tends to be greater than required by plants, this may lead to accumulation of P in the soil over time (Chang et al., 1991; Dormaar and Chang, 1995). Repeated application of manure can increase the concentration of other nutrients in surface soil (Qian et al., 2003). For example, Olson et al. (1998) reported that a medium-textured soil receiving four annual applications (120 Mg ha^{-1}) of cattle feedlot manure had eleven times higher K than where no manure was applied. Similarly, Chang et al. (1991) observed increased levels of soluble Ca, Mg, Na, Cl, and Zn in soil after eleven annual applications of cattle manure.

Subhumid Prairie (Diverse Annual-Perennial Crop Rotations of Cereals, Oilseeds, Grain Legumes— Pulses and Forage Legumes/Grasses)

Moisture availability is much less restrictive in the subhumid prairie than in the semiarid prairie regions. Crop yield potential is also substantially higher in this region than in the semiarid prairies, so higher nutrient inputs are generally used to optimize crop productivity. Summer fallow plays a minor role in crop production in this area, with continuous cropping using diverse annual and perennial crop rotations being much more common than fallow-based systems (Campbell et al., 2002). Livestock populations also tend to be higher in this region than in the semiarid prairies. Adoption of conservation tillage practices is not as widespread as in the semiarid prairies, as yield increases from moisture conservation are not as consistent in the subhumid region. Impact of the diversified rotations, the inclusion of livestock manures and perennial forages in the cropping system, and efficient management of relatively high nutrient inputs are key management aspects in this region.

Crop Rotations

Oilseed crops, particularly canola, comprise a significant proportion of the cropping area on the subhumid prairies. Nutrient management practices may need to be altered to optimize yield and quality of an oilseed crop such as canola as compared to a cereal crop such as wheat. For example, the S requirement of canola is substantially greater than that of cereal crops (Grant and Bailey, 1993) and yield loss due to S deficiency may occur in canola grown on soils with S levels that may be sufficient for optimum production of wheat. Furthermore, S depletion by canola may hasten the occurrence of S deficiency in subsequent wheat crops on soils containing marginal levels of available S. Conversely, if canola is grown with adequate S to optimize crop yield, cereal crops following the canola may be adequately supplied with S on marginal soils by mineralization of S from the high S-containing canola residue. This could translate into higher grain protein content/quality for cereal crops. Boron requirements of both canola and seed alfalfa are greater than those of cereal crops (Grant and McCaughey, 1991; Grant and Bailey, 1993). Corn, beans, and flax tend to have higher Zn requirements than small-grain crops such as wheat or barley.

Optimal placement of nutrients differs among the diverse crops grown in the subhumid prairie. For example, wheat and canola are more efficient than flax at accessing P from fertilizer granules due to their ability to proliferate roots when in contact with a high P reaction zone (Strong and Soper, 1974a,b). Small-seeded oilseed crops, such as canola and flax, tend to be more sensitive to damage from seed-placed fertilizers than are cereals (Nyborg and Hennig, 1969). Pulse crops, such as field pea, may also be sensitive to seed-placed fertilizer (Henry et al., 1995). Thus, rates of seed-placed fertilizer must be reduced for sensitive crops in order to avoid the risk of seedling damage and consequent loss in crop competitiveness, resulting in late maturity, lower yield, and reduced grade. To optimize crop yield and quality, the specific nutritional needs of each crop in the rotation must be carefully considered.

Inclusion of Legume Crop

Cropping systems in the subhumid prairies region were traditionally based on cereals and oilseeds, but legumes and forage grasses have played an important part of crop rotations in these regions for many years (Campbell et al., 1990). Inclusion of legumes in rotations can offer significant advantages for INM in this region (Izaurralde et al., 1993). Legumes provide available N to subsequent crops (Zentner et al., 1990; Entz et al., 1995), improve soil physical conditions (Toogood and Lynch, 1959), increase soil organic matter, and reduce potential for erosion (Campbell et al., 1990), suppress weeds (Dryden et al., 1983), and disrupt plant disease cycles (Campbell et al., 1990). Compared to wheat-fallow, continuous wheat and wheat-pea sequences had greater organic matter and microbial biomass in soil

(Zentner et al., 2001), while accumulation of nitrate-N was reduced (Campbell et al., 1994; Entz et al., 2001; Malhi et al., 2002; Table 4.5).

In the subhumid prairie as in the semiarid prairie, legume crops in rotation increase N availability for the following non-legume crop. Izaurralde et al. (1990) and Izaurralde, Choudhary et al. (1995) reported an improvement in yield and protein content of cereal seed following legumes compared to cereal-cereal cropping sequence. Under no-till on these soils, yields of unfertilized barley following legumes were similar to those of fertilized continuous barley and the equivalent value of legume residue was 19 kg N ha^{-1} (Izaurralde, Choudhary et al., 1995). Beckie and Brandt (1997) suggested that a credit of 32 kg N ha^{-1} should be given to crops that follow field pea in the Black soil zone. Bullied et al. (2002) reported that wheat yield and N uptake, averaged 2,955 kg ha^{-1} and 76.1 kg N ha^{-1}, respectively, following chickling vetch and lentil, and 2,456 kg ha^{-1} and 56.4 kg N ha^{-1} following single-year hay legume, compared to only 1,706 kg ha^{-1} and 37.9 kg N ha^{-1} following canola. Growing legumes in rotation also improved organic matter, mineral N content, and N mineralization potential in soil as compared to nonlegume crops (Campbell, Biederbeck et al., 1991; Campbell, Bowren et al., 1991; Campbell, Lafond et al., 1991). Other non-N benefits from including a legume or an oilseed crop in rotation can improve productivity of the following cereal crops (Beckie and Brandt, 1997; Stevenson and van Kessel, 1997). Therefore, crop rotations that include legumes are recommended for sustaining productivity, quality, and fertility of soils in the subhumid region (Wani et al., 1991, 1994b).

Forages in Rotation

Available moisture in the subhumid prairies is usually adequate to support the inclusion of forage crops in the cropping systems (Campbell et al., 1990) which, in turn, can play an important role in INM in this region. Forage legumes help to reduce reliance on external inputs, conserve energy, increase sustainability of crop production and economic returns, protect soil from erosion, improve soil quality, and minimize environmental damage (Morrison and Kraft, 1994; Entz et al., 2002). A five-year rotation in a gray luvisol at Breton, Alberta including forages had 38 percent more total N and 117 percent more microbial N in the top 15 cm of the soil than did a two-year wheat-fallow rotation (McGill et al., 1986). There was also a higher concentration of total C and N in soil in the five-year rotation with forages, than two-year rotations with only cereals (Juma et al., 1997), due to the lower rate of C input to soil in the two-year rotation compared to the five-year rotation. Grain yields were enhanced when alfalfa or alfalfa-grass

TABLE 4.5. Influence of alfalfa on extracting subsoil nitrate-N after six years in Manitoba

	Concentration of nitrate-N (mg N kg^{-1}) at various soil depths (cm)									
Crop rotation[a]	0-30	30-60	60-90	90-120	120-150	150-180	180-210	210-240	240-270	270-300
Continuous alfalfa	8.5	6.2	4.7	2.7	2.0	1.2	4.0	5.5	7.2	8.0
Continuous fallow	23.7	7.5	8.5	9.7	10.5	10.7	11.0	10.2	9.7	8.5
WPBWWW rotation + N	9.5	3.7	2.2	2.0	3.5	7.7	9.0	9.7	8.2	7.7

Source: Adapted from Entz, Bullied et al. (2001).

[a]W, P, and B in rotation refer to wheat, field pea, and barley.

mixtures were included in rotations (Ferguson and Gorby, 1971; Hoyt and Leitch, 1983; Baddarudin and Meyer, 1990; Hoyt, 1990). Wani et al. (1994b) reported that barley grain yield was greater in the crop rotation that included a forage legume without applied N (4.3 Mg ha^{-1}) than in the continuous cereal rotation with 90 kg ha^{-1} of applied N (3.6 Mg ha^{-1}). They also reported that legume-based forage rotations had greater total soil N, mineral N accumulation in soil under incubation, potentially mineralizable N in soil, and plant N uptake compared to continuous cereal rotation.

The method of termination of the forage stands influenced their impact on nutrient dynamics, soil quality, and yield of subsequent crops. Termination of perennial forage stands with tillage increased loss of available soil moisture (Clayton, 1982) and risk of soil erosion (Entz et al., 1995). Bullied and Entz (1999) found that termination with herbicide alone resulted in higher levels of ground cover, water use efficiency, and grain yields compared to tillage treatments, but Biederbeck and Slinkard (1988) reported that termination by desiccation resulted in less N supplied by the legume to the following cereal crop. The available soil N in spring was also reduced when termination of alfalfa was delayed, with N higher in tilled plots than no-till plots (Mohr et al., 1999).

Deep-Rooted Crops in Rotation

Loss of nitrate-N by leaching may result in environmental damage and at economic cost to the farmer. Nutrients present in the subsoil may be used by deep taproot plants and become available again in the surface soil after crop residues are returned to the soil. This can improve economic productivity when surface soil has low fertility and enhances retention of nutrients and minimizes environmental damage under high fertility. The excess nitrate-N in the soil profile could be decreased by growing deep-rooted perennial forages in the crop sequence (Campbell et al., 1994; Izaurralde, Feng et al., 1995). Entz et al. (2001) observed that annual cereal and oilseed crops extracted nitrate-N only from 0 to 120 cm of soil, while alfalfa extracted significant amounts of nitrate-N to a soil depth of 270 cm (Table 4.5). In contrast, deep-rooted legumes, such as sweet clover (*Melilotus officinalis* L.), green manure, and alfalfa-bromegrass hay crops increased nitrate leaching and sodium bicarbonate-extractable P (Olsen-P) in the subsoil layers (15-120 cm depth) of a thin Black Chernozem at Swift Current, possibly through root decomposition in situ (Campbell et al., 1993). Malhi et al. (2002) found that high input and low crop diversity tended to result in higher nitrate-N at a depth of 0 to 90 cm compared to low input and high crop diversity treatments (Table 4.6).

TABLE 4.6. Interaction effects of cropping diversity and input level on nitrate-N in 0-90 cm soil, averaged across six crop phases, at Scott, Saskatchewan from 1995 to 2000

Year	Input level[b]	Nitrate-N (kg ha^{-1}) under various in-cropping diversities[a]		
		LOW	DAG	DAP
1995	ORG	52	39	31
	RED	62	49	52
	HIGH	76	56	53
	LSD$_{(0.05)}$		14	
1996	ORG	49	47	40
	RED	45	39	41
	HIGH	67	39	45
	LSD$_{(0.05)}$		9	
1997	ORG	60	54	51
	RED	61	66	54
	HIGH	90	78	63
	LSD$_{(0.05)}$		17	
1998	ORG	56	56	45
	RED	72	68	54
	HIGH	101	74	71
	LSD$_{(0.05)}$		14	
1999	ORG	30	33	37
	RED	40	33	31
	HIGH	76	39	42
	LSD$_{(0.05)}$		7	
2000	ORG	52	64	35
	RED	49	42	38
	HIGH	66	44	42
	LSD$_{(0.05)}$		7	

Source: Adapted from Malhi, Brandt et al. (2002).

[a]LOW, DAG, and DAP refer to low diversity, diversified annual grains, and diversified annual grains-perennial forages in six-year rotations.

[b]ORG, RED, and HIGH refer to organic, reduced, and high inputs.

Legume Intercropping

Intercropping of legume and nonlegume crops may be a method of providing biologically fixed N in the cropping system, if moisture is adequate. Use of legumes as intercrops with cereals may be a viable alternative to continuous cereal cropping because annual legumes contribute N through BNF, reduce weed competition, and increase input of root mass to soil (Izaurralde et al., 1993). In studies conducted in Alberta, barley-pea intercropping improved N yield but not seed yield; the barley-pea residues returned 22 percent more N to soil than barley-barley residues (Izaurralde et al., 1990). However, intercropping of annual crops is not common in the subhumid prairies. Relay cropping and double crops in winter cereal-based cropping systems were also considered agronomically feasible in southern Manitoba, because the cover crops did not have any detrimental effect on the grain yield of the main crops (Thiessen, Martens, and Entz, 2001; Thiessen et al., 2001), but such systems are not generally used by producers.

Intercropping is more common in forage mixtures than in annual cropping. Grasses grown in association with legumes use some of the N fixed by legumes, leading to higher forage dry matter and protein yield (Tomms et al., 1995; Walley et al., 1996). Recycling of N derived from symbiotic N_2^- fixation to grasses improves the sustainability of forage production and can result in lower fertilizer N requirements.

Management of fertilizer N for mixed grass-legume stands is complex, because N application stimulates the grass component while reducing the ability of the legume component to survive and to contribute to yield in future years (Cooke et al., 1973; Nuttall et al., 1980; Harapiak and Flore, 1984). The dry matter yield response of mixed stands to applied N and net returns from N fertilization were influenced by percentage of alfalfa in forage stands, initial soil nitrate-N level, soil type, and forage species (Nuttall et al., 1980, 1991; Broersma et al., 1989; Fairey, 1991; Malhi et al., 1993). Forage stands containing greater than 50 percent alfalfa showed little response to applied N, with dry matter yield increases and net returns from applied N being highest on soils with low percentage (36 percent) of alfalfa in the stands, and low levels of nitrate-N (6 kg N ha^{-1}) in soil (Webster et al., 1976). Malhi et al. (2002) also found a much greater increase in dry matter yield from N application to pure bromegrass than bromegrass-alfalfa mixtures (Table 4.7). In comparison to zero-N control for example, application of 200 kg N ha^{-1} increased dry matter yield by 9.68 Mg ha^{-1} for pure bromegrass, and only by 3.52 Mg ha^{-1} for 1:1 bromegrass-alfalfa mixture.

Growing bromegrass-alfalfa rather than pure bromegrass increased protein concentration by 18-46 g kg^{-1}, net economic returns by \$446-498 ha^{-1},

TABLE 4.7. Dry matter yield (DMY) of hay from various bromegrass-alfalfa compositions, treated annually with different rates of ammonium nitrate at Lacombe and Eckville in central Alberta (average of 1993-1995 data at two sites)

Composition	DMY (Mg ha^{-1}) at N rates (kg N ha^{-1})				
	0	50	100	150	200
Pure bromegrass	4.95	7.61	10.41	12.22	14.63
Bromegrass:alfalfa (2:1)	10.98	12.76	13.57	14.24	15.09
Bromegrass:alfalfa (1:1)	11.30	12.84	13.38	14.54	14.81
Bromegrass:alfalfa (1:2)	11.30	13.03	14.15	14.49	15.15
Pure alfalfa	10.47	10.48	10.97	10.44	10.53

Source: Adapted from Malhi, Zentner, and Heier (2002).

and energy output (EO - MJ ha^{-1}) to energy input (EI - MJ ha^{-1}) ratio (EO/EI) by 26.3-30.4, and could save about 100 kg N ha^{-1} (Malhi et al., 2002). Adding N at 200 kg N ha^{-1} to bromegrass caused a decline in EO/EI ratio from eighty-two to seventeen, while including alfalfa in the stand improved EO/EI ratio from 82 to 105. Compared to pure grass pastures, alfalfa-grass mixtures had 6-11 percent higher daily weight gains of calves (Kopp et al., 1997), had higher seasonal mineral N supply, and tended to have greater total soil C, N, and organic C (Chen et al., 2001). The findings suggested that in grazed systems the use of alfalfa in mixed stands would be more profitable (by reducing N fertilizer requirements in the long term), and more sustainable than using fertilizer N on pure grass stands, but grazed swards with a high percentage of legumes could also increase the risk of nitrate-N loss into the environment when the swards are terminated.

Organic Nutrient Resources

Manures. Manures can serve as slow release fertilizers for crop production (Hoyt and Rice, 1977; Spratt and McIver, 1979; Freeze and Sommerfeldt, 1985; Stevenson et al., 1998; Mooleki et al., 2001), increase the supply of N and P to crops (Qian et al., 2000a,b), and improve the physical, chemical, and biological conditions of the soil (Campbell et al., 1986; Sommerfeldt et al., 1988). Schoenau et al. (2000) reported that about 20 to 30 percent of the N contained in hog manure effluent was taken up by the crop in the year of application, and availability of N (of which about 50 percent of the N was present as ammonium) ranged from 60 to 70 percent of that obtained for urea applied at equivalent rates of N. They observed equivalent or greater increases in yields of canola, barley, and crested wheatgrass (*Agropyron cristotatum* L.) with hog manure effluent than with an equivalent rate of N supplied as urea (Table 4.8).

In a long-term crop rotation study at the Breton plots in Alberta, manure plots had much higher concentrations of total N, water soluble C, and microbial C in soil than plots treated with N, P, K, and S synthetic fertilizers (McGill et al., 1986; Juma et al., 1997). In the same study, model results suggested that the low rates of C input in the two-year rotation (wheat-fallow) led to organic matter decline and addition of manure was needed to maintain soil organic C in the 0-15 cm soil depth. However, soil C levels did not decline in the five-year rotation (wheat-oat-barley-forage-forage).

Combined application of organic manures and synthetic fertilizers usually produce greater crop yields than when each is applied alone. Supplementing manure with commercial fertilizer is beneficial in achieving the desired balance of available nutrients needed by the crop. For example, some liquid swine manure sources in western Canada are low in available S relative to N, and crops grown on S deficient soils were found to benefit from the addition of fertilizer sulfur along with the manure (Schoenau et al., 2004). Addition of fertilizer N with solid manures of low short-term N availability due to high C:N ratio will also result in better yield and crop utilization of other manure nutrients like P.

Additions of N and P fertilizers plus farm yard manure to eroded soil produced greater increase in wheat yield, and total and light fraction organic C and N than either of these amendments alone, but their levels on

TABLE 4.8. Yield of canola, barley, and crested wheatgrass from injected swine manure effluent and urea N fertilizer in east-central Saskatchewan, Canada (Farm Facts, 2000. Managing manure as a fertilizer. Saskatchewan Agriculture and Food, November 2000)

	Seed yield (mg ha^{-1})		Crested wheatgrass
Treatment	Canola	Barley	Hay yield (Mg ha^{-1})
Control (no fertilizer)	0.56	2.04	1.06
37,059 L ha^{-1} (84 kg N ha^{-1})	1.29	4.03	2.72
74,118 L ha^{-1} (168 kg N ha^{-1})	1.74	4.30	4.99
148,236 L ha^{-1} (336 kg N ha^{-1})	1.62	3.98	4.89
Urea (112 kg N ha^{-1})	1.46	4.09	

Source: Adapted from Schoenau et al. (2000).

eroded soil did not reach that obtained in noneroded soils (Izaurralde, Solberg et al., 1998; Malhi et al., 1999; Table 4.9). This suggests that the physical condition of soil must be improved for satisfactory production when top soil layers are eroded (Izaurralde, Nyborg et al., 1998).

Green manures. Because of their ability to fix atmospheric N_2, green manure (GM) legumes offer opportunities to increase and sustain productivity, and improve grain quality (Zentner et al., 2004). In the subhumid prairies, as available moisture increases, the risk of moisture depletion reducing yield in the crop following the green manure is reduced, so green manure is a viable option for INM. In northern Alberta, Clayton and Austenson (1992) reported that legume green manure can provide 60 kg N ha^{-1} year^{-1}. Green manure also increases availability of P, K, S and some micronutrients through its favorable impact on biocycling, soil properties,

TABLE 4.9. Effect of erosion level (artificial topsoil removal) and amendments on seed yield of wheat (average of 1991 to 1994), and total organic and light fraction organic C and N in soil (0-20 cm at Cooking Lake, Alberta, Canada)

Treatments		Seed yield (mg ha^{-1})	TOC[a] (mg C ha^{-1})	LFOC[a] (mg C ha^{-1})	TN[a] (mg N ha^{-1})	LFN[a] (kg N ha^{-1})
Erosion (cm)	Amendments					
0	Control	2.05	67.72	2.94	7.07	150
	Fertilizer (F)	3.32	75.83	3.47	7.67	180
	Manure (M)	3.07	76.34	4.62	7.97	280
	M + F	3.75	82.92	5.75	8.43	360
10	Control	0.81	39.77	0.91	4.83	40
	Fertilizer	2.66	47.42	1.86	5.30	90
	Manure	2.27	55.47	3.00	6.12	190
	M + F	3.20	59.11	4.38	6.32	270
20	Control	0.32	29.96	0.34	3.99	10
	Fertilizer	1.85	31.27	0.77	4.07	40
	Manure	1.18	39.98	1.42	4.84	100
	M + F	2.32	43.90	1.56	5.15	90
LSD$_{0.05}$			9.06	1.13	0.66	80

Source: Adapted from Izaurralde, Solberg et al. (1998).

[a]TOC, LFOC, TN, and LFN refer to total organic C, light fraction organic C, total organic N, and light fraction organic N, respectively.

oxidation reduction regime, pH, and increased chelation capacity. In addition, green manure plays an important role in the reclamation of salt affected soils and the improvement of physical and biological properties of soil (Bremer and van Kessel, 1990).

In an alternative cropping systems experiment, Wani et al. (1994a) reported that barley grain yield after fababean green manure with no applied N was greater (4.2 Mg ha^{-1}) than continuous cereals with 90 kg N ha^{-1} (3.6 Mg ha^{-1}). Also, potentially mineralizable N in soil, and plant N uptake as well as total soil N were greater with green manure compared to continuous cereal rotation. Campbell, Biederbeck et al. (1991), Campbell, Bowren et al. (1991), Campbell (1993), and Campbell et al. (1993) demonstrated that properly inoculated legume green manure in rotation in thin Black Chernozem at Indian Head, Saskatchewan, maintained crop production and soil quality similar to that provided by the use of synthetic fertilizers. Because of their ability to contribute N to the soil, forage legumes significantly reduce the fertilizer N requirements of subsequent crops (Rice and Biederbeck, 1983). However, because they increase the N supply capacity of the soil, they can increase the risk of nitrate leaching in subhumid environments if N release is out of synchrony with crop uptake (Hoyt and Leitch, 1983).

Incorporation of green manure legume can supply large quantities of biologically fixed N_2 to crops, reduce reliance on nonrenewable energy, and improve soil quality and productivity. However, the use of green manures in cropping systems in the Canadian prairies is limited. There is a need for critical analysis of the potential and limiting factors of green manure in cropping systems in order to assess its future scope.

Crop residues. Accurate assessment of the contribution of crop residues to the nutrient supply of the growing crop is an important step in effective nutrient management. This is of particular importance in the subhumid prairies, as the extended crop rotations and relatively high level of residue returned to the soil will increase the level of nutrients supplied from residue mineralization (Grant et al., 2002). Of the nutrients taken up by the cereal crops, approximately 25 percent of N and P, 50 percent of S, and 75 percent of K are retained in crop residues. When crop residues are returned to soil, they can contribute appreciable amounts of plant nutrients and energy for the growth and activity of soil microorganisms to maintain or enhance soil organic matter level, microbial and faunal activity, and soil productivity. The residues also help to improve soil structure, reduce bulk density, and increase the porosity and infiltration rate of soil (Singh and Malhi, 2005). As discussed previously, legume crop residues in particular may benefit the following crops both by increasing N supply and by improving soil conditions.

However, incorporation of residues having wide C:N ratio (e.g., cereal straw) can increase immobilization of inorganic N and restrict crop growth and yield due to deficiency of N in the following crop, for the initial years of residue incorporation (Nyborg, Solberg, Izaurralde et al., 1995).

In long-term experiments on two different soils, incorporation of straw increased organic C and total N contents in the soil with low organic matter (Nyborg, Solberg, Malhi et al., 1995; Solberg et al., 1998; Table 4.10). Returning straw to Black Chernozem and Gray Luvisol soils for five years improved soil physical properties including increased soil aggregation, and reduced bulk density and penetration resistance (Singh and Malhi, 2005). Seed yield of canola was highest in plots where straw was returned to soil under no tillage (Malhi, Lemke et al., 2003), and barley grain yield was greater when crop residues were returned (3.6 Mg ha^{-1}) than removed (1.9 Mg ha^{-1}) Wani et al. (1994b). Campbell et al. (1998) reported that prolonged burning of crop residues may have negative consequences for soil conservation as it usually decreases soil organic C and N content. This may also be a problem if residues are removed for industrial uses (Malhi, Lemke et al., 2003).

TABLE 4.10. Effect of straw management and N fertilizer application (1983-1995) on total organic and light fraction organic C and N in soil at Breton, Alberta, Canada

Treatments		0-30 cm Soil		0-15 cm Soil	
N rate (kg N ha^{-1})	Straw	TOC[a] (mg C ha^{-1})	TN[a] (mg Nha^{-1})	LFOC[a] (mg C ha^{-1})	LFN[a] (kg N ha^{-1})
0	Removed	36.8	4.24	0.75	26
	Retained	34.9	4.12	0.82	32
25	Removed	33.5	3.89	1.02	37
	Retained	36.9	4.11	1.16	49
50	Removed	38.4	4.31	1.17	49
	Retained	42.6	4.59	1.38	60
75	Removed	42.7	4.62	1.33	57
	Retained	43.0	4.52	1.54	67
LSD$_{0.05}$		6.6	0.71	0.37	15

Source: Adapted from Solberg et al. (1998).

[a]TOC, LFOC, TN, and LFN refer to total organic C, light fraction organic C, total organic N, and light fraction organic N, respectively.

Central Canada

Dairy Farming Mixed Cropping Systems

Dairy-based cropping systems occupy a large proportion of the agricultural land in central Canada and are present on various soil types from the Podzolic sandy soils of the Laurentians or the Appalachians to the Gleysolic clay soils of southwestern Ontario and the St. Lawrence Lowlands. These systems generally involve a range of annual and perennial crops grown in rotation. Sources of nutrients include animal manure and mineral fertilizers. As perennial legumes are generally present in the rotation, N fixation is also a major source of N for the crops.

Manure. Management practices for animal manure application vary widely in this region. According to Beauchamp (1998), about one half of the manure produced in Ontario is applied in the late summer or autumn of the year previous to the growing season for which the manure is intended for crop uptake. While much of the manure application is on forage crops, only a few studies have looked at the response of perennial forage crops to manure application in eastern Canada. In a field study with timothy grass in Québec, Warman (1986) noted that dairy manure applied for two years at 90 Mg ha^{-1} wet weight outyielded all other treatments including mineral fertilizer and chicken manure. Chantigny et al. (2004) concluded that for an equivalent total N application rate, timothy yields in a single year or over several years with liquid hog manure were 86 percent of those of mineral fertilizer. Treating manure in various ways (anaerobic digestion, flocculation) prior to field application increased crop yield response to levels almost equal to those of the fertilizer.

Tran et al. (1996) showed that repeated application of solid manure resulted in greater yield and uptake of N, P, K, Ca, and minor elements by silage corn compared to equivalent rates of mineral fertilizer. Because all aboveground crops are removed for silage, external or recycled organic matter inputs are highly desirable to maintain soil quality (N'dayegamiye and Angers, 1990). However, repeated applications of manure at excessively high rates can adversely affect soil quality and crop yield, and lead to large amounts of ammonia volatilization (Beauchamp et al., 1982) and other N losses.

In addition to providing plant nutrients, cattle manure can also provide large quantities of organic matter or organic C to soil. For instance, the application of 40 Mg ha^{-1} of solid cattle manure with a dry matter content of 25 percent will supply about 4 Mg ha^{-1} of C. When applied year after year on the same field, such C input will induce large changes in soil physical and biological properties. In a silage corn system in Quebec, five biannual applications of solid dairy manure over ten years resulted in significant improvements in soil aggregate stability, soil porosity, and a range of biological properties such as microbial populations and N mineralization potential (N'dayegamiye and Angers, 1990). Changes were generally directly proportional to manure application rates. Since dairy-based cropping systems usually involve various combinations of tillage, crops, and manure management practices, interactions between these factors can be expected. For example, in a long-term study, liquid dairy manure had little to no effect on surface soil C in an annually tilled barley system whereas significant increases in soil C, microbial biomass, and soil aggregate stability were observed in the barley/forage rotation which was only tilled every three years (Bissonnette et al., 2001).

The fate of nutrients applied with manure on hayland needs to be quantified. Chantigny et al. (2004) found that about 1 percent of the applied manure N was lost as N_2O during the six week period following application, which is about twice the losses following N mineral fertilizer; ammonia losses represented up to 15 percent of the applied manure N. Repeated manure application can also result in an accumulation of elements in soil such as P. Long-term solid dairy cattle manure application at a low rate (20 Mg ha^{-1} wet weight every two years) on a Gleysolic silt loam in Quebec significantly increased the labile P fractions in a soil cropped to silage corn (Tran and N'dayegamiye, 1995). Labile P pools in soil were significantly correlated with silage corn P uptake. In a companion study with higher application rates (>100 Mg ha^{-1} every two years), soil P levels were considered excessively high according to local recommendations (Tran, Côté, and N'dayegamiye, 1996). Simard et al. (1995) studied the P status of soils under dairy production in a watershed from Québec and found that these soils were generally highly enriched in labile P compared with forest soils. Zheng et al. (2003) found that applying dairy manure on a perennial crop-based rotation system resulted in higher concentrations of organic P fractions than mineral fertilizers. The converse was true for the inorganic fractions which showed higher values with mineral fertilizers.

Crop rotation and tillage. In most cropping systems, plant residues supply most of the organic matter to soils. In a mathematical analysis of C cycling in a dairy system and a swine system, Beauchamp and Voroney (1994) clearly showed the important contribution of crop roots and other residue to soil C maintenance in dairy systems, primarily because of the inclusion of perennial legume/grass in the rotation. The use of perennial crops in a rotation system has two major implications. First, generally high quantities of C are returned to the soil in such systems (Beauchamp and Voroney, 1994; Bolinder et al., 2002). Second, perennial crops are only tilled every three to seven years, which will reduce C loss due to decomposition. Zheng et al. (2001) showed that the combined use of dairy manure, reduced tillage, and including perennial forages in rotation resulted in high accumulation of labile P in the soil, and cautioned against the risk of water contamination in those situations. In a subsequent study, Zheng et al. (2003) found that including perennial crops in the rotation resulted in greater amounts of labile P fractions in the 30-60 cm soil layers than under monoculture. This was attributed to deeper placement of root material by the perennials and/or to higher C inputs in the rotation.

The impact and management of soils under dairy-based cropping systems have not been studied as much as other cropping systems. This brief overview shows that the biological, chemical, and physical properties of these soils can respond greatly to changes in management practices. Two main management factors that contribute to these changes are the use of perennial crops in the rotation and the application of animal manure. These two factors can interact in their effect on soil properties and nutrient cycling, and these effects are largely dependent on the amount of organic matter and nutrients that they provide to the soil. More long-term sites on a range of soils, involving various crop rotations and manure management practices, are needed to fully evaluate the impact of these systems on crop production, and on soil, water, and air quality.

Annual Cropping Systems: Corn, Soybeans, Cereals, Beans, and Potatoes

A majority of the soils in southeastern Ontario and the St. Lawrence Lowlands of Quebec are fairly flat and have clay and clay loam textures. For example, in the southernmost counties (Kent and Essex) of southwestern Ontario, where cash crops predominate, 80 percent of field crop production is on the Brookston soil which has a texture ranging from clay to clay loam. Similar trends are observed in southwestern Quebec where cash crops are generally grown on gleysolic clay soils. The rainfall in these areas varies between 800 to 1,000 mm (Martin et al., 2000). The soils are imperfectly drained, and typically have fairly low surface organic matter (~2 percent C). To overcome drainage problems, the fields are systematically tile drained, which enables producers to plant earlier in the spring and harvest later in the fall.

Annual cropping systems are dominated by cash crops and intensive management, including tillage, nutrient addition (fertilizers or manures), herbicide application, and residue management (especially when conservation tillage is employed). These annual cropping systems have contributed to a decline in soil quality (MacDonald et al., 1995). Intensive tillage, including use of moldboard plowing and arable crop production practices have decreased the organic matter content of agricultural surface soils by 15-30 percent relative to uncultivated soils (Gregorich et al., 1995).

Since corn (grain and fodder) is a dominant field crop in this region and it requires a large amount of N fertilizer (typical recommendation rates of about 150-170 kg N ha^{-1}), there is a concern that some of this applied N may be lost from the fields during wet periods via nitrate leaching and/or denitrification (Reynolds et al., 2003). These N losses pose environmental and economic problems. For example, in a study conducted near Harrow, Ontario, under conventional management practices, the flow weighted mean nitrate concentration of tile drainage water was 10.6 mg N L^{-1}, which exceeded the water quality guideline (10 mg N L^{-1}) and resulted in an annual loss of 25.8 kg NO_3-N ha^{-1} (Drury et al., 1996). Water table management practices have been found to reduce nitrate concentrations to acceptable levels and to reduce nitrate losses through tile drainage by up to 50 percent (Tan et al., 1993; Drury et al., 1996). Under anaerobic conditions, nitrate in the soil is converted to nitric oxide, nitrous oxide, and dinitrogen by denitrification, causing economic loss and potential environmental contamination (Drury et al., 1992; McKenney and Drury, 1997).

Since N is such an important crop nutrient in cash crop systems, soil, crop, and water management practices are being developed to overcome these agronomic and environmental limitations and to enable producers to utilize the N more efficiently. These management practices include innovative crop rotations, fertilizer application methods, tillage systems, and compost application to enhance nutrient utilization, increase crop productivity, and reduce environmental pollution by using both inorganic and organic forms of N more efficiently.

Crop Rotation

A long-term study initiated in 1958 near Harrow, Ontario compared continuous corn (with/without synthetic fertilizer) to corn grown in rotation with oats and two years of alfalfa (Drury and Tan, 1995). During forty-five years, fertilized corn grown in the four-year rotation had increasing yield trends whereas the fertilized continuous corn treatments showed no appreciable yield increases, in spite of the use of improved corn cultivars (Figure 4.2). Increased yield potential from use of improved corn cultivars were therefore probably offset by soil deterioration. When no fertilizer was applied, the corn grown in rotation had moderately declining crop yields whereas the continuous corn yield decreased further, to unsustainable levels (between 0

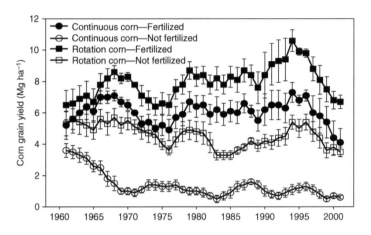

FIGURE 4.2. Corn grain yields (five year averages) from fertilized or nonfertilized continuous corn or corn grown in rotation with oats and two years of alfalfa at Harrow, Ontario. *Source:* Adapted from Drury and Tan (1995).

and 2 Mg ha^{-1}). Soil C was greatest with the fertilized and nonfertilized rotation corn treatments (2.1-2.2 percent C), followed by the fertilized continuous corn treatment (1.86 percent C), and much lower (1.6 percent C) in the nonfertilized continuous corn treatment (Drury et al., 1998). Crop rotation increased C storage below the plow layer by 40 percent compared to monoculture corn, likely due to the alfalfa component and to this C being more biologically stable (Gregorich et al., 2001). However, despite many advantages from crop rotations, including the N fixing ability of leguminous crops, supplemental fertilizer was required to optimize crop yields.

A three-year crop rotation involving winter wheat-corn-soybeans was compared to one in which red clover was underseeded in the winter wheat preceding corn and soybean in Quebec. In addition to the N benefits of underseeding the red clover, red clover accelerated the decomposition of the wheat straw, and enhanced seed germination and crop growth (Drury et al., 1999). No-tillage management resulted in yield reductions as a result of wet and cool seedbed conditions in the spring (Fortin and Pierce, 1990; Drury et al., 1999). The inclusion of red clover in the crop rotation increased corn grain yields for zone tillage treatments in four out of six years (average of 9 percent), but had little impact on conventional tilled corn.

Cover crops have been investigated as a means of retaining residual N in soil as well as fixing atmospheric N$_2$ by legumes. Red clover underseeded in a spring wheat crop in the St. Lawrence lowlands of Quebec provided the succeeding wheat crop with 82 kg N ha^{-1} and increased wheat yields in two out of three years (Garand et al., 2001). However, the red clover was not very effective in removing residual N in the soil. In another study, seeding of rye in August in corn reduced nitrate leaching from 8.8 to 4.3 mg L^{-1} during the fallow period up to the following spring (Ball-Coelho et al., 2004). Vyn et al. (2000) observed that cover crops including red clover, rye, oilseed radish, and oats established after winter wheat, were found to be equally effective in lowering the residual nitrate levels in soil, while red clover was the only cover crop which increased N availability to the following corn crop. Forage radish, canola, and barley cover crops planted as trap crops following sweet corn in Quebec also reduced soil nitrate-N levels (Isse et al., 1999).

Intercropping can also reduce the risk of nitrate loss. For example, when corn was intercropped with Italian ryegrass, the nitrate-N content of the top 100 cm of soil was 47 percent lower than with monoculture corn (Zhou et al., 1997).

Tillage

Tillage practices have been found to affect the C and N distribution in a soil profile. No-till and conventional tilled soils in Quebec and in southwestern Ontario had similar C and N storage in a 60 cm profile under cereal crops and corn, respectively; however, the distribution in the profile differed, with more C and N in the surface layers of a no-till soil compared to conventional tillage (Angers et al., 1997). Tillage can also affect crop growth and nutrient uptake by affecting soil physical properties (soil temperature, moisture, and structure), and soil biological properties such as earthworms (Drury et al., 2004). Under no-till, soils tend to be cooler and wetter in the spring, which can result in slower emergence and reduced plant populations as compared to conventional tillage (Drury et al., 1999). The three-year average corn grain yields were 13 percent lower for no-till, compared to conventional tillage, which would also affect nutrient uptake and the amount of residual N in the soil. In a long-term study (twenty years) on a clay loam soil in southwestern Ontario, corn grain yields from moldboard plow tillage, no-till and ridge tillage were compared (Drury et al., 2004). Both no-till and ridge till performed about as well as moldboard plow under normal to high summer rainfall conditions, but not as well as moldboard plow under summer drought conditions. Averaged over the entire twenty-year period, corn grain yield from moldboard plow tillage (6.51 t ha^{-1}) was 5.3 percent greater than no-till (6.18 t ha^{-1}) and 1.4 percent greater than ridge tillage (6.42 t ha^{-1}). Alternative conservation tillage systems such as zone tillage have been developed to improve the seedbed conditions in the crop row and thereby increase crop yields and nutrient uptake while maintaining some of the soil quality benefits of no-till (Drury et al., 2003).

Changes in soil moisture and C distribution with tillage and N placement can also affect biological processes such as N mineralization and denitrification. Plant-available N was found to be greater in a silt loam soil under conventional tillage than under no-till during the first two months after tillage in a variable landscape in southern Ontario (Dharmakeerthi et al., 2004). Changes in soil physical properties, such as temperature, appeared to contribute to these differences in mineralization rates between tillage treatments. Nitrous oxide emissions measured from April to October in three growing seasons under conventional tillage, zone tillage, and no-till systems in a clay loam soil in southwestern Ontario were found to be very high for all treatments in 2000, as a result of the wet growing season conditions (Drury et al., 2003, 2006). Spikes in N_2O emissions occurred following N application to corn in mid-June of each year (i.e., most of the N was applied to corn when it reached the six leaf stage). These N_2O spikes were most prominent with the deep N placement treatment. On average, shallow N placement (2.83 kg N ha^{-1}) decreased N_2O emissions from the soil by 26 percent compared to deep N placement (3.83 kg N ha^{-1}). These decreased emissions may be due to the drier soil conditions surrounding the fertilizer in the shallow N placement treatment compared to the wetter soil conditions surrounding the fertilizer in the deep N placement treatment. In the deep N placement treatments, averaged over the three growing seasons, the N_2O emissions from zone-tillage treatment (2.98 kg N ha^{-1} year^{-1}) were 20 percent lower than from no-till (3.71 kg N ha^{-1} year^{-1}), and 38 percent lower than those from moldboard plow tillage (4.81 kg N ha^{-1} year^{-1}) (Drury et al., 2006). The pattern of N_2O emissions with zone tillage < no-till < conventional tillage may have stemmed from a combination of relatively low C inputs at depth with the zone tillage and no-till treatments, compared to conventional tillage and relatively dry soil conditions with the zone tillage and conventional tillage treatments in the N application zone, compared to no-till treatment. Further, yields and N uptake were greater in the conventional and zone tillage treatments than in the no-till treatment which reduces the amount of N available for denitrification. However, in southwestern Quebec, no-till was associated with higher N_2O emissions compared to conventional tillage (Fan et al., 1997; MacKenzie et al., 1998).

The improved soil structure and increased crop residue on the surface of no-till soils can increase the amount of water entering the soil profile (Drury et al., 1999; Tan et al., 2002). In a five-year period on a clay loam soil in southwestern Ontario, long-term no-till resulted in 48 percent more tile drainage flow than conventional tillage. The nitrate loss through tile drainage was greater with no-till (16.5 kg N ha^{-1} year^{-1}) than conventional tillage (12.7 kg N ha^{-1} year^{-1}), primarily as a result of the increase in drainage volume (Tan et al., 2002). However, due to a dilution effect with the greater drainage volumes in no-till, the flow weighted nitrate concentration in tile drainage water was lower for the no-till (11.8 mg N L^{-1}) than conventional tillage (13.5 mg N L^{-1}). Both of these nitrate concentrations exceeded the Canadian drinking water guidelines (10 mg N L^{-1}).

Tillage practices also affect the distribution and fate of other nutrients such as P. In two Québec soils under continuous corn production, Weill et al. (1990) showed that available P increased at the soil surface under no-till as compared to moldboard plowing. Further analysis by O'Hallaran (1993) confirmed the accumulation of labile P forms after seven years no-till at that site. The accumulation of labile P at or near the soil surface under NT might have an impact on P losses to water. In Brookston in Ontario, Gaynor and Findlay (1995) measured P losses from surface runoff and tile drainage. Overall, conservation tillage increased P loss. Dissolved P accounted for >80 percent of total P loss.

Composts and Manures

Improvements in crop yields can occur with INM, using synthetic fertilizers and manures in combination. In a continuous corn study with compost and fertilizer in southwestern Ontario, highest yields were obtained with a fertilized pig manure straw compost (75 Mg ha^{-1} dry weight) treatment, and the lowest when no fertilizer or compost was applied (Figure 4.3). The yield advantage with one compost application lasted for three or four years when fertilizer was also added. With compost alone, the yields were increased compared to the unfertilized control but were considerably lower compared to the fertilized treatments. Both compost and fertilizer were required to optimize yields and the effects of fertilizer and compost were additive. In addition to the nutrient supply from the composted material, there were soil physical quality benefits such as lower bulk density, and higher organic C contents when pig manure straw compost, and yard waste compost were added to soils (Reynolds et al., 2003).

Nitrous oxide emissions were increased over seven-fold when nitrate was added to liquid pig manure amended soils under aerobic conditions in a laboratory incubation study compared to liquid pig manure amended soils with no added nitrate (Yang et al., 2003). However, when nitrate was applied to soils amended with pig manure, straw compost, or yard waste compost under similar incubation conditions, N$_2$O emissions were decreased by

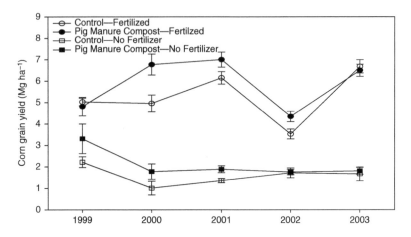

FIGURE 4.3. Corn grain yields from a fertilized or nonfertilized study either with a pig manure compost application once in the autumn of 1998 or with a control treatment (no compost amendment).   Error bars are standard error (n = 4). *Source:* C. F. Drury, unpublished results.

93 and by 50 percent. Further, compared to a control, ryegrass growth in twenty weeks was increased by 164, 40, and 64 percent with liquid pig manure, wheat straw + liquid pig manure compost, and yard waste compost, respectively (Yang et al., 2004). However, seventy-six and fifty times greater N_2O emission occurred with liquid pig manure as a nutrient source compared to control and wheat straw + liquid pig manure compost, respectively, compromising its sustainability. Hence it appears that there are many benefits associated with the use of composted material as a way of improving soil and environment quality compared to untreated organic amendments.

Ball-Coelho et al. (2004) observed that corn yield and nitrate leaching were similar with inorganic fertilizer and injected liquid hog manure. They found that splitting nutrient applications and overseeding a rye cover crop could help reduce the residual N after harvest and thereby minimize the risk of nitrate leaching and groundwater contamination. In a long-term P fertilization study in southwestern Quebec, soil test P increased linearly with time when fertilizer P exceeded crop removal and it decreased linearly with time when no additional P was applied (Zhang et al., 2004). Therefore, manure and fertilizer application rates should be adjusted in fields which have received manure in past years to ensure that P addition and P removal by crops are balanced. There is a need to account for these nutrient sources and more closely match nutrient addition with crop nutrient requirements, thereby minimizing potential for environmental contamination.

Reducing N Losses with Nitrification Inhibitors

Nitrogen inhibitors have been tried as a means of enhancing N uptake and decreasing N leaching losses. In a study in central Ontario, when the nitrification inhibitor dicyandiamide was added with urea-ammonium nitrate fertilizer (UAN), nitrate leaching was decreased by about 50 percent compared to urea alone (Ball-Coelho and Roy, 1999). However, there were minimal and inconsistent effects of the nitrification inhibitor and UAN solution on crop yields compared to UAN alone.

Manure plays an important role as a nutrient source in both dairy-based and other intensive farming systems. Problems with use of manure include accumulation of nutrients in soil, water pollution from nitrate leaching and P in runoff, and air pollution from N_2O emissions and odor. Improved management practices to handle manure application to soil are being investigated. Use of INM, optimum manure and fertilizer management, cover crops, legumes, and diverse crop rotations are key components of sustainable cropping systems in central Canada and Quebec.

Atlantic Maritime

Cropping systems in the Atlantic Maritime region involve vegetable, horticulture, and forage cropping but are largely centered on potato production. Fertilization of potatoes is necessary for maximum economic yield. Arsenault and Malone (1999) showed variable response of potato yield to fertilizer N rates. Of three potato varieties used in the study, one showed no response to added N while the yield of the other two varieties increased with rates up to 134 kg N ha^{-1}. Phosphorus fertilization was identified as an important management tool in potato production, contributing to improvement in both yield and quality (Sanderson, 2003). Monoammonium phosphate was superior as a P source on limed soils, while diammonium phosphate produced the largest yield increases on unlimed soils (Sanderson, 2003).

Fruit production is important in some areas of the Maritimes, such as the Annapolis valley. In Nova Scotia, optimum apple production requires soil pH to be maintained around 5.5 to 6.5, which can be adjusted by using dolomitic limestone if both soil pH and magnesium levels are low. Generally, fertilizer is spread out to about a one meter radius around the trunk of the tree, and a N-P_2O_5-K_2O blend of 10-10-10 is used with a typical rate of 0.02 kg N-P_2O_5-K_2O per tree per year of age for the first ten years, with periodic applications of boron as well (AFHRC, 2003). Too much fertilizer N can result in rank shoot growth and poor fruit color. For several decades now, the area devoted to apple production has decreased while overall production per hectare has tripled due to high density plantations (1,000 trees per hectare) with better weed control and fertility management (AFHRC, 2003). For blueberry and raspberry production in Nova Scotia, municipal compost is an effective source of nutrients (Warman, 2003). Since nutrients are released more slowly from compost than from synthetic fertilizer, with less opportunity for leaching, they may be of less environmental concern than soluble commercial fertilizers.

For successful production of vegetable crops such as potatoes in many regions of the Atlantic Maritimes, particularly in Prince Edward Island with its highly weathered acidic Podzolic soils, applications of lime and gypsum are frequently made to increase pH and supply Ca and S (Sanderson and Carter, 2002). Although the surface soil pH may be readily increased through liming, the deeper region of the soil profile tends to remain very acidic and acts as a limitation for deep tap root system crops like alfalfa. However, deep tillage and amelioration have been shown to be helpful in preventing chemical and physical soil problems (Carter and Richards, 2000). For crops like rutabagas, P is generally applied preplant and N may be broadcast prior to planting and also side-dressed in early summer. On sandy soils with low organic matter content, S may be recommended, with gypsum as the source. Cruciferous vegetables like cabbage and broccoli have quite high nutrient requirements and prefer a soil pH of 5.8 to 6.5 that usually necessitates application of lime. For these crops, N may be applied preplant as well as side-dressed or foliar applied; P and K may be preplant broadcast and incorporated. To avoid Mg deficiency, dolomitic limestone is applied, or Mg may be added to the fertilizer, or applied as a foliar spray of magnesium sulfate. Beans, as a legume, require no N but may receive up to 30 kg N ha^{-1} while 50 kg P ha^{-1} is usually recommended (APASCC, 1997).

In forage grass production, considerable emphasis is placed on fertilizer management for maximum quality of forage grasses like timothy, and balanced application of N, P, and K are required to ensure long-term productivity of the stand (Belanger and Richards, 1998). It would be desirable to increase the efficiency of N uptake by forage grasses and there have been attempts to improve N use efficiency by selecting genotypes of timothy that will produce greater yields with less N fertilizer while still maintaining plant N and protein content for optimal production of ruminant animals (Michaud et al., 1998). Copper limitations in cattle diets have been addressed through copper fertilization. Gupta (1989) reported that while copper sulfate addition did not affect timothy or alfalfa yields in Prince Edward Island, it did increase plant copper concentrations, with foliar application being more effective than soil application.

Due to the livestock industry in the Atlantic Maritimes, particularly dairy production, manures are readily available for land application and dairy slurries are commonly applied to forage fields. In a survey conducted in New Brunswick and Nova Scotia, Rodd et al. (2002) reported that 80 percent of the cattle manure was applied to forage land (either hay or pasture) and the rest was plowed into crop land. They found that manure application tended to be equally distributed between spring, summer, early autumn, and late autumn. Summer and early autumn applications are done after hay cuts. Timing of manure application does not appear to influence forage yield substantially. They also noted that forage yield was directly proportional to the rate of addition of dairy manure and N appeared to be 50% available as synthetic N fertilizer. Recent studies have also looked at the effects of liquid hog manure application on hay production. Chicken manure and compost were also identified as useful soil amendments on mixed forage stands in Nova Scotia (Warman and Cooper, 1999). The poultry manure served as an effective source of calcium and magnesium on these soils, eliminating the need for lime application.

Short-term (i.e., two years) potato rotations have become more common in recent years and concerns have risen about deterioration of soil quality in these rotations. Sandy soils, high moisture, and shallow water tables all contribute to these concerns. Losses of nutrient by leaching and the threat of water contamination by nonpoint source leaching of nitrate has been identified as an issue in various cropping systems (Fairchild et al., 2004). The highest concentrations of nitrate in drainage waters from fields in New Brunswick and Prince Edward Island were from those under potato production, although the values were still below the accepted tolerance limit (Fairchild et al., 2004). Nitrate is mainly leached in autumn, winter, and early spring when there is high rainfall, low evapotranspiration and no crop. Use of cover crops and keeping soil nitrate levels low by avoiding early autumn tillage of forage crops is recommended (Sanderson et al., 1999; Wallace, 2002). Legumes, grasses, and barley are identified as effective crops that can be grown in between potato cropping years to add biomass to soil and improve soil tilth (Carter et al., 2003). Surface applications of commercial fertilizer and animal manures also increase the risk of losses of nutrient by runoff as well as by volatile losses of N to the atmosphere. A single compost application in a potato-barley-red clover rotation combined with different rates of N application in the potato crop increased potato productivity and improved physical soil properties such as water-holding capacity (Carter et al., 2004).

For the potato-based cropping system, organic nutrient sources (manures, composts, and legumes), synthetic fertilizers, crops with high biomass production (grasses, barley, etc.), and cover crops can form key components of INM. These practices would help offset soil deterioration from intensive tillage operations use in potato production and thereby result in environmentally sustainable crop production. Liming and fertilizer applications are important for apple and vegetable production on acid soils. Application of N, P, and K improves longevity and productivity of grasses and Cu has been used to address Cu deficiency in animals. Intercropping of legumes in forage fields may also be worth considering.

Pacific Maritime and Moran Cordillera

Predominant production systems in British Columbia include agroforestry, horticultural crops, dairy, and poultry industries. Short supply of high quality arable land has resulted in increased intensity of agricultural operations, with 53 percent of total agricultural product sales being produced on less than 10 percent of the provincial agricultural land (Kowalenko, 2000). In the past few years, there is increased land devoted to small fruit and intensive animal production in British Columbia (Zebarth et al., 1998; Neilsen and Neilsen, 2002). As in the Atlantic Maritime Provinces, there has been a move to high density orchards with several thousand trees per ha (Neilsen et al., 1998).

Accompanying the increased intensity of agriculture are concerns about volumes and over-application of manure created in the livestock industry resulting in nitrate leaching into groundwater and aquatic environments (Zebarth et al., 1998; Kowalenko, 2000) as well as negative effects on crops (Bittman and Kowalenko, 1998). A small land base for dairy operations means that careful plant nutrient management is needed to produce maximum quantity of high quality feeds economically (Bittman and Kowalenko, 1998). Animal manure slurries represent an important source of nutrients for crop production, particularly in the production of perennial forage crops (Bittman et al., 1999). A long growing season and generally ample rainfall in this region allows for high production potentials for grass, with upward of 16 Mg ha^{-1} per year of biomass production capability. With the high production potential due to a favorable climate, grasses can use large quantities of nutrients and reduce risk of leaching or runoff as the grasses provide ground cover. For application of manure slurry, the splash plate applicator is typically used and applications are made as uniformly as possible to avoid burning of the crop (Bittman et al., 1999). The risk of N loss by volatilization and air quality concerns have led to increased interest in ways of reducing ammonia volatilization losses. The losses are difficult to predict as they are dependent on factors such as temperature, wind speed, pH and texture of soil, and fertilizer placement.

As in the Atlantic Maritimes, management of a broad diversity of relatively intensive agricultural systems in British Columbia results in a wide range of fertilization practices. The addition of fertilizer with irrigation water has been employed for horticultural crops (Haynes et al., 1985). Kowalenko (2000) has reported that fertigation increased both leaching and depletion of nutrients within the root zone, and that applying N through trickle irrigation to raspberries resulted in lower yields compared to a granular N fertilizer application. Others have noted that roots are shallower under irrigation and this can result in the root system becoming very dependent on outside sources of nutrients and water (Neilsen et al., 1998). Generally, broadcasting of fertilizer and manure tends to be most common, with potential for high N losses through volatilization. It has been suggested that up to 80 percent of the ammonium N in broadcast slurry manure could be lost (Amberger, 1990). It is estimated that over 7,000 Mg per year of ammonia is lost from agriculture land in the Fraser valley of British Columbia, with 30 percent of the losses associated with the crop stage at which fertilizer application occurs (Zebarth et al., 1999). Banding may be more effective than broadcasting, but it depends on the N fertilizer form and crop. Fairey and Lefkovitch (1998) found no difference in effectiveness between broadcast ammonium nitrate and injected UAN solution in a tall fescue stand. For hybrid poplar production, banding of N was not as effective as placing N fertilizer in the hole at a point (van-den Driessche, 1999).

Applications of large amounts of nutrients as fertilizer or manure have raised concerns regarding potential impacts on the environment (Bittman and Kowalenko, 1998; Zebarth et al., 1998). Leaching occurs mainly during the winter months when precipitation is high and evaporation rates are low (Kowalenko, 1991). Ideally, the application of manure on bare ground during autumn should be avoided (Paul and Zebarth, 1997). Large manure applications in the spring prior to seeding could also produce high levels of residual nitrate in the autumn (Zebarth et al., 1996). Forages that are over-fertilized also tend to contain high nitrate levels, thereby lowering the feed quality (Bittman and Kowalenko, 1998). Planting of trees or shrubs near riparian zones may help to reduce the entry of nitrate from croplands into river and stream beds.

As a result of increasing intensity of land use in the Abbotsford region, nitrate levels in the Abbotsford-Sumas aquifer that extend through British Columbia in Canada and Washington state in the United States are of concern (Zebarth et al., 1998). Land use above the aquifer includes intensive poultry operations as well as berry production and forage and pastures. An increase in manure production coupled with increased intensive fruit production resulted in a nitrogen surplus and the conclusion that the contamination of the aquifer is caused by application of nitrogen in excess of crop demand and mineralization of N during land use changes (Zebarth et al., 1998).

High intensity and diversity of cropping systems in part of this region presents unique opportunities and challenges. Addition of large quantities of manures and fertilizers, particularly to maximize yield of high value crops, is of particular concern for pollution of soil, water, and air.

RESEARCH ACCOMPLISHMENTS, GAPS, AND FUTURE NEEDS

Misuse of synthetic fertilizers and manures in modern agriculture has been implicated as a causal factor for increased nitrate-N concentration in groundwater, P movement to surface waters, and gaseous emissions to the atmosphere. However, most issues surrounding nutrient loading and risk of escape to the water or atmosphere can be addressed through sound nutrient and agronomic management. For any nutrient source, either organic or inorganic, the keys for maximizing crop nutrient recovery and minimizing potential losses to the environment are the establishment and utilization of proper nutrient rates, placement, and timing for a cropping system and region. The combined use of inorganic fertilizers and organic manures will often improve nutrient use efficiency and reduce nutrient loss by ensuring balanced nutrient availability. INM improves efficiency by synchronizing the supply of nutrients from soil, biological N_2 fixation, manures, and crop residue as well as from inorganic fertilizers, with the nutrient demand of crops. Further research is needed in various agroecological regions of Canada on the effects of INM on crop yield, soil quality, supply of plant-available nutrients, accumulation and distribution of nitrate-N and other nutrients in the soil profile, and greenhouse gas emissions. Moreover, INM must involve overall agronomic management of the entire production system, including crop rotation, moisture management, tillage system, and control of insects, weeds, and diseases, rather than consideration of nutrient management in isolation.

Crops grown in diversified rotations generally produce greater yields of higher quality grain and often have a lower N requirement for optimal grain yield as compared to crops grown in monoculture (Grant et al., 2002). Quality of eroded soil can also be improved by growing deep-rooted perennial forages in the crop rotation where feasible (Franzluebbers and Stuedemann,

2003). Cropping systems that reduce nutrient losses to surface/groundwater are important for INM and sustainable agriculture production. In the Prairie region, legumes such as field pea, alfalfa, clover, and lentil are more commonly being used to replace summer fallow in the traditional cereal-fallow and cereal-oilseed rotations. Use of legumes as mixture in grass hay fields and pastures is also increasing. More information from western Canada is needed on the benefits of increased yield and protein concentration of the grass component when alfalfa is included in the mixture. The reduction in N fertilizer requirement by the presence of legumes is often attributed to reduced mineralization or the additional N_2 fixed that is released through decomposition of roots and residues for succeeding crops (Soper and Grenier, 1987; Beckie and Brandt, 1997; Beckie et al., 1997). There is considerable interest in the inclusion of forage crops, especially legumes, in cropping systems. Management practices to maximize the benefits of legumes need to be developed to achieve a reduction in fertilizer N input and environmental problems through increased efficiency in the use of fertilizer and energy use.

Use of green manure and cover crops to replace fallow, a special feature of crop rotations in the drier regions of the prairie provinces, have implications for soil conservation, reductions in inputs, and sustainability of crop production. Green manure legumes can supply biologically fixed N_2 to following crops, reduce the reliance on nonrenewable energy, and improve soil quality and productivity. However, they can increase the risk of nitrate leaching. Fitting green manure legumes into cropping systems has not been easy, particularly in semiarid areas, as they compete with cash crops for space, time, water, and other inputs. Because of this and other factors like cost of green manure production and incorporation into soil, the adoption of green manure technology in Canada is limited. There is a need for critical analysis of the potential and limiting factors of green manure technology in order to assess the future scope for green manure in Canada.

Cover crops traditionally have been grown to prevent soil erosion and conserve soil moisture, reduce loss of available N from soil, and minimize nitrate-N entering the groundwater. The return of plant residue over time may increase organic matter in soil, which in turn can increase the N supplying capacity of soil and yield of subsequent crops. Cover crops have also been used for weed control, thereby leaving more nutrients in soil for crop uptake.

Cropping systems on mixed animal and crop farms generally involve an INM approach as both organic manure and inorganic fertilizers are used. The changes in soil organic matter and other soil properties due to application of organic nutrient sources are slow. The impact of manure or straw retention inputs and straw removal on soil organic matter and other properties may not be measured effectively in short-term experiments. Also, the net release of nutrients from organic sources is a function of their chemical composition and physical form of residues, soil and climatic conditions, and cultural practices. Until now, very little attention has been paid to the chemical composition or quality of organic nutrient sources. So, quality of organic sources of nutrients should be given due attention in the INM program.

There is a critical need for more information on the impact of INM under livestock-based cropping systems. Available information shows that the biological, chemical, and physical properties of soils can respond greatly to changes in management practices. Two main management factors that contribute to these changes are the use of perennial crops in the rotation and the application of animal manure. These two factors can interact in their effect on soil properties, and these effects are largely dependent on the amount of organic matter and nutrients that they provide to the soil. More long-term sites on a range of soils and involving various crop rotations and manure management practices are needed to fully evaluate the impacts of these systems on crop production, and on soil, water, and air quality. There is a specific need to further evaluate the atmospheric N losses, and surface, and groundwater P losses following cattle manure application both on cultivated and hay land.

The demand for organically grown food and fiber products is increasing in Canada and internationally, and the price premiums on organically-grown products make their production attractive. Researchers have reported a decline in crop yields during transition from a conventional to an organic system (Clark et al., 1999; Entz et al., 2001) or in the absence of N or P fertilizer (Campbell et al., 1993), and reduction (by half) in available P in soil (Entz et al., 2001). Organic and low input-cropping systems are being promoted to reduce the use of nonrenewable resources and environmental degradation, while sustaining productivity and profitability. Low nutrient inputs often result in lower nitrate-N leaching than more intensive high input conventional cropping systems. But, in soils lacking other essential nutrients such as P, K, S, or micronutrients, deep leaching of nitrate-N in the soil profile were found to be equal to or greater than high input-cropping systems (Malhi et al., 2002).

In organic farming and low input agriculture systems, low soil organic matter coupled with low concentration of total and available nutrients in soil can be major constraints to high-yielding crop production. Application of P-enriched compost, prepared by aerobic decomposition of crop/organic residues with low grade rock phosphate and phosphate-solublizing microorganisms increased the concentration of available P in soil and sustained high crop yield in studies conducted in India (Manna et al., 2001). However, there is limited research information regarding this topic in Canada.

More work to identify organic and natural sources, and techniques to increase their effectiveness in organic cropping systems are needed.

While information from other countries indicates that biofertilizers can play an important role in sustaining productivity of soils through biological N_2 fixation and enhancing native P availability to crops, there has been limited investigation of the use of biofertilizers under Canadian conditions. In rice cropping systems in warm, humid regions, inoculation of *Rhizobium, Azospirillum, Azotobacter,* blue green algae (BGA) and *Azolla* can increase N supply through BNF and has resulted in savings of fertilizer N in rice cropping systems (Singh, 1981). *Azolla* (a water fern) is used as a green manure organic fertilizer in rice production systems in many Asian countries. However, in Canada, labor requirements, the cold, a short growing season, lack of ability to over-winter, common usage of pesticides, and the need for fertilizer P are likely to limit adoption on a commercial scale.

If P is lacking in soil on organic farms, it has to be applied from outside sources. As there is not enough manure to apply on all farm fields, rock phosphate fertilizer can be used as a natural source of P to correct P deficiency. However, the amount of P that becomes available from the rock phosphate fertilizer to the crop in the growing season is very limited (maybe less than 5 percent of the applied P as rock phosphate), especially on high pH soils. Phosphorus solubilizing microorganisms can increase availability of sparingly soluble P in soil through secretion of organic acids that lower soil pH. *Penicillium bilaii* has been shown to increase the amount of available P in soil and its use to increase effectiveness of natural rock phosphate may prove useful for organic farming. In Canada, producers often use *Rhizobium* inoculum for legumes, but there is low level of acceptance of other biofertilizers or microbial inoculants by producers because of their low visible impact on crop production and the lack of information to demonstrate their benefits. Research to explore the potential of these organic nutrient sources is therefore suggested.

Agroforestry can combine production of trees with other crops or animal production systems (Small Woodlands Program of BC, 2004). Products of such systems include conventional timber and livestock/crop production as well as many non-timber forest products including mushrooms, florals, medicinals, herbs and craft products. Both shade systems (shade tolerant plants grown under a closed tree canopy) and sun systems (with open canopies and crops grown between trees) are utilized. Knowledge on the interactions of crops and trees for INM may be helpful to determine nutrient input rates and times to crops and trees.

Much of the manure and some of the fertilizer in Canada are still being applied using broadcast or broadcast-incorporation methods. This method of application increases the susceptibility of nutrients to loss by volatilization of ammonia and transport off the field by runoff water. Research is underway on the development and evaluation of improved low disturbance equipment for subsurface placement of manure and fertilizer in bands in the soil. In areas of high potential nutrient loss by leaching, tools and techniques such as soil, plant, and manure testing must be refined to avoid nutrient over-application and nutrient surpluses. In the production of high value crops where the cost of the fertilizer is small relative to the value of the product, the tendency is to apply for maximum yield which can lead to nutrient escape when the requirements are overestimated. Future research should look toward developing better nutrient budgets for the diverse array of high value horticultural crops, such as berries, that are replacing traditional crops in these areas.

In summary, long-term studies are needed to investigate the beneficial effects of INM on physical, chemical, and biological properties of soil, soil organic matter dynamics, and nutrient use efficiency, as well as effects on crop productivity, and crop quality in different cropping systems. Use of cover crops may prove particularly useful in situations where organic manures with high proportion of inorganic N (e.g., liquid pig manure) or a high rate of synthetic N and P fertilizers are applied and a fallow period follows the crop. Runoff and erosion can be reduced by elimination of tillage during fallow, with a consequent improvement in soil fertility/quality and productivity by improvement of soil organic matter content as well, producing more even water distribution across the landscape. To maintain or enhance organic matter in soil, major strategies can include increasing biomass production and return of crop/root residues to soil per unit area, by balanced use of manures, and inorganic fertilization, and decreasing loss of soil organic matter by eliminating tillage.

CONCLUSIONS

Sustainable crop production requires that nutrient inputs are managed effectively. Shortages of one or more nutrients can lead to accelerated loss of soil organic matter, soil erosion, and nutrient depletion. Soil nutrients supplied in excess of crop requirements can lead to accumulation in the soil, pollution of groundwater and surface water, and N_2O emissions to the atmosphere. Judicious use of fertilizers and other nutrient sources is therefore critical for long-term sustainability of agricultural systems, and INM is one of the important strategies to meet this goal. INM implies consideration of both inorganic and organic nutrient sources in a diverse crop production system for sustainability of soil quality, high crop yields and quality, and minimal impact on environment. Important considerations in INM are the selection of appropriate nutrient rates, placements, and timings that will promote maximum nutrient recovery by the crop. Tools that aid in this process include soil, plant, and manure testing along with nutrient recommendation systems that are specific to the region. Use of INM in combination with appropriate diverse crop rotations, conservation tillage, and other management practices will play a critical role in the long-term sustainability of Canadian agriculture.

REFERENCES

AAFC. (2005). A national ecological framework for Canada. http://sis.agr.gc.ca/cansis/nsdb/ecostrat (accessed August 31, 2005).

Abul-Kashem, M.D., O.O. Akinremi, and G.J. Racz. (2004). Phosphorus fractions in soil amended with organic and inorganic phosphorus sources. *Canadian Journal of Soil Science.* 84: 83-90.

AFHRC. (2003). Producing apples in eastern and central Canada, available at http://res2.agr.ca/kentville/pubs/pub1899, accessed June 15, 2004, Agriculture and Agri-Food Canada, Ottawa, Canada.

Amberger, A. (1990). Ammonia emissions during and after land spreading of slurry. In Nelson, V.C., J.H. Voorburg, and P. L'Hermite (eds.) *Odor and Ammonia Emissions from Livestock Farming.* pp. 126-131. London, UK: Elsevier.

Angers, D.A., M.A. Bolinder, M.R. Carter, E.G. Gregorich, C.F. Drury, B.C. Liang, R.P. Voroney, et al. (1997). Impact of tillage practices on organic carbon and nitrogen in cool, humid soils of Eastern Canada. *Soil and Tillage Research.* 41: 191-201.

APASCC. (1997). Beans-vegetable production guide for the Atlantic Provinces, available at http://www.nsac.ns.ca/lib/apascc/acv, accessed June 15, 2004.

Arsenault, W.J. and A. Malone. (1999). Effects of nitrogen fertilization and in-row seedpiece spacing on yield of three potato cultivars in Prince Edward Island. *American Journal of Potato Research.* 76: 227-229.

Asea, P.E.A., R.M.N. Kucey, and J.W.B. Stewart. (1988). Inorganic phosphate solubilization by two *Penicillium* species in solution culture and soil. *Soil Biology and Biochemistry.* 20: 459-464.

Bailey, L.D. (1990). The effects of 2-chloro-6(trichloromethyl)-pyridine (N-Serve) and N fertilizers on the growth, yields, protein and oil content of Canadian rape. *Canadian Journal of Plant Science.* 70: 979-986.

Ball-Coelho, B.R. and R.C. Roy. (1999). Enhanced ammonium sources to reduce nitrate leaching. *Nutrient Cycling in Agroecosystems.* 54: 73-80.

Ball-Coelho, B.R., R.C. Roy, and A.J. Bruin. (2004). Nitrate leaching as affected by liquid swine manure and cover cropping in sandy soil of southern Ontario. *Canadian Journal of Soil Science.* 84: 187-197.

Beauchamp, E.G. (1983). Response of corn to nitrogen in preplant and sidedress applications of liquid dairy cattle manure. *Canadian Journal of Soil Science.* 63: 377-386.

Beauchamp, E.G. (1998). *Time of Application of Manures and Crop Response to N. A Review of Literature and a Series of Research Reports.* 79p. Ontario, Canada: Department of Land Resource Science, University of Guelph, Guelph.

Beauchamp, E.G., G.E. Kidd, and G. Thurtell. (1982). Ammonia volatilization from liquid dairy cattle manure in the field. *Canadian Journal of Soil Science.* 62: 11-19.

Beauchamp, E.G. and R.P. Voroney. (1994). Crop carbon contribution to the soils with different cropping and livestock systems. *Journal of Soil and Water Conservation.* 49: 205-209.

Beckie, H.J. (1997). *Field Pea Magic. Research Letter # 97-01.* Saskatoon, Canada: Agriculture and Agri-Food Canada, Research Centre.

Beckie, H.J. and S.A. Brandt. (1997). Nitrogen contribution of field pea in annual cropping systems: I. Nitrogen residual effect. *Canadian Journal of Plant Science.* 77: 311-322.

Beckie, H.J., S.A. Brandt, J.J. Schoenau, C.A. Campbell, J.L Henry, and H.H. Janzen. (1997). Nitrogen contribution of field pea in annual cropping systems. 2. Total nitrogen benefit. *Canadian Journal of Plant Science.* 77: 323-331.

Beckie, H., J.D. Schlechte, A.P. Moulin, S.C. Gleddie, and D.A. Pulkinen. (1998). Response of alfalfa to inoculation with *Penicillium bilaii* (Provide). *Canadian Journal of Plant Science.* 78: 91-102.

Belanger, G. and J.E. Richards. (1998). Relationship between P and N concentrations in timothy. *Canadian Journal of Plant Science.* 79: 65-70.

Biederbeck, V.O., O.T. Bouman, J. Looman, A.E. Slinkard, L.D. Bailey, W.A. Rice, and H.H. Janzen. (1993). Productivity of four annual legumes as green manure in dryland cropping systems. *Agronomy Journal.* 85: 1035-1043.

Biederbeck, V.O. and A.E. Slinkard. (1988). Effect of annual green manures on yield and quality of wheat on a brown loam. *Proceedings of Soils and Crops Workshop.* pp. 345-361. February 1988, Saskatoon, Saskatchewan, Canada.

Bissonnette, N., D.A. Angers, R.R. Simard, and J. Lafond. (2001). Interactive effects of management practices on water-stable aggregation and organic matter of a Humic Gleysol. *Canadian Journal of Soil Science.* 81: 545-551.

Bittman, S. and C.G. Kowalenko. (1998). Whole season grass response to and recovery of nitrogen applied at various rates and distributions in a high rainfall environment. *Canadian Journal of Plant Science.* 78: 445-451.

Bittman, S., C.G. Kowalenko, D.E. Hunt, and O. Schmidt. (1999). Surface banded and broadcast manure effects on tall fescue yield and nitrogen uptake. *Agronomy Journal.* 91: 826-833.

Blackshaw, R.E., J.R. Moyer, C. Doram, A.I. Boswall, and E.G. Smith. (2001). Suitability of undersown sweetclover as a fallow replacement in semiarid cropping systems. *Agronomy Journal*. 93: 863-868.

Bolinder, M.A., D.A. Angers, G. Bélanger, R. Michaud, and M.R. Laverdière. (2002). Root biomass and shoot to root ratios of perennial forage crops in eastern Canada. *Canadian Journal of Plant Science*. 82: 731-737.

Brandt, S.A. (1990). Indianhead lentil as a green manure substitute for summer-fallow. *Proceedings of Soils and Crops Workshop*. pp. 222-230. February 22-23, 1990, University of Saskatchewan, Saskatoon, Saskatchewan, Canada.

Brandt, S.A. (1996). Alternatives to summerfallow and subsequent wheat and barley yield on a dark Brown soil. *Canadian Journal of Plant Science*. 76: 223-228.

Brandt, S.A. (1999). Management practices for black lentil green manure for the semi-arid Canadian prairies. *Canadian Journal of Plant Science*. 79: 11-17.

Bremer, E. and C. van Kessel. (1990). Availability of N to plants from legume and fertilizer sources: Which is greater? *Proceedings of Soils and Crops Workshop*. pp. 262-268. February 22-23, 1990, Saskatoon, Saskatchewan, Canada.

Broersma, K., A.R. Yee, and E.D. Pickering. (1989). Yield and quality of some grass-legume mixtures in response to N, P and K fertilizers in the central interior of British Columbia. *Technical Bulletin 1989-2E*. 29 pp. Research Branch, Agriculture Canada, Information Division, Ottawa, Ontario, Canada.

Bullied, W.J. and M.H. Entz. (1999). Soil water dynamics after alfalfa as influenced by termination technique. *Agronomy Journal*. 91: 294-305.

Bullied, W.J., M.H. Entz, S.R. Jr. Smith, and K.C. Bamford. (2002). Grain yield and N benefits to sequential wheat and barley crops from single-year alfalfa, berseem and red clover, chickling vetch and lentil. *Canadian Journal of Plant Science*. 82: 53-65.

Campbell, C.A. (1978). Soil organic carbon, nitrogen and fertility. In M. Schnitzer and S.U. Khan (eds.). *Soil Organic Matter*. pp. 173-272. Developments in Soil Science. Amsterdam, The Netherlands: Elsevier.

Campbell, C.A., R.P. Biederbeck, R.P. Zentner, and G.P. Lafond. (1991). Effect of crop rotations and cultural practices on soil organic matter, microbial biomass and respiration in a thin Black Chernozem. *Canadian Journal of Soil Science*. 71: 363-376.

Campbell, C.A., K.E. Bowren, M. Schnitzer, R.P. Zentner, and L. Townley-Smith. (1991). Effect of crop rotations and fertilization on soil organic matter and some biochemical properties of a thick Black Chernozem. *Canadian Journal of Soil Science*. 71: 377-387.

Campbell, C.A., G.P. Lafond, V.O. Biederbeck, and G.E. Winkleman. (1993). Influence of legumes and fertilization on deep distribution of available phosphorus (Olsen-P) in a thin Black Chernozem. *Canadian Journal of Soil Science*. 73: 555-565.

Campbell, C.A., G.P. Lafond, A.J. Leyshon, R.P. Zentner, and H.H. Janzen. (1991). Effect of cropping practices on the initial potential rate of N mineralization in a thin Black Chernozem. *Canadian Journal of Soil Science*. 71: 43-53.

Campbell, C.A., G.P. Lafond, and R.P. Zentner. (1993). Spring wheat yield trends as influenced by fertilizer and legumes. *Journal of Production Agriculture.* 6: 564-568.

Campbell, C.A., G.P. Lafond, R.P. Zentner, and Y.W. Jame. (1984). Effect of cropping, summerfallow and fertilizer nitrogen on nitrate-nitrogen lost by leaching on a Brown Chernozemic loam. *Canadian Journal of Soil Science.* 64: 61-74.

Campbell, C.A., G.P. Lafond, R.P. Zenter, and Y.W. Jame. (1994). Nitrate leaching in an Udic Haploboroll as influenced by fertilization and legumes. *Journal of Environmental Quality.* 23: 195-201.

Campbell, C.A., M. Schnitzer, J.W.B. Stewart, V.O. Biederbeck, and F. Selles. (1986). Effects of manure and P fertilizer on properties of Black Chernozem in southern Saskatchewan. *Canadian Journal of Soil Science.* 66: 601-613.

Campbell, C.A., F. Selles, G.P. Lafond, B.G. McConkey, and D. Hahn. (1998). Effect of crop management on C and N in long-term crop rotations after adopting no-tillage management: Comparison of soil sampling strategies. *Canadian Journal of Soil Science.* 78: 155-162.

Campbell, C.A., J.S.J. Tessier, and F. Selles. (1988). Challenges and limitations to adoption of conservation tillage—Soil organic matter, fertility, moisture, and soil environment, Land Degradation and Conservation Tillage. *Proceedings of the 34th Annual Canadian Society of Soil Science/Agricultural Institute of Canada Meeting.* pp. 140-185. Calgary, Alberta, Canada.

Campbell, C.A. and R.P. Zentner. (1993). Soil organic matter as influenced by crop rotations and fertilization. *Soil Science Society of America Journal.* 57: 1034-1040.

Campbell, C.A., R.P. Zentner, S. Gameda, B. Blomert, and D.D. Wall. (2002). Production of annual crops on the Canadian prairies: Trends during 1976-1998. *Canadian Journal of Soil Science.* 82: 45-57.

Campbell, C.A., R.P. Zentner, H.H. Janzen, and K.E. Bowren. (1990). Crop rotation studies on the Canadian prairies. *Publication 1841/E.* 133pp. Communications Branch, Agriculture Canada, Ottawa, Ontario, Canada.

Campbell, C.A., R.P. Zentner, F. Selles, V.O. Biederbeck, B.G. McConkey, R. Lemke, and Y.T. Gan. (2004). Cropping frequency effects on yield of grain, straw, plant N, N balance and annual production of spring wheat in the semiarid prairie. *Canadian Journal of Plant Science.* 84: 487-501.

Campbell, C.A., R.P. Zentner, F. Selles, P.G. Jefferson, B.G. McConkey, R. Lemke, and B.J. Blomert. (2005). Long-term effects of cropping systems and nitrogen and phosphorus fertilizer on production and nitrogen economy of grain crops in a Brown Chernozem. *Canadian Journal of Plant Science.* 85: In Press.

Canadian Soil Classification System. (1978). *The System of Soil Classification for Canada.* Ontario, Canada: Agriculture and Agri-Food Canada, Ottawa.

Carefoot, J.M., M. Nyborg, and C.W. Lindwall. (1990). Tillage-induced soil changes and related grain yield in a semi-arid region. *Canadian Journal of Soil Science.* 70: 203-214.

Carter, M.R., H.T. Kunelius, J.B. Sanderson, J. Kimpinski, H.W. Platt, and M.A. Bolinder. (2003). Productivity parameters and soil health dynamics under one-term 2 year potato rotations in Atlantic Canada. *Soil and Tillage Research.* 72: 153-168.

Carter, M.R. and J.E. Richards. (2000). Soil and alfalfa response after amelioration of subsoil acidity in a fine sandy loam Podzol in Prince Edward Island. *Canadian Journal of Soil Science.* 90: 607-614.

Carter, M.R., J.B. Sanderson, and J.A. McLeod. (2004). Influence of compost on the physical properties and organic matter fractions of a fine sandy loam throughout the cycle of a potato rotation. *Canadian Journal of Soil Science.* 84: 211-218.

CFI. (2003). *Food For our Future. Canadian Fertilizer Information System 2001— Statistical Compendium.* Vancouver, Canada: Canadian Fertilizer Institute.

Chambers, P.A., M. Guy, E.S. Roberts, M.N. Charlton, R. Kent, C. Gaganon, G. Grove, and N. Foster. (2001). Nutrients and their impact on the Canadian environment, *Technical Bulletin C2001-980118-1.* 241p. Agriculture and Agri-Food Canada, Environment Canada, Fisheries and Oceans Canada, Health Canada and Natural Resources Canada, Ottawa, Ontario, Canada.

Chang, C. and T. Entz. (1996). Nitrate leaching losses under repeated cattle feedlot manure applications in southern Alberta. *Journal of Environmental Quality.* 25: 145-153.

Chang, C., T.G. Sommerfeldt, and T. Entz. (1990). Rates of soil chemical changes with eleven annual application of cattle feedlot manure. *Canadian Journal of Soil Science.* 70: 673-681.

Chang, C., T.G. Sommerfeldt, and T. Entz. (1991). Soil chemistry after eleven annual applications of cattle feedlot manure. *Journal of Environmental Quality.* 20: 475-480.

Chantigny, M.H., D.A. Angers, P. Rochette, G. Bélanger, J. Tremblay, D. Massé, and D. Côté. (2004). Valorisation agronomique sur cultures fourragères de lisiers de porc pré-traités et réduction des impacts environnementaux consécutifs aux épandages. *Rapport final de recherche. Centre de Recherche sur les Sols et les Grandes Cultures.* 88p. Agriculture et Agroalimentaire Canada, Sainte-Foy, Québec, Canada.

Chen, W., W.P. McCaughey, C.A. Grant, and L.D. Bailey. (2001). Pasture type and fertilization effects on soil chemical properties and nutrient redistribution. *Canadian Journal of Soil Science.* 81: 395-404.

Clark, S., K. Klonsky, P. Livingston, and S. Temple. (1999). Crop-yield and economic comparisons of organic, low-input, and conventional farming systems in California's Sacramento Valley. *American Journal of Alternative Agriculture.* 14: 109-121.

Clayton, G.W. (1982). Zero tillage for cereal production on forage sods in Manitoba. MSc Thesis, University of Manitoba, Winnipeg, Manitoba, Canada.

Clayton, G.W. and H.M. Austenson. (1992). Barley yield and nitrogen utilization following short-term green-manured legumes. *Proceedings of Alberta Soil Science Workshop.* pp. 201-207. February 1992, Lethbridge, Alberta, Canada.

Clayton, G.W., W.A. Rice, N.Z. Lupwayj, A.M. Johnston, G.P. Lafond, C.A. Grant, and F. Walley. (2004a). Inoculant formulation and fertilizer nitrogen effects on field pea: Crop yield and seed quality. *Canadian Journal of Plant Science.* 84: 79-88.

Clayton, G.W., W.A. Rice, N.Z. Lupwayj, A.M. Johnston, G.P. Lafond, C.A. Grant, and F. Walley. (2004b). Inoculation formulation and fertilizer nitrogen effects on field pea: Nodulation, N_2 fixation and nitrogen partitioning. *Canadian Journal of Plant Science.* 84: 89-96.

Cooke, D.A., S.E. Beacom, and J.A. Robertson. (1973). Comparison of continuously grazed bromegrass-alfalfa with rotationally grazed crested wheatgrass-alfalfa. *Canadian Journal of Animal Science.* 53: 423-429.

Cox, W.J., R.W. Zobel, H.M. van Es, and D.J. Otis. (1990). Tillage effects on some soil physical and corn physiological characteristics. *Agronomy Journal.* 82: 806-812.

Dean, J.R. and K.W. Clark. (1977). Nodulation, acetylene reduction and yield of faba beans as affected by inoculum concentration and soil nitrate level. *Canadian Journal of Plant Science.* 57: 1055-1061.

Dharmakeerthi, R.S., B.D. Kay, and E.G. Beauchamp. (2004). Effect of soil disturbance on N availability across a variable landscape in southern Ontario. *Soil and Tillage Research.* 79: 101-112.

Doran, J.W. and M.S. Smith. (1987). Organic matter management and utilization of soil and fertilizer nutrients. In *Soil Fertility and Organic Matter as Critical Components of Production Systems.* Special Publication No. 19, pp. 53-72. Madison, WI: Soil Science Society of America.

Dormaar, J.F. and C. Chang. (1995). Effect of 20 annual applications of excess feedlot manure on labile soil phosphorus. *Canadian Journal of Soil Science.* 75: 507-512.

Drury, C.F., D.J. McKenney, and W.I. Findlay. (1992). Nitric oxide and nitrous oxide production from soil: Water and oxygen effects. *Soil Science Society of America Journal.* 56: 766-770.

Drury, C.F., T.O. Oloya, D.J. McKenney, E.G. Gregorich, C.S. Tan, and C.L. van Luyk. (1998). Long-term effects of fertilization and rotation on denitrification and soil carbon. *Soil Science Society of America Journal.* 62: 1572-1579.

Drury, C.F., W.D. Reynolds, C.S. Tan, T.W. Welacky, W. Calder, and N.B. McLaughlin. (2006). Emissions of nitrous oxide and carbon dioxide: Influence of tillage type and nitrogen placement depth. *Soil Science Society of America Journal.* 70: 570-581.

Drury, C.F. and C.S. Tan. (1995). Long-term (35 years) effects of fertilization, rotation and weather on corn yields. *Canadian Journal of Plant Science.* 75: 355-362.

Drury, C.F., C.S. Tan, J.D. Gaynor, T.O. Oloya, and T.W. Welacky. (1996). Influence of controlled drainage-subirrigation on surface and tile drainage nitrate loss. *Journal of Environmental Quality.* 25: 317-324.

Drury, C.F., C.S. Tan, W.D. Reynolds, C.A. Fox, N.B. McLauglin, and S. Bittman. (2004). Tillage practices for corn production. In S. Bittman and C.G. Kowalenko (eds.) *Advanced Silage Corn Management.* pp. 72-79. British Columbia, Canada: Pacific Field Corn Association.

Drury, C.F., C.S. Tan, W.D. Reynolds, T.W. Welacky, S.E. Weaver, A.S. Hamill, and T.J. Vyn. (2003). Impacts of zone tillage and red clover on corn performance and soil quality. *Soil Science Society of America Journal.* 67: 867-877.

Drury, C.F., C.S. Tan, T.W. Welacky, T.O. Oloya, A.S. Hamill, and S.E. Weaver. (1999). Red clover and tillage influence soil temperature, moisture and corn emergence. *Agronomy Journal.* 91: 101-108.

Dryden, R.D., L.D. Bailey, and C.A. Grant. (1983). Crop rotation studies, 1893-1982 (Mostly field crops, Canada). *Canadian Agriculture.* 29: 18-21.

Entz, M.H., V.S. Baron, P.M. Carr, D.W. Meyer, S.R. Jr. Smith, and W.P. McCaughey. (2002). Potential of forages to diversify cropping systems in the Northern Great Plains. *Agronomy Journal.* 94: 240-250.

Entz, M.H., W.J. Bullied, D.A. Foster, R. Gulden, and K. Vessey. (2001). Extraction of subsoil nitrogen by alfalfa, alfalfa-wheat, and perennial grass systems. *Agronomy Journal.* 93: 495-503.

Entz, M.H., W.J. Bullied, and F. Katepa-Mupondwa. (1995). Rotational benefits of forage crops in Canadian prairie cropping systems. *Journal of Production Agriculture.* 8: 521-529.

Entz, M.H., R. Guilford, and R. Gulden. (2001). Crop yield and nutrient status on 14 organic farms in the eastern portion of the northern Great Plains. *Canadian Journal of Plant Science.* 81: 351-354.

Fairchild, G.L., D.A. Barry, M.J. Goss, A.S. Hamill, P. Lafrance, P.H. Milburn, R.R. Simard, and B.J. Zebarth. (2004). *Groundwater Quality.* Available at http://res2 .agr.ca/publications/hw/pdf, accessed June 15, 2004.

Fairey, N.A. (1991). Effects of nitrogen fertilizer, cutting frequency, and companion legume on herbage production and quality of four grasses. *Canadian Journal of Plant Science.* 71: 717-725.

Fairey, N.A. and L.P. Lefkovitch. (1998). Effects of method, rate and time of application of nitrogen fertilizer on seed production of tall fescue. *Canadian Journal of Plant Science.* 78: 453-458.

Fan, M.X., A.F. MacKenzie, M. Abbott, and F. Cadrin. (1997). Denitrification estimates in monoculture and rotation corn as influence by tillage and nitrogen fertilizer. *Canadian Journal of Soil Science.* 77: 389-396.

Ferguson, W.S. and B.J. Gorby. (1971). Effect of various periods of seed-sown to alfalfa and bromegrass on soil nitrogen. *Canadian Journal of Soil Science.* 51: 65-73.

Fortin, M.C. and F.J. Pierce. (1990). Developmental and growth effects of crop residues on corn. *Agronomy Journal.* 82: 710-715.

Franzluebbers, A.J. and J.A. Stuedemann. (2003). Bermudagrass management in the southern Piedmont, USA. III. Particulate and biologically active soil carbon. *Soil Science Society of American Journal.* 67: 132-138.

Freeze, B.S. and T.G. Sommerfeldt. (1985). Breakeven hauling distance for beef feedlot manure in southern Alberta. *Canadian Journal of Soil Science.* 65: 687-693.

Gangbazo, G., G.M. Barnett, A.R. Pesant, and D. Cluis. (1999). Disposing hog manure on inorganically-fertilized corn and forage fields in southeastern Quebec. *Canadian Agricultural Engineering.* 41: 1-12.

Garand, M.J., R.R. Simard, A.F. MacKenzie, and C. Hamel. (2001). Underseeded clover as a nitrogen source for spring wheat on a Gleysol. *Canadian Journal of Soil Science.* 81: 93-102.

Gauer, L.E., C.F. Shaykewich, and E.H. Stobbe. (1982). Soil temperature and soil water under zero tillage sin Manitoba. *Canadian Journal of Soil Science.* 62: 311-327.

Gaynor, J.D. and W.I. Findlay. (1995). Soil and phosphorus loss from conservation and conventional tillage in corn production. *Journal of Environmental Quality.* 24: 734-741.

Gleddie, S.C., G.L. Hnatowich, and D.R. Polonenko. (1991). A summary of wheat response to Provide™ (*Penicillium bilaii*) in western Canada. *Proceedings of Alberta Soil Science Workshop.* pp. 306-313. February 19-21, 1991, Lethbridge, Alberta, Canada.

Gludden, R.H. and J.K. Vessey. (2000). *Penicillium bilaii* inoculation increases root hair production in field pea. *Canadian Journal of Plant Science.* 80: 801-804.

Government of British Columbia. (2004). About the agriculture industry. Available at http://www.agf,gov.bc.ca/aboutind/index.htm, accessed June 15, 2004.

Grant, C.A. and L.D. Bailey. (1993). Fertility management in canola production, *Canadian Journal of Plant Science.* 73: 651-670.

Grant, C.A., S. Bittman, M. Monreal, C. Plenchette, and C. Morel. (2005). Soil and fertilizer phosphorus: Effects on plant P supply and mycorrhizal development. *Canadian Journal of Plant Science.* 85: 3-14.

Grant, C.A., K.R. Brown, G.J. Racz, and L.D. Bailey. (2001). Influence of source, timing and placement of nitrogen on grain yield and nitrogen removal of Sceptre durum wheat under reduced- and conventional-tillage management. *Canadian Journal of Plant Science.* 81: 17-27.

Grant, C.A., K.R. Brown, G.J. Racz, and L.D. Bailey. (2002). Influence of source, timing and placement of nitrogen fertilizer on seed yield and nitrogen accumulation in the seed of canola under reduced-and conventional-tillage management. *Canadian Journal of Plant Science.* 82: 629-638.

Grant, C.A., S. Jia, K.R. Brown, and L.D. Bailey. (1996). Volatile losses of NH_3 from surface applied urea and urea ammonium nitrate with and without the urease inhibitor NBPT. *Canadian Journal of Soil Science.* 76: 417-419.

Grant, C.A. and W.P. McCaughey. (1991). Nutrient requirements for alfalfa production. In E.G. Siemer (ed.) *Proceedings of Intermountain Meadow Symposium.* pp. 133-150. July 1-3, 1991, 3rd, Steamboat Springs, CO, USA. Fort Collins, CO: Colorado State University.

Grant, C.A., G.A. Peterson, and C.A. Campbell. (2002). Nutrient considerations for diversified cropping systems in the northern Great Plains. *Agronomy Journal.* 94: 186-198.

Gregorich, E.G., D.A. Angers, C.A. Campbell, M.R. Carter, C.F. Drury, B.H. Ellert, P.H. Groenvelt, et al. (1995). Changes in soil organic matter. In D.F. Acton and L.J. Gregorich (eds.). *The Health of Our Soils: Toward Sustainable Agriculture in Canada.* pp. 41-50. Centre for Land and Biological Resources Research, Research Branch. Ontario Canada: Agriculture and Agri-Food Canada, Ottawa.

Gregorich, E.G. and C.F. Drury. (1996). Fertilizer increases corn yield and soil organic matter. *Better Crops with Plant Food.* 80(4): 3-5.

Gregorich, E.G., C.F. Drury, and J. Baldock. (2001). Changes in soil carbon under long-term maize in monoculture and legume-based rotation. *Canadian Journal of Soil Science.* 81: 21-31.

Gupta, U.C. (1989). Copper nutrition of cereals and forages grown in Prince Edward Island. *Journal of Plant Nutrition.* 12: 53-64.

Halvorson, A.D., A.L. Black, J.M. Krupinsky, S.D. Merrill, and D.L. Tanaka. (1999). Sunflower response to tillage and nitrogen fertilization under intensive cropping in a wheat rotation. *Agronomy Journal.* 91: 637-642.

Halvorson, A.D., A.L. Black, J.M. Krupinsky, and S.D. Merrill. (1999). Dryland winter wheat response to tillage and nitrogen within an annual cropping system. *Agronomy Journal.* 91: 702-707.

Harapiak, J.T. and N.A. Flore. (1984). Behavior of fertilized mixed forage stands. *Proceedings of Alberta Soil Science Workshop.* pp. 256-263. February 21-22, 1984, Edmonton, Alberta, Canada.

Haynes, R.J., P.L. Carey, and K.M. Goh. (1985). Strawberries: Effects of nitrogen and potassium fertilizers. *New Zealand Commercial Grower.* 40: 44.

Henry, L., J. Harapiak, H. Ukrainetz, and B. Green. (1995). Revised guidelines for safe rates of fertilizer applied with the seed. *Farm Facts.* 6p. 10M01/95 ISSN0840-9447 SCR-1194. Saskatchewan Agriculture, Regina, Saskatchewan, Canada.

Holmstrom, D.A., H.T. Kunelius, and J.A. Ivany. (2000). Forages underseeded in barley for residue management for potatoes. *Canadian Journal of Plant Science.* 81: 205-210.

Hoyt, P.B. (1990). Residual effects of alfalfa and bromegrass cropping on yields of wheat grown for 15 subsequent years. *Canadian Journal of Soil Science.* 70: 109-113.

Hoyt, P.B. and R.H. Leitch. (1983). Effects of forage legume species on soil moisture, nitrogen and yield of successive barley crops. *Canadian Journal of Soil Science.* 63: 125-136.

Hoyt, P.B. and W.A. Rice. (1977). Effects of high rates of synthetic fertilizer and barnyard manure on yield and moisture use of six successive barley crops grown on three gray Luvisolic soils. *Canadian Journal of Soil Science.* 57: 425-435.

Hynes, R.K., K.A. Kraig, D. Covert, R.S. Smith, and R.J. Rennie. (1995). Liquid rhizobial inoculants for lentil and field pea. *Journal of Production Agriculture.* 8: 547-552.

Isse, A.A., A.F. MacKenzie, K. Stewart, D.C. Cloutier, and D.L. Smith. (1999). Cover crops and nutrient retention for subsequent sweet corn production. *Agronomy Journal.* 91: 934-939.

Izaurralde, R.C., M. Choudhary, N.G. Juma, W.B. McGill, and L. Haderlein. (1995). Crop and nitrogen yield in legume-based rotations practiced with zero tillage and low-input methods. *Agronomy Journal.* 87: 958-964.

Izaurralde, R.C., Y. Feng, J.A. Robertson, W.B. McGill, N.G. Juma, and B.M. Olson. (1995). Long-term influence of cropping systems, tillage methods, and N sources on nitrate leaching. *Canadian Journal of Soil Science.* 75: 497-505.

Izaurralde, R.C., N.G. Juma, and W.B. McGill. (1990). Plant and nitrogen yield of barley-field pea intercrop in cryoboreal-subhumid central Alberta. *Agronomy Journal.* 28: 295-301.

Izaurralde, R.C., N.G. Juma, W.B. McGill, D.S. Chanasyk, S. Pawluk, and M.J. Dudas. (1993). Performance of conventional and alternative cropping systems in cryoboreal subhumid central Alberta. *Journal of Agricultural Science.* 120: 33-41.

Izaurralde, R.C., M. Nyborg, E.D. Solberg, H.H. Janzen, M.A. Arshad, S.S. Malhi, and M. Molina-Ayala. (1998). Carbon storage in eroded soils after five years of reclamation techniques. In R. Lal, J.M. Kimble, R.F. Follett, and B.A. Stewart (eds.). *Soil Processes and the Carbon Cycle.* pp. 369-385. Boca Raton, FL: CRC Press.

Izaurralde, R.C., E.D. Solberg, M. Nyborg, and S.S. Malhi. (1998). Immediate effects of topsoil removal on crop productivity loss and its restoration with commercial fertilizers. *Soil and Tillage Research.* 46: 251-259.

Jakobsen, I., L.K. Abbott, and A.D. Robson. (1992). External hyphae of vesicular arbuscular mycorrhizal fungi associated with *Trifolium subterraneum* L. I. Spread of hyphae and phosphorus inflow in roots. *New Phytology.* 120: 509-516.

Janzen, H.H. and R.M.N. Kucey. (1988). Carbon, nitrogen, and sulfur mineralization of crop residues as influenced by crop species and nutrient regime. *Plant and Soil.* 106: 35-41.

Jensen, E.S. (1986). The influence of rate and time of nitrate application supply on nitrogen fixation and yield in pea (*Pisum sativum* L.). *Fertilizer Research.* 10: 193-202.

Johnston, A.M., F.J. Larney, and C.W. Lindwall. (1995). Spring wheat and barley response to long-term fallow management. *Journal of Production Agriculture.* 8: 264-268.

Juma, N., J.A. Robertson, R.C. Izaurralde, and W.B. McGill. (1997). Crop yield and soil organic matter over 60 years in a Typic Cryoboralf at Breton, Alberta. In E.A. Paul, K.H. Paustian, E.T. Elliott, and C.V. Cole (eds.) *Soil Organic Matter in Temperate Agroecosystems—Long-Term Experiments in North America.* pp. 273-281. Boca Raton, FL: CRC Press.

Kopp, J.D., W.P. McCaughey, and K.M. Wittenberg. (1997). Calf-cow production response to pasture forage species. *Proceedings of the 18th International Grassland Congress.* pp. 37-38. Session 29. Saskatoon/Winnipeg, Canada.

Kowalenko, C.G. (1991). Fall versus spring sampling for calibrating nutrient applications on individual fields. *Journal of Production Agriculture.* 4: 322-329.

Kowalenko, C.G. (2000). Nitrogen pools and processes in agricultural systems of coastal British Columbia—A review of published research. *Canadian Journal of Plant Science.* 80: 1-10.

Kucey, R.M.N. (1983). Phosphate-solubilizing bacteria and fungi in various cultivated and virgin Alberta soils. *Canadian Journal of Soil Science.* 63: 671-678.

Kucey, R.M.N. (1987). Increased phosphorus uptake by wheat and field beans inoculated with a phosphorus-solubilizing *Penicillium bilaii* strain and with vesicular-arbuscular mycorrhizal fungi. *Applied Environmental Microbiology.* 53: 2699-2703.

Kucey, R.M.N. (1988). Effect of *Penicillium bilaii* on the solubility and uptake of P and micronutrients from soil by wheat. *Canadian Journal of Soil Science.* 68: 261-270.

Kucey, R.M.N. and M.E. Leggett. (1989). Increased yield and phosphorus uptake by Westar canola (*Brassica napus* L.) inoculated with a phosphorus-solubilizing isolate of *Penicillium bilaii* strain and with vesicular-arbuscular mycorrhizal fungi. *Canadian Journal of Soil Science.* 69: 425-432.

Kutcher, H.R., G. Lafond, A.M. Johnston, P.R. Miller, K.S. Gill, W.E. May, T. Hogg, E. Johnson, V.O. Biederbeck, and B. Nybo. (2002). Rhizobium inoculant and seed-applied fungicide effects on field pea production. *Canadian Journal of Soil Science.* 82: 645-651.

Lafond, G.P., H. Loeppky, and D.A. Derksen. (1992). The effects of tillage systems and crop rotations on soil water conservation, seedling establishment and crop yield. *Canadian Journal of Plant Science.* 72: 103-125.

Larney, F.J., E. Bremer, H.H. Janzen, A.M. Johnston, and C.W. Lindwall. (1997). Changes in total, mineralizable and light fraction soil organic matter with cropping and tillage intensities in semiarid southern Alberta, Canada. *Soil and Tillage Research.* 42: 229-240.

Larney, F.J. and H.H. Janzen. (1996). Restoration of productivity to a desurfaced soil with livestock manure, crop residue, and fertilizer amendments. *Agronomy Journal.* 88: 921-927.

Larney, F.J. and C.W. Lindwall. (1995). Rotation and tillage effects on available soil water for winter wheat in a semi-arid environment. *Soil and Tillage Research.* 36: 111-127.

Lemke, R.L., R.C. Izaurralde, M. Nyborg, and E.D. Solberg. (1999). Tillage and N-source influence soil-emitted nitrous oxide in the Alberta Parkland region. *Canadian Journal of Soil Science.* 79: 15-24.

Li, G.C., R.L. Mahler, and D.O. Everson. (1990). Effects of plant residues and environmental factors on phosphorus availability in soils. *Communications in Soil Science and Plant Analysis.* 21: 471-491.

Lukima, E.V., K.W. Freeman, K.J. Wynn, W.E. Thomason, R.W. Mullen, A.R. Klatt, G.V. Johnson, et al. (2001). Nitrogen fertilization optimization algorithm based on in-season estimates of yield and plant nitrogen uptake. *Journal of Plant Nutrition.* 24: 885-898.

Lupwayi, N.Z., G.W. Clayton, K.N. Harker, T.K. Turkington, W.A. Rice, and A.M. Johnston. (2003). Impact of crop residue type on phosphorus release. *Better Crops with Plant Food.* 87: 4-5.

MacDonald, K.B., W.R. Fraser, F. Wang, and G.W. Lelyk. (1995). A geographic framework for assessing soil quality. In D.F. Acton and L.J. Gregorich (eds.). *The Health of Our Soils: Toward Sustainable Agriculture in Canada.* pp. 19-30. Ottawa, Ontario: Agriculture and Agri-Food Canada.

MacKenzie, A.F., M.X. Fan, and F. Cadrin. (1998). Nitrous oxide emission in three years as affected by tillage, corn-soybean -alfalfa rotations and nitrogen fertilization. *Journal of Environmental Quality.* 27: 698-703.

Malhi, S.S., S.A. Brandt, D. Ulrich, R. Lemke, and K.S. Gill. (2002). Accumulation and distribution of nitrate-nitrogen and extractable phosphorus in the soil profile under various alternative cropping systems. *Journal of Plant Nutrition.* 25: 2499-2520.

Malhi, S.S., J.T. Harapiak, R. Karamanos, K.S. Gill, and N. Flore. (2003). Distribution of acid extractable P and exchangeable K in a grassland soil as affected by long-term surface application of N, P and K fertilizers. *Nutrient Cycling in Agroecosystems.* 67: 265-272.

Malhi, S.S., J.T. Harapiak, M. Nyborg, and N.A. Flore. (1991). Soil chemical properties after long-term fertilization of bromegrass: Nitrogen rate. *Communications in Soil Science and Plant Analysis.* 22: 1447-1458.

Malhi, S.S., D.H. Laverty, J.T. Harapiak, L.M. Kryzanowski, and D.C. Penney. (1993). Fertilizer management for forage crops in central Alberta. *Technical Bulletin 1993-3E.* 17p. Agriculture Canada, Research Branch, Ottawa, Ontario, Canada.

Malhi, S.S., R. Lemke, R. Farrell, and D. Leach. (2003). Influence of tillage and crop residue management on crop yield, greenhouse gas emissions and soil quality. *Proceedings of 16th Conference of International Soil Tillage Research Organization.* pp. 690-695. July 13-18, 2003, Brisbane, Australia.

Malhi, S.S. and M. Nyborg. (1986). Increase in mineral N in soils and loss of mineral N during early spring in north-central Alberta. *Canadian Journal of Soil Science.* 66: 397-409.

Malhi, S.S. and M. Nyborg. (1987). Influence of tillage on nitrate-N in soil. *Canadex (Tillage/Soil Fertility) 516.530.* Communications Branch, Agriculture Canada, Ottawa, Ontario, Canada.

Malhi, S.S., E.D. Solberg, R.C. Izaurralde, and M. Nyborg. (1999). Effects of simulated erosion and amendments on grain yield and quality of wheat. pp. 297-300. In D. Anac and P. Martin-Prevel (eds). *Improved Crop Quality by Nutrient Management.* Dordrecht, The Netherlands: Kluwer Academic Publishers.

Malhi, S.S., R.P. Zentner, and K. Heier. (2002). Effectiveness of alfalfa in reducing fertilizer N input for optimum forage yield, protein concentration, returns and energy performance of bromegrass-alfalfa mixtures. *Nutrient Cycling in Agroecosystems.* 62: 219-227.

Manna, M.C., P.K. Ghosh, B.N. Ghosh, and K.N. Singh. (2001). Comparative effectiveness of phosphate-enriched compost and single superphosphate on yield, uptake of nutrients and soil quality under soybean-wheat rotation. *Journal of Agricultural Science, Cambridge.* 137: 45-54.

Martin, F.R.J., A. Bootsam, D.R. Coote, B.G. Fairley, L.J. Gregorich, J. Lebedin, P.H. Milburn, B.J. Stewart, and T.W. Van der Gulik. (2000). Canada's Rural Water Resources. In D.R. Coote and L.J. Gregorich (eds.). *The Health of Our Water-Toward Sustainable Agriculture in Canada.* pp. 5-14. Research Planning and Coordination Directorate, Research Branch. Ontario, Canada: Agriculture and Agri-Food Canada, Ottawa.

McConkey, B.G., D. Curtin, C.A. Campbell, S.A. Brandt, and F. Selles. (2002). Crop and soil nitrogen status of tilled and no-tillage systems in semiarid regions of Saskatchewan. *Canadian Journal of Soil Science.* 82: 489-498.

McGill, W.B., K.R. Cannon, J.A. Robertson, and F.D. Cook. (1986). Dynamics of soil microbial biomass and water-soluble organic C in Breton L after 50 years of cropping to two rotations. *Canadian Journal of Soil Science.* 66: 1-19.

McGill, W.B., J.F. Dormaar, and E. Reinl-Dwyer. (1988). New perspectives on soil organic matter quality, quantity, and dynamics on the Canadian Prairies. *Proceedings of Canadian Society of Soil Science and Canadian Society of Extension Joint Symposium, Land Degradation: Assessment and Insight into a Western Canadian Problem.* pp. 30-48. August 23, 1988, Calgary, Alberta, Canada: Agricultural Institute of Canada.

McKenney, D.J. and C.F. Drury. (1997). Nitric oxide production in agricultural soils. *Global Change Biology.* 3: 317-326.

McKenzie, R.H., A.B. Middleton, E.D. Solberg, J. DeMulder, N. Flore, G.W. Clayton, and E. Bremer. (2001). Response of pea to rhozobia nitrogen and starter nitrogen in Alberta. *Canadian Journal of Soil Science.* 81: 637-643.

McKenzie, R.H., J.W.B. Stewart, J.F. Dormaar, and G.B. Schaalje. (1992a). Long-term crop rotation and fertilizer effects on phosphorus transformations: I. In a Chernozemic soil. *Canadian Journal of Soil Science.* 72: 569-579.

McKenzie, R.H., J.W.B. Stewart, J.F. Dormaar, and G.B. Schaalje. (1992b). Long-term crop rotation and fertilizer effects on phosphorus transformations: I. In a Luvisolic soil. *Canadian Journal of Soil Science.* 72: 581-589.

Meek, B., L. Graham, and T. Donovan. (1982). Long-term effects of manure on soil nitrogen, phosphorus, potassium, sodium, organic matter and water infiltration rate. *Soil Science Society of American Journal.* 46: 1014-1019.

Michaud, R., G. Belanger, A. Bregard, and J. Surrenant. (1998). Selection for nitrogen efficiency and N concentration in timothy. *Canadian Journal of Plant Science.* 78: 611-613.

Miller, P.R., Y. Gan, B.G. McConkey, and C.L. McDonald. (2003a). Pulse crops for the northern Great Plains: I. Grain productivity and residual effects on soil water and nitrogen. *Agronomy Journal.* 95: 972-979.

Miller, P.R., Y. Gan, B.G. McConkey, and C.L. McDonald. (2003b). Pulse crops for the northern Great Plains: II. Cropping sequence effects on cereal, oilseed and pulse crops. *Agronomy Journal.* 95: 980-986.

Mohan, J.D. and J.J. Child. (1979). Growth response of inoculated peas (*Pisum sativum* L.) to combine nitrogen. *Canadian Journal of Botany.* 57: 1687-1693.

Mohr, R.M., M.H. Entz, H.H. Janzen, and W.J. Bullied. (1999). Plant-available N supply as affected by method and timing of alfalfa termination. *Agronomy Journal.* 91: 622-630.

Mooleki, S.P., J.J. Schoenau, G. Hultgreen, G. Wen, and J.L. Charles. (2001). Crop response to liquid swine effluent and solid cattle manure over four years in east-central Saskatchewan. *Proceedings of Soils and Crops Workshop.* pp. 291-297. February 2001, University of Saskatchewan. Saskatoon. Saskatoon, Canada.

Morrison, I.N. and D. Kraft. (1994). *Sustainability of Canada's Agri-Food System: A Prairie Perspective.* International Institute for Sustainable Development: Manitoba, Canada.

Moyer, J.R., R.E. Blackshaw, E.G. Smith, and S.M. McGinn. (2000). Cereal cover crops for weed suppression in a summer fallow-wheat sequence. *Canadian Journal of Plant Science.* 80: 441-449.

Muldoon, J.F., D.J. Hume, and W.D. Beversdorf. (1980). Effects of seed- and soil applied *Rhizobium japonicum* inoculants on soybean in Ontario. *Canadian Journal of Plant Science.* 60: 399-410.

N'dayegamiye, A. and D.A. Angers. (1990). Effets de l'apport prolongé de fumier de bovins sur quelques propriétés physiques et biologiques d'un loam limoneux Neubois sous culture de maïs. *Canadian Journal of Soil Science.* 70: 259-262.

Neilsen, D. and G.H. Neilsen. (2002). Efficient use of nitrogen and water in high density apple orchards. *Hort Technology.* 12: 19-25.

Neilsen, D., G.H. Parchomchuk, G.H., Neilsen, and E.J. Hogue. (1998). Using soil solution monitoring to determine the effects of irrigation management and fertigation on nitrogen availability in high-density apple orchards. *Journal of the American Society of Horticultural Science.* 123: 706-713.

Nuttall, W.F., D.A. Cooke, J. Waddington, and J.A. Robertson. (1980). Effects of nitrogen and phosphorus fertilizers on a bromegrass-alfalfa mixture grown under two systems of pasture management. I. Yield, percentage legume in sward, and soil tests. *Agronomy Journal.* 72: 289-294.

Nuttall, W.F., D.H. McCartney, S. Bittman, P.R. Horton, and J. Waddington. (1991). The effect of N, P, S fertilizer, temperature and precipitation on the yield of bromegrass and alfalfa pasture established on a Luvisolic soil. *Canadian Journal of Plant Science.* 71: 1047-1055.

Nyborg, M. and A.M.F. Hennig. (1969). Field experiments with different placements of fertilizers for barley, flax and rapeseed. *Canadian Journal of Soil Science.* 49: 79-88.

Nyborg, M. and S.S. Malhi. (1989). Effect of zero and conventional tillage on barley yield and nitrate nitrogen content, moisture and temperature of soil in north-central Alberta. *Soil and Tillage Research.* 15: 1-9.

Nyborg, M., S.S. Malhi, E.D. Solberg, and R.C. Izaurralde. (1999). Carbon storage and light fraction C in a grassland dark Gray chernozem soil as influenced by N and S fertilization. *Canadian Journal of Plant Science.* 79: 317-320.

Nyborg, M., E.D. Solberg, R.C. Izaurralde, S.S. Malhi, and M. Molina-Ayala. (1995). Influence of long-term tillage, straw and N fertilizer on barley yield, plant-N uptake and soil-N balance. *Soil and Tillage Research.* 36: 165-174.

Nyborg, M., E.D. Solberg, S.S. Malhi, and R.C. Izaurralde. (1995). Fertilizer N, crop residue, and tillage alter soil C and N content in a decade. In R. Lal, J.M. Kimble, E. Levine, and B.A. Stewart (eds.). *Soil Management and Greenhouse Effect.* pp. 93-99. Boca Raton, FL: Lewis Publication.

O'Hallaran, I.P. (1993). Effect of tillage and fertilization on inorganic and organic soil phosphorus. *Canadian Journal of Soil Science.* 73:359-369.

Olson, B.M., E.R. Bennett, R.H. McKenzie, T. Ormann, and R.P. Atkins. (1998). *Manure Nutrient Management to Sustain Ground Water Quality Near Feedlots.* Alberta, Canada: CAESA, Lethbridge.

Ontario Ministry of Food and Rural Affairs. (2004). http://www.omafra.gov.on.ca/english/stats/crops/estimate_metric.html, accessed September 13, 2005.

Padbury, G., S. Waltman, J. Caprio, G. Coen, S. McGinn, D. Mortensen, G. Nielsen, and R. Sinclair. (2002). Agroecosystems and Land Resources of the Northern Great Plains. *Agronomy Journal.* 94: 251-261.

Paul, J.W. and B.J. Zebarth. (1997). Denitrification and nitrate leaching during the fall and winter following dairy cattle slurry application. *Canadian Journal of Plant Science.* 77: 231-240.

Peterson, G. (1996). N: The vital nutrient in the Great Plains. *Fluid Journal.* 4: 19-21.

PPI. (2002). Plant nutrient use in North American agriculture. *Technical Bulletin* No. 2002-1. Potash and Phosphate Institute, Norcross, GA.

Qian, P. and J.J. Schoenau. (2000a). Effect of swine manure and urea on soil phosphorus supply to canola. *Journal of Plant Nutrition.* 23: 381-390.

Qian, P. and J.J. Schoenau. (2000b). Use of ion exchange membrane to assess soil N supply to canola as affected by addition of liquid swine manure and urea. *Canadian Journal of Soil Science.* 80: 213-218.

Qian, P., J.J. Schoenau, T. Wu, and S.P. Mooleki. (2003). Copper and zinc amounts and distribution in soil as influenced by application of animal manures in east-central Saskatchewan. *Canadian Journal of Soil Science.* 83(2): 197-202.

Raun, W.R., J.B. Solie, G.V. Johnson, M.L. Stone, R.W. Mullen, K.W. Freeman, W.E. Thomason, and E.V. Lukima. (2002). Improving nitrogen use efficiency in cereal grain production with optical sensing and variable rate application. *Agronomy Journal.* 94: 815-820.

Reynolds, W.D., X.M. Yang, C.F. Drury, T.Q. Zhang, and C.S. Tan. (2003). Effects of selected conditioners and tillage on the physical quality of a clay loam soil. *Canadian Journal of Soil Science.* 83: 381-393.

Rice, W.A. and V.O. Biederbeck. (1983). The role of legumes in the maintenance of soil fertility. *Proceedings of Alberta Soil Science Workshop.* pp. 35-42. February 22-23, 1983, Edmonton, Alberta, Canada.

Rice, W.A., G.W. Clayton, P. Olsen, and N.J. Lupwayi. (2000). Rhizobial inoculant formulations and soil pH influence field pea nodulation and nitrogen fixation. *Canadian Journal of Soil Science.* 80: 395-400.

Rice, W.A. and P. Olsen. (1992). Effects of inoculation method and size of Rhizobium meliloti population on nodulation of alfalfa. *Canadian Journal of Soil Science.* 72: 57-68.

Rodd, A.V., J. MacLeod, P. Warman, and R. Gordon. (2002). Effect of timing and rate of surface applied manure addition on the yield of grassland agri-ecosystems. Poster: *Atlantic Agricultural Science and Technology Workshop.* November 14-15, 2002. Truro, Nova Scotia, Canada.

SAF, Saskatchewan Agriculture and Food. (1999). Nutrient values of manure. *Farm Facts,* ISSN 0840-9447.

Sanderson, J.B. (2003). *Efficient Phosphorus Management for Potatoes.* QC-05. Charlottetown, Prince Edward Island, Canada: Agriculture and Agri-Food Canada.

Sanderson, J.B. and M.R. Carter. (2002). Effect of gypsum and elemental sulfur on calcium and sulfur content of rutabagas in Podzolic soils. *Canadian Journal of Plant Science.* 82: 785-787.

Sanderson, J.B., J.A. MacLeod, and J. Kimpinski. (1999). Glyphosate application and timing of tillage of red clover affects potato response to N, soil N profile, and root and soil nematodes. *Canadian Journal of Soil Science.* 79: 65-71.

Schoenau, J.J. (1997). Soil fertility benefits from swine manure addition. *Proceedings of Saskatchewan Pork Industry Symposium.* pp. 59-63. Saskatoon, Saskatchewan, Canada.

Schoenau, J.J., K. Bolton, and K. Panchuk. (2000). Managing manure as a fertilizer. *Farm Facts.* Saskatchewan Agriculture and Food, ISSN 0840-9447.

Schoenau, J.J. and C.A. Campbell. (1996). Impact of crop residues on nutrient availability in conservation tillage systems. *Canadian Journal of Plant Science.* 76: 621-626.

Schoenau, J.J., J. Charles, G. Wen, P. Qian, and G. Hultgreen. (1999). Effect of hog and cattle manure additions on soil fertility and crop growth in east-central Saskatchewan. *Proceedings of a Tri-Provincial Conference on Manure Management.* pp. 66-85. Manure Management 1999. Saskatoon, Saskatchewan, Canada.

Schoenau, J.J., S.P. Mooleki, S.S. Malhi, and G. Hultgreen. (2004). Strategies for maximizing crop recovery of nutrients applied as liquid swine manure. *Proceedings of Great Plains Soil Fertility Conference.* pp. 8-13. March 2004, Denver, CO, USA.

Selles, F., C.A. Campbell, and R.P. Zentner. (1995). Effect of cropping and fertilization on plant and soil phosphorus. *Soil Science Society of America Journal.* 59: 140-144.

Simard, R.R., D. Cluis, G. Gangbazo, and S. Beauchemin. (1995). Phosphorus status of forest and agricultural soils from a watershed of high animal density. *Journal of Environmental Quality.* 24: 1010-1017.

Singh, B. and S.S. Malhi. (2006). Response of soil physical properties to tillage and straw management on two contrasting soils in a cryoboreal environment. *Soil and Tillage Research.* 85: 143-153.

Singh, P.K. (1981). Use of *Azolla* and blue-green algae in rice cultivation in India. In P.V. Bose and A.P. Ruschel (eds.). *Associative N_2-Fixation.* pp. 236-242. Boca Raton, FL: CRC. Press.

Slinkard, A.E., V.O. Biederbeck, L.D. Bailey, P. Olson, W.A. Rice, and L. Townley-Smith. (1987). Annual legumes as a fallow substitute in northern Great Plains of Canada. In J.F. Power (eds.). *The Role of Legumes in Conservation Tillage Systems.* pp. 6-7. Ankley, IA: Soil Conservation Society of America .

Small Woodlands Program of B.C. (2004). Available at http://www.woodlot.bc.ca/swp/html/agro/agro_guide,accessed June 15, 2004.

Smith, E.G., H.H. Janzen, T. Entz, and J.M. Carefoot. (2004). Productivity and profitability of straw-tillage and nitrogen treatments on irrigation in southern Alberta. *Canadian Journal of Plant Science.* 84: 411-418.

Solberg, E.D., M. Nyborg, R.C. Izaurralde, S.S. Malhi, H.H. Janzen, and M. Molina-Ayala. (1998). Carbon storage in soils under continuous cereal grain cropping: N fertilizer and straw. In R. Lal, J.M. Kimble, R.F. Follett, and B.A. Stewart (eds.). *Management of Carbon Sequestration in Soil.* pp. 235-254. Boca Raton, FL: CRC Press.

Sommerfeldt, T.G., C. Chang, and T. Entz. (1988). Long-term annual manure applications increase soil organic matter and nitrogen and decrease carbon to nitrogen ratio. *Soil Science Society of American Journal.* 52: 1667-1672.

Soper, R.J. and M.R. Grenier. (1987). Fertility value of annual legume in a crop rotation. *Manitoba Agronomy Forum-Winnipeg.* pp. 7-12. Winnipeg, Manitoba, Canada: Manitoba Institute of Agrologists.

Spratt, E.D. and R.N. McIver. (1979). The effect of continued use of P fertilizer or barnyard manure on yield of wheat and the fertility status of a clay Chernozem soil. *Canadian Journal of Soil Science.* 59: 451-454.

Stevenson, F.C., A.M. Johnston, H.J. Beckie, S.A. Brandt, and L. Townley-Smith. (1998). Cattle manure as a nutrient source for barley and oilseed crops in zero and conventional tillage systems. *Canadian Journal of Plant Science.* 78: 409-416.

Stevenson, F.C. and C. van Kessel. (1997). Nitrogen contribution of pea residue in a hummocky terrain. *Soil Science Society of American Journal.* 61: 494-503

Strong, W.M. and R.J. Soper. (1974a). Phosphorus utilization by flax, wheat, rape, and buckwheat from a band or pellet-like application. 1. Reaction zone root proliferation. *Agronomy Journal.* 66: 597-601.

Strong, W.M. and R.J. Soper. (1974b). Phosphorus utilization by flax, wheat, rape, and buckwheat from a band or pellet-like application. 2. Influence of reaction zone phosphorus concentration and soil phosphorus supply. *Agronomy Journal.* 66: 601-605.

Tan, C.S., C.F. Drury, J.D. Gaynor, and T.W. Welacky. (1993). Integrated soil, crop and water management system to abate herbicide and nitrate contamination of the Great Lakes. *Water Science and Technology.* 28: 497-507.

Tan, C.S., C.F. Drury, W.D. Reynolds, J.D. Gaynor, T.Q. Zhang, and H.Y. Ng. (2002). Effect of long-term conventional tillage and no-tillage systems on soil and water quality at the field scale. *Water Science and Technology.* 46: 183-190.

Thiessen, H., J.R. Martens, and M.H. Entz. (2001). Availability of late season heat and water resources for relay and double cropping with winter wheat in prairie Canada. *Canadian Journal of Plant Science.* 81: 273-276.

Thiessen, H., J.R. Martens, J.W. Hoeppner, and M.H. Entz. (2001). Legume cover crops with winter cereals in southern Manitoba: Establishment, productivity, and microclimate effects. *Agronomy Journal.* 93: 1086-1096.

Tomm, G.O., F.L. Walley, F.L., C. van Kessel, and A. Slinkard. (1995). Nitrogen cycling in an alfalfa and bromegrass sward via litterfall and harvest losses. *Agronomy Journal.* 87: 1078-1085.

Toogood, J.A. and D.L. Lynch. (1959). Effect of cropping systems and fertilizers on mean-weight diameter of aggregates of Breton plot soils. *Canadian Journal of Soil Science.* 39: 151-156.

Townley-Smith, L., A.E. Slinkard, L.D. Bailey, V.O. Biederbeck, and W.A. Rice. (1993). Productivity, water use and nitrogen fixation of annual-legumes green-manure crops in the Dark Brown soil zone of Saskatchewan. *Canadian Journal of Plant Science.* 73: 139-148.

Tran, T.S., D. Côté, and A. N'dayegamiye. (1996). Effets des apports prolongés de fumier et de lisier sur l'évolution des teneurs du sol en éléments nutritifs majeurs et mineurs. *Agrosol.* 9: 21-30.

Tran, T.S. and A. N'dayegamiye. (1995). Long-term effects of fertilizers and manure application on the forms and availability of soil phosphorus. *Canadian Journal of Soil Science.* 75: 281-285.

Van-den Driessche, R. (1999). First-year growth response of four Populus tricho-carpa × Populus deltoides clones to fertilizer placement and level. *Canadian Journal of Forest Research.* 29: 554-562.

Vessey, J.K. and K.G. Heisinger. (2001). Effect of *Penicillium bilaii* inoculation and phosphorus fertilization on root and shoot parameters of field-grown pea. *Canadian Journal of Plant Science.* 81: 361-366.

Vigil, M.F. and D.C. Nielsen. (1998). Winter wheat yield depression from legume green fallow. *Agronomy Journal.* 90: 727-734.

Vyn, T.J., J.G. Faber, K.G. Janovicek, and E.G. Beauchamp. (2000). Cover crop effects on nitrogen availability to corn following wheat. *Agronomy Journal.* 92: 915-924.

Wallace, J. (2002). Uses of Cover Crops. NSOGA, available at http://www.gks .com/nccrp/usesofcc.phps, accessed June 15, 2004.

Walley, F.L., G.O. Tomm, A. Matus, A.E. Slinkard, and C. van Kessel. (1996). Allocation and cycling of nitrogen in an alfalfa-bromegrass sward. *Agronomy Journal.* 88: 834-843.

Wani, S.P., W.B. McGill, K.L. Haugen-Kozyra, and N.G. Juma. (1994). Increased proportion of active soil N in Breton loam under cropping systems with forages and green manures. *Canadian Journal of Soil Science.* 74: 67-74.

Wani, S.P., W.B. McGill, K.L. Haugen-Kozyra, J.A. Robertson, and J.J. Thurston. (1994). Improved soil quality and barley yields with fababeans, manure, forages and crop rotation on a Gray Luvisol. *Canadian Journal of Soil Science.* 74: 75-84.

Wani, S.P., W.B. McGill, and J.A. Robertson. (1991). Soil N dynamics and N yield of barley grown on Breton loam soil using N from biological fixation or fertil-izer. *Biology and Fertility of Soils.* 12: 10-18.

Warman, P.R. (1986). The effect of fertilizer, chicken manure and dairy manure on timothy yield, tissue composition and soil fertility. *Agriculture Wastes.* 18: 289-298.

Warman, P.R. (2003). Using municipal compost on blueberry and raspberry crops, available at http://www.nsac.ns.ca/rgs/research/newsletter winter 2003.pdf, accessed June 15, 2004.

Warman, P.R. and J.M. Cooper. (1999). Fertilization of a mixed forage crop with fresh and composted chicken manure and NPK fertilizer: Effects on soil and tissue Ca, Mg, S, B, Cu, Fe, Mn and Zn. *Canadian Journal of Soil Science.* 80: 345-351.

Webster, G.R., D.K. Mcbeath, L.A. Heapy, H.C. Lore, U.M. Von Maydell, and J.A. Robertson. (1976). *Influence of Fertilizers, Soil Nutrients and Weather on Forage Yield and Quality in Central Alberta.* Edmonton, Alberta, Canada Alberta Institute of Pedology Publication Number M-76-10.

Weill, A.N., G.R. Mehuys, and E. McKyes. (1990). Effect of tillage reduction and fertilizer type on soil properties during corn (*Zea mays* L.) production. *Soil and Tillage Research.* 17:63-76.

Weinhold, B.J. and A.D. Halvorson. (1999). Nitrogen mineralization responses to cropping, tillage, and nitrogen rate in the northern Great Plains. *Soil Science Society of America Journal.* 63: 192-196.

Williams, C.H. and C.H. David. (1973). The effect of superphosphate on the Cd content of soils and plants. *Australian Journal of Soil Research.* 11: 43-56.

Yanan, T., O. Emteryd, L. Dianqing, and H. Grip. (1997). Effect of organic manure and chemical fertilizer on N uptake and nitrate leaching in a Eum-orthic Anthrosol profile. *Nutrient Cycling Agroecosystems.* 48: 225-229.

Yang, X.M., C.F. Drury, W.D. Reynolds, and C.S. Tan. (2004). Nitrogen mineralization and uptake by ryegrass in a clay loam soil amended with composts or liquid pig manure. *Canadian Journal of Plant Science.* 84: 11-17.

Yang, X.M. C.F. Drury, W.D. Reynolds, C.S. Tan, and D.J. McKenney. (2003). Interactive effects of composts and liquid pig manure with added nitrate on soil carbon dioxide and nitrous oxide emissions from soil under aerobic and anaerobic conditions. *Canadian Journal of Soil Science.* 83: 343-352.

Zebarth, B.J., B. Hii, H. Liebscher, K. Chipperfield, J.W. Paul, G. Grove, and S.Y. Szeto. (1998). Agricultural land use practices and nitrate contamination in the Abbotsford aquifer, British Columbia, Canada. A*griculture, Ecosystems and Environment.* 69: 99-112.

Zebarth, B.J., J.W. Paul, O. Schmidt, and R. McDougall. (1996). Influence of the time and rate of liquid manure application on yield and nitrogen utilization of silage corn in south coastal British Columbia. *Canadian Journal of Plant Science.* 76: 153-164.

Zebarth, B.J., J.W. Paul, and R. Van Kleek. (1999). The effect of nitrogen management in agricultural production on water and air quality: Evaluation on a regional scale. *Agriculture, Ecosystems and Environment.* 72: 35-52.

Zentner, R.P. and C.A. Campbell. (1988). First 18 rears of a long-term crop rotation study in southwestern Saskatchewan—Yield, grain protein, and economic performance. *Canadian Journal of Plant Science.* 68: 1-21.

Zentner, R.P., C.A. Campbell, V.O. Biederbeck, P.R. Miller, F. Selles, and M.R. Fernandez. (2001). In search of a sustainable cropping system for the semiarid Canadian prairies. *Journal of Sustainable Agriculture.* 18: 117-136.

Zentner, R.P., C.A. Campbell, V.O. Biederbeck, and F. Selles. (1996). Indianhead black lentil as green manure crop for wheat rotation in the Brown soil zone. *Canadian Journal of Plant Science.* 76: 417-422.

Zentner, R.P., C.A. Campbell, V.O. Biederbeck, F. Selles, R. Lemke, P.G. Jefferson, and Y. Gan. (2004). Long-term assessment of management of an annual legume green manure crop for fallow replacement in the Brown soil zone. *Canadian Journal of Plant Science.* 84: 11-22.

Zentner, R.P., C.A. Campbell, H.H. Janzen, and K.E. Bowren. (1990). *Benefits of Crop Rotation for Sustainable Agriculture in Dryland Farming.* Publication 1839/E. Edmonton, Alberta, Canada: Communications Branch, Agriculture Canada.

Zentner, R.P., C.A. Campbell, F. Selles, B.G. McConkey, P.G. Jefferson, and R. Lemke. (2003). Cropping frequency, wheat classes and flexible rotations: Effects on production, nitrogen economy, and water use in a Brown Chernozem. *Canadian Journal of Plant Science.* 83: 667-680.

Zentner, R.P., D.D. Wall, C.N. Nagy, E.G. Smith, D.L. Young, P.R. Miller, C.A. Campbell, et al. (2002). Economics of crop diversification and soil tillage opportunities in the Canadian prairies. *Agronomy Journal.* 94: 216-230.

Zhang, T.Q., A.F. MacKenzie, B.C. Liang, and C.F. Drury. (2004). Soil test phosphorus fractions with long-term phosphorus addition and depletion. *Soil Science Society of America Journal.* 68: 519-528.

Zheng, Z., J.A. MacLeod, J. Lafond, J.B. Sanderson, and A.J. Campbell. (2003). Phosphorus status of a Humic Gleysol after 10 years of cultivation under contrasting cropping practices. *Canadian Journal of Soil Science.* 83: 537-545.

Zheng, Z., R.R. Simard, J. Lafond, and L.E. Parent. (2001). Changes in phosphorus fractions of a Humic Gleysol as influenced by cropping systems and nutrient sources. *Canadian Journal of Soil Science.* 81: 175-183.

Zhou, X., A.F. MacKenzie, C.A. Madramootoo, J.W. Kaluli, and D.L. Smith. (1997). Management practices to conserve soil nitrate in maize production systems. *Journal of Environmental Quality.* 26: 1369-1374.

Chapter 5

Integrated Nutrient Management: The European Experience

Paolo Sequi
A. E. Johnny Johnston
Rosa Francaviglia
Roberta Farina

INTRODUCTION

Integrated nutrient management (INM) when applied to plant nutrition implies that farmers should consider each year, the total amount of nutrients that are required for a crop to attain its economic optimum yield on each field of the farm. They should then consider the sources from which these nutrients could come, including the supply from the soil, and the need to supplement this supply with fertilizers and manures. The two most important reasons for encouraging the adoption of INM are to use all nutrient inputs efficiently, which should increase the profitability of a farm, and minimize losses of nutrients, especially nitrogen (N) and phosphorus (P), from the plant-soil system, which has environmental benefits. Theoretically such an approach is applicable to all farms, but the adoption of INM is likely to vary from region to region depending on a number of factors. These factors include the intensity of crop production, the willingness of farmers to consider using INM, and whether there is any external pressure for them to do so.

This chapter discusses agricultural policy from a Pan-European perspective in relation to developing INM practices for farmers to optimize productivity and economic viability, while at the same time minimizing any adverse impact of agriculture on the environment.

BACKGROUND

Among the nonfarming community in some parts of Europe, protection of the environment receives greater attention than does sustainable crop production. In part, this is because media coverage emphasizes what it perceives to be serious adverse effects of the intensification of farming on things such as the nitrate content of water, eutrophication of surface water bodies, and loss of amenity through changes in the landscape such as the enlargement of fields and loss of hedgerow habitats for wildlife. While in these circumstances pressure for change has come from urban communities whose concern is with the environment, farmers themselves need to be aware that decreasing nutrient inputs can cause a loss of yield. We see the uptake of INM in those parts of Europe where intensive agriculture is practiced as a way of maintaining crop yields while minimizing the loss of nutrients from the soil-plant system. Thus we discuss briefly nutrient losses in relation to soil and climate. The uptake of INM will probably vary within Europe partly because of external pressure, partly because agriculture is extremely diverse, ranging from large, highly intensive, high input specialized farms, mainly in northern Europe, to subsistence farming systems relying on little or no external inputs (traditional systems) mainly in the much drier southern Europe. Throughout Europe, however, maintaining soil fertility and achieving sustainable crop production is a common goal, but the scale of the inputs and therefore possible environmental impact will vary greatly.

Within Europe there are groupings of countries with somewhat different philosophies toward farming and environmental issues.

1. The original European Union of fifteen countries (EU-15) includes Austria, Belgium, Luxembourg, Denmark, Finland, France, Germany, Greece, Ireland, Italy, the Netherlands, Portugal, Spain, Sweden, and the United Kingdom. In these countries the EU Common Agricultural Policy (CAP) has been one of the most important driving forces for agricultural intensification and specialization. Based on the premise that food security and maintenance of a viable rural economy were paramount, the provisions of the CAP that have encouraged increased production have also been blamed for the adverse impacts of those production systems on the environment. Market pressures and technological developments that have contributed to the greatly increased productivity have rarely encouraged recycling, for example, of pig and poultry wastes, unless given support from public funds. However, public concern related to production methods and environmental impact is bringing some pressure on farmers for change, and some reorientation of the CAP could create opportunities for them to do so through agri-

environment schemes and changes in farming practice, for example adoption of organic farming. The full implication of public supported agri-environment schemes within the EU-15 has yet to be realized, and this is discussed later.

2. The European Free Trade Area (EFTA) includes the EU-15, together with Iceland, Norway, and Switzerland. Policies developed in these countries, such as Norway and Switzerland have been strongly influenced by national concern to ensure agricultural self-efficiency and the protection of the rural environment. These countries often provide a higher level of public support to agriculture than do many other countries in the EU.

3. The European Accession Countries (EAC-10) comprise the Czech Republic, Cyprus, Estonia, Hungary, Latvia, Lithuania, Malta, Poland, Slovakia, and Slovenia (members since May 2004). Candidate countries to the EU are Bulgaria, Romania (members since September 2007), and Turkey.

4. The Russian Federation and the Western EECCA (Eastern Europe, Caucasus, and Central Asia) include Belarus, Moldova, Russian Federation, and Ukraine. The Balkan Countries are Albania, Bosnia-Herzegovina, Croatia, Macedonia, Serbia, and Montenegro.

In most of the countries that comprised the Central and Eastern European (CEE) group, agricultural policies were developed historically within the framework of central planning, with an emphasis on collective farming and its growth as a public industry, rather than a collection of private farms as in the EU. Significant environmental damage has often been associated with agriculture in these CEE countries where the exploitation of natural and agricultural resources has been excessive in the past. The recent decline in resource use in these countries has been largely due to economic restructuring and constraints, rather than policy measures, consumer needs, technological development, or response to environmental pressures. Moreover, the abandonment of land, too little grazing of pastures, and lack of capital to maintain or improve farm infrastructure are creating additional environmental pressures.

As some of the EAC-10 countries become EU members, agriculture is likely to intensify when they have full access to the CAP. However, there is a special accession program for agriculture that will be applied to new Member States, the aim being to reduce the incentives for increasing production, and therefore any associated environmental risk. There will be a need for agriculture in these countries to embrace INM as their farm economies improve and they once again start to use fertilizers as an input to increase productivity.

MAJOR SOIL CLASSES, CLIMATIC ZONES, AND AGRICULTURAL PRODUCTION SYSTEMS

Soil Types

Farming systems depend on climate and soil type. The major soil types in Europe range in texture from sandy to clayey, from well drained to imperfectly drained, from acid to calcareous, and include semiarid and salt affected soils. Some information about these soil types is given in Appendix 1. Arable agriculture is usually found on the deeper well-drained soils, animal farming on wetter, heavier-textured soils not usually suitable for arable cropping. Nutrient losses by leaching will be greater on free-draining soils while losses of N by denitrification will be larger where soil conditions can become anaerobic. Land use, soil management, and degradation hazards differ considerably within the European regions because, even within a small area, there may be a wide variety of soils depending on the local climate, geology, relief, altitude, exposure, slope, vegetation, hydrology, and human influence. A qualitative appraisal of the vulnerability of the main soil types to threats such as erosion, acidification, pollution, and compaction is given in Table 5.1.

Climatic Zones

Differences in climate both within and between northern and southern Europe are dramatic. The climate varies from subtropical in the very far south to polar in the far north. The Mediterranean climate is dry and warm, while

TABLE 5.1. Vulnerability of European soil types to degradation

Soil type	Erosion	Compaction	Acidification	Pollution
Black earths	Low			
Sands	Moderate	High	High	High
Acid loams	High	High	High	High
Nonacid loams	High	High	Moderate	Moderate
Clays	Low	Moderate	Low	Low
Acid shallow	High		High	High
Nonacid shallow	High			Low
Semiarid	High			
Salt affected	Moderate			
Imperfectly drained		Moderate	Low	High

Source: Adapted from Fraters (1994).

the western and northwestern regions have a mild, generally humid climate, influenced by the North Atlantic Drift. In central and eastern Europe, the climate is of the humid continental-type with cool summers. In the far northeast, the climate is subarctic and tundra. All of Europe is subject to the moderating influence of prevailing westerly winds from the Atlantic Ocean, and consequently, similar climatic conditions are found at higher latitudes in Europe than in other continents.

Possible nutrient losses in drainage depend on the balance between rainfall and evapotranspiration (ETP). This balance differs greatly at different latitudes in Europe (Figure 5.1). In humid environments, such as Aberdeen (United Kingdom) and Eelde (The Netherlands), the difference between rainfall and evaporation is close to the amount of drainage water, so there is a risk of transfer of dissolved and particulate material to aquifers. Thus the application of fertilizers and manures must be regulated, as much as possible, to just satisfy plant nutrient requirements. In Mediterranean environments, or under more temperate and less humid climates, for example in parts of northern Italy, more intense ETP limits water moving to groundwater to a short period of time, generally at the end of the cold season. But in the semiarid region of southern Italy and in warmer climates there is insufficient rainfall to cause drainage. In these latter regions, real water tables do not exist and cannot therefore be polluted by drainage from agricultural soils. In Sicily, for example, there are dry areas with an arid climate, yet salinity in some rivers is greater than that in the sea despite the fact that chemical inputs to agriculture in Sicily are very small. In northern Europe on the other hand, excess rainfall leads to through drainage in most years, and the enrichment of this drainage water with nitrate leads to environmental problems. In both northern and southern Europe, intense and prolonged rainfall can lead to surface water runoff that can take entrapped soil or organic manure particles with it to rivers. This can result in the adverse effects of eutrophication, but the extent of eutrophication will depend on the fertility of the soil, or the amount of organic manure recently applied to the land. Despite such differences, the CAP appears to favor a common approach to solving all such pollution problems in Europe. We consider that INM might be a better approach. One approach that is common to both is the use of nutrient management plans.

In the drier areas, growing arable crops is impossible without irrigation, and such a situation also influences environmental issues. For these drier areas, codes of good agricultural practice must (1) relate the irrigation water supply with soil water retention capacity and crop demand for water, and (2) set the maximum amount of nutrients that can be applied to soil as fertilizers to avoid plant stress caused by too little nutrients or a surplus of

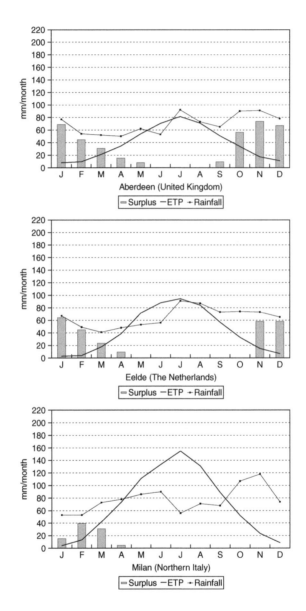

FIGURE 5.1. Some typical trends of precipitation and evapotranspiration through-out the year. *Source:* Sequi (2003).

FIGURE 5.1 *(continued)*

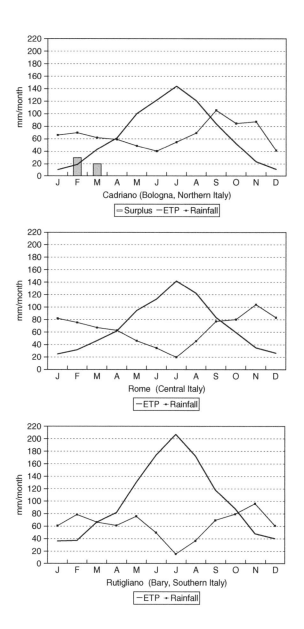

nutrients causing salinity. Although this stress may seem rather unlikely, in these conditions water usually moves upward, and therefore salts tend to accumulate in the upper part of the profile where there are more roots.

Agricultural Production Systems

There have been many changes in the structure of agricultural production systems in Europe since World War II. These include developments in international trade and agricultural policies, technological developments, and greater availability and affordability of many of the inputs necessary to produce crops. Each of these has influenced yield, labor productivity and efficiency, and has led to some significant changes in the underlying structure of European agriculture.

Agricultural land accounts for about 45 percent of the total land area in the EU-15, although the proportion varies between 4 and 8 percent in Finland and Sweden, and up to 60 percent or more in Denmark, Greece, Ireland, Spain, and the United Kingdom. The share of agricultural land is more than 55 percent in some accession countries, for example, Czech Republic, Hungary, and Poland (OECD, 2001).

The total number of farms has declined all over Europe, but perhaps unexpectedly, the proportion of small farms to the total number of farms is increasing.

In the eastern and southern regions for which there are data, there are still a large number of holdings of fewer than 10 hectares (ha), while in most countries in the western region the average size of holdings is larger, with fewer farms. This second group includes countries such as Denmark, the United Kingdom, The Netherlands, Belgium, and France. Here the trend over time has been an increase in the average size of holdings while the number of farms has been decreasing. This suggests that it is the medium-sized farms that have been amalgamated into larger farms with a corresponding increase in the size of larger farms. Such a change is driven by increases in labor costs. This change in farm size has been accompanied by an increasing specialization in arable or livestock husbandry and with this intensification in land use there has been an increased use of inputs of fertilizers, pesticides, feedstuffs, and energy. Where these inputs are not being used efficiently there are increased risks of adverse effects on the environment.

In the eastern region, there has been a slower increase in the average size of holdings because the structure of the agricultural sector was greatly influenced by the policies of the former centrally planned economies. A major exception was Poland, with 72 percent of farms privately owned, and with the farms tending to be small because of the arrangements for land tenure and state aid. These small farms were maintained under communism rather than being replaced by the large industrial-type farms common to a number of other countries with centrally planned economies.

Five major agricultural regions can be discerned in Europe (Kostrowicki, 1991). These regions are determined by both environmental and socioeconomic factors. Zone 1 is the northern region, and includes Denmark, Finland, Great Britain, Iceland, Ireland, Norway, and Sweden, where agriculture is mainly market oriented, with extensive cattle raising. Zone 2 includes the Atlantic and central continental regions, Austria, Belgium, France, the western part of Germany, Luxembourg, Switzerland, and the northern part of Italy, characterized mainly by intensive and market-oriented agriculture. Zone 3, which includes the Mediterranean countries, Greece, southern France, peninsular Italy, Malta, Portugal, and Spain, has different patterns of agriculture, ranging from a market-oriented agriculture mainly with fruit trees, olive and grapes to considerable areas of traditional low-input production systems. Zone 4 is the eastern region, which includes Albania, Bosnia-Herzegovina, Bulgaria, Croatia, Czech Republic, the eastern part of Germany, Hungary, Macedonia, Montenegro, Poland, Serbia, Slovakia, Slovenia, and Romania. Here traditional agriculture still predominates, but the proportion of marked-oriented and specialized agriculture has been increasing rapidly. Zone 5 is the European part of the former USSR, Belarus, Estonia, Latvia, Lithuania, Moldova, Russian Federation, and Ukraine, which was dominated by large-scale socialized agriculture, but which is now slowly adapting to a more quality-oriented agriculture.

Highly productive "industrialized" systems in western Europe fall into two broad categories. First, there are areas of intensive field-crop farming dominated by large holdings, for example in France (the Loire valley to Calais region), eastern England, eastern Denmark, northern Germany, and much of The Netherlands. Second, there are areas of very intensive animal production (dairy products and meat) and/or fruit and vegetable growing. Included in this area are the coastal and southern areas of The Netherlands, northern Belgium (Flanders), western Denmark, parts of Germany, Provence (France), the area along the Mediterranean coast of Spain, and central and northern Italy.

AGRICULTURAL PRODUCTION, NUTRIENT USE, AND ISSUES

Agricultural production, both crops grown and yields achieved per hectare, varies greatly throughout Europe not only due to soil and climate, but also for economic and political reasons.

Crop Yields

Mean yields of different crops grown in selected regions are given in Table 5.2. In western Europe, total agricultural production has increased significantly since the 1950s. In the EU-15 in particular, increases in production led to an oversupply of some products such as cereals, meat, and dairy products, although these surpluses are now declining.

Yields, measured in tonnes of harvested crop per hectare (t ha^{-1}) of agricultural land, have increased due to genetic improvements in the yield potential of a number of crops, larger nutrient inputs, and the ability to limit the adverse effects of weeds, pests, and diseases. For example, the total European production of cereals increased from 199 million tonnes (Mt) in 1970 to 283 Mt in 1990 (i.e., by 42 percent; world production increased by 62 percent over the same period). Against this, the area of cereals harvested in Europe declined from 72 million hectares (Mha) in 1970 to 66 Mha in 1990. This increase in production per unit land area is significant because it suggests that either nutrient inputs per unit area of land have increased, or the efficiency of nutrient use has increased. Increasing nutrient inputs highlights the need to get farmers involved with the principles and use of INM to use nutrients more efficiently, and improve the financial viability of their farms. Productivity per unit area of land can also be used to assess the quality of agro-ecosystems. If agricultural activity is modifying the ecosystem in such a way that yields cannot be maintained in the long-term, then the activity can be defined as "unsustainable."

Fertilizer Use

The consumption of mineral fertilizers in kg per hectare (kg ha^{-1}) of agricultural land in recent years is compared in Tables 5.3 through 5.6.

TABLE 5.2. Mean yield of different crop types (t ha^{-1}) in the European agricultural regions (1980-2000)

Regions	Cereals	Oil crops	Root tuber crops	Vegetables	Fruit crops
1	5.06	0.86	24.5	26.2	9.36
2	6.44	0.83	27.6	22.5	18.28
3	3.44	0.41	14.1	21.0	8.58
4	3.29	0.53	17.7	18.1	6.07
5	2.27	0.34	20.3	15.9	3.69

Source: Adapted from FAOSTAT, Statistic Division, Food and Agriculture Organization of the UN. Reprinted with permission.

Data availability and quality are not homogeneous in the different countries, due in part to the level of information that farmers are required to submit or are collected at national level.

Using the zones identified earlier in the section titled "Agricultural Production Systems," the following changes in total fertilizer consumption, expressed as $N + P_2O_5 + K_2O$ in kg ha^{-1} agricultural land, can be seen.

Western Europe. Total fertilizer consumption increased greatly during 1961-1971 followed by a smaller increase in 1971-1981. These increases were followed by a decline of about 17 percent, on average, in the period 1991-2001. While these trends are similar for all three fertilizers supplying N, P, and K, phosphate fertilizers showed the largest decrease (29 percent) in the decade 1991-2001. In 2001, the average total consumption was 111 kg ha^{-1}, with northern European countries having the largest consumption.

European Accession Countries. (EAC-10). In these countries, there has been a slight increase (13 percent) in fertilizer consumption in the period 1991-2001, with the exception of Estonia and Latvia. In both Hungary and Poland there was a sharp decrease in the decade 1981-1991. The largest consumption in 2001 was in Cyprus, Slovenia, Czech Republic, and Poland.

EU Candidate Countries. In these countries, fertilizer consumption decreased sharply both in 1981-1991 and 1991-2001. This decrease followed a peak in consumption in 1961-1971 and 1971-1981 in Bulgaria and Romania; the average total consumption was 30 kg ha^{-1} in 2001.

Russian Federation and the Western EECCA. Here there was a very sharp decrease (64 percent on average) in fertilizer consumption in the decade 1991-2001; in 2001 it was only 26 kg ha^{-1}, although in Belarus it was 84 kg ha^{-1}. In the Balkan countries, the largest decreases have been in Albania (41 percent) and Macedonia (46 percent). The average total consumption in 2001 was 35 kg ha^{-1} with a peak of 68 kg ha^{-1} in Croatia.

Nutrient Balances

The increase in agricultural production in many parts of Europe has been accompanied by a surplus of nutrients in the soil at the farm level in some countries (e.g., in northwest Europe), and these mainly originate from the disposal of animal wastes. Where excess nutrients have been lost to the freshwater and marine environments this has contributed to problems of eutrophication, and there have been increased losses of N to the air as ammonia and nitrous oxide. The presence of large nutrient balances indicates where the principles of INM should be applied to try to decrease the size of these positive nutrient balances.

TABLE 5.3. Total fertilizer consumption

Countries	Total N + P$_2$O$_5$ + K$_2$O (kg ha^{-1})					% change			
	1961	1971	1981	1991	2001	1961-1971	1971-1981	1981-1991	1991-2001
Austria	62	113	107	82	67	82	-5	-23	-18
Belgium-Luxembourg	206	282	270	240	187	37	-4	-11	-22
Denmark	136	214	213	210	118	58	-1	-2	-43
Finland	87	187	182	140	134	115	-3	-23	-4
France	70	152	176	183	141	117	16	4	-23
Germany	169	259	263	173	153	53	2	-34	-11
Greece	18	38	63	71	49	113	66	13	-31
Ireland	36	74	103	150	130	106	40	45	-13
Italy	42	82	117	124	109	95	43	6	-12
Netherlands	203	283	330	282	212	39	17	-15	-25
Portugal	36	34	69	66	55	-5	103	-5	-16
Spain	22	43	44	62	74	94	3	40	20
Sweden	71	140	132	87	91	96	-6	-34	5
United Kingdom	71	103	126	120	113	46	21	-5	-6
EU-15	**88**	**143**	**157**	**142**	**117**	**63**	**10**	**-9**	**-18**
Iceland	5	11	13	10	9	111	15	-24	-7
Norway	143	224	269	206	183	57	20	-24	-11
Switzerland	49	69	84	79	58	41	21	-6	-26
EFTA	**66**	**102**	**122**	**98**	**83**	**54**	**20**	**-20**	**-15**
Western Europe	**84**	**136**	**151**	**134**	**111**	**62**	**11**	**-11**	**-17**

Countries	Total N + P_2O_5 + K_2O (kg ha^{-1})					% change			
	1961	1971	1981	1991*	2001	1961-1971	1971-1981	1981-1991	1991-2001
Czech Republic				72	92				29
Cyprus	82	112	92	138	194	37	-18	50	41
Estonia				77	47				-39
Hungary	30	139	225	50	55	358	62	-78	9
Latvia				66	26				-61
Lithuania				44	46				5
Malta	24	41	26	54	70	71	-37	110	30
Poland	44	148	177	60	85	237	20	-66	41
Slovakia				39	49				24
Slovenia				112	142				26
European AC-10	**45**	**110**	**130**	**71**	**81**	**144**	**18**	**-45**	**13**
Bulgaria	25	106	169	77	25	318	60	-55	-67
Romania	23	42	108	31	22	88	155	-71	-30
Turkey	2	13	34	44	43	530	160	32	-3
Candidates	**17**	**54**	**104**	**51**	**30**	**222**	**93**	**-51**	**-41**

Countries	Total N + P_2O_5 + K_2O (kg ha^{-1})					
	1961	1971	1981	1991**	2001	% change 1991-2001
Belarus				149	84	-43
Moldova, Republic of				53	2	-96
Russian Federation				25	7	-70
Ukraine				64	11	-82

TABLE 5.3 *(continued)*

Countries	Total N + P$_2$O$_5$ + K$_2$O (kg ha^{-1})					% change 1991-2001
	1961	1971	1981	1991**	2001	
Russian Federation and the western EECCA				73	26	−64

Countries	Total N + P$_2$O$_5$ + K$_2$O (kg ha^{-1})					% change 1991-2001
	1961	1971	1981	1991***	2001	
Albania				28	16	−41
Bosnia-Herzegovina				8	18	120
Croatia				83	68	−18
Macedonia, FYR of				45	24	−46
Serbia and Montenegro				35	48	36
Balkan Countries				40	35	10

Source: FAOSTAT, Statistic Division, Food and Agriculture Organization of the UN. Reprinted with permission.

*1991 data for Cyprus, Hungary, Malta, Poland, Bulgaria, Romania, and Turkey; 1992 data for Estonia, Latvia, Lithuania, and Slovenia; 1993 data for Czech Republic and Slovakia.

**1992 data.

***1991 data for Albania; 1994 data for Croatia, and Macedonia; FYR of 1995 data for Bosnia-Herzegovina;1996 data for Serbia and Montenegro.

TABLE 5.4. Nitrogen fertilizers consumption

Countries	Total N (kg ha^{-1})					% change			
	1961	1971	1981	1991	2001	1961-1971	1971-1981	1981-1991	1991-2001
Austria	13	36	44	38	37	177	22	-14	-1
Belgium-Luxembourg	58	98	127	123	107	69	29	-3	-13
Denmark	42	104	130	134	77	147	24	3	-42
Finland	21	67	73	69	74	217	8	-6	8
France	18	47	69	84	81	158	48	22	-4
Germany	45	90	112	100	105	99	25	-11	5
Greece	9	22	41	45	31	140	80	10	-31
Ireland	5	17	48	78	80	239	177	63	2
Italy	17	35	56	56	54	111	59	0	-5
Netherlands	105	176	237	197	150	67	35	-17	-24
Portugal	18	23	36	36	28	30	58	-1	-21
Spain	10	20	26	33	38	107	28	25	15
Sweden	26	62	67	55	64	136	9	-18	16
United Kingdom	25	49	76	75	72	97	53	-1	-4
EU-15	**30**	**61**	**82**	**80**	**71**	**105**	**35**	**-2**	**-11**
Iceland	3	6	7	5	6	80	14	-20	7
Norway	48	88	114	110	97	81	30	-4	-12
Switzerland	7	17	31	31	31	132	82	0	-2
EFTA	**20**	**37**	**51**	**49**	**44**	**88**	**37**	**-4**	**-9**
Western Europe	**28**	**56**	**76**	**75**	**67**	**103**	**35**	**-2**	**-11**

TABLE 5.4 *(continued)*

Countries	Total N (kg ha^{-1})					% change			
	1961	1971	1981	1991*	2001	1961-1971	1971-1981	1981-1991	1991-2001
Czech Republic				47	73				53
Cyprus	39	60	52	76	100	53	−14	46	32
Estonia				27	32				19
Hungary	13	57	85	41	38	328	49	−52	−7
Latvia				25	15				−40
Lithuania				25	29				17
Malta	22	35	18	3	30	55	−49	−83	900
Poland	15	46	64	36	47	216	38	−43	29
Slovakia				27	33				26
Slovenia				85	68				−20
European AC-10	**22**	**50**	**55**	**39**	**47**	**121**	**10**	**−28**	**19**
Bulgaria	13	54	84	61	24	315	57	−27	−61
Romania	19	29	59	19	18	54	104	−69	−3
Turkey	1	7	21	28	29	402	176	33	6
Candidates	**11**	**30**	**55**	**36**	**24**	**171**	**82**	**−34**	**−34**

Countries	Total N (kg ha^{-1})					% change 1991-2001
	1961	1971	1981	1991**	2001	
Belarus				27	31	18
Moldova, Republic of				25	2	−92

	Total N (kg ha^{-1})					
	1961	1971	1981	1991***	2001	% change 1991-2001
Russian Federation				12	5	−57
Ukraine				32	9	71
Russian Federation and the western EECCA				**24**	**12**	**−50**
Countries	**1961**	**1971**	**1981**	**1991*****	**2001**	**% change 1991-2001**
Albania				20	9	−53
Bosnia-Herzegovina				3	10	235
Croatia				41	38	−7
Macedonia, FYR of				28	12	−58
Serbia and Montenegro				26	37	43
Balkan Countries				**24**	**21**	**−19**

Source: FAOSTAT, Statistic Division, Food and Agriculture Organization of the UN. Reprinted with permission.

*1991 data for Cyprus, Hungary, Malta, Poland, Bulgaria, Romania, and Turkey; 1992 data for Estonia, Latvia, Lithuania, and Slovenia; 1993 data for Czech Republic and Slovakia.

**1992 data.

***1991 data for Albania; 1994 data for Croatia, and Macedonia, 1995 data for Bosnia-Herzegovina;1996 data for Serbia and Montenegro.

TABLE 5.5. Phosphate fertilizer consumption

Countries	Total P$_2$O$_5$ (kg ha^{-1})					% change			
	1961	1971	1981	1991	2001	1961-1971	1971-1981	1981-1991	1991-2001
Austria	25	35	25	20	14	42	−28	−21	−31
Belgium-Luxembourg	54	86	60	44	28	60	−30	−27	−35
Denmark	36	45	36	27	13	24	−20	−25	−52
Finland	38	68	57	35	23	77	−16	−39	−32
France	28	59	53	41	26	111	−11	−22	−38
Germany	44	70	61	30	18	59	−12	−50	−39
Greece	7	14	18	19	12	84	33	7	−37
Ireland	17	31	25	31	20	88	−21	26	−34
Italy	19	33	40	41	31	70	24	3	−25
Netherlands	44	48	40	38	27	9	−16	−6	−27
Portugal	15	9	21	19	16	−43	144	−9	−15
Spain	9	14	11	17	21	54	−24	51	24
Sweden	25	41	33	14	12	64	−19	−59	−10
United Kingdom	23	28	24	20	17	19	−12	−17	−16
EU-15	27	41	36	28	20	51	−13	−21	−29
Iceland	1	3	4	3	2	126	15	−24	−36
Norway	43	61	67	34	29	41	10	−50	−14
Switzerland	21	22	21	18	9	7	−8	−11	−49
EFTA	22	29	30	18	13	32	5	−40	−27
Western Europe	26	39	35	27	19	48	−10	−24	−29

Countries	Total P$_2$O$_5$ (kg ha^{-1})					% change			
	1961	1971	1981	1991*	2001	1961-1971	1971-1981	1981-1991	1991-2001
Czech Republic				12	12				−1
Cyprus	39	43	32	50	58		−27	59	17
Estonia				18	8				−56
Hungary	11	37	60	4	8	230	65		107
Latvia				18	6			−94	−65
Lithuania				5	6				12
Malta	1	1	4	3	20	39	356	−36	643
Poland	12	37	43	8	17	220	17	−81	114
Slovakia				7	8				24
Slovenia				28	33				19
European AC-10	**16**	**29**	**35**	**15**	**18**	**88**	**19**	**−56**	**16**
Bulgaria	11	44	66	10	1	306	50	−85	−94
Romania	3	12	37	10	3	275	206	−73	−66
Turkey	1	5	13	15	12	894	149	22	−21
Candidates	**5**	**20**	**39**	**12**	**5**	**320**	**89**	**−70**	**−54**

Countries	Total P$_2$O$_5$ (kg ha^{-1})					% change 1991-2001
	1961	1971	1981	1991**	2001	
Belarus				31	5	−83
Moldova, Republic of				18	0	−100
Russian Federation				7	1	−79
Ukraine				13	1	−88

TABLE 5.5 *(continued)*

Countries	Total P_2O_5 (kg ha^{-1})						% change 1991-2001
	1961	1971	1981	1991**	2001		
Russian Federation and the western EECCA				17	2	−88	

Countries	Total P_2O_5 (kg ha^{-1})					% change 1991-2001
	1961	1971	1981	1991***	2001	
Albania				7	6	−18
Bosnia-Herzegovina				3	4	26
Croatia				22	14	−34
Macedonia, FYR of				4	9	115
Serbia and Montenegro				3	4	25
Balkan Countries				**8**	**7**	**23**

Source: FAOSTAT, Statistic Division, Food and Agriculture Organization of the UN. Reprinted with permission

*1991 data for Cyprus, Hungary, Malta, Poland, Bulgaria, Romania, and Turkey; 1992 data for Estonia, Latvia, Lithuania, and Slovenia; 1993 data for Czech Republic, and Slovakia.

**1992 data.

***1991 data for Albania; 1994 data for Croatia, and Macedonia, 1995 data for Bosnia-Herzegovina;1996 data for Serbia and Montenegro.

TABLE 5.6. Potash fertilizer consumption

Countries	Total K$_2$O (kg ha^{-1})					% change			
	1961	1971	1981	1991	2001	1961-1971	1971-1981	1981-1991	1991-2001
Austria	24	42	38	25	16	72	-10	-35	-35
Belgium-Luxembourg	94	98	83	73	52	4	-15	-13	-29
Denmark	57	65	47	49	28	14	-27	3	-42
Finland	27	52	53	37	36	90	1	-31	-1
France	24	46	54	57	34	92	16	7	-40
Germany	80	99	90	43	30	24	-9	-53	-30
Greece	1	2	4	7	7	80	125	66	-9
Ireland	14	25	31	40	30	79	21	31	-27
Italy	6	14	21	26	25	130	46	26	-5
Netherlands	55	59	53	47	34	9	-11	-10	-28
Portugal	3	2	11	10	10	-17	411	-8	-1
Spain	3	8	7	13	16	181	-13	79	28
Sweden	20	37	31	18	15	82	-14	-42	-16
United Kingdom	22	26	26	24	23	18	-3	-5	-4
EU-15	**31**	**41**	**39**	**33**	**25**	**34**	**-5**	**-15**	**-24**
Iceland	1	2	3	2	2	213	18	-32	-7
Norway	52	76	89	62	58	48	17	-30	-8
Switzerland	21	30	32	29	18	44	7	-9	-39
EFTA	**24**	**36**	**41**	**31**	**26**	**48**	**14**	**-24**	**-17**
Western Europe	**30**	**40**	**40**	**33**	**25**	**36**	**-2**	**-16**	**-23**

TABLE 5.6 (continued)

Countries	Total K$_2$O (kg ha^{-1})					% change			
	1961	1971	1981	1991*	2001	1961-1971	1971-1981	1981-1991	1991-2001
Czech Republic				12	7				-37
Cyprus	4	9	8	11	35	141	-9	41	216
Estonia				33	8				-76
Hungary	6	45	79	6	9	669	76	-93	61
Latvia				24	5				-80
Lithuania				14	11				-19
Malta	1	5	4	3	20	650	-22	-16	505
Poland	18	65	70	15	20	266	7	-78	33
Slovakia				6	7				16
Slovenia				40	41				2
European AC-10	**7**	**31**	**40**	**16**	**16**	**344**	**30**	**-59**	**0**
Bulgaria	2	8	19	5	0	425	137	-72	-92
Romania	1	2	12	3	1	139	722	-77	-79
Turkey	0	0	0	1	2	634	-32	424	48
Candidates	**1**	**3**	**11**	**3**	**1**	**347**	**221**	**-70**	**-71**

Countries	Total K$_2$O (kg ha^{-1})					% change 1991-2001
	1961	1971	1981	1991**	2001	
Belarus				91	48	-47
Moldova, Republic of				9	0	-100

Countries	1961	1971	1981	1991***	2001	% change 1991-2001
Russian Federation				6	1	−85
Ukraine				20	1	−96
Russian Federation and the western EECCA				**31**	**12**	**−61**

Total K$_2$O (kg ha^{-1})

Countries	1961	1971	1981	1991***	2001	% change 1991-2001
Albania				1	1	32
Bosnia-Herzegovina				3	4	26
Croatia				21	16	−25
Macedonia, FYR of				11	4	−63
Serbia and Montenegro				6	6	6
Balkan Countries				**8**	**6**	**−5**

Source: FAOSTAT, Statistic Division, Food and Agriculture Organization of the UN. Reprinted with permission.

*1991 data for Cyprus, Hungary, Malta, Poland, Bulgaria, Romania, and Turkey; 1992 data for Estonia, Latvia, Lithuania, and Slovenia; 1993 data for Czech Republic and Slovakia.

**1992 data.

***1991 data for Albania; 1994 data for Croatia, and Macedonia, 1995 data for Bosnia-Herzegovina; 1996 data for Serbia and Montenegro.

The OECD (2001) has introduced nutrient balance indicators for N, P, and K to calculate for each nutrient the difference between total inputs and outputs to assess changes over a one-year period. A positive (surplus) or negative (deficit) balance indicates, respectively, a potential risk of pollution of soil, water, and air from agricultural systems or a depletion of soil fertility. For example, for its N balance the OECD includes the following elements:

Nitrogen inputs
- inorganic or chemical N fertilizers;
- net livestock manure N production;
- biological N fixation;
- atmospheric deposition of N;
- N from recycled organic matter; and
- N contained in seeds and planting materials.

Nitrogen outputs, or nitrogen uptake
- crop and fodder production; and
- grass from temporary and permanent pasture.

The size of each of the inputs and outputs varies considerably and in consequence so does the N balance in European agriculture (Tables 5.7 and 5.8). In general however, countries with high livestock densities and intensive farming systems have the largest N surpluses (e.g., The Netherlands, Belgium, and Denmark). Nitrogen efficiency, that is the ratio of output to input, is generally better in countries where both livestock and farming systems are less intensive.

FARMING PRACTICES AND ENVIRONMENTAL ISSUES

Whereas in developing countries deterioration of soil quality is mainly due to exploitation, in Europe two main issues can be identified. In areas with high livestock densities, the amount of manure produced is in excess of that which can be used efficiently on the land belonging to the farm, and soils have been receiving excessive amounts of manure for many years. Where this occurs the practice has resulted in large accumulations of organic matter and nutrients beyond the level that can be retained by the soil, causing ground- and surface-water pollution and other environmental problems. On the other hand, in southern Europe there is an emerging issue of the depletion of soil organic matter (SOM).

TABLE 5.7. Soil surface nitrogen balance estimates (1995-1997)

Country	N input (1,000 t)	N output (1,000 t)	N efficiency (output/ input) (%)	N balance 1,000 t	N balance kg ha^{-1} of total agricultural land
Netherlands	960	447	47	513	262
Belgium	443	196	44	247	181
Denmark	611	287	47	323	118
United Kingdom	2,865	1,387	48	1,478	86
Ireland	878	480	55	397	79
Norway	206	131	63	75	73
Portugal	384	120	31	264	66
Finland	272	134	49	138	64
Germany	3,442	2,390	69	1,052	61
Switzerland	251	155	62	96	61
Czech Republic	558	325	58	233	54
France	4,550	2,965	65	1,585	53
Spain	2,086	885	42	1,202	41
Greece	653	457	70	195	38
Sweden	373	268	72	105	34
Italy	1,909	1,424	75	485	31
Poland	1,881	1,348	72	533	29
Austria	364	269	74	95	27
Turkey	2,716	2,216	82	500	12
Iceland	34	21	61	13	7
Hungary	446	537	120	−91	−15
EU-15	**19,789**	**11,709**	**59**	**8,080**	**58**

Source: Adapted from Annex Table 1. Soil surface nitrogen balance estimates: 1985-1987 to 1995-1997, *Environmental Indicators for Agriculture: Methods and Results* Volume 3, © OECD 2001.

Nutrient Inputs in Fertilizers and Manures

Specialization in agriculture, especially the separation of arable cropping and livestock farming tends to be regionalized on the basis of soil type, climate, and topography. This has often led to a disturbance of nutrient cycles that were typical of traditional, integrated farming systems. Where arable

TABLE 5.8. Composition of nitrogen inputs and outputs in national soil surface nitrogen balances (1995-1997)

	N inputs from			N outputs (uptake) from	
Country	Inorganic fertilizers	Net livestock manure	Other*	Harvested crops	Pasture
	Share of total inputs (%)			Share of total outputs (%)	
Netherlands	40	49	11	20	72
Belgium	38	50	12	35	53
Denmark	49	37	14	64	19
United Kingdom	47	28	25	35	64
Ireland	47	45	8	9	91
Norway	51	42	7	20	32
Portugal	37	54	10	46	54
Finland	67	24	9	55	8
Germany	51	27	22	49	38
Switzerland	22	46	32	19	25
Czech Republic	44	21	35	51	14
France	54	26	20	58	33
Spain	44	35	21	54	31
Greece	53	35	13	51	47
Sweden	53	29	18	44	15
Italy	46	28	27	65	20
Poland	46	27	27	53	36
Austria	35	34	31	38	50
Turkey	41	39	19	39	59
Iceland	33	34	33	0	62
Hungary	45	23	32	84	4
EU-15	**49**	**31**	**20**	**49**	**40**

Source: Adapted from Annex Table 2. Composition of nitrogen inputs and outputs (uptake) in national soil surface nitrogen balances: 1985-1987 to 1995-1997, *Environmental Indicators for Agriculture: Methods and Results* Volume 3, © OECD 2001.

*Includes mainly biological nitrogen fixation, nitrogen recycled from organic matter, nitrogen contained in seeds and planting materials, and atmospheric deposition of nitrogen.

cropping predominates, fertilizer use has increased to make up for diminishing recycling of nutrients through animal manure. With the intensification of livestock production and a larger number of animals being grazed on less pasture, or being kept in stalls, or in open yards, much greater quantities of excreta must be disposed of, usually on a limited area of land. Where intensification has led to a greater use of feedstuffs, the composition of excreta is often different to that of animals on open grazing (see "Water Quality"). As a consequence of intensification, there has been a significant increase in the quantity of excreta produced in some countries (Figure 5.2). The N supply from fertilizers and the total N supply in fertilizers and manure are shown in Figures 5.3 and 5.4, respectively.

Many factors concur in the changing availability, composition, and actual use of organic fertilizers and amendments, or biosolids, to use a generic term. Animal wastes are the largest proportion of all biosolids. The nutrient content often depends not only on that of the forage and feed, but on many

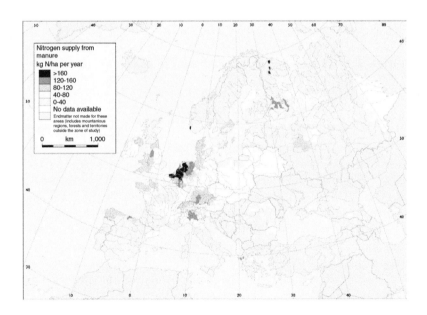

FIGURE 5.2. Nitrogen supply from manure in agricultural soils in Europe. *Source:* European Environmental Agency. Europe's Environment, The Dobris assessment, second edition 1995, Chapter 22, Agriculture, http://reports.eea .europa.eu/92-826-5409-5/en. Reprinted with permission by the European Environment Agency.

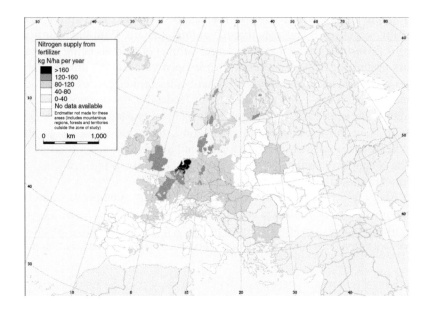

FIGURE 5.3. Nitrogen supply from fertilizers in agricultural soils in Europe. *Source:* European Environmental Agency. Europe's Environment, The Dobris assessment, second edition 1995, Chapter 22, Agriculture, http://reports.eea .europa.eu/92-826-5409-5/en. Reprinted with permission by the European Environment Agency.

other factors, including the addition of feed supplements. To reduce the environmental impact of stock farming from pollution by N, P, copper (Cu), and zinc (Zn), and emissions of methane, animal diets can be manipulated (Mordenti et al., 1991). For example, the P content of the excreta of monogastric animals largely depends on the digestibility of phytic P compounds in the feed and the addition of P supplements. It is possible to decrease the excretion of P up to 50 percent or more by adding the enzyme phytase in order to improve the digestibility of phytic compounds. In addition, some inorganic P compounds that are more readily digested than calcium phosphate could be used as a dietary supplement.

With sewage sludge, there have been impressive changes in its use/disposal. Human excrement has been used to improve soil fertility from ancient times, long before the testimony of the Roman historians. For example, some forty years ago it was still used to fertilize the entire area south of Milan in Italy, increasing grass harvests from three or four to ten or more per year, and,

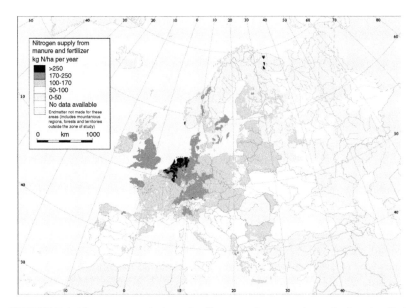

FIGURE 5.4. Nitrogen supply from fertilizers and manure in agricultural soils in Europe. *Source:* European Environmental Agency. Europe's Environment, The Dobris assessment, second edition 1995, Chapter 22, Agriculture, http://reports .eea.europa.eu/92-826-5409-5/en. Reprinted with permission by the European Environment Agency.

in addition, animal breeding was very successful. Nowadays, applying human excrement directly to the soil is prohibited and the fate of P in the raw sewage varies. In the larger sewage treatment works in the EU, soluble P in the effluent must be removed before it is discharged to rivers but this requirement does not apply to smaller works. The recycling of sludge to soil in Europe is now controlled by various EU Directives.

Returning treated urban sewage to a limited area of land near sewage treatment works also disturbs nutrient cycles, because a large percentage of the nutrient content does not originate in the area served by the treatment works. Food now often originates in different continents so that there is an intercontinental nutrient transfer. Food may also contain supplements, including metals, as can proprietary medicines sold by chemists for human consumption. The number of food supplements is very large, recently, for example, about 600 food supplements containing chromium were found in Italy (Sequi, 2003). Metals in food supplements together with those from industry can be discharged into the sewage system, and some, especially from industry, would be generally classified as undesirable.

Water Quality

Some of the N surplus in both intensive arable and animal farming systems is lost to groundwater, a major source of drinking water all over Europe. Thus the quantity and quality of groundwater is of vital importance. Attention has been focused on the concentration of nitrate in view of the 50 mg L^{-1} nitrate limit in drinking water in the EU Drinking Water Directive. Levels of nitrate exceeding the limit are found in about a third of the groundwater bodies for which information is currently available, and this is often related to the intensity of agricultural production in the area, although other sources such as discharges from waste water treatment works can be significant. For example, concentrations of nitrate are much smaller and are around background levels in those areas in the northern regions where the landscape is largely forest and there is little nitrate leaching. Specifically in Sweden, nitrate leaching from agricultural soils is above background levels, for example, 10-15 kg ha^{-1} arable land as compared with 5-7 kg ha^{-1} from forested areas. Nitrate levels have remained fairly constant during the 1990s in the rivers in the northern accession countries and western Europe. In the central accession and Balkan countries, industrial production together with polluting discharges decreased in the 1990s and there was a dramatic reduction in pesticide and fertilizer use in agriculture. Consequently, pollution in surface waters has eased considerably. Although in many places river quality has improved there are still many polluted stretches in rivers, however, this pollution is not always due to agriculture.

Soil Organic Matter

Concern is now being expressed about the level of soil organic matter (SOM) in many soils in Europe. According to the European Soil Bureau's (Zdruli et al., 2004) preliminary estimates, 74 percent of the soils in southern Europe have less than 2 percent organic carbon (C) (less than 3.4 percent organic matter) in the topsoil (0-30 cm). This is due to excessive soil cultivation, and has now become a major issue in land degradation. Such estimates of the C content of soils are probably optimistic, if we consider that in Italy the average C content of soil is now about 1 percent, half what it was fifty years ago (Sequi, 2003).

Factors affecting soil organic matter status are both natural and human-induced. The most important natural factors are: climate, soil type, topography, and the vegetation covering the soil. Human induced factors are mainly land use and management. In Mediterranean areas, the combination of high temperature in summer, rainfall largely in autumn and winter, deep ploughing (up to 0.5-0.6 m), and soil erosion deplete SOM. When carbon stocks in the soil are in decline, the process could be defined as "soil nutrient mining" (Zdruli et al., 1998). Thus strategies to maintain or even enhance soil C are required for the sustainable use of land resources. Organic human and animal waste together with any other organic matter available, for example, vegetal biomass from agro-industrial processes, plant residues, etc., can be used to increase soil C provided appropriate management techniques are used (Sequi and Benedetti, 1995).

Applying organic manures increases the amount of SOM. This improves the soil's physical properties, either directly or by supporting the living organisms in the soil, giving a better soil structure, better water-holding capacity and soil aeration, and providing protection to the soil surface by the presence of a surface layer (mulch) of organic material. In terms of its influence on chemical properties, SOM retains nutrients in readily plant-available forms, and supplies nutrients during its decomposition. Organic acids produced during decomposition aid the dissolution of soil minerals. Nutrients can be fixed in organic complexes (mainly a negative influence for a shorter or longer period). Soil organic matter may also have effects on the accumulation of growth inhibitors where crops are grown in monoculture, and the production of antibiotics that protect against some bacterial diseases.

INTEGRATED NUTRIENT MANAGEMENT AND AGRI-ENVIRONMENTAL POLICIES

Changes in and adoption of a range of agri-environmental policies within Europe may persuade farmers to think more carefully about nutrient inputs and consider applying the concepts of INM because agricultural activities properly managed can minimize many adverse environmental effects. For example, INM can minimize surface and groundwater pollution from excessive nutrient inputs from both fertilizers and manures. Strategies adopted by policy makers to protect the environment by requiring action to be taken by farmers must be critically examined before they are implemented. At the same time they must also be appreciated by the nonfarming community as a possible way of improving the general environment if undertaken appropriately (Sequi and Indiati, 1996).

The European Union's (EU) agri-environmental policy uses a combination of voluntary, regulatory, and cross-compliance programs to achieve environmental goals. The European Commission (EC) does not design or run the day-to-day operations of most agri-environmental programs. Instead it establishes guidelines for three broad categories into which both EU-wide and EU member state programs are placed. The amount and source of compensation, if any, which is paid to farmers depends upon the category under which the agri-environment program falls. This arrangement gives greater latitude to member states in the design and implementation of its agri-environmental programs than is found in other countries. However, some "EU-wide and EU Member Country Programmes" in other countries do conform to current EU agri-environmental legislation.

Basic legal standards apply to all EU member states and their farmers. In consequence, farmers must comply with these environmental regulations without receiving any compensation. The EU Nitrate Directive is an example of a basic legal standard that applies specifically to agriculture in both "EU-wide and EU Member Country Programmes."

Good farming practices, which farmers are expected to observe without receiving direct compensation from the state, are intended to achieve basic environmental standards. However, unlike basic legal standards, the Commission does not mandate good farming practices, but allows each member state to decide what good farming practice is and set appropriate standards. Member states can make good farming practices mandatory or cross-compliant by tying the adoption of such practices to state payments. Prior to the 2003 reform of the CAP, only a few EU member states had such programs in place.

The June 2003 CAP reform agreement provided further incentives for producers to observe environmental rules by tying payments to producers to comply with statutory environmental standards, as well as food safety and animal health and welfare standards. In addition, there is a new requirement that producers maintain land in "good agricultural and environmental condition" to receive payments. These reforms make compulsory the cross-compliance provision that was previously optional to member states. The 2003 CAP reform also added several new agri-environmental measures. It provided for increased funding for projects to promote environmentally beneficial actions, and allowed member states to make new payments to producers to support agricultural activities that are important for protecting the environment. It also allowed member states to offer temporary support to help producers adapt to new environmental standards (European Commission, 2003).

Adoption of agri-environmental measures by farmers is strictly voluntary but subsidies are paid only if they are adopted. The EU subsidizes most measures that fall under a set of broad policy objectives, listed under the

EU-wide programs in Appendix 2. In return for adopting such measures, EU producers receive a payment calculated on the basis of income foregone and the financial incentive needed for adoption. Payments are limited to 450-900 euros per hectare depending on the type of land use. Producers in the EU can choose specific agri-environmental measures best suited to their production system under "EU Member Country Programmes." Most EU agri-environmental programs, while funded at the EU level, are administered by the member states. Therefore, it is difficult to break down EU expenditure on specific programs. However, perhaps the relevant difference between the different EU agri-environmental programs is not the level of funding, but the type of program that is considered to be "agri-environmental." Agri-environmental measures or goals provided by the EC, divided into three broad categories, are listed in Appendix 3.

The EU uses environmental protection as a rationale for the continued government support of agriculture as a whole and has a wide range of measures it considers to be environmentally related (point 3 in Appendix 2). Some examples would be the protection of farm incomes and employment, promotion of rural development, and the upkeep of woodland.

Agri-environmental policy in the EU is now a part of rural development policy and the two can be difficult to distinguish. For example, the compensation payment scheme in EC 1257/99 highlights one of the explicit goals of EU agri-environmental policy not found elsewhere, namely preventing the abandonment of land. The EC has stated that keeping large numbers of family farmers on the land is necessary to preserve the natural environment in the EU and prevent land being abandoned (CEC, 1991). The 2003 CAP reform agreement gave member states broad discretion to maintain product-specific support in order to prevent land abandonment and cessation of production (European Commission, 2003). The proximity of rural and urban areas in the EU, combined with the fact that it is difficult to find "natural" landscapes in many parts of Europe, especially areas near large populations, may cause land abandonment concerns to play a more prominent role in directing EU agri-environmental policy.

INTEGRATED NUTRIENT MANAGEMENT

Integrated nutrient management seeks to both increase agricultural production and safeguard the environment for future generations (Gruhn et al., 2000). In the Introduction, we expanded this definition as follows: INM when applied to plant nutrition implies that farmers should consider each year the total amount of nutrients that are required for a crop to attain its economic optimum yield on each field of the farm. They should then consider the sources from which these nutrients could come, including the supply from the soil and the need to supplement this supply with fertilizers and manures.

It is easy to define INM conceptually here, but it is not so readily put into practice with our current state of knowledge and technology transfer. Besides relating directly to nutrient inputs, crop nutrient requirements depend on other interrelated factors such as crop rotation, the use of cover crops, and improvements in plant yield potential, which together will enhance soil fertility and achieve optimum yield. There is scope for using INM for the great majority of crops that take up their nutrients from the soil solution, but the concept must be approached with realism, and without assuming that it is easy or will be a panacea to minimize the transport of nutrients from the soil-plant-animal system to the wider environment where some elements, especially N and P, may have an adverse impact on the quality of the environment.

The integrated use of different sources of nutrients for plants has been well known to farmers in Europe for many centuries, although not called INM. Since their introduction, the use of mineral fertilizers increased slowly initially, often as a supplement to organic manures, and then more quickly. From the 1950s their preeminence in ensuring adequate plant nutrition has permitted the intensification and specialization of farm practices discussed previously. One consequence has been a considerable decrease in integrated farming with both crop and animal production on the same farm (Shroeder, 2005). The effect has been to disrupt nutrient flows creating an imbalance in nutrient inputs and outputs. The cost of transporting organic manure is very large; because of its bulky nature and low nutrient content large quantities have to be added to get adequate yields. Besides, for farmers, mineral fertilizers are usually readily available, and very easy to use, transport, and handle, and plant response is predictable. However, the price paid in terms of sustainability, and the environmental impact of agriculture on a local and national scale may be large, especially in terms of soil and water quality. In this respect, INM strategy is very interesting, because it could permit a combination of old and new knowledge on plant nutrition to reach environmental and crop productivity goals.

However, INM that seeks to make efficient use of nutrients in organic manures and fertilizers is in no way similar to organic farming, which refuses the use of any chemical product, except under very special circumstances. INM, in contrast, relies on nutrient application and conservation, new technologies to increase nutrient availability to plants, and the dissemination of knowledge between farmers and researchers (Gruhn et al., 2000).

The appropriate use of INM combines both the readiness of plant response to mineral fertilizer and the positive effect of organic manures on nutrient supply and soil properties. INM is concerned with the conservation and best management of endogenous SOM as well as the use of exogenous organic matter (EOM). In Mediterranean areas, where water shortage is the main factor limiting productivity, addressing the problem of SOM depletion in soil is a scientific, political, and social priority. In areas with intensive agriculture, in northern Europe, it appears more important to address the issue of excess of nutrients in soils, mainly N and P. In both situations, INM could be a solution. In the first case, application of diverse sources of nutrients via organic materials could achieve major fertility goals; in the intensive farming areas the results of INM could be a reduction in nutrient leaching to groundwater.

The management of soil fertility involves not only the application of different fertilizers but regulation of nutrient flows to boost their efficiency. A traditional example is the use of crop rotations in which no crop should ever follow itself. Particular sequences confer particular benefits to long- and short-term soil fertility and to pest management. Types of crops that can be rotated include cereals and legumes, and deep-rooted and shallow-rooted plants. Rotation is normally used in organic agriculture with good results, especially the use of forage crops for animal feed and the application of farmyard manure to arable crops. Of course rotation usually implies having both crops and animals on the same farm.

Where all-arable cropping is practiced, the crops are usually rotated, but it is possible to maintain soil fertility by incorporating crop residues and using cover crops as green manure. Cover crops are important to retain nutrients from one season to the next. Their benefit varies considerably with species in terms of nutrient supply and organic matter build-up. In irrigated Mediterranean areas, where spring crops such as maize and vegetables are harvested by the end of summer or early autumn, a winter cover crop, usually a legume, could be grown and incorporated into soil. Three goals are achieved: reduction of soil erosion, reduction of N losses, and an increase in SOM where crop residues are incorporated into the soil. Unnecessary soil cultivations should be avoided because they tend to reduce SOM through increased soil aeration, aiding the decomposition of organic matter. In areas where soil erosion is a problem, it is often better to leave crop residues on the soil surface as protective mulch. If burning of crop residues is allowed, this may be acceptable if there is a serious problem with pests.

The use of mineral fertilizers in agriculture plays a key role in obtaining both crop yield and quality that contribute to the economic viability of the farm and the fertility of the soil. The correct use of inorganic fertilizers, as a supplement to the supply of nutrients from the soil and organic manures, requires management decisions on the type of fertilizer, optimal rate, and time and method of application.

The following criteria must be gathered to assess the amount of fertilizer to be applied:

- the nutrient requirement of a crop to achieve the desired yield;
- the nutrient supply from the soil, which can be estimated by different methods;
- where applicable, the availability of the nutrients from any organic manure that can be applied; and
- the rate at which the nutrients become available during the growing season of the crop.

There are many uncertainties about the quantity of each nutrient a crop requires, its supply from the soil, and the interaction with many other factors that can affect final yield. Among the many factors that affect the total nutrient requirement of a crop are the yield potential of the cultivar grown, the soil, climate and water supply, and the control of pests, diseases, and weeds. The supply of nutrients from soil reserves is frequently difficult to predict both in terms of the amount and rate of release from inorganic and organic reserves. This is especially so for nutrients mineralized from SOM because this microbially mediated process is affected by many soil and environmental factors. Soil analysis by appropriate methods can be used to estimate the amount of a nutrient that is probably available; this and the importance of soil structure will be discussed later.

The daily uptake rate for each nutrient from the soil supply varies during growth but there is a maximum daily uptake which must be met if the crop is to achieve its maximum economic yield. For example, Figure 5.5 shows the daily uptake of P by spring barley grown on soil with and without an adequate supply of plant-available P, measured as Olsen P. On the soil with an adequate supply of Olsen P, daily P uptake increased initially, reached a peak at about 0.6 kg P ha^{-1} and then declined. On the soil with too little Olsen P, the maximum daily uptake rate was less than 0.2 kg ha^{-1}. As a result of this difference in P availability, the crop adequately supplied with P yielded 6.4 t ha^{-1} grain, and that with too little P only 2.9 t ha^{-1}. As in this example with P, for all nutrients the initial question must be what is the maximum daily uptake rate for each nutrient for a crop to achieve its optimum yield and how this can be met by uptake by the roots from the soil solution.

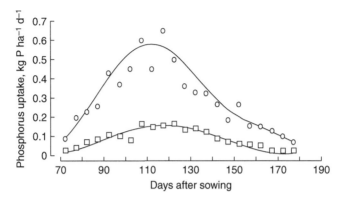

FIGURE 5.5. Daily uptake of P by spring barley on soil with and without adequate supply of available P. *Source:* Johnston (2005). Reprinted with permission.

Some examples of guidelines used in Europe are given hereafter for N and other nutrients.

Principles of Nitrogen Fertilization and Recommendations

How to Estimate the Soil N Supply is the First Key Question in Fertilization

Two main sources of N can be identified, soil mineral nitrogen (SMN), directly available at the beginning of the growing season, and organic nitrogen, not immediately available but able to be released at different rates during the season. In Europe, in less advanced agricultural areas, N fertilizer applications are based solely on experience and this often leads to an oversupply of N. When N recommendations are related to the expected yield there can be an appreciable improvement in N management, especially when combined with additional correction factors such as previous crop, previous climatic conditions (rain and temperature), green or farmyard manure applied, and so on.

Modern methods, mainly used in western European countries, to estimate the N required by a crop are based on calculation, using, for example, tables of reference, expert systems, or simulation methods. Tables giving the N requirement for different crops are based on results from field experiments but their reliability is limited because of spatial and temporal variability between experimental sites. More advanced methods are based on soil and plant analysis. Expert systems and simulation models are much more precise in estimating N fertilizer requirements, but they need a large amount of data. Expert systems are now easier because they are computer aided. Several of them are available for farmers directly from regional extension services. Simulation models are not suitable for farmers use, but they are very helpful for planning and political decisions at local, regional, and national levels, and are very useful for research purposes.

The principles and practice of INM come together in the balance method, as suggested for use in several European countries, such as Italy and Great Britain. It is easy to use and is flexible. It requires an estimate of the amount of N required to reach the optimum yield or specific quality goals and then considers sources and sinks of N in soil, to quantify the amount of N to be added as fertilizer and/or manure.

Nitrogen inputs from soil can be divided into the following:

- Soil Mineral Nitrogen represented by nitrate (NO_3-N) and ammonia (NH_4-N). The amount of SMN, which depends on the previous crop, possible application of manure, etc., can be measured or estimated. In Mediterranean soils, the amount of NH_4-N "fixed" in the lattices of layer silicates may be equal to or exceed organic N (Vittori Antisari and Sequi, 1988; Vittori Antisari et al., 1992).
- Nitrogen mineralized from organic matter is usually taken as a percentage of total soil organic nitrogen. It depends on the C:N ratio of the organic material, soil texture, rainfall, temperature, type of tillage, etc.
- Nitrogen from atmospheric deposition: the amount varies greatly between regions and may be large in parts of northern Europe but very small in rural areas. It should be estimated for a given region.
- Nitrogen from organic sources: the amount is very variable depending on the source. Some of this N is immediately available and some may become available during the crop's growing season. This latter quantity is important but difficult to quantify in many cases.

Nitrogen losses are mainly as follows:

- Leaching of nitrate below the depth to which crop roots grow.
- Denitrification of nitrate under anaerobic conditions leading to losses of nitrous oxide and nitrogen gas to the atmosphere.
- Ammonia volatilization when organic manures and urea are applied, especially to calcareous soils.

The amount of N fertilizer to be applied is the difference between plant need (uptake) and N inputs less N losses. In addition, the efficiency with which crops use N fertilizer is variable and depends on soil type.

Decisions to be made by farmers pertain not only to the amount of N fertilizer, but also timing, type of fertilizer, and economic considerations. Timing of the application must take into account the life cycle of the plant. For example, winter wheat in Mediterranean areas receives nitrate in two to three applications, when vegetative growth is very rapid and N needs are very large. If these needs are not satisfied, yield can be dramatically reduced. Sometimes correct timing, and a split between early application and later topdressings, is even more important than the total amount of N applied. In several European countries the means of production and farming practices are mandatory, including fertilizer management, for crops with Protected Designation of Origin (PDO), Protected Geographical Indication (PGI), or Traditional Speciality Guaranteed (TSG).

Considering all of the points set out previously, the amount of fertilizer N to be applied in any one growing season can be decided best by implementing INM to the full. Currently the most difficult and sometimes neglected aspect is the availability of N from organic sources such as farmyard manure and sewage sludge. This is because there is too little reliable information on the rate of mineralization of the organic N in these manures. Improving the use of such manures has benefit not only in nutrient cycling and the supply of organic matter but it would also minimize the cost of disposal.

Principles of Fertilization and Recommendations for Other Nutrients

Reserves of plant-available P can accumulate in many soils, and the size and availability of the reserves must be allowed for in deciding on the amount of P to be applied as fertilizer. Applying P in organic manures has the same problem as in applying N, namely the amount and rate of release by mineralization. There is much evidence that plants rarely take up more than 25 percent of their total P requirement from the P fertilizer applied to them. In consequence, the remainder of the P in the crop must have come from soil reserves. Soil analysis using an appropriate method can be used to estimate the size of the available reserves. Research at Rothamsted on a range of soils has shown that Olsen P is a good predictor of plant-available P, and when yield is related to increasing levels of Olsen P, yield increases until an asymptote is reached. The level of Olsen P at which the asymptote is reached can be considered as the critical value, and for most crops applying additional P at soil P values above the critical value does not further increase yield. Examples can be found in Figure 5.6. The management of P fertilization, therefore, requires that the critical value is determined, and that the soil P level is initially increased to, and then maintained

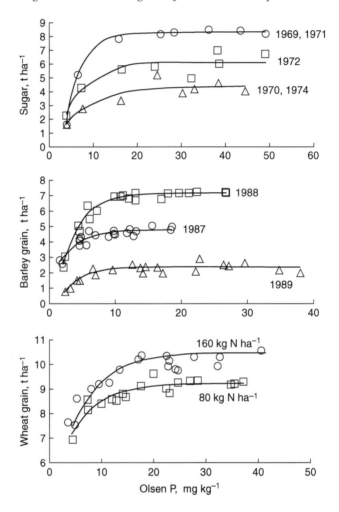

FIGURE 5.6. Olsen P critical values above which any further application does not increase yield. *Source:* Johnston (2005). Reprinted with permission.

at this value. This can often be done by replacing the amount of P removed from the field in harvested crops. This approach has the advantage that, where organic manures are applied frequently it is not necessary to know the immediate availability of the P to crops, but rather whether the P in the manure will maintain the plant-available P at the critical value. Both this and the effectiveness of applying replacement applications of P in fertilizers can be checked by sampling and analyzing soil every four to five years. This approach also requires that soil and fertilizer/manure management practices minimize soil losses by erosion and nutrient losses in overland surface flow from excess rainfall.

The same principle can be applied to K fertilization. Plant-available soil K reserves are usually estimated by determining the exchangeable K content of a soil and critical K concentrations can be determined; see Figure 5.7 for an example. Applications of K in fertilizers or organic manures can be based on replacing the amount taken off the field in the harvested crop, and whether this is maintaining the critical value can be checked by periodic soil analysis.

For both P and K, the critical value can vary depending on soil conditions which affect the ability of roots to grow through the soil to find the nutrients they require to optimize their aboveground production. For example, Table 5.9 shows for three crops grown in the field, how the critical Olsen P value varied with soil C percent, which was related to soil structure, and that the percentage variance accounted for was greater where there was more C. This demonstrates the fact that the success of INM will depend on many different factors.

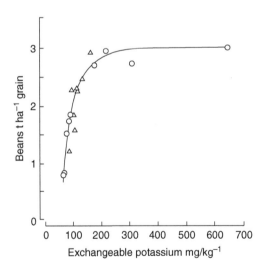

FIGURE 5.7. Critical value of exchangeable K for *Vicia faba* beans. *Source:* Johnston, A.E., personal communication.

TABLE 5.9. Effect of soil organic matter on the relationship between the yield of three arable crops and Olsen P in a silty clay loam soil, Rothamsted

Crop	Soil organic matter (%)	Yield at 95% of the asymptote	Olsen P, mg kg^{-1}, associated with the 95% yield	Percentage variance accounted for
Field experiments				
Spring barley	2.4	5.00	16	83
Grain, t ha^{-1}	1.5	4.45	45	46
Potatoes	2.4	44.70	17	89
Tubers, t ha^{-1}	1.5	44.10	61	72
Sugar beet	2.4	6.58	18	87
Sugar, t ha^{-1}	1.5	6.56	32	61
Pot experiments in the glasshouse				
Grass dry matter	2.4	6.46[a]	23	96
g pot^{-1}	1.5	6.51	25	82

Source: Adapted from Johnston and Poulton (2005).

[a]The response curves at the two levels of soil organic matter were not visually different.

Even if INM allows us to estimate the amount of a particular nutrient to be applied with some accuracy, the efficiency with which the nutrient is used will depend, in part, on how the fertilizer is applied. The greatest opportunity for successfully meeting a crop's nutrient supply is by fertigation (IPI, 2002). At its simplest, for a crop grown under controlled conditions on an inert substrate, it has been shown that nutrient supply can be adjusted to meet the changing needs of a growing crop to achieve a given level of production. Where crops are grown in the field the most efficient way to supply nutrients would also be by fertigation, but in the absence of irrigation it is necessary to consider whether fertilizers can be applied placed in bands close to the seed or whether they are broadcast over the soil surface and worked into the soil. Band placement works best with broad row crops where the placement, especially of top dressings after crop emergence, does little or no damage to plant roots. Any form of placement requires appropriate equipment, and this is often expensive especially for crops such as cereals that are drilled in narrow rows often only 12.5 cm apart. Another important factor is that the placement of large quantities of fertilizer should not create problems with salt effects and high osmotic potentials in the soil solution in the root zone. In northern Europe, where autumn sown crops feature largely in crop rotations, the window of opportunity for drilling under the best soil conditions is limited and farmers usually prefer to drill as quickly as possible. To achieve this they tend to maintain their soils at satisfactory levels of available P and K and apply N in the spring. It is in adjusting the amount of N to be applied according to the principles of INM that there is the greatest opportunity to put INM into practice.

Legislative Constraints and Gaps

Although agriculture has changed dramatically, especially in many parts of Europe (see "Agriculture Production, Nutrient Use, and Issues"), nutrient applications should be based on INM criteria in all agricultural systems. These include specialized horticultural crops, forestry management, recreational areas, and a range of others, and not just conventional broad acre agriculture. In addition, radical changes in food chains have broken traditional nutrient cycles (see the section titled "Nutrient Balances"). Thus any legislative rules must take into consideration structural changes, not limited to agricultural practice but extended to all human activities, in order to be effective. Notwithstanding this, INM presents an opportunity to preserve environmental equilibria as much as possible. Farmers (1) must make use of any available waste resource for environmental reasons, (2) must return to soil sufficient amounts of any nutrient to replace those withdrawn, but (3) must avoid excess of both nutrients and undesired elements added by human activities.

Waste resources are usually organic and can supply both nutrients and organic matter but such amendments often do not fit the needs of modern agriculture. Agriculture must readapt itself to the use of organic amendments where they can be used effectively and efficiently. Rather than use organic amendments, farmers may prefer to manage the organic matter levels in their soils simply by proper crop rotations or reducing the intensity of tillage (Sequi, 1996, 1999). In addition, there are increasingly very rigid legislative constraints on the use of many organic amendments, especially sewage sludge in agriculture, that seek to protect the soil in particular and the environment in general. Such constraints sometimes appear irrational, partly because policymakers in different countries interpret the results of experiments in different ways even when the results do not appear to conflict. When results do not agree, perhaps because of differences in experimental techniques, there are even greater differences in interpretation. For example, there are often great differences between countries in the maximum allowable levels of undesired substances or elements in organic wastes applicable to the soil. The limited scientific reliability of legislative rules is shown by comparing the allowable levels established by the Environmental Protection Agency of the United States (U.S. EPA), and those already decided or under consideration by the EC, or in some European countries. For example, for sewage sludge to be applied to agricultural soils in the EC, the upper limit for zinc is 2,000 to 3,000 mg Zn kg^{-1} and for dioxins is some hundreds of nanograms, while the U.S. EPA has established a cautionary limit of 7,500 mg Zn kg^{-1} and does not consider dioxins for all biosolids.

A second consideration concerns the levels of possible pollutants which are expressed in total rather than available amounts, both in organic wastes and in soils to which wastes may be applied. For example, for zinc (Zn) no data exist as to whether a waste or a soil containing 300 or 3,000 mg Zn kg^{-1}, can be considered zinc deficient or dangerous with respect to plant nutrition and the amount of Zn reaching the human food chain.

Irrational rules are common in a generalized way throughout the world, but the negative effects are mainly evident in Europe. Lack of appropriate data has hindered and continues to hinder application of useful wastes to the soil, but it has also favored and favors the disposal to soil of dangerous wastes. Such irrational regulations may involve both excessive inputs and legalized pollution simultaneously (Sequi, 1990). Legislative rules are often applied stringently because of possible criminal proceedings and may discourage plant managers, public administrators, and of course farmers from recycling organic wastes.

Getting INM into Farm Practice

Although the principles of INM are admirable, research on the subject in European countries is minimal probably because there is little history of a holistic approach to crop nutrition. Nonetheless, results of some field studies, not specifically designed to study INM, have given information on its possible application in Europe.

Some Data from Italy

Combined use of mineral fertilizers and animal manures. An experiment, established in 1962 at the experimental farm of Padova University (Nardi et al., 2004), evaluated the effect of mineral, organic and mixed mineral organic fertilizers/manures on soil properties for forty years. The results were as follows:

1. Soil N content was higher with organic fertilization and positively correlated with the total organic carbon applied. N from mineral

fertilizer was readily available to plants and microorganisms, but leaching was greater.
2. Total soil organic carbon was maintained only with farmyard manure; it declined over forty years by 23 percent with liquid manure and mixed fertilization, 43 percent with mineral fertilizer, and 51 percent in the untreated control.

The results have been confirmed in a similar experiment lasting thirty years in the Emilia Romagna region (Toderi et al., 1999). Other studies in Italy confirm that the use of animal manure is very important to maintain soil quality. In general, the results suggest that a combination of animal manure and mineral fertilizer applied just to meet plant need is a reasonable compromise, and could help to save money, and protect the environment.

Use of green manure. An experiment (Caporali et al., 2004) in Central Italy using winter green manure reduced the need for mineral fertilizer, protected the soil from erosion, and generally improved soil quality. Italian ryegrass (*Lolium multiflorum* Lam.), subterranean clover (*Trifolium subterraneum* L.), and hairy vetch (*Vicia villosa* Roth) were compared with natural weed cover and amounts of N were tested on the maize that followed. Maize yield varied with both fertilizer and cover crop. Differences between ryegrass and weeds, compared to the two legume crops, were larger at low rates of applied N. Soil N content was always greater with legume cover crop. The results showed that soil N inputs for maize after subterranean clover or hairy vetch green manure were equivalent to an application of 100 to 200 kg ha^{-1} of mineral fertilizer N, much more than the recommended rates for fertilizer N in central Italy. The nonleguminous cover crop had little effect on the subsequent maize yield. The authors concluded that green cover crops were advisable when no animal manure is available, especially in those Mediterranean areas where erosion is an issue.

Waste from food processing industry. The use of agri-industrial wastes in agriculture may be an opportunity for both industry and agriculture, because it solves both disposal problems and an opportunity for recycling nutrients, that is, agronomic, environmental, and social benefits. However, different problems, both agronomic and environmental, must be solved before the application of wastes to soils. A large variety of wastes coming from the food industry and characteristic of Mediterranean areas have been the subject of field studies for nutrient supply in recent years, namely olive oil milling, cheese making, winery residues, and from the preparation of tomatoes, fruit juices, etc. For example, winery sludge consists of grape stalks, lees, marc, and washing water. It has a large content of organic C and N and a small content of P and K. Several studies have proved its value in improving soil structure and increasing the yield of spring crops (Masoni et al., 2000) but nitrate leaching was also increased. For winter crops, such as wheat (Masoni et al., 2002), the yield increase was very large and the risk of nitrate leaching less than with spring crops. Though losses of N were not negligible, they were not large enough to discourage the use of such waste in agriculture.

Other organic materials, such as distillery and winery vinasses, obtained after anaerobic digestion of winery sludge by a process developed by Anagni, Italy, have been tested since 2000. They have been used in field experiments in a national official research program,[1] and are extensively used by farmers at present. Such vinasses have been recently classified as an organic N fertilizer in Italy.[2]

Integrated Nutrient Management in the United Kingdom

The British Survey of Fertiliser Practice assesses annually, and on a random selection of farms, the use of fertilizers by type, amount and crop. Farmers are also asked whether or not they apply organic manures and if so, how much, and to what crop. For arable farms in particular, the results of the survey show that only a few farmers make any allowance for the addition of nutrients in any organic manure they apply. Thus, such farmers could be considered not to practice INM.

This is in spite of the fact that data on the average nutrient content and nutrient availability in a range of organic manures are given in an official bulletin, Fertiliser Recommendations for Agricultural and Horticultural Crops (RB 209) published by the Ministry of Agricultural, Fisheries and Food (MAFF, 2000). Farmers can therefore, if they wish, make an allowance for the nutrients in organic manures. There is no immediately obvious reason why many farmers apparently do not make the appropriate allowances. It may be a lack of awareness of the savings that can be made.

A number of organizations in the United Kingdom have a role in technology transfer, one such is Linking Environment And Farming (LEAF),* which has introduced a whole farm business and environmental management tool through an on-farm audit. This is a comprehensive guide to aid a farmer in implementing a whole farm management system. The LEAF audit seeks not only to help improve a farm as a business but to do this in an

---

*LEAF is a UK-based organization, LEAF, NAC, Stoneleigh, Warwickshire. CV 8 2LZ; e-mail: audit@leafuk.org; available online: www.leafuk.org. LEAF works closely with similar organizations in Europe.

environmentally and socially acceptable way. There is no requirement for farmers to use the audit, but having completed it and returned it to LEAF, the farmer is told where improvements are possible. LEAF also promotes a number of demonstration farms throughout the country where the benefits from using the audit can be shown to other farmers. These benefits are often both financial and environmental, the latter especially on farm wild life.

In the latest edition (2003), the audit is divided into sections: organization and planning; soil management and fertility; crop health and protection; animal husbandry; pollution control and by-product management; energy efficiency; and additional considerations for landscape and nature conservation and community issues.

The section on soil management and fertility is divided into sections: site; soil erosion and cultivation techniques; crop nutrition; organic manure; inorganic fertilizer; and irrigation. In each of these sections, questions are asked and labeled boxes have to be ticked to indicate how far the farmer thinks he has achieved the aim posed in the question. The boxes are: A, fully achieved; B, considerable progress; C, some progress; D, not started; and E, not applicable.

Questions asked, for example, in the section on site are: Have the soil types on the farm been identified and mapped? In the section on soil erosion: Have areas prone to erosion been identified and mapped? Is there a general policy to conserve and build up SOM? In the section on crop nutrition, among the questions asked are:

1. Do you have a nutrient management plan integrated with a livestock manure management plan?
2. Are soils regularly analyzed for soil nutrient status and pH?
3. Are nutrient balance sheets used as part of the nutrient management plan?
4. Are nutrient contributions from different crops in the rotation and N mineralization from cultivations considered?
5. Are fertilizers only applied where required?
6. Is the timing of fertilizer applications such as to minimize loss?
7. Is a FACTS registered agronomist used to give crop nutrition advice?*

---

*FACTS (Fertiliser Advisory Certification and Training Scheme) is a fertilizer industry requirement. All those selling or advising on fertilizer use must be FACTS registered. A FACTS-registered fertilizer salesperson or an agronomist must demonstrate by examination, competence to give appropriate advice and to show each year that he or she is maintaining/developing his or her professional competence.

In the section on organic manure, the questions include: Is there a live-stock manure plan integrated with a nutrient management plan? Is manure and slurry applied to minimize pollution? In the section on inorganic fertil-izers, the questions asked include: Are all inorganic fertilizer and manure applications recorded? Are estimates of soil N supply made (to adjust fertil-izer N applications)? Various questions about application techniques are also included. The section on irrigation is concerned with scheduling and amounts applied and the possible use of fertigation.

This is a whole-farm, not an individual field approach but it should raise an awareness in the farmer's mind of ways for improving all those tech-niques related to managing the farm that could improve profitability, mini-mize waste, and contribute to better landscape management.

APPENDIX 1

Soils

Fraters (1994) grouped the European soil types in classes of broadly sim-ilar vulnerability to different degradation processes as follows.

Well-Drained Soils

These soils cover 44 percent of the European land area, and occur gener-ally in gently undulating relief. Their formation was conditioned by climate and natural vegetation; hence they show a zonal distribution from north to south in relation to variation in rainfall and temperature. They differ from each other by texture, SOM (frequently referred to as humus) content and acidity (pH), and five soil categories can be identified: black earths, sandy soils, acid loamy soils, nonacid loamy soils, and clayey soils.

Black earths. Agriculturally, the black earths, or chernozems, are of major importance. They are characterized by a humus-rich topsoil, 50 or 100 cm deep, overlying a lighter textured horizon containing calcium carbonate. They occur in the continental climate zone of the central European plain, the steppe, in a belt extending from Poland and Hungary to the Urals, and cover about 9 percent of Europe's land area. They are very fertile and suit-able for arable farming. However, erratic rainfall often limits yields. In Russia and the Ukraine, black earths are notable for extensive cereal pro-duction.

Sandy soils. These are mainly podzols and develop under the influence of slowly decomposing acid humus, the dissolved organic material moving through the soil with percolating rainwater. This organic material precipitates in the subsurface soil layers in combination with iron and aluminum oxides to produce an impermeable layer that limits rooting depth. In Europe, such sandy soils cover over 9 percent of the land. In northern Europe, they occur mainly under forests but in western Europe they are currently used for intensive agriculture with the application of fertilizers or manures to overcome their inherently low fertility. The sandy soils also include arenosols, which are confined to limited areas, mainly in Poland, Spain, and the United Kingdom.

Acid loamy soils. These are represented mainly by the podzoluvisols, and they occur most extensively under the cold continental conditions of the central taiga of the north and center of Russia and Moldova, and cover about 14 percent of the land area. In more central areas, grassland and crops occupy a large portion of the land, but low fertility and a short growing season limit production.

Nonacid loamy soils. These are widespread in western and central Europe, and cover about 8 percent of the land. They comprise the orthic luvisols on loessic parent materials, which are among the most productive agricultural soils. In more northern latitudes, in Ireland and Sweden, they also include cambisols, on which livestock production predominates, and the haplic and luvic phaeozems.

Clayey soils. These soils cover about 4 percent of Europe's land area, and in the Mediterranean area they are red and reddish-brown soils of heavy texture developed over limestone, the chromic cambisols and the chromic luvisols. Despite their large clay content, these soils are well drained because of their texture and the presence of a permeable subsoil. They are associated in level areas and in depressions in the landscape with heavy textured dark clay soils, the vertisols, which although fertile, present management problems due to their unfavorable physical properties. They are preferably used for the production of cereals. The luvisols are sensitive to erosion, and are used mainly for vineyards, olives, and citrus orchards.

Shallow and Stony Soils

These soils—lithosols, cambisols, and rendzinas—cover 30 percent of the total land area of Europe, and usually occur on steep slopes. They predominate in the major mountain and hill ranges of Europe, such as the Pyrenees, the French Massif Central, the Alps, the Apennines, the Carpathians, the Scandinavian mountain ranges, and the Caucasus. These soils are not suitable for intensive arable farming and the use of heavy machinery; they are used mainly for nonintensive grazing and forestry for wood production.

In this group of soils, the acid, shallow, and stony soils that occur on non-calcareous parent material (such as in Scandinavia and Germany) are more vulnerable to acidification than the nonacid shallow and stony soils on calcareous bedrock, for example in France, Greece, Italy, and Spain.

Semiarid and Salt-Affected Soils

In specific areas, such as the eastern part of southern Europe, excessive evaporation because of the high temperatures, results in soil salinization, which leads to the formation of solonchaks, with excess of salts, or solonetz, with excess of sodium (Na). These semiarid and salt-affected soils cover 9 percent of Europe, mostly in the southern part of Russia, the Ukraine, and Romania. The lack of rainfall combined with the excess of salt or Na permits only nonintensive agriculture. The semiarid but not salt-affected soils (xerosols and kastanozems) do have potential for more intensive agriculture provided irrigation is available.

Imperfectly Drained Soils

These soils cover 17 percent of the land surface of Europe. The imperfect drainage can be due to surface water stagnation (e.g., with gleyic luvisols and planosols) or to fluctuating groundwater (e.g., with gleysols and fluvisols). Wet soils, occurring in level relief and in areas with a high groundwater table, are very extensive in the northern part of Europe (Russia, Scotland, and Ireland), where they occur in association with peat soils, the frequent periods of anaerobic conditions favoring the formation and retention of peat. Imperfectly drained soils due to surface waterlogging are extensive in England and in Germany. Organic soils—or histosols, that is, peat and muck—are developed under the influence of high water tables and low temperatures, and are found in the boreal parts of Russia, Finland, Ireland, and the uplands of Britain. Remnants of former extensive peat areas occur elsewhere, for example in the Netherlands and northern Germany. The prolonged waterlogging of these soils strongly influences their use and management. Depending on the climate under which they occur, they are used either for extensive wood production and/or grazing (such as in the boreal parts of Russia and Finland), or, after reclamation by installing artificial drainage systems, for arable cropping, dairy farming, or horticulture, as in western Europe.

A qualitative appraisal of the vulnerability of the main soils types to threats such as erosion, acidification, pollution, and compaction is given in Table 5.1.

APPENDIX 2

EU-wide and EU Member Country Programs Conforming to EC 2078/92 (examples)

EU-Wide Programs

The Nitrate Directive (EC 91/676/EEC) seeks to reduce pollution caused by nitrates from agricultural sources by requiring member states to implement action programs in areas identified as being vulnerable to pollution. Under the directive, the application of livestock manure is to be limited to 170 kg N ha^{-1} by 2003.

Under EC Reg. 1257/99, support can be given to farmers who for at least five years use production methods designed to protect the environment and maintain the countryside, in order to promote farming methods which promote the protection and improvement of the environment (which includes the landscape and its features, natural resources, the soil, and genetic diversity), environmental planning in farming practice, extensification, the conservation of farmed environments of high natural value, and the upkeep of the landscape.

EC Reg. 1257/99 allows for compensatory payments to farmers who produce in less-favored areas such as mountainous areas, areas threatened with abandonment, or areas in which "the maintenance of agriculture is necessary to ensure the conservation or improvement of the environment, the management of the landscape, or its tourism value."

As part of the EU's Agenda 2000 agricultural reform, farmers are expected to observe basic environmental standards, also known as good farming practices, without direct compensation. Good farming practices are not legal texts mandated by the Commission but rather the Commission allows each member state to decide what good farming practice is.

Environmental programs under commodity support regimes. Certain of the EU, commodity regimes provide payments for implementing environmental practices or require that producers implement environmental practices as a condition for receiving payments.

Producers of beef cattle must not exceed a maximum stocking density (livestock units per hectare) as a condition of eligibility for payments. In addition, producers who observe lower stocking densities are eligible for an extensification premium.

Producers of arable crops producing more than ninety-two tonnes of arable crops are required to set aside a portion of their land. The base level set-aside requirement is 10 percent. Member states are required to introduce measures that ensure that set-aside land is maintained so as to protect the environment. Some examples of recommended practices relate to the use of field margins, choice of set-aside cover, timing of cutting, cultivation, and the spreading of animal manure.

Examples of EU Member State Programs

United Kingdom. In England, the Countryside Stewardship makes payments to farmers and other land managers to enhance and conserve agricultural landscapes as well as associated wildlife and history, and to improve opportunities for public access. Grants are available toward capital works such as hedge laying and planting, repairing dry stone walls, etc. The Organic Farming Scheme provides payments to farmers for adopting authorized conservation practices above those set out under the minimum standard of "good farming practices."

Italy. Sicily's "Plan for Rural Development" has provisions that include: (1) providing payments for adopting "integrated production methods" (or organic methods) to reduce nitrogen and phosphorus applications by at least 25 percent over the levels required under "good farming practices" for specific crops; (2) making payments for activities that maintain scenic aspects of agriculture; and (3) providing financial aid for maintaining olive trees in excess of 100 years old, and nuts and chestnuts on terraces at more than 300 m altitude.

Germany. Measures established at the state level include those that promote extensive farming by reducing inputs to arable land, organic farming, and support the rearing of local breeds of animals in danger of extinction. For example, the Schleswig-Holstein region offers a twenty-year set-aside for arable land. North Rhine/Westphalia provides incentives for the conservation of fruit trees and wetlands, as well as arable land set-asides. Rheinland-Palatina provides incentive measures to preserve traditional agriculture activities such as wine growing in hill areas.

Greece. Farmers are required to rotate cotton with cereals and to limit the application of nitrate fertilizer to specified low levels. Irrigation systems that reduce nitrogen leaching and erosion are promoted.

France. All agri-environmental programs in France are administered under the auspices of the Land Management Contract (Contrat Territorial d'Exploitation, or CTE). The program funds project-specific contracts between individual producers and the government. Projects may cover a broad set of objectives, including environmental protection. The program is cofunded by EU (Guarantee Fund) and the Government of France. Support is given to producers for project startup expenses, in addition to annual aid for up to five years for the increased cost of production resulting from the project. The share of expenses compensated varies. Examples of environmental projects include: rehabilitation and upkeep of irrigation, conversion to organic agriculture, replacing chemical herbicides with mechanical weed control, planting natural grasses between rows, replacing chemical fertilizers with compost, establishing, and maintaining grassland (pasture), etc. About half the contracts have gone to livestock operations.

Ireland. The basic program is the Rural Environment Protection Scheme (REPS). A farmer who joins the scheme enters into an environmental management agreement comprising a series of twelve obligations that must be fulfilled on all parts of the farm. The obligations include drawing up and following a plan for protection of water, nutrient management, stock management, hedge and stonewall repair, and habitat protection.

APPENDIX 3

EU Agri-environmental Goals (by category)

1. *Environmentally beneficial productive farming*
   * input reduction
   * organic farming
   * extensification of livestock production
   * conversion of arable land to grassland and rotation measures
   * undersowing and cover crops, strips, preventing erosion, and fire
   * preserving areas of special biodiversity/nature interest
   * maintenance of existing sustainable and extensive systems
   * preserving farmed landscape

2. *Nonproductive land management*
   * set-aside
   * upkeep of abandoned land and woodland
   * maintenance of the countryside and landscape features
   * maintaining public access

3. *Socioeconomic measures and impacts*
   * training
   * supporting farm incomes
   * employment
   * societal attitudes

Source: European Commission (1998).

NOTES

1. Italian Ministry of Agricultural and Forestry Policies, PARSIFAL Project.
2. Italian law 748/84 "New rules for fertiliser regulation."

REFERENCES

Caporali, Fabio, Campiglia, Enio, Mancinelli, Roberto, and Roberto Paolini (2004). Maize performances as influenced by winter cover crop green manuring. *Italian Journal of Agronomy.* 8: 37-45.

Commission of the European Communities (CEC). The Development and Future of the CAP. COM 91 (100), Office for Official Publications of the European Communities, Luxembourg, 1991.

European Commission (1998). "State of Application of Regulation (EEC) No. 2078/92: Evaluation of Agri-Environment Programmes." DGVI Commission Working Document VI/7655/98.

European Commission (2003). "CAP Reform Summary." <http://europa.eu.int/comm/agriculture/mtr/sum_en.pdf>, July 2003.

European Environmental Agency (1995). Agriculture. In Europe's environment, the Dobris Assessment. State of Environment report No 1/1995, Office des Publications, Luxembourg, 676p.

FAOSTAT, Statistic Division, Food and Agriculture Organization of the UN. < http://www.fao.org/waicent/portal/statistics_en.asp>.

Fraters, Dico (1994). Generalized soil map of Europe: Aggregation of the FAO-UNESCO soil units based on the characteristics determining the vulnerability to soil degradation processes. Report 71240300, RIVM, Bilthoven, The Netherlands.

Gruhn, Peter, Goletti, Francesco, and Montague Yudelman (2000). *Integrated Nutrient Management, Soil Fertility and Sustainable Agriculture. Current Issues and Future Challenges.* 2033 K Street, N.W. Washington, DC. 20006 USA, International Food Policy Research Institute, 38p.

IPI (2002). *Fertigation: Fertilisation Through Irrigation.* Research Topics No. 23, International Potash Institute, Horgen, Switzerland, 81p.

Johnston, A.E. (2005). Phosphorus nutrition of arable crops. In Sims, J.T. and A.N. Sharpley (eds.) *Phosphorus: Agriculture and the Environment.* pp. 495-520. ASA, CSSA, SSSA, Madison, Wisconsin, USA.

Johnston, A.E. and P.R. Poulton, (2005). Soil organic matter: its importance in sustainable agricultural systems. *Proceedings 565, The International Fertilizer Society,* York, UK. 46p.

Kostrowicki, Jerzy (1991). Trends in the transformation of European agriculture. In Brouwer, F.M., A.J. Thomas, and M.J. Chadwick (eds.) *Land Use Changes in Europe.* pp. 21-47. The GeoJournal Library Vol 18. Dordrecht: Kluwer Academic Publishers.

MAFF (2000). *Fertiliser Recommendations for Agricultural and Horticultural Crops RB209.* Ministry of Agriculture, Fisheries and Food, The Stationery Office, London, UK.

Masoni, Alessandro, Mariotti, Marco, Arduini, Iduna, and Laura Ercoli (2002). Distribution of winery sludge on wheat. *Italian Journal of Agronomy.* 6: 85-95.

Masoni, Alessandro, Mariotti, Marco, and Laura Ercoli (2000). Distribuzione dei fanghi di cantina al mais. Produzione granellare, assorbimento e lisciviazione dell'azoto e del fosforo. *Rivista di Agronomia.* 34: 234-245.

Mordenti, Archimede, Piva, Gianfranco, and G.A. Fleming (1991). Animal nutrition. Environment interactions reduction of nitrogen, phosphorus, heavy metal and drug pollution via diet manipulation. In Fleming, G.A. and Archimede Mordenti (eds.) *The Production of Animal Wastes.* pp. 42-81. European Conference "Environment, Agriculture, Stock Farming in Europe," Mantua, Italy: Ministry of Agriculture and CCIAA.

Nardi, Serenella, Morari, Francesco, Berti, Antonio, Tosoni, Moira, and Luigi Giardini (2004). Soil organic matter properties after 40 years of different use of organic and mineral fertilisers. *European Journal of Agronomy.* 21: 357-367.

Organisation for Economic Co-operation and Development (2001). *Environmental Indicators for Agriculture. Methods and Results.* Vol. 3. Paris, France.

Sequi, Paolo (1990). The role of agriculture in nutrient cycling. *Alma Mater Studiorum.* 3(2): 155-190.

Sequi, Paolo (1996). The role of composting in sustainable agriculture. In de Bertoldi, Marco, Paolo Sequi, Bert Lemmes, and Tiziano Papi (eds.) *The Science of Composting.* pp. 23-29. Glasgow, UK: Blackie Academic & Professional.

Sequi, Paolo (1999). Impact of agriculture on the environment. In Brufau, J. and A. Tacon (eds.) *Feed Manufacturing in the Mediterranean Region. Recent Advances in Research and Technology. Cahiers Options Méditerranéens.* 37: 223-228.

Sequi, Paolo (2003). The viewpoint of a soil scientist. In Langenkamp, Heinrich and Luca Marmo (eds.) *Biological Treatment of Biodegradable Waste. Technical Aspects.* pp. 159-184. Brussels, European Commission, Joint Research Centre, EUR 20517 EN.

Sequi, Paolo and Anna Benedetti (1995). Management techniques of organic materials in sustainable agriculture. In Dudal, R. and R.N. Roy (eds.) *Integrated Plant Production Systems. FAO Fertilizer and Plant Nutrition Bulletin.* 12: 139-154.

Sequi, Paolo and Roberto Indiati (1996). Minimizing surface and ground water pollution from fertilizer application. In Rosen, David, Elisha Tel-Or, Yitzhak Hadar, and Yona Chen (eds.) *Modern Agriculture and the Environment.* pp. 147-158. Lancaster, UK: Kluwer Academic Publisher.

Shroeder, Jaap (2005). Revisiting the agronomic benefits of manure: a correct assessment and exploitation of its fertilizer value spares the environment. *Bioresource Technology.* 96: 253-261.

Toderi Giovanni, Giordani, Gianni, Comellini, Franca, and Marina Guermandi (1999). Effetti di un trentennio di apporto di materiali organici di diversa origine e della concimazione azotata su alcune componenti della fertilità del terreno. *Rivista di Agronomia.* 33: 1-7.

Vittori Antisari, Livia, Marzadori, Claudio, Ciavatta, Claudio, and Paolo Sequi (1992). Influence of cultivation on soil nitrogen pools. *Communications in Soil Science and Plant Analysis.* 23: 585-599.

Vittori Antisari, Livia and Paolo Sequi (1988). Comparison of total nitrogen by four procedures and sequential determination of exchangeable NH_4, organic N and fixed NH_4 in soil. *Soil Science Society of America Journal.* 52: 1020-1023.

Zdruli, Pandi, Almaraz Russel, and Hari Eswaran (1998). Developing land resources information for sustainable land use in Albania. CATENA VERLAG, *Advances in GeoEcology.* 31: 153-159.

Zdruli, Pandi, Jones Robert, and Luca Montanarella (2004). Organic matter in the soils of Southern Europe. European Soil Bureau Research Report No.15. Official publication of the European Communities, Luxembourg. EUR 21083 EN.17p.

Chapter 6

Integrated Nutrient Management: Experience and Concepts from New Zealand

Antony H. C. Roberts
Tony J. van der Weerden
Douglas C. Edmeades

INTRODUCTION

New Zealand's land-based animal and crop production need to be conducted in a manner that is both economically and environmentally sustainable. Soil fertility management on pastoral farms, now and in the future, revolves around supplying the quantity and quality of pasture and forage crops to suit the requirements for animal production and health, and limit avoidable environmental degradation. The legume is critical to pastoral farming in that it supplies the most deficient plant nutrient, that is, nitrogen (N), through biological fixation, as well as providing high quality forage for animals. Similarly, a full understanding of arable crop requirements, and the impact of cultural practices on soil and water quality are required to develop an integrated nutrient management (INM) program. Agriculture is under scrutiny because of its impact on soil condition and contamination, water quality, and biodiversity. These resource management issues will also impact the profitability of New Zealand agriculture, and the long-term marketing of produce, thus giving land managers a strong incentive to adopt or develop solutions to these issues.

This chapter briefly discusses agricultural production in New Zealand, issues surrounding fertilizer use, agricultural sustainability, and environmental concerns. The extent to which nutrient management is integrated in New Zealand agriculture, and the soil and nutrient management tools developed and implemented over the last ten years are discussed including case studies. Emerging issues and research gaps are identified.

MAJOR SOIL AND CLIMATIC REGIONS AND MAJOR CROPPING SYSTEMS

New Zealand lies to the southwest of Australia in the Pacific Ocean. It consists of three major islands, which stretch some 1,600 km, over 14 degrees of latitude. The three islands embrace such climatic extremes as subtropical Northland, the cold uplands of the Alpine regions, the semiarid basins of Central Otago, and the very wet mountains and lowlands of Westland. The topography is varied, with around 50 percent being steep, 20 percent moderately hilly, and less than 33 percent is either rolling or flat (Taylor and Pohlen 1968). The sharp relief of the islands is sufficient to produce a significant range of temperature from north to south and with altitude. Mean air temperatures at sea level are approximately 15°C in the north, 12°C in the center (about Cook Straight), and 9.5°C in the south, and these fall by about 1.5°C for each 305 m of elevation. The prevailing westerly winds are forced up and sideways by the mountain ranges, causing those areas directly exposed to the westerly airstreams, and of higher altitudes to have the greatest rainfall. This tendency for wetness in the west and dryness in the east is most extreme in the southern South Island, where more than 5,000 mm of rain falls on the western-southern Alps, compared to less than 500 mm in the Central Otago basin on the eastern side (Taylor and Pohlen, 1968).

The geology underlying the soil mantle varies in texture and composition and includes igneous rocks ranging from ultrabasic to acidic (e.g., basalt, andesites, and rhyolite), metamorphic rocks, and sedimentary rocks (e.g., conglomerates, sandstones, mudstones, and limestones). Based on their parent materials, the soils of New Zealand may be broadly classified into three major "groups": sedimentary (sandstone, siltstone, and mudstone), ash (andesite, basalt), and pumice (rhyolite) soils (Roberts and Morton, 1999). Sedimentary soils include the following soil orders: brown, sands, gley, melanic, pallic, recent, semiarid, and most ultic soils. Ash soils include the allophanic and granular orders, and most oxidic soil, while pumice soils are composed of the pumice soil order.

New Zealand's largest industries are concerned with biomass production, and processing, and 60-70 percent of export earnings derive from these industries. Agriculture is at the forefront of these industries. Annual agricultural production in 2001/2002 was worth $16.61 billion. Pastoral farming, based primarily on grass/legume pastures, is the dominant enterprise with 13.5 Mha in pasture. Some 39 m sheep and 4.5 million cattle are wintered on around 33,101 properties (Meat and Wool Innovation April 2003), and 3 million dairy cattle are carried on 12,751 properties (LIC Dairy Statistics 2003/04). In addition, there are 2,115 deer farms (NZ Statistics Department, 1993).

Arable crop production in New Zealand covers approximately 300,000 hectares (ha) of land cultivated for arable and vegetable crops, with the concentration of cropping activity occurring in Canterbury. The main crops grown are wheat, barley, oats, peas, maize, grass and clover seeds, potatoes, sweetcorn, and onions (Williams, 1998). These are typically grown as part of a mixed cropping rotation, where land is rotated between cropping and pasture phases, and the latter may be utilized for either seed production or grazing.

In terms of INM, the country's critically important production agriculture focuses on meeting crop and animal nutrient requirements by applying nutrients that are deficient, or unavailable from natural soil reserves, as multi-nutrient fertilizer materials in order to maximize the quantity and quality of the harvested product. Most fertilizer nutrients used in New Zealand are supplied using imported raw materials for example, phosphate rock and elemental sulfur, to manufacture single superphosphate or by importing already manufactured compound fertilizers, for example, diammonium phosphate (DAP) and urea. Given that increasing, or even just maintaining, the current soil nutrient status requires more nutrients imported than amounts exported in products, New Zealand agriculture will always be a net sink for imported nutrients, at current production levels. Therefore, New Zealand agriculture can only be as sustainable as the capacity of the earth to continue to supply fertilizer raw materials.

AGRICULTURAL PRODUCTION
AND NUTRIENT BALANCES

Globally, continual agricultural technology development has allowed greater production from the same land area (Isherwood, 1998; Lomburg 2001; Borlaug 2003). The same is true for New Zealand. For example, over the past forty years to the mid-1990s, land under cultivation has declined by

6 percent to 16.6 Mha, of which 13.5 Mha is in pasture (Whittington, 1995). Yet, during this time, production has increased as follows (Anonymous, 1998):

- the number of grazing animals (sheep, dairy, beef, pig, deer, and goat) by 58 percent to 58 million
- wool production by 4 percent to 197,000 tonnes (t) per annum
- meat production by 71 percent to 1.8 million tonnes (Mt) per annum
- dairy production by 131 percent to 10.65 million liters per annum
- crop production by 58 percent to 950,000 t per annum

More recent detailed major agricultural production outputs and national fertilizer sales since 1993/1994 (Table 6.1) show a trend for increasing productivity from the same land area but declining livestock numbers which can be attributed to increasing production efficiency. Production in the dairy sector has approximately doubled since 1990, while wheat production has increased by 50 percent. Beef numbers have remained constant while sheep numbers have declined by 40 percent. Estimated macronutrient fertilizer sales for New Zealand are also shown (Table 6.1). Phosphate fertilizer sales have doubled while N fertilizer inputs have increased by approximately 400 percent since 1990. Much of this increase is presumably associated with the doubling in dairy production, although increased production of arable and vegetable crops and lamb meat also influences the fertilizer sales.

On a national basis, fertilizer nutrient imports into the country exceed the export of nutrients in finished products due mainly to the fact that a proportion of the fertilizer nutrients applied remains on the farm, or is lost via leaching runoff, or transfer to nonproductive areas. This is consistent with the fact that some nutrients are accumulating under some farm types and soil groups, as discussed later. However, while fertilizers are usually the most dominant source on nutrients entering the farm, they are not the only ones. On some soils, nutrients are released (weathered) from soil minerals, for example, K and P, and some nutrients come from rainfall, for example, S and N, especially if the farm is close to the coast. Irrigation water, farm dairy effluent, organic manures, and supplementary feed are other sources of nutrients to the farm. Nutrient balances are useful tools to assist in developing a profitable and sustainable nutrient management program, providing they take account of all inputs and outputs. A recent examination, using national agricultural statistics, with a computerized nutrient budget program OVERSEER Nutrient Budgets 2, highlights some interesting points.

TABLE 6.1. New Zealand fertilizer sales (NPK only), livestock numbers, milksolids production, and wheat grain production since 1993/1994

Year	1993/1994	1994/1995	1995/1996	1996/1997	1997/1998	1998/1999	1999/2000	2000/2001	2001/2002	2002/2003	2003/2004
No. sheep[a] (000)	50,298	49,466	48,816	47,349	46,077	44,400	44,190	42,260	40,033	39,546	39,688
No. cattle[a] (000)	4,758	5,048	5,183	4,852	4,775	4,379	4,283	4,276	4,345	4,495	4,614
No. dairy cattle[b] (000)	3,550	3,839	4,090	4,165	4,398	4,516	4,543	4,794	5,038	4,495	4,614
Tonnes milksolids[b] (000)	736	733	788	880	893	880	981	1,096	1,152	1,191	1,254
Tonnes Wheat (000)	2,419	2,452	2,770	NC	NC	NC	NC	NC	3,015	3,203	2,710
Tonnes Barley (000)	3,955	3,028	3,672	NC	NC	NC	NC	NC	4,409	3,783	2,980
Tonnes N[c] (000)	1,057	1,256	1,555	1,400	1,389	1,535	1,800	2,307	2,691	3,230	3,480
Tonnes P[c] (000)	1,690	1,700	1,706	1,694	1,668	1,668	1,773	2,009	2,113	2,081	2,177
Tonnes K[c] (000)	1,127	1,121	1,157	1,121	1,077	1,096	1,104	1,041	1,160	1,344	1,328

Note: NC = data not collected.

[a]Meat and Wool New Zealand Farm Statistics.

[b]Livestock Improvement Corporation Dairy Statistics 1998-99 and 2003-04.

[c]New Zealand Fertiliser Manufacturer's Association Statistics.

257

Dairy Farm Nutrient Budgets

Nine regions represent about 80 percent of the total dairy industry with respect to total area and number of cows. Estimated nutrient balances as determined by OVERSEER for each region and nutrient are given in Table 6.2. The inputs are based on current typical fertilizer inputs used by farmers within their respective regions, and therefore the predicted balances are representative of current practice.

These results suggest that inputs of phosphate (P), potassium (K), and calcium (Ca) exceed the losses of these nutrients under dairying in all regions. The average positive balances are 19, 37, and 196 kg ha^{-1} year for P, K, and Ca, respectively. In contrast, four regions are in negative balance for magnesium (Mg), namely Taranaki, South Auckland, West Coast, and Bay of Plenty (Table 6.2). These same four regions, together with the Central Plateau, Southland, and Northland, are in negative balance with respect to sodium (Na). While this will have no impact on pasture growth, declining soil Na levels could lead to sodium deficiency in milking cows. Assuming that the soil fertility levels for the nutrients P, K, Ca, and Mg are optimum for pasture production (which will not always be the case), then it could be

TABLE 6.2. Dairy farm nutrient balances (inputs from all sources minus outputs from all sources) for each nutrient and region

Region	Nutrient balances (kg nutrient/ha/year^{-1})						
	N	P	K	S	Ca	Mg	Na
Northland	0	26	43	0	206	4	−10
South Auckland	0	16	39	0	155	−19	−33
Bay of Plenty	0	5	33	0	204	−2	−50
Central Plateau	0	6	31	0	209	16	−71
Taranaki	0	13	45	0	164	−23	−44
Wairarapa	0	13	18	0	165	8	5
North Canterbury	0	23	43	0	230	10	25
West Coast	0	27	42	0	237	−16	−25
Southland	0	23	37	0	196	3	−25

Source: D.C. Edmeades, unpublished data.

Note: For N and S the models are constrained such that the balance is always zero. In the case of N, any deficit of N is balanced by an increase in symbiotic N from clover, and for S, any surplus is assumed to be leached.

expected that soil test levels for P, K, and Ca are increasing and for Mg declining. An analysis of around 250,000 advisory soil test levels between 1988 and 2001 (Wheeler et al., 2004) supports the nutrient balance in so far as Olsen P levels have increased in the dairy sector and Mg levels have slowly declined.

Sheep/Beef Farm Nutrient Budgets

Nutrient balances were also derived for the sheep/beef farming sector for seven (of eight) of the farm Classes used by the Meat and Wool Board Economic Service, carrying 95 percent and 98 percent of all sheep and beef, respectively (Table 6.3), using the methodology described. The balances were estimated using typical fertilizer inputs used by farmers within their respective regions and therefore are representative of current practice.

For P, all Classes are in a positive balance, except for the High Country (Class 1) of the South Island, noting that the positive balance for Class 6 (Canterbury Finishing) is small. The K balances are positive for all Classes despite the fact that significant fertilizer K is applied only on Class 5 (Northland/Waikato Finishing). Except for the High Country, all Classes are in positive balance with respect to Ca. Class 5 (NI Intensive Finishing

TABLE 6.3. Nutrient balances (kg nutrient/ha) for each nutrient and sheep/beef farm class

Class	N	P	K	S	Ca	Mg	Na
1	0	−6	15	0	−1	5	2
2	0	10	19	0	36	8	5
3	0	14	31	0	45	2	−11
4	0	13	22	0	46	8	−13
5	0	13	18	0	36	−21	−31
6	0	4	16	0	33	9	5
7	0	10	18	0	46	8	−1

Source: D.C. Edmeades, unpublished data.

Notes: Class 1: South Island High Country; Class 2: South Island Hill Country; Class 3: North Island Hard Hill Country; Class 4: North Island Hill Country; Class 5: North Island Intensive Finishing; Class 6: South Island Finishing/Breeding; Class 7: South Island Intensive Finishing.

For N and S the models are constrained such that the balance is always zero. In the case of N, any deficit of N is balanced by an increase in symbiotic N from clover, and for S, any surplus is assumed to be leached.

on volcanic soils) is in negative balance with respect to Mg, and Classes 3 (NI East Coast), 4 (Wanganui/Manawatu), and 5 (NI Intensive Finishing) are in a negative balance with respect to Na (Table 6.3).

The mostly positive P balances are consistent with the soil test summary data (Wheeler et al., 2004) that shows increasing soil Olsen P levels on sheep/ beef farms also. Current soil K levels are within the biological optimal range even though no fertilizer K is being applied. This is understandable given that sedimentary soils have significant reserve K (it is generally accepted that economically it is prudent to mine these soil K reserves), and sheep and beef cattle are more efficient recyclers of pasture K on the farm. The exception is Class 5, where the current average soil K level for these mostly volcanic soils is also within the optimal range but is being achieved with significant fertilizer K inputs. Current soil Mg levels on both sedimentary and volcanic soils are above the minimum level from optimal pasture production. For most Classes there is a slight positive Mg balance suggesting that this situation can be sustained for some time. Soil Mg reserves are sufficiently high on sedimentary soils to fully meet medium term needs, but the Class 5 volcanic soils are in a negative balance indicating that they could become Mg deficient in time.

In New Zealand agriculture, very little nutrients are purposefully returned or added to the soil as organic nutrient sources. In the grazed pasture situation, 80 to 90 percent of the nutrients animals ingest from forage are returned as dung and urine. On dairy farms, dairy effluent is applied to a proportion of the farm, and small amounts of poultry manure (actually poultry manure and bedding material such as wood shavings) are applied to pastoral and cropping farms in close proximity to broiler and layer chicken facilities.

FERTILIZERS AND SUSTAINABLE DEVELOPMENT

Pasture and Crop Productivity in New Zealand

In order to increase or maintain both pasture and crop productivity, major elements are routinely applied (primarily nitrogen [N], phosphate [P], sulfur [S], and potassium [K]); soil pH is adjusted by the addition of lime ($CaCO_3$) application and trace elements (usually cobalt [Co], copper [Cu], selenium [Se], boron [B], and/or molybdenum [Mo]) are applied to create a soil environment which encourages legume growth and function (biological N fixation), as well as forage with the required mineral content for animal health. However, there are sixteen known essential elements for plant and animal health and production and not all are presently applied as fertilizer or soil amendments. Agriculture is depleting soil reserves of these nutrients through the net export of nutrients as product or as indirect soil losses.

To date, fertilizer recommendations for optimizing pasture production have been based on the premise that only those macronutrients that are deficient in soil are applied, typically N, P, K, and S. Lime is applied as required to alter the soil pH and thus calcium (Ca), magnesium (Mg), and sodium (Na) are rarely recommended. For this reason, currently, pastoral farming is for some regions and farming types a net exporter of these nutrients, as shown in Tables 6.2 and 6.3. This nutrient depletive philosophy is not sustainable and at some time in the future, soils will be unable to supply sufficient quantities of these essential nutrients (e.g., Ca, Mg, and Na and possibly some trace elements) to sustain high levels of pasture and animal production. At present, it is not known how long our soils will continue to supply sufficient levels of these nutrients, and it is possible that consumer perception may determine that soil nutrient reserve depletion does not fit in with their concept of "sustainable agriculture."

It is possible then to surmise that future fertilizer nutrient requirements will include at least two additional essential cations, namely Ca and Mg. Fortuitously, the prevalence of calcium phosphate fertilizer (single superphosphate) use in New Zealand pastoral agriculture has meant that sufficient Ca has been unconsciously added to replace losses, occurring particularly on well buffered soils where lime application has been sporadic. In the early 1990s, a marked swing away from these "traditional" fertilizers to the use of high analysis NPK types (such as DAP) may have resulted in Ca depletion to the extent that animal production may have suffered. Losses of Ca from dairy soils can be alarmingly high. Recent work (Rajendram et al., 1998) has measured cation leaching (as a function of counterion movement due to nitrate leaching) under grazed dairy pastures. Bearing in mind that this is representative of intensive dairying and leaching, measurements include, a very high drainage year, inputs of Ca are almost in balance with outputs, that is, $+10$ kg ha^{-1} Ca (Figure 6.1). However, should a fertilizer be used which contains little or no Ca, then Ca depletion could be potentially quite large. Edmeades and Perrott (2004) also concluded that New Zealand's current farming practices are sustainable with respect to Ca, providing the use of superphosphate and lime continues.

There is little evidence of widespread application of Mg fertilizer to enhance pasture production or herbage Mg content (for animal requirements). The exception is on undeveloped pumice soils that may be deficient in Mg for grass/legume pasture growth. Most soils are not deficient in Mg for pasture growth and so typically Mg does not form part of the fertilizer program. Hypomagnesemia remains an issue in many areas, particularly before and

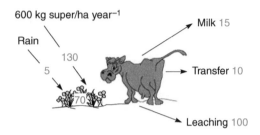

FIGURE 6.1. Annual inputs and outputs of calcium (kg/ha) at the Dairying Research Corporation (DRC) No. 2 Dairy with no N fertilizer applied. *Source:* S. F. Legard, unpublished data.

after calving or lambing. Despite research showing Mg fertilizer as effective in preventing or reducing this metabolic disease (O'Connor et al., 1987), most farmers directly supplement their animals during risk periods. As indicated in the pastoral nutrient budgets, Mg balances are frequently negative implying soil Mg reserves are declining. In his review, Edmeades (2004) suggested that in the absence of fertilizer Mg inputs, current reserves of soil Mg will be sufficient for a further twenty years.

Sodium is not required for plant growth but is essential for optimal animal production. Edmeades and O'Connor (2003) in a recent review identified the central regions of both Islands as becoming Na deficient for grazing animals and noted that it was likely the Na would become a regular fertilizer input in these regions in the next few decades.

Environmental Concerns: Grazed Pastures

Phosphate runoff (surface water quality) and nitrate leaching (groundwater quality) are the two main issues relating to nutrient management under pastoral systems.

Phosphorus. Unlike nitrate (NO_3^2) and sulphate (SO_4^2), P leaching is not an issue in New Zealand except for minor areas of unmineralized peat soils, and some highly weathered podzolized sedimentary soils under high rainfall, and with low phosphate retention. This is because the binding capacity for P (phosphate retention) is much higher than other anions, and most New Zealand soils have a large P sorbing capacity due to large quantities of short-range order iron and aluminum compounds in soils (McLaren and Cameron, 1990). However, it has been estimated that 45 percent to 80 percent (McColl, 1982; Gregg et al., 1993) of P inputs to surface water arise from diffuse agricultural sources such as runoff.

There have been many studies (n = 27) in New Zealand which have measured P losses from a variety of land uses (Ministry for the Environment, 2002). Mean export coefficients are; 1.0, 1.98, 0.46, 0.39, and 0.35 kg P ha^{-1} per year from dairy land, hill soils, low intensity pastoral use, native catchments, and exotic forests respectively. Thus there should be no doubt that pastoral farming increases P loadings on waterways relative to the natural environment. Gillingham and Thorrold (2000) in a review, reported that about 90 percent of the total P entering waterways comes from diffuse sources (runoff and sediment), and that most of this is in the form of sediment (particulate associated P or PAP), the balance being dissolved P in runoff water. For this reason they emphasized management practices to reduce sediment entering the waterways (stream fencing and riparian planting, and control of active erosion) as the primary tools to reduce P losses from pastures. Other simple techniques include, avoiding direct application of fertilizer to waterway, ensuring adequate pasture cover during periods of heavy rain, and avoiding the application of fertilizer prior to the rainy season.

Nitrogen. Although New Zealand has sufficient good quality water resources there is increasing concern regarding the impact of agricultural activities on water quality. The New Zealand Ministry of Health (MoH) has adopted the World Health Organization's maximum acceptable standard of 11.3 mg NO_3-N L^{-1} for drinking water (Ministry of Health, 1995). The concentration of NO_3^2 in some shallow aquifers frequently exceeds the MoH limit for drinking water (Burden, 1982; Dillon et al., 1989). For example, in the Hamilton basin, over 30 percent of the shallow wells (<15 m deep) have NO_3-N levels >10 mg L^{-1} and shallow bore water is not recommended for feeding infants (Hoare, 1986). Grazed pasture and cropland are the major sources of this contamination, although in specific circumstances, point sources such as dairy factory and piggery waste and market gardening have been identified. Nitrate contamination of surfaces waters (e.g., rivers and lakes) has resulted in eutrophication causing undesirable growth of aquatic plants and algae leading to loss of fisheries, reduced aesthetic appeal, and increases in the cost of water abstraction (Cameron and Haynes, 1986).

The nitrate contamination derives primarily from the urine of grazing animals that can contain up to the equivalent of 1,000 kg N ha^{-1} for dairy cattle (Ledgard et al., 2000) and 500 kg N ha^{-1} for sheep. The soil/plant system cannot retain these high amounts of N once the urea has been biologically converted through ammonium ions (strongly retained) to nitrate because the retention capacity for this anion by New Zealand soils is low and the nitrate prone to leaching. In New Zealand, most leaching occurs over the winter and early spring when pasture N uptake is low, soils are at or near field capacity, and drainage is occurring. Studies under grazed dairy pastures in the

Waikato region of New Zealand have shown that up to around 200 kg N ha^{-1} of fertilizer N can be applied annually before the drainage water exceeds the MoH drinking water standard and that very little N is directly lost from the fertilizer up to this application rate (Ledgard et al., 1999).

The use of nitrification inhibitors, such as dicyandiamide (DCD), as a tool to assist in reducing the impact of urine patches on nitrate leaching has been investigated in New Zealand for the past five years (Di and Cameron, 2002, 2003, 2004a,b,c, 2005). Lysimeter studies on irrigated soils formed from sedimentary rocks show the potential to decrease nitrate leaching by 59 to 76 percent under dairy cow urine patches equivalent to 1,000 kg N ha^{-1} (Di and Cameron, 2002, 2004, 2005), with concomitant decreases in cation leaching that is, Ca, K, and Mg, ranging from 31 to 65 percent (Di and Cameron, 2004a, 2005). Pasture production responses have also been measured following DCD application (Di and Cameron, 2004, 2005), which means that farmers could substitute DCD for some of the N fertilizer they would otherwise apply, and maintain pasture production, while reducing nitrate and cation leaching. The research has resulted in DCD formulations being released commercially, either as a fine particle suspension (Di and Cameron, 2005) or coated on inert granules by the two major fertilizer companies in New Zealand.

Greenhouse gases (GHG). The main contributor to GHG emissions from grazed pasture relate to methane production from grazing ruminants and nitrous oxide emissions from excreted N. The smallest contribution will come from fertilizer N and can be minimized by applying the N fertilizer in environmental conditions to promote maximum plant growth rather than environmental loss. An exciting by product of the nitrification inhibitor work has been the discovery that applying DCD to grazed pasture soils also significantly decreases nitrous oxide, the most destructive of GHG emissions from urine patches. Reductions of N$_2$O between 76 and 82 percent have been measured on urine patches treated with DCD (Di and Cameron, 2002, 2003).

Environmental Concerns: Arable Crops

While nitrate leaching and P runoff are the two main issues relating to nutrient management under pastoral systems, it is only the former that has been a major concern under crop production. Within the mixed cropping system the pastoral phase is composed of two species: ryegrass and clover. This provides significant amounts of N via fixation by the *Rhizobium* bacteria associated with clover nodules. White clover seed crops are also included in many rotations, which can result in 160-220 kg N ha^{-1} being supplied to soils (Williams and Wright, 1997). Consequently, cultivation of land out of clover seed production or pasture results in high rates of soil mineralization. Much of this soil mineral N is potentially available for uptake by the subsequent crop (e.g., wheat, potatoes). However, to maximize this utilization requires knowledge of the behavior of soil N under various farm management practices, including timing of cultivation, effects of rainfall, cover crops, planting dates, and soil types. Additional crop N requirements are met with fertilizer inputs, typically urea. These are applied in a timely manner to match crop demand. If the N contribution from the soil is underestimated, there is a risk of N fertilizer being over prescribed. By harvest, residual N may remain in the soil from several sources: soil mineralization, crop residue decomposition, and unused fertilizer. Some of this N can subsequently be lost via nitrate leaching during the winter drainage season (approximately May to September).

Although crops such as winter potato production cover a very small land area, they receive high rates of N fertilizer. Research has shown that these cropping systems result in high rates of nitrate leaching (Francis et al., 2003). In New Zealand, there is very little input of manures compared to many other countries.

Greenhouse gas emissions from the cropping sector are more closely associated with the cultivation of land than with fertilizer and manure applications. Significant amounts of carbon dioxide and nitrous oxide can be emitted from soils that have been ploughed (Crush et al., 1992; van der Weerden et al., 1999).

INTEGRATED NUTRIENT MANAGEMENT

Pastoral Soils

Until relatively recently New Zealand pastoral farmers relied almost exclusively on N fixation by legumes such as white clover (*Trifolium repens*) in the sward. A well-established grass/clover pasture (around 80 percent and 20 percent respectively) growing in fertile soils with adequate moisture could contribute approximately 200 kg N ha^{-1} annually to pasture nutrition. Fertilizer nutrient applications in this system are predicated on ensuring the optimum supply of major and trace elements to support legume growth, as the nitrogen fixation function is directly proportional to clover growth. In addition, as the mixed pasture grown is usually the major livestock forage for pastoral farms, the nutritional quality of this forage, particularly with respect to trace elements essential for animals but not plants, such as cobalt (Co) and selenium (Se), can be manipulated in part in the fertilizer program.

In recent years, the amount of N fertilizer used on pastoral farms has increased dramatically (Roberts et al., 1992) as farmers have pushed the productivity of their farm systems higher, and found that tactical applications of N fertilizer usefully increase the amount and change the distribution of pasture growth to better fit animal demand throughout the year.

Grazed pastoral soils are net accumulators of organic matter, and so, the use of organic amendments on pastoral soils are either not required or are not economical to apply, even if the amendment itself is free, as there is considerable expense involved in cartage and spreading. In general, INM for pastoral farms involves the use of traditional fertilizer nutrients, the appropriate rates of which take account of nutrients supplied by the soil and those brought in as supplementary feed from outside the farm gate, or redistributed within the farm such as by the land application of farm dairy effluent.

There are a host of organic amendments and biostimulants on the New Zealand market ranging from seaweed, fish, or animal rendering waste material-based liquid "fertilizers" through vermicompost, to various living microorganism cultures, and so called elicitor compound solutions. Unfortunately, many of the often large production benefits claimed by the zealous marketers of these products are not supported by any credible scientific proof of efficacy (Edmeades, 2002). Most permanent pasture soils in New Zealand have a high level of indigenous vesicular-arbuscular mycorrhiza (VAM) fungi (Crush, 1975; Powell, 1977) leading to an established sward with abundant roots already highly colonized by indigenous VAM. There have been no experiments in New Zealand which have showed productivity gains from introduced VAM inoculation (Powell, 1984).

New Zealand's long-term cropping soils contain, on average, 3-5 percent organic carbon. The requirement for organic matter inputs through manure applications is low, as a mixed cropping rotation is practiced in most arable regions. Farmers that do apply manures such as poultry and pig manure to cropped land, have in the past often overlooked the nutrient inputs added as part of the manure application. However, there is increased awareness of the nutritional values of the waste by-products, with companies actively promoting their products on the basis of organic matter inputs and nutrient inputs.

A mixed cropping rotation provides an opportunity to rest land from continuous cultivation. Research has shown that continuous cropping using conventional cultivation practices reduces soil stability, which in turn can lower crop yields (Francis et al., 1998). Poorly structured soils typically require increased mechanical cultivation, while the incidence of surface crusting and soil erosion is also increased. Pasture will help restore soil structure in addition to increasing microbial activity (Fraser et al., 1996; Haynes and

Tregurtha, 1999) through inputs of organic matter. Cultivation of pasture leads to significant losses in organic matter, particularly in the first year. Preliminary results from a trial on organic matter management has shown that 11 percent of soil C in the top 30 cm can be lost in the first eight months following cultivation of a long-term pasture (M. Beare, personal communication). In this trial, cultivation was conducted by ploughing to 20 cm depth, followed by two passes of a maxi-till and grubber to 10 cm depth. There is increasing interest in minimal tillage and no-tillage cropping which help to maintain organic matter levels, reduce soil erosion, and more importantly, in dry regions such as Canterbury, aid in conserving soil moisture. Long-term estimates suggest that regions such as Canterbury will become increasingly drier in the future.

TECHNICAL REQUIREMENTS FOR INM

Pastoral

Many factors influence nutrient requirements on individual farms. The best fertilizer management practices should consider agronomic, environmental, and economic factors as they all underpin sustainable agriculture. Economic factors to consider include the cost of fertilizer, transport, and spreading; the pasture and animal production response; the returns from the livestock enterprise; the ability to finance fertilizer purchases; the opportunity cost of money spent on fertilizer; and the farmer's goals and planning horizon. The high residual value of fertilizer application makes a long-term approach essential when making fertilizer use decisions. Short-term changes in soil fertility are of special interest when fertilizer is withheld or large capital applications are made to increase soil fertility.

A good starting point to developing a profitable strategy for fertilizer nutrient application is to measure the level of soil fertility, in terms of pH, P, K, S, Ca, and Mg, that is current on the farm. These tests, in conjunction with past fertilizer history, will assist in establishing where on the pasture development/maintenance sequence the farm sits. However soil tests, like all biological measurements, are variable and therefore a single soil test taken at one time is of limited value.

Maximum advantage from soil analysis will be achieved by repeated testing over a number of years. In this way, a picture of trends in soil fertility status of the farm is built up and can be compared relative to target soil test ranges. Thus, in the long-term, regular soil sampling is required to monitor an increase in soil nutrient levels from capital fertilizer inputs or to fine-tune maintenance requirements. Herbage analysis should also be used to complement the soil sampling program and can be sampled on the same transects as the soils. Pasture samples are useful for fine-tuning the major nutrient requirements and are essential for determining trace element requirements for both pasture growth and animal health.

The fertilizer nutrients required are determined for each individual farm based on knowledge of the farm's soils, animal production, and farm management system. For example, what materials the soils are formed from determine how much P fertilizer is required to increase soil test levels, how well it retains sulphate S against leaching, and whether or not there is any K mineralized from the soil. The amount of animal product going off the farm and the stocking rate, milking times, effluent management, forages, and supplementary feed used all affect additions, losses, and movement of nutrients onto, off, and around the farm.

There are rules of thumb available, created from experience with nutrient flows on farms as well as scientific knowledge, to assist in determining nutrient requirements. For example, on sedimentary soils, approximate maintenance fertilizer is equivalent to 0.5-0.7 kg ha^{-1} 15 percent potassic superphosphate or equivalent for every 1 kg ha^{-1} milksolids produced from pasture (Roberts and Morton, 1999) and for sheep/beef farms it is the equivalent of 20 kg superphosphate/ha for every stock unit wintered. On these soils it takes on average 5 kg P ha^{-1} (over and above maintenance P) to raise the Olsen test by 1 unit; 125 kg K ha^{-1} to raise soil test K 1 unit, and 30-40 kg S ha^{-1} to overcome an S deficiency (Roberts and Morton, 1999).

The drawback with these approximations is that while they are quick and easy to use, they will not necessarily give you the most profitable fertilizer program for your farm business. Software-based decision support systems (DSS) incorporating decades of agronomic research into soil/plant nutrient interactions and knowledge of grazed pasture systems have been developed. The DSS, called OVERSEER3.0 nutrient requirement software (Metherell et al., 1997) and OVERSEER Nutrient Budgets 2 software, are designed to help farmers and consultants optimize their nutrient requirements to match their production objectives without leading to depletion or excessive buildup of nutrients, by taking into account nutrient additions from nonfertilizer sources such as farm dairy effluent, supplementary feed, soil reserves, nutrients in irrigation water, and atmospheric contributions. Use of the software helps identify farms that may save money by reducing fertilizer expenditure and those that could increase profit by increasing then maintaining or increasing soil fertility levels. The uses of the DSS are demonstrated in the section "Technical Requirements for INM" in example case study farms.

Arable

Fertilizer recommendations for arable cropping take a similar approach to that of the pastoral sector. Soil and plant analysis are increasingly used as the basis for establishing a fertilizer recommendation and correcting nutrient deficiencies. Where information exists on the optimum soil fertility status for various crops, recommendations by fertilzer and other agricultural consultants aim to increase the soil fertility to the optima. High value crops usually require soils to be at optimum soil fertility to maximize crop yields. As the crop value decreases and/or the cost of fertilizer inputs increase, the economically optimum rate of fertilizer declines.

How does one determine the economically optimum fertilizer rate? Recently, New Zealand scientists have developed a yield response model called Parjib (Reid, 2002). This model was used as the basis for a decision support tool called Parjib-Express, which provides farmers with transparent information on yield and economic responses to fertilizer inputs, thus enabling them to maximize their profit, as opposed to maximizing yield (Reid et al., 2005). Soil test results are required as an input to this tool, where the soil N supply is estimated using the anaerobically mineralizable N test (Keeney and Bremner, 1966).

In an attempt to improve N fertilizer recommendations to some broadacre crops, there has been increasing interest in collecting soil samples to 60-90 cm depth for determining soil mineral N content. Sampling is conducted in the late winter-early spring period, prior to spring N fertilizer dressings. The test results are used to help better determine the rate of N required, rather than estimating soil N supply on the basis of paddock history, crop rotation, soil type, and climate. While providing a snapshot of the mineral N status at the time of sampling, there is an associated practical challenge to this type of sampling.

In 2005, agronomy scientists released a new decision support tool for wheat production called the Wheat Calculator. This software program uses inputs such as deep soil mineral N content, along with other key information including cultivar, sowing date, soil type, and climate data, to help a farmer or consultant manage their N and irrigation inputs to optimize wheat yields while minimizing the amount of nitrate leached (Jamieson et al., 2003). Similar calculators are being developed for potato and maize production.

ACTUAL IMPLEMENTATION AND INM

At the time of writing, in New Zealand, two companies supply 95 percent of the fertilizer sold and used. Both companies are 100 percent farmer-owned cooperatives, and thus aim to recommend the most suitable fertilizer at the appropriate rate to help maximize shareholder profits rather than Company profits. Fertilizer suitability will be determined by several factors such as crop type, method of establishment, soil conditions, climatic conditions, and time of year. Fertilizer consultants are equipped with computer-based software tools to assist with determining the most appropriate rate of nutrients, based on soil test results, fertilizer history, and yield potential. Indeed, there are many instances where consultants have tried to persuade farmers to reduce their fertilizer rates, for two reasons: (1) to maximize profit, and (2) to reduce the potential of environmental contamination. However, farmers do not always heed the advice of the fertilizer company staff as there is a perception that the advice is not given independently from trying to close a sale. In addition, there is a host of qualified and unqualified operators in the field, all influencing the decisions farmers make about fertilizer programs that makes servicing the 50,000-55,000 farmers, through the network of less than 100 private consultants, and the 120 or so fertilizer company field officers quite challenging. In order to help farmers achieve greater efficiency in fertilizer use on farms, as well as minimize off-farm impacts of fertilizer use on water quality, a number of initiatives have been taken. The New Zealand Fertiliser Manufacturer's Association has developed a voluntary Fertiliser Code of Practice for all agricultural sectors, and Fonterra, the largest dairy farm cooperative in the country has developed a Clean Stream Accord with the New Zealand Government. In addition, there are some well-qualified private consultants who use science and science-based decision aides to give farm-specific nutrient advice including economic and environmental assessments of fertilizer policies.

There has been a gradual increase in the land area dedicated to organic farming systems. Organically certified produce receive premium prices, both domestically and internationally. Fertilizer companies supply a range of fertilizers that are either permitted for use by organic farmers or have restricted use where dispensation is required by the certification organization such as Bio-Gro and Certenz. Most nutrients can be supplied using permitted products, however the major limitation is nitrogen. Material available that does contain N (such as certified fish and bone meal) is often too costly to apply at a broadacre scale.

A recently formed New Zealand No-Tillage Association (NZNTA) is a consequence of growing interest in the use of no-tillage practices for crop production. The association is now helping to increase further interest in no-tillage practices. In 2000, 4 percent of all seeding in New Zealand was conducted using no-tillage methods: by 2005 the proportion of seeding by no-tillage methods had increased to approximately 15-20 percent

(C. J. Baker, NZNTA president, personal communication). The NZNTA is currently exploring the potential for carbon trading on the international market. From a nutrient supply point of view, where no-tillage is practiced on soils with low fertility, fertilizer may need to be applied down the spout with seed. However, farmers need to exercise caution to minimize the risk of germination injury, which is influenced by soil moisture, fertilizer type and rate, fertilizer placement, and seed type. There are some no-tillage drills (e.g., Cross Slot) that place fertilizer and seed apart in the soil, thereby reducing the risk of germination injury.

DISCUSSION OF CASES

The following two cases describe real farm situations, one a pastoral dairy farm and the other a more extensive hill country sheep and beef farm, where farmers have sought advice on nutrient management from trained professionals using science-based tools and information.

Case 1: Bay of Plenty, Dairy Farm

This high-producing dairy farm comprises 125 ha of free draining yellow brown pumice (rhyolitic ash) soils. It is currently producing 1,400 kg milksolids (MS) ha^{-1} per year and the goal is to achieve 1,500 kg MS ha^{-1}. The current average Olsen P level is ninety and the estimated economic optimal Olsen P level (the level required for maximum profitability) is in the range 40 to 45, as determined by using OVERSEER 3 nutrient requirement decision support software. Currently, the effluent from the dairy enters an anaerobic/aerobic two-pond system and the sludge is periodically removed and applied to an area (19 ha^{-1}) close by. There are no permanent streams, but three dams have been installed to reduce flooding at times of intense rainfall, and there are significant riparian plantings. There is a "runoff" that is used to grow supplementary feed and where the cows are wintered.

The farm is close to Lake Rotorua, a major tourist attraction, and there is intense pressure to reduce nutrient loadings to this lake. For this reason, professional advice was sought to improve nutrient use efficiency on the farm and reduce nutrient losses from the farm to the lake.

Table 6.4 sets out the options identified on this farm to reduce nutrient loadings of N and P and a qualitative estimate of the costs and benefits of each option.

The most obvious changes in farm management were to reduce fertilizer P inputs and mine the soil P levels back to the economically optimal range.

TABLE 6.4. Management options identified to reduce nutrient loading—Case 1

Option	Cost	Priority	Benefits
Withhold fertilizer P	Nil	1	Saving in fertilizer costs, reduce P loading
Increase effluent area from 19 to 39 ha	Pipes and pumps	1	Reduce fertilizer costs, reduce N and P loadings
Reduce proportion of fertilizer N applied in May, June, July from 50 percent to 0 percent	More winter supplements	2	Reduce N loading
Reduce fertilizer N inputs from 200 to 100 kg N/ha/year	More winter feed	2	Reduce N loading
Riparian/wetlands around two existing dams	Fencing and planting	3	Reduce N and P loadings, improve landscape
Riparian/wetland at boundary outfall	Fencing and planting, small loss in productive land	3	Reduce N and P loadings, improve landscape
Reduce stocking rate	Loss in production	4	Reduce N loading
Reduce stocking rate but increase per cow production	Cost of improved animal genetics and winter supplements	4	Reduce N loading

Source: D.C. Edmeades, Unpublished data.

This would greatly reduce fertilizer costs without any loss in production and therefore was in the farmer's economic interests. This farm produced about $4,000 worth of nutrients in the farm dairy effluent. The current practice was to apply this to 19 ha. Simple calculation showed that the nutrient inputs per hectare were well above agronomic requirements. Therefore increasing the effluent area to 39 ha made more efficient use of the nutrients, and hence reduced nitrate leaching in particular, and saved further fertilizer expenditure—no further fertilizer inputs were required on the effluent area.

This farm is using high inputs of fertilizer N to achieve the desired production goals. Restricting fertilizer N inputs over the winter period when the soils are already at field water capacity was calculated to reduce annual nitrate leaching, without a significant loss in production. While reducing total N fertilizer inputs would reduce nitrate leaching, it would also reduce total production, and so was not an economically viable option. Similarly reducing stocking rate would decrease nitrate leaching by reducing the number of urine patches per unit area, but this option was not preferred by the farmer unless per cow production could be increased.

The farm already has three dams, originally installed to reduce soil erosion. The farmer was prepared to consider further planting of trees and shrubs around these areas to improve their riparian buffer effect.

Of crucial importance in terms of encouraging changes in behavior, was that the farmer wanted more information particularly to quantify the costs and the benefits of the various options. The point here is that the farmer was more than happy to make decisions in favor of the environment if at the same time it increased the farm's profitability.

Case 2: Hill Country Sheep and Beef Farm

This farm comprises 585 ha of rolling to steep pumice soils within the Lake Okareka catchment, a pristine tourist attraction. The farm is currently under development and has low soil nutrient levels and there is considerable pressure on the landowner to return the land to forestry. However the owner has taken the decision to make the farm more profitable but at the same time meet the requirement to minimize nutrient loading. Fencing and watering is complete and professional advice was sought to develop a fertilizer management plan consistent with the goals of the farm including minimizing avoidable nutrient losses. Current soil test levels indicated gross P and S deficiency.

A capital fertilizer program was recommended coupled with a combination of other options (see Table 6.5) to reduce nutrient loadings from the farm. Although the farm has already some significant riparian plantings it was recommended that these be extended and improved. Similarly, there are a number of natural ponds on the farm and it was recommended that these be upgraded into constructed wetlands. The farmer was more than happy to retire the less productive steep hillsides that show evidence of active erosion. This would have the added advantage of improving the aesthetic value of the land. The farmer was not prepared to reduce stock numbers unless this could be coupled with an increase in per animal production.

Once again the farmer was very enthusiastic for more information on the costs and benefits of the various options.

TABLE 6.5. Management options identified to reduce nutrient loading—Case 2

Option	Cost	Priority	Benefits
Riparian/wetland plantings within farm	Loss in productive land, fencing, and planting	1	Reduce N and P loadings
Retirement of worst hill sides	Loss in production, fencing, and planting	2	Reduce N and P loadings, improve aesthetics
Reduce stocking rate but increase per animal performance	Cost of improved animal genetics and winter supplements.	3	Reduce N loading
Reduce stocking rate	Loss in production	4	Reduce N loading

Source: D.C. Edmeades, unpublished data.

RESEARCH GAPS AND FUTURE RESEARCH NEEDS

As the day approaches when nutrient budgeting as a way of minimizing the offsite impacts of farm nutrients is a requirement of farming in all sectors, there is an increasing need for practical tools to help farmers achieve this. Computer software such as the OVERSEER suite of programs provides a tool for pastoral farmers and consultants alike. Apart from calculating nutrient balances, this tool is capable of estimating the drainage water nitrate concentration based on several input variables. This provides an estimate of the long-term effects of various farming practices including the influence of livestock management, cover crops, time of cultivation, fertilizer timing and rate, soil type, and rainfall. At the time of writing, the OVERSEER model was being reexamined to improve its accuracy for arable, vegetable, and horticultural production systems. Other decision support systems are also being developed that will provide similar outcomes.

A common challenge with these tools is the need for a measure or estimate of the soil N concentration, particularly for arable and horticultural production. This is often determined by measuring the anaerobically mineralizable N or, more recently, deep soil mineral N content where soil cores are sampled to 60-90 cm depths. Each method measures a different part of the soil nitrogen pool, and both methods have their limitations. The former provides a gross estimate of the potentially available N supply via soil mineralization, while the latter provides a practical challenge. Although development of software tools such as the Wheat Calculator can help to minimize the amount of potentially leachable N by the end of the growing season, determination of soil N supply will need to be simplified. This problem is likely to be overcome in time with the development of an N index system, where estimated N content is based on variables such as cropping history, fertilizer history, soil type and depth, irrigation and rainfall, and crop type and yield. An N index was in use in New Zealand in the early 1990s for cereal production, but is in need of updating due to changes in farming practices and higher yielding cereal varieties. Such an index is in use in the United Kingdom, resulting in a decline in physical soil sampling for soil N content. With the establishment of an N index system, farmers and consultants will be able to predict the additional N requirement from the fertilizer bag with greater assurance.

Current research efforts by various science providers revolve around the use of remote sensing technology such as electromagnetic sensors placed on farm equipment to "map" soil properties and estimate forage yields throughout paddocks and farms. In addition, the New Zealand dairy industry is actively involved in the Australian CSIRO's "Pastures from Space" program, whereby satellite imagery is used to estimate pasture and crop yields, and predicts future production also. The use of these technologies in pastoral agriculture, and even for arable field cropping, is nascent in New Zealand. Much more work needs to be undertaken to determine the value proposition of such technologies before farmers will adopt these systems.

As the processes and tools for nutrient management are integrated together, we predict there will be a rapidly emerging requirement for Whole Farm Nutrient Management Plans to be produced for each farm, particularly those in sensitive catchments with respect to water quality. It is envisaged that these plans will involve written records with a structure similar to that set out in Exhibit 6.1.

Much information will be captured using ortho-corrected farm maps (Figure 6.2) to provide permanent proof of placement of fertilizer products, rates, and dates of application (Figure 6.3). This technology is already in place in some ground spreading fertilizer equipment as well as aerial topdressing airplanes.

Linking Geographic Information System and Global Positioning System (GIS and GPS) technology together allows the actual path of travel of either ground spread trucks or airplanes, and the swath width of the fertilizer spread to be recorded on farm maps also (Figure 6.4). In a simple picture, there is powerful evidence of where the fertilizer was placed and where native vegetation, or surface waterways were purposefully avoided.

There are several challenges facing future production: (1) maintaining the balance between economic sustainability and environmental sustainability;

EXHIBIT 6.1. Structure of Whole Farm Nutrient Management Plans

1. Nutrient management objectives

Example: Our nutrient management objective is to maintain soil fertility to optimize pasture productivity while taking all practical steps to avoid nutrient losses to water.

2. Land management units

The following land management units have been identified on this property.

Unit	Description	Approximate area (ha)
A	Steep hill country	200
B	Rolling downlands	150
C	Flats	55
D	Flats (Effluent block)	18

3. Soil/herbage/animal test results

(a) Soil (date of sampling   /   /   )

Transect No.	N	Olsen P	Mg	S	K	Ca
1						
2						
3						
4						

Comments
Graphs of soil test trends

(b) Herbage tests (date of sampling   /   /   )
(c) Animal tissue tests (date of sampling   /   /   )

4. Recommended fertilizer program (including costs/economics)

(a) Insert fertilizer plan here.

5. Nutrient budget analysis

(a) Insert whole farm nutrient budget.
(b) Insert Effluent block nutrient budget (for dairy farm).

6. Environmental risks

The following environmental risks have been identified for these land management units.

Risk factor	Unit A	Unit B	Unit C	Unit D
Nitrate leaching				
P runoff				
Extreme soil P status				
Effluent Block N loading				
Effluent Block K loading				
Riparian margins				
Stock access to waterways				

7. Revised recommended fertilizer program

Insert new fertilizer plan which addresses risk factors identified above.

8. New nutrient budget outcomes

Comments
Include comments on how this program contributes to NM objective above, reasons for choice of products, etc.

9. Management practices

The following management practices will be used to achieve the objective for nutrient management as listed above.

	Checklist	
Specific Industry requirements	**Yes**	**No**
Annual nutrient budget to be prepared for each block??		
Proof of placement		

	Checklist	
Specific Regional Council requirements	**Yes**	**No**
Compliance with the code of practice		
Apply no more than 150 kg N		

(continued)

EXHIBIT 6.1 *(continued)*

In addition we will implement our own management policies to achieve the above objectives including:

Management practices	Undertaken Yes/No	Verification
Will use only Spreadmark Certified operators		
Annual soil tests on each block		
Timing		

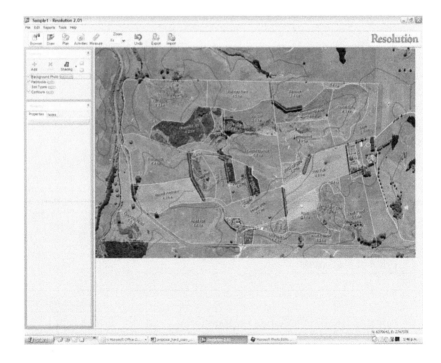

FIGURE 6.2. An example of a farm map showing paddock fences and other details. *Source:* A. H. C. Roberts, Ravensdown Fertiliser Co-Operative, unpublished data.

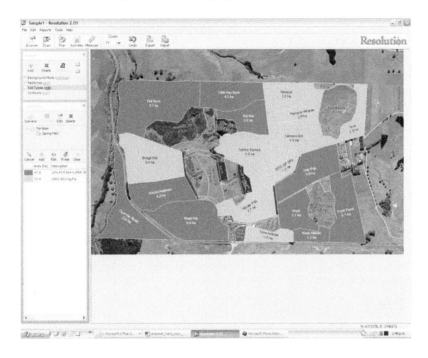

FIGURE 6.3. Map showing paddocks with different fertilizer products and rates applied. *Source:* A. H. C. Roberts, Ravensdown Fertiliser Co-Operative, unpublished data.

(2) the need for continuing investment in fundamental research in areas such as crop physiology and land and environmental management, as these are the cornerstones for the development of decision support tools; and (3) the need to integrate decision support tools for farmers and consultants to reduce the level of duplicate data entry.

SUMMARY AND CONCLUSION

Earlier, it was stated that farming in the future requires both an environmentally sustainable and economically sustainable approach. With increased understanding of crop requirements together with the evolution of decision support tools, consultants and agronomists will be able to provide better fertilizer advice for sustained production. There are several challenges facing the viability of future production: (1) balancing economic and environmental

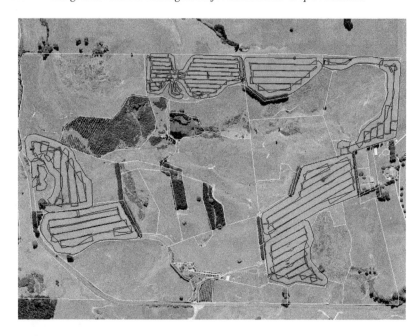

FIGURE 6.4. Map showing path of travel and swath width of fertilizer applications. *Source:* A. H. C. Roberts, Ravensdown Fertiliser Co-Operative, unpublished data.

sustainability; (2) continuing investment in fundamental research in crop physiology and land management research, as these are the cornerstones for the development of decision support tools; and (3) integrating decision support tools for farmers and consultants to reduce the level of duplicate data entry. To meet these challenges, the farming industry, regional, and central governments need to work closely together to ensure that New Zealand's major export earning industry is able to continue to utilize the land resource into the future, while ensuring that other stakeholders in the country's wider environment have their needs met also.

REFERENCES

Anonymous. (1998). *New Zealand Official Yearbook.*
Borlaug, N.E. (2003). Feeding a world of 10 billion people: The miracle ahead. In Bailey R. (ed.) *Global Warming and Other Eco-Myths: How the Environmental Movement Uses False Science to Scare Us to Death.* pp. 29-59. Muscle Shoals, AL: IFDC.

Burden, R.J. (1982). Nitrate contamination of New Zealand aquifers. *New Zealand Journal of Science.* 25: 205-220.

Cameron, K.K.C. and R.J. Haynes. (1986). Retention and movement of nitrogen in soils. In Haynes R.J. (ed.) *Mineral Nitrogen in the Plant-Soil System.* pp. 166-241. Orlando: Academic Press.

Crush, J.R. (1975). Occurrence of endomycorrhizas in soils of MacKenzie Basin, Canterbury, New Zealand. *New Zealand Journal of Agricultural Research.* 18: 361-364.

Crush, J.R., G.C. Waghorn, and M.P. Rolston. (1992). Greenhouse gas emissions from pasture and arable crops grown on a Kairanga soil in the Manawatu, North Island, New Zealand. *New Zealand Journal of Agricultural Research.* 35: 253-257.

Di, H.J. and K.C. Cameron. (2002). The use of a nitrification inhibitor, dicyandiamide (DCD), to decrease nitrate leaching and nitrous oxide emissions in a simulated grazed and irrigated grassland. *Soil Use and Management.* 18: 395-403.

Di, H.J. and K.C. Cameron. (2003). Mitigation of nitrous oxide emissions in spray-irrigated grazed grassland by treating the soil with dicyanamide, a nitrification inhibitor. *Soil Use and Management.* 19: 284-290.

Di, H.J. and K.C. Cameron. (2004a). Effects of a nitrification inhibitor dicyandiamide on potassium, magnesium and calcium leaching in grazed grassland. *Soil Use and Management.* 20: 2-7.

Di, H.J. and K.C. Cameron. (2004b). Effects of temperature and application rate of a nitrification inhibitor, dicyandiamide (DCD), on nitrification rate and microbial biomass in a grazed pasture soil. *Australian Journal of Soil Research.* 42: 927-932.

Di, H.J. and K.C. Cameron. (2004c). Treating grazed pasture soil with a nitrification inhibitor, eco-n, to decrease nitrate leaching in a deep sandy soil under spray irrigation—A lysimeter study. *New Zealand Journal of Agricultural Research.* 47: 351-361.

Di, H.J. and K.C. Cameron. (2005). Reducing environmental impacts of agriculture by using a fine particle suspension nitrification inhibitor to decrease nitrate leaching from grazed pastures. *Agriculture Ecosystems and Environment.* 109: 202-212.

Dillon, P.J., M.E. Close, and R.J. Scott. (1989). Diffuse source nitrate contamination of groundwater in New Zealand and Australia. *Hydrology and Water Resources Symposium,* Christchurch. pp. 351-355.

Edmeades, D.C. (2002). The effects of liquid fertilisers derived from natural products on crop, pasture, and animal production: A review. *Australian Journal of Agricultural Research.* 53: 956-976.

Edmeades, D.C. (2003). The long-term effects of manures and fertilisers on soil productivity and quality: A review. *Nutrient Cycling in Agroecosystems.* 66: 165-180.

Edmeades, D.C. (2004). Magnesium requirements of pastures in New Zealand. *New Zealand Journal of Agricultural Research.* 47: 363-380.

Edmeades, D.C. and M.B. O'Connor. (2003). Sodium requirements for temperate pastures in New Zealand: A review. *New Zealand Journal of Agricultural Research.* 46: 37-47.

Edmeades, D.C. and K.W. Perrott. (2004). The calcium requirements of pastures in New Zealand: A review. *New Zealand Journal of Agricultural Research.* 47: 11-21.

Francis, G.S., F.J. Tabley, and K.M. White. (1998). Soil structural changes in New Zealand mixed cropping rotations. *Proceedings of the World Congress of Soil Science.* 16: 47.

Francis, G.S., L.A. Trimmer, C.S. Tregurtha, and P.H. Williams. (2003). Winter nitrate leaching losses from three land uses in the Pukekohe area of New Zealand. *New Zealand Journal of Agricultural Research.* 46: 215-224.

Fraser, P.M., P.H. Williams, and R.J. Haynes. (1996). Earthworm species, population size and biomass under different cropping systems across the Canterbury Plains, New Zealand. *Applied Soil Ecology.* 3: 49-57.

Gillingham, A.G. and B.T. Thorrold. (2000). A review of New Zealand research measuring phosphorus in runoff from pasture. *Journal of Environmental Quality.* 29: 88-96.

Gregg, P.E.H., M.J. Hedley, A.B. Cooper, and J. Cooke. (1993). Transfer of nutrients from grazed pasture to aquatic surface waters and their impact on water quality. *Proceedings of the New Zealand Fertiliser Manufacturer's Conference.* 22: 64-75.

Haynes, R.J. and R. Tregurtha. (1999). Effects of increasing periods under intensive arable vegetable production on biological, chemical and physical indices of soil quality. *Biology and Fertility of Soils.* 28: 259-266.

Hoare, R.A. (1986). *Groundwater Nitrate in the Hamilton Basin.* Waikato Valley Authority. Technical Report 1986/16. 18p.

Isherwood, K.F. (1998). *Fertiliser Use and the Environment.* Paris, France: International Fertiliser Industry Association.

Jamieson, P.D., T. Armour, and R.F. Zyskowski. (2003). On-farm testing of the Sirius Wheat Calculator for N fertiliser and irrigation management. In "Solutions for a better environment." *Proceedings of the 11th Australian Agronomy Conference,* 2-6 Feb. 2003, Geelong, Victoria. Australian Society of Agronomy. ISBN 0-9750313-0-9.

Keeney, D.R. and J.M. Bremner. (1966). Comparisons and evaluation of laboratory methods of obtaining an index of soil nitrogen availability. *Agronomy Journal.* 58: 498-503.

Ledgard, S.F., C.A.M. de Klein, J.R. Crush, and B.S. Thorrold. (2000). Dairy farming, nitrogen losses and nitrate-sensitive areas. *Proceedings of the New Zealand Society of Animal Production.* 60: 256-260.

Ledgard, S.F., J.W. Penno, and M.S. Sprosen. (1999). Nitrogen inputs and losses from clover/grass pastures grazed by dairy cows, as affected by nitrogen fertilizer application. *Journal of Agricultural Science, Cambridge.* 132: 215-225.

Lomburg, B. (2001). *The Skeptical Environmentalist: Measuring the Real State of the World.* Cambridge: Cambridge University Press.

McColl, R.H.S. (1982). Water quality in agricultural areas—The prospects for management of diffuse pollution sources. *Waters in New Zealand's Future. Proceedings of the National Water Conference.* 4: 153-160.

McLaren, R.G. and K.C. Cameron. (1990). *Soil Science: An Introduction to the Properties and Management of New Zealand Soils.* Auckland: Oxford University Press. 249p.

Meat and Wool Innovation. (2003). *Compendium of New Zealand Farm Production Statistics.* 13th Edition. Wellington, NZ: MWI Economic Service.

Metherell, A.K., B.S. Thorrold, S.J.R.Woodward, D.G. McCall, P.R. Marshall, J.D. Morton, and K.L. Johns. (1997). A decision support model for fertiliser recommendations for grazed pasture. *Proceedings of the New Zealand Grassland Association.* 59: 137-140.

Ministry for the Environment. (2002). *Lake Managers Handbook. Land-Water Interactions.* Wellington, NZ: Ministry for the Environment, June 2002.

Ministry of Health. (1995). *Drinking Water Standards for New Zealand 1995.* Wellington, NZ: Ministry of Health, 87p.

New Zealand Statistics Department. (1993). Agriculture. Wellington, NZ: Government Printer. 86p.

O'Connor, M.B., M.G. Pearce, I.M. Gravett, and N.R. Towers. (1987). Fertilizing with magnesium to prevent hypomagnesaemia (grass staggers) in dairy cows. *Proceedings of the Ruakura Farmer's Conference.* 39: 47-49.

Powell, C.L. (1977). Mycorrhizas in hill country soils. Growth responses in ryegrass. *Canadian Journal of Soil Science.* 20: 495-502.

Powell, C.L. (1984). Field Inoculation with VA Mycorrhiza Fungi. In Powell C.L. and D.J. Bagyaraj (eds.). *VA Mycorrhiza.* pp. 205-222. Boca Raton: CRC Press.

Rajendram, G.S., S.F. Ledgard, J.W. Penno, M.S. Sprosen, and L.Ouyang. (1998). Effect of rate of nitrogen fertiliser on cation and anion leaching under intensively grazed dairy pasture. In Currie L.D. and P. Loganathan (eds.). *Environmental Long-term Nutrient Needs for New Zealand's Primary Industries: Global Supply, Production Requirements and Constraints.* pp. 67-73. Occasional Report No. 11. Fertiliser and Lime Research Centre, Massey University, Palmerston North.

Reid, J.B. (2002). Yield response to nutrient supply across a wide range of conditions. 1. Model derivation. *Field Crops Research.* 77: 161-171.

Reid, J.B., T.J. van der Weerden, and M.W. Willimott. (2005). Parjib_Express-Software Providing an Economic Basis for Fertilizer Recommendations for Root and Tuber Crops. In Nichols, M.A. (ed). *Proceedings of the 1st International Symposium on Root & Tuber Crops: "Food Down Under."* pp. 143-150, Acta Horticulturae 670. International Society of Horticultural Science, ISBN 978-90-66055-68-1.

Roberts, A.H.C., S.F. Ledgard, M.B. O'Connor, and N.A. Thomson. (1992). Effective use of N fertiliser—Research and practice. *Proceedings of the Ruakura Dairy Farmers' Conference.* 44: 77-83.

Roberts, A.H.C. and J.D. Morton (1999). *Fertiliser Use on New Zealand Dairy Farms.* Auckland: New Zealand Fertiliser Manufacturer's Association.

Taylor N.H. and I.J. Pohlen. (1968). Soils of New Zealand Part 1. *New Zealand Soil Bureau Bulletin.* 26: 7-14.

van der Weerden, T.J., R.R. Sherlock, P.W. Williams, and K.C. Cameron. (1999). Nitrous oxide and methane oxidation by soil following cultivation of two different leguminous pastures. *Biology and Fertility of Soils.* 30: 52-60.

Wheeler, D.M., G.P. Sparling, and A.H.C. Roberts. (2004). Trends in some soil test data over a 14 year period in New Zealand. *New Zealand Journal of Agricultural Research.* 47(2): 155-166.

Whittington, B. (1995). *Situation and Outlook for New Zealand Agriculture.* Wellington, NZ: Ministr y of Agriculture and Fisheries. 133p.

Williams, P.H. and C.E. Wright. (1997). Effect of short term pastures on soil nitrogen status under contrasting management practices. *Proceedings of the New Zealand Agronomy Society.* 27: 15-18.

Chapter 7

Integrated Nutrient Management: Experience from South Asia

Milkha S. Aulakh
Guriqbal Singh

INTRODUCTION

Nutrient management has played, and will continue to play an important role in increasing the production of crops to a great extent. Over the years the ever-increasing human population, especially in developing countries, required more and more foodgrains, edible oils, fibers, and other products. Similarly, more and more feed and fodder were required for animals to meet requirements of milk and milk products, and meat for humans. All these materials are obtained directly or indirectly from plants (field crops), which, of course, need nutrients.

The rice (*Oryza sativa* L.)-wheat (*Triticum aestivum* L.) cropping system in the Indo-Gangetic Plains covers 10.3 million hectare (Mha) in India, 1.5 Mha in Pakistan, and 0.5 Mha each in Nepal and Bangladesh (Pande et al., 2000). According to an estimate, to produce 8 t ha^{-1} rice and 5 t ha^{-1} wheat grains, the crops require 285 kg N, 58 kg P, 349 kg K, 48 kg S, 5 kg Fe, 6 kg Mn ha^{-1}, apart from many other nutrients (Tandon and Narayan, 1990). So there is a need to supply the crops with sufficient nutrients. Not only are the total amounts of nutrients important but also balanced fertilization. For example, continuous production of high-yielding varieties of rice and wheat supplied with only N, P, and K has resulted in the appearance of Zn and Fe deficiency in rice and Mn deficiency in wheat, especially in coarse-textured soils. Sulfur deficiency is also widespread in Indian soils

285

(Tandon, 1995) and is predicted to increase further due to the application of non–sulfur-containing fertilizers (Aulakh, 2003a).

It has been observed that, in general, soil fertility is declining in South Asia. This could be due to overexploitation of soil reserves and/or under-fertilization. On the other hand, a substantial proportion of farmers over-fertilize, resulting in nutrient loss, environmental pollution, and increasing production costs. Due care is not taken in balanced fertilization. Only the major nutrients (mainly N and P; K to a lesser extent) are being applied and not in the required ratio. For example, the optimum ratio of $N:P_2O_5:K_2O$ is considered to be 4:2:1, however, the prevalent ratio in India is 6.5:2.5:1 (FAI, 2004). Mostly chemical fertilizers are applied, giving little or no attention to organic sources like farmyard manure (FYM), green manure (GM), crop residues (CR), biofertilizers (BF), and biosolids (by-products of agro-industry, sewage sludge, or other industries). Fertilizers may have residual effects on the succeeding crops, so it is always desirable to monitor the nutrient requirements of crops throughout different cropping systems rather than solely on individual crops.

Integrated nutrient management (INM) could be defined as "the maintenance/adjustment of soil fertility to an optimum level for crop productivity to obtain the maximum benefit from all possible sources of plant nutrients—organics as well as inorganics—in an integrated manner." Emphasis on the importance and use of INM practices in sustainable agriculture in South Asia has been given in many studies reported from India (Bahl et al., 1986, 1988; Bhandari et al., 1992; Meelu, 1996; Tilak and Singh, 1996; Aulakh and Pasricha, 1998; Mani and Yadav, 2000; Prasad, 2000; Aulakh et al., 2000, 2001a; Bhandari et al., 2002; Yadav et al., 2002; Kabba and Aulakh, 2004; Aulakh et al., 2004), Bangladesh (Islam and Saha, 1998; Saha et al., 1998; Islam, 2001; Panaullah et al., 2001), Pakistan (Zia et al., 1992; Ahmad and Muhammad, 1998), Nepal (Sherchan et al., 1995; Gurung et al., 1996; Pilbeam et al., 1999; Brown and Schreier, 2000; Manandhar, 2001; Regmi et al., 2002), and Sri Lanka (De et al., 1993).

This chapter synthesizes the information on INM in field crops in South Asia, covering India, Pakistan, Bangladesh, Nepal, Sri Lanka, Afghanistan, Bhutan, and the Maldives. Major emphasis is given on the climate and major crops of these countries, agricultural production and nutrient balances, positive and negative aspects of mineral fertilizer use, integrated use of chemical and organic nutrient sources, and research gaps and future research needs. In order to illustrate different trends obtained in different countries of South Asia, most examples have been cited from large countries such as India as such work is often not available in small countries.

CLIMATE AND MAJOR CROPS

This chapter covers eight countries of South Asia, namely India, Pakistan, Bangladesh, Nepal, Sri Lanka, Afghanistan, Bhutan, and the Maldives. South Asia has temperate, subtropical, and tropical regions; however, a major portion of the agricultural land falls in the subtropical region. The subtropical region has summer and winter crop-growing seasons where summer (May-September) is characterized by high temperature and rainfall (monsoons); the winter (November-March) is often dry with low temperatures (Figure 7.1). Irrigated areas are under annual double cropping systems where summer crops such as rice, maize (*Zea mays* L.), groundnut (*Arachis hypogaea* L.), and soybean [*Glycine max* (L.) Merrill] are followed by wheat, maize, rapeseed, and mustard (*Brassica* spp.), or vegetables in winter.

In tropical regions, the temperature remains warm with few fluctuations throughout the year. Even the coolest month has an average temperature of more than 18°C in tropical climates. However, soil moisture may vary among wet and dry seasons. Thus, response of crops to INM often remains unaffected under irrigated conditions but may vary significantly under rainfed conditions.

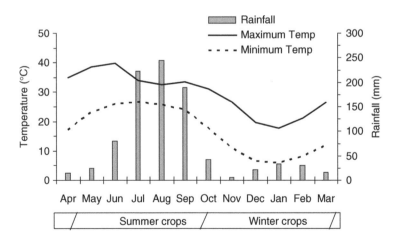

FIGURE 7.1. Monthly minimum and maximum temperature, rainfall and crop growing seasons in the subtropical region of South Asia. *Source:* Aulakh, Singh, and Sadana (2004).

AGRICULTURAL PRODUCTION
AND NUTRIENT CONSUMPTION

It has been estimated that nutrient uptake by major cereals (wheat, rice, and maize), in general, is 20-27 kg N, 10-12 kg P_2O_5, and 20-35 kg $K_2O\ t^{-1}$ of grain harvest (Table 7.1). The corresponding values for pulses and oilseeds are 2-5 and 40-50 kg N, 8-13 and 14-27 kg P_2O_5, and 6-16 and 20-30 kg $K_2O\ t^{-1}$, respectively. The use of fertilizers is higher in cereal crops than in other field and vegetable crops and horticultural plants. The rapid increase in the consumption of fertilizers (N + P_2O_5 + K_2O) in South Asia

TABLE 7.1. Average nutrient uptake by major cereals, pulses, oilseeds, and other principal crops

Crop	N (kg t^{-1})	P_2O_5 (kg t^{-1})	K_2O (kg t^{-1})	S (kg t^{-1})	Zn (g t^{-1})
	Nutrient uptake by crop[a]				
Wheat	22.3	10.6	34.8	3.5	39
Rice	21.6	10.3	32.4	3.5	30
Maize	26.9	11.7	23.9	3.5	69
Barley	7.3	3.9	16.9	3.5	25
Chickpea	50.2(5.0)[b]	8.0	25.9	6.1	30
Mungbean	51.1(5.1)[b]	12.6	6.2	4.1	41
Rapeseed and mustard	44.1	21.3	30.5	12.8	65
Sunflower	33.3	14.2	34.0	5.3	36
Groundnut	59.9(6.0)[b]	19.0	27.5	6.8	73
Sesamum	42.3	26.6	23.6	5.5	87
Linseed	51.7	17.2	23.6	4.2	35
Potato	4.4	2.0	7.8	1.4	18
Cotton	22.2	14.1	37.4	3.0	16
Sugarcane	1.8	0.7	2.5	0.5	10

Source: Adapted from Aulakh and Bahl (2001).

[a]Nutrient uptake per tonne of produce, which includes total of grain, seed, tuber, or cane and their proportionate straw or leaves.

[b]After deducting 90 percent of N uptake, which is contributed by biological nitrogen fixation in legumes. P and K uptake data converted to P_2O_5 and K_2O by multiplying the P and K content with 2.29 and 1.205, respectively.

resulted in a dramatic increase in the total cereal production (Figure 7.2). According to an estimate, by 2010 AD about 246 million tonnes (Mt) of foodgrains will be needed in India annually and to produce this about 25 ± 2 Mt of plant nutrients ($N + P_2O_5 + K_2O$) as mineral fertilizer would be needed (Prasad, 2000). In 2025 AD, total nutrient ($N + P_2O_5 + K_2O$) consumption is estimated to be 30-35 Mt (Pasricha et al., 1996), whereas the present consumption is around 18 Mt.

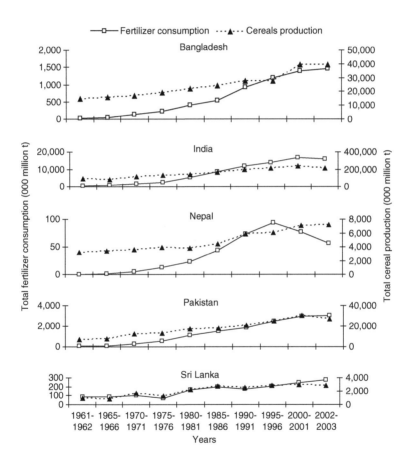

FIGURE 7.2. Trends of consumption of fertilizers ($N + P_2O_5 + K_2O$) and total cereal production in different countries of South Asia. *Source:* Adapted from data of FAI (2004).

BENEFITS AND LIMITATIONS
OF CHEMICAL FERTILIZERS

Over the past five decades, the role of fertilizers in augmenting foodgrain production has been widely recognized. Increased use of chemical fertilizers no doubt helped to increase the production of foodgrains, resulting in improved food security. High-yielding–semi-dwarf varieties of crops were more responsive to chemical fertilizers than the traditional tall varieties. The green revolution of the late 1960s and early 1970s helped the South Asian region to achieve self-sufficiency in foodgrains. During the 1960s when the food situation became very grim, emphasis was placed on increasing wheat and rice production. Adoption of high-yielding varieties, improved management technologies, expansion of irrigation facilities, and assured procurement of foodgrains at remunerative prices resulted in a quantum leap in food production. For instance, the productivity of wheat and rice in India was 663 kg ha^{-1} and 668 kg ha^{-1} in 1950-1951, which increased to 2,708 kg ha^{-1} and 1,901 kg ha^{-1} in 2000-2001, respectively. The cropping intensity has also increased substantially.

To obtain high productivity, more and more emphasis has been placed on chemical fertilizers (Table 7.2). The dramatic increase in fertilizer consumption and resultant increased agricultural productivity has continued to meet the food requirement of an increasing population despite a minimal change in the cultivated area over the past fifty years. In his keynote address at the 15th World Soil Science Congress, Borlough rightly stated that most populous countries like India and China would have required to put two to three times more land under cereal crops to meet the food needs of 1991, if they had not increased the input of fertilizers, and had continued to use the technology of 1960 (Borlough and Dowswell, 1994). He further said that even the high-yielding varieties or "miracle seeds" would not have been able to create the miracle of high yields without the use of fertilizers. Fertilizers are also the means of saving land through increase in land productivity.

In the mid-sixties, fertilizer application was limited to N. With the prolonged use of fertilizer N alone, crop yields obtained in N-treated plots were even inferior to that obtained in minus N-treatments, as illustrated by a long-term study in India (Table 7.3). Intensive continuous cropping resulted in the deficiency of P, K, Zn, and S in that sequence and their application became necessary to obtain optimum yields. However, where FYM was used in combination with chemical fertilizers, not only were the levels of crop yields maintained over a long period, but it also resulted in a significant improvement in overall soil productivity. Similarly, in a twenty-year study in Nepal, the mean response of crops in rice-rice-wheat rotation was

TABLE 7.2. Evolution of human population, cultivated area under foodgrains, fertilizer consumption, grain production, and cereal productivity in India

	1950-1951	1960-1961	1970-1971	1980-1981	1990-1991	2000-2001	2001 vs. 1951 (%)
Human population (m)	361.1	439.2	548.2	683.3	846.3	1,027.0	+184
Area (Mha)	97.3	115.6	124.3	126.7	127.8	121.0	+24
Fertilizer Consumption (000 t N + P_2O_5 + K_2O)	69.8	293.8	2,256.6	5,515.6	12,546.2	16,702.3	+23,929
Cereal productivity (kg grain ha^{-1})[a]	626	930	1,236	1,375	1,846	2,144	+142
Foodgrain production (Mt)	50.8	82.0	108.4	129.6	176.4	196.8	+285

Source: FAI (2004).

[a]Mean of three major cereal crops (wheat, rice, and maize).

TABLE 7.3. Long-term effects of chemical fertilizers alone or integrated with organics (FYM) on the grain yield of maize (kg ha^{-1}) on an alfisol in India

Treatment	1-12 years	13-18 years	19-24 years	25-28 years
Check	830	830	550	530
N	1,700	1,150	40	0
N + P	2,400	2,850	400	0
N + P + K	3,170	3,230	820	120
N + FYM	2,270	2,590	2,750	2,590

Source: Adapted from Lal and Mathur (1988).

spectacularly enhanced with the optimum use of N + P or N + P + K through chemical fertilizers or FYM (Table 7.4). It is surmised that FYM, and for that matter any other organic nutrient source, can play an additional role beyond its capacity to contribute nutrients.

The fertilizer-use efficiency (FUE) of N fertilizers is only 30-50 percent (Aulakh and Singh, 1997). Thus, chemical fertilizers, applied in high amounts

TABLE 7.4. Effect of chemical fertilizers and FYM on the twenty-year mean grain yield in a long-term experiment on rice-rice-wheat rotation (1978-1998) experiment, Bhairhawa, Nepal

Treatment	First rice N	First rice P	First rice K	Second rice N	Second rice P	Second rice K	Wheat N	Wheat P	Wheat K	Grain yield First rice	Grain yield Second rice	Grain yield Wheat
	kg ha^{-1}									t ha^{-1}		
Control	0	0	0	0	0	0	0	0	0	0.399	1.066	0.532
N	100	0	0	100	0	0	100	0	0	0.719	1.330	0.588
NP	100	13	0	100	13	0	100	18	0	2.577	2.465	1.200
NK	100	0	25	100	0	25	100	0	25	0.630	1.300	0.611
NPK	100	13	25	100	13	25	100	18	25	2.760	3.082	2.301
FYM[a]	80	18	40	80	18	40	80	18	40	2.797	3.138	2.202

Source: Adapted from Regmi et al. (2002).

[a]FYM applied at the rate of 4 t ha^{-1} (dry weight basis) to each crop. Total N, P, and K contents in FYM are on an average basis.

on large acreage, resulted in a large amount of nutrients that are not immediately utilized by the crop. Movement of these nutrients off-field has led to pollution of air, and surface and groundwater. The adverse effects of fertilizers include eutrophication of surface waters, accumulation of nitrates in ground and surface waters, emission of greenhouse gases (CO_2, N_2O, and CH_4), depletion of the ozone layer, and heavy metal accumulation in soils (Williams, 1992; Prasad and Katyal, 1992; Pathak et al., 2002; Singh and Sekhon, 2002). Depending on the environment and cropping practices, about 14-15 percent of applied fertilizer N (FN) may leach below 150 cm depth (Arora et al., 1980; Aulakh et al., 1992, 2000, 2001; Aulakh, 1994; Aulakh and Singh, 1997). Aulakh and Singh (1997) noted that NO_3^- enrichment of groundwater beneath soils was evident from the tubewell waters. The leached nitrates may contaminate household and livestock water coming from shallow wells (Singh and Sekhon, 1977). Nitrogen movement below the root zone and into the groundwater (Olson et al., 1970; Spalding and Kitchen, 1988) can cause human and animal health problems (USEPA, 1985). Nitrite, the reduced form of nitrate, can cause methemoglobinemia, a disease affecting the mechanisms of oxygen exchange in blood, in the fetus, and in infants (Sarkar, 1990). Nitrate content exceeding 10 mg N l^{-1} in drinking water is considered harmful for health. In some parts of India, groundwater has nitrate content much higher than 10 mg NO_3-N l^{-1} (Handa, 1987; Aulakh, 1994). However, such effects are in specific areas of high fertilizer use. For instance, compared to the country's average annual consumption of 90 kg ha^{-1} of N + P_2O_5 + K_2O in India, consumption in individual states varies from 794 kg ha^{-1} in Pondicherry, and 175 kg ha^{-1} in Punjab, to as low as 3 and 2 kg ha^{-1} in Arunachal Pradesh and Nagaland, respectively (FAI, 2004). Otherwise, in most South Asian countries, fertilizer consumption is relatively low; thus such adverse effects of fertilizers are neither expected nor have been reported on a large scale.

Continued production of high-yielding crop varieties removes large amounts of both micro- and macronutrients. If these crops are grown with inputs of only macronutrients and with no or very limited use of organics crop yields, quality and FUE can decrease and micronutrient deficiencies can appear in the crops (Balasubramanian et al., 2004; Wassmann et al., 2004; Aulakh and Malhi, 2004, 2005). Deficiency of Zn, Fe, Mn, Cu, B, and Mo in India (Benbi and Brar, 1992; Prasad, 2000) and of Zn, B, and Cu in Bangladesh (Ahmed and Rahman, 1992; Islam and Saha, 1998) is quite common. Animals fed on fodder grown on micronutrient-deficient soils are showing micronutrient deficiencies. Similar effects are also expected in humans.

The above-mentioned negative effects of fertilizers could be minimized or decreased with balanced fertilizer use and by meeting crop nutrient requirements with well-managed organic sources of nutrients. The nutrient release from organic sources is generally slower than from fertilizers; the release could be closely matched to crop uptake, thus reducing the risk of environmental pollution. However, one problem is that not enough organic sources are available to meet the nutrient demand for sustained production of foodgrains and other products and that the distribution of these organic sources does not necessarily match the areas where they are required. According to an estimate, the annual potential organic resources in India generated through excreta of livestock and human beings, and from CR, compost, sewage sludge, and BFs is about 17 Mt of N + P_2O_5 + K_2O (Pasricha et al., 1996). About 380 Mt of CR are produced in India, which have the potential of about 7.3 Mt of NPK. However, only 33 percent of these residues are available for direct use, the others being used as animal feed. Animal dung and night soil have the potential of about 7 Mt and 5 Mt of NPK, respectively (Pasricha et al., 1996). Further, 3.2 Mt of pressmud generated from more than 350 sugar mills can provide 0.17 Mt of NPK. Thus, only 25 percent of the nutrient needs of Indian agriculture can be met by utilizing various organic sources such as FYM, CR, urban and rural wastes, green manuring and BFs (Prasad, 2000). Therefore, balanced fertilization through INM is the best method of sustaining crop yield and quality without adversely affecting the environment (Mishra et al., 1991; Biswas et al., 1992).

INTEGRATED NUTRIENT MANAGEMENT PRACTICES

The previous section has clearly pointed out the need for INM. Excellent reviews on INM involving chemical fertilizers, BFs (N-fixing bacteria in legumes and nonlegumes, phosphorus solubilizing microorganisms, mycorrhizae, etc.), and organic manures (GM, FYM, and CR) have been published (Meelu et al., 1992; Mishra et al., 1991; Meelu, 1996; Pasricha et al., 1996; Tilak and Singh, 1996). Therefore, in this section various practices and components of INM are discussed citing a few salient examples.

Crop Rotation

The concept of crop rotation is perhaps as old as agriculture itself. Rotations of shallow-rooted crops with deep-rooted ones, cereals and nonlegumes with legumes, or high nutrient demanding ones with low nutrient demanding ones have been followed for centuries, and of course, will continue to be followed.

Fertilizers are generally applied on an individual crop basis. It is well-known that nutrient use efficiency is very low, that is, N 30-50 percent, P 10-20 percent, K < 80 percent, Zn 2-5 percent, Fe 1-2 percent, and Cu 1-2 percent (Prasad, 2000); some portion is lost from the soil in various forms and some portion remains in the soil in inorganic or organic forms. A part of this residual fertilizer nutrient, depending upon its availability, could be utilized by succeeding crop(s). So it is desirable that fertilizer recommendations be made on a cropping systems basis rather than on an individual crop basis.

The rooting pattern of crops exerts a profound influence in controlling the NO_3^2 distribution pattern and leaching in the soil profile. A four-year field study by Aulakh et al. (2000) with rice-wheat cropping system has demonstrated how shallow-rooted rice followed by deep and extensively-rooted wheat could minimize nitrate leaching to lower soil layers and potential groundwater contamination (Figure 7.3a). While there was a substantial downward movement of NO_3-N to the 60-cm depth for the shallow-rooted rice crop, NO_3^2 from deeper soil layers was utilized by the subsequent wheat crop. Therefore, shallow-rooted rice grown in rotation with deeper and more extensive rooting crops appears a promising approach to efficiently utilize residual N and minimize leaching of NO_3^2 to greater soil depths in irrigated soils.

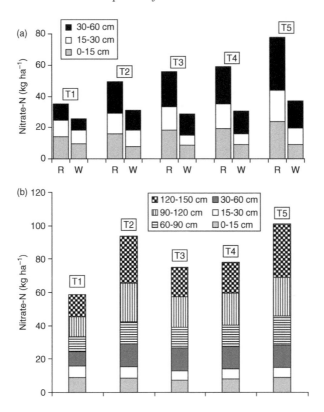

FIGURE 7.3. Nitrate-N in different layers of the soil profile, (a) after two years of rice (R) and wheat (W), and (b) after four years of FN and GM treatments in rice-wheat rotation. Treatment $T_1 = GM_0 FN_0$ (Rice) $- FN_0$ (Wheat), $T_2 = GM_0 FN_{120} - FN_{120}$, $T_3 = GM_{20} FN_0 - FN_0$, $T_4 = GM_{20} FN_{60} - FN_{90}$, and $T_5 = GM_{40} FN_{120} - FN_{90}$. *Source:* Prepared from data of Aulakh, Khera, Doran, Kuldip-Singh, et al. (2000). Subscripts for GM denote 0 and 20 t ha^{-1} of Sesbania aculeate GM and for FN denote 0, 60, 90, and 120 kg fertilizer N ha^{-1}.

In semiarid subtropical regions, the hot summer and the cold winter are the two main crop-growing seasons. Seasonal conditions exert an enormous influence on the response of crops to fertilizer P. In general, winter season crops are more responsive to fertilizer P application compared to summer crops. Summer-grown crops, both cereals and legumes, often do not respond to directly summer-applied P where the preceding winter crops had received the recommended amount of fertilizer P (Vig et al., 1999; Bahl and Aulakh,

2003; Aulakh, 2003b). In rice-wheat rotation, when the recommended dose of 26 kg P ha^{-1} has been applied to wheat, P application to rice can be omitted (Gill and Meelu, 1983) without any loss in the grain yield of either crop. The pearl millet (*Pennisetum typhoides* L.)-mungbean [*Vigna radiata* (L.) Wilczek] rotation enhances the nutrient pool in the soil compared to pearl millet-pearl millet cropping system (Yadav et al., 2002). The differential response of winter and summer crops is due to two reasons. First, during summer, the temperature is relatively high (26-45°C), compared to 5-20°C in winter (Figure 7.1), and this, coupled with high soil moisture status in the rainy season (74 to 85 percent of annual 600 to 1,200 mm rainfall occurs during July-September period), enhance the solubility of native soil P and residual fertilizer P. Second, legumes can better utilize less soluble P than cereals; their effects, when grown in rotation, are discussed in the following section.

Grain and Forage Legumes

Grain and forage legumes are grown for producing their specific end products, that is, grain and forage, respectively. Though some amount of N accumulated by these legumes by biological, N fixation (BNF) is consumed by the crop itself, and a substantial amount is left in the soil, plant roots, and stubble, which can be utilized by the succeeding crop in the cropping system.

After the grain or pods of pulse crops are harvested, the straw can be incorporated into the soil. At maturity, the straw of different grain legumes contains considerable amounts of nutrients (N, P, K, S, and others). Incorporating the straw of these grain legumes into the soil not only provides nutrients but also improves physical and biological properties of the soil. Inclusion of short-duration pulse crops like summer mungbean in between wheat and rice has shown enormous beneficial effects. In situ growth of mungbean and incorporation of its residues (after picking of pods) before transplanting rice supplied 60 kg (Rekhi and Meelu, 1983) to 85 kg N ha^{-1} (Antil et al., 1989) to rice crop. In the case of the summer mungbean-maize system, the benefit in maize was of 40 kg N ha^{-1} (Ali and Mishra, 2000). Similarly, in tropical regions of India, incorporation of mungbean residues resulted in an additional 15.6 percent yield of winter sorghum [*Sorghum bicolor* (L.) Moench] and supplied 50 percent of the NPK requirement of winter sorghum (Shinde et al., 1996). Other studies also support that after harvesting the grain, incorporating of mungbean straw can add 60-80 kg N ha^{-1} (Prasad et al., 1999; Prasad, 2000).

In a four-year study, the incorporation of cowpea [*Vigna unguiculata* (L.) Walp] and mungbean in a maize-wheat rotation increased the productivity of maize + wheat from 9,630 kg ha^{-1} to 9,830 and 10,210 kg ha^{-1}, respectively; grain yields of cowpea (450 kg ha^{-1}) and mungbean (1,100 kg ha^{-1}) were an additional benefit (Randhawa, 1992). Similarly, in a pigeonpea [*Cajanus cajan* (L.) Millsp.]-wheat system, substantial increases in wheat yield have been reported (Sharma and Pal, 1986; Singh and Kalra, 1989). Several studies with different cropping systems where N-equivalent benefits were estimated revealed that pulse crops may contribute 18-70 kg N ha^{-1} to soil, which could be utilized by succeeding crops (Bahl et al., 1986, 1988; Srinivasan et al., 1991; Acharya and Biswas, 2002). In a five-year field study at Ludhiana, India, Bahl and Pasricha (2000) demonstrated enormous benefits of including fieldpea (*Pisum sativum* L.) in the rotation and incorporating fieldpea residues. As compared to maize succeeding wheat, maize grown after fieldpea produced an additional 130 and 310 kg grain ha^{-1} with and without the incorporation of fieldpea residues, respectively (Table 7.5). These benefits were further enhanced when the recommended dose of FN was used along with fieldpea residue incorporation. The enhanced maize production was due mainly to the supply of N in the soil by biological N fixation (BNF), and mineralization of fieldpea residue N (and perhaps other nutrients) as was evident from the uptake of 26 and 87 kg N ha^{-1} over and above N taken from the applied FN (Table 7.5). Significantly greater N uptake with the combined use of fieldpea residues and FN as compared to the sum of their individual effects suggested synergistic improvement in

TABLE 7.5. Grain yield and N uptake by maize grown in wheat-maize and fieldpea-maize cropping systems in a five-year study at Ludhiana, India

	Crop rotations followed		
Treatment	Wheat-maize	Fieldpea (without residues)-maize	Fieldpea (with residues)-maize
Maize grain yield (kg ha^{-1})			
Without fertilizer N	440	570	750
With 120 kg fertilizer N ha^{-1}	1,280	1,860	2,520
N uptake by maize (kg ha^{-1})			
Without fertilizer N	31	40	68
With 120 kg fertilizer N ha^{-1}	49	75	136

Source: Adapted from Bahl and Pasricha (2000).

nutrient use efficiency. Thus, inclusion of legumes in the crop rotations both increases cereal yields, and supplements legume grain production, while improving soil properties and nutrient use efficiency.

Application of P to wheat rather than to groundnut in a cropping sequence has been reported to be as efficient as phosphorus application to both these crops (Pasricha et al., 1980; Goswami and Pasricha, 1992). Significantly higher response of legumes such as pigeonpea and soybean than wheat to residual fertilizer P has been reported under subtropical semiarid soils (Bahl and Pasricha, 1998; Aulakh et al., 2003). In the slightly alkaline semiarid subtropical soils, most of the inorganic P exists as calcium-bound P (Pasricha et al., 2002), which legumes often can utilize more effectively along with residual soil P compared to nonleguminous plants (Kalra and Soper, 1968; Hundal and Sekhon, 1976). Legumes have been found to be very effective in solubilizing Ca-P and improving the available P status of soils (Reddy and Surekha, 1999), presumably due to their genetic abilities such as development of proteoid roots and mobilization of nutrients in the soil with the help of specific root exudates (Gardner et al., 1983; Marschner, 1995). The investigations with different cropping systems confirmed that groundnut, pigeonpea, and mungbean grown in the summer season could solubilize sufficient residual soil P to meet their full P requirement. Residual P could partially fulfill the demand of soybean, which has a high requirement of 80 kg P_2O_5 ha^{-1} for optimum yield (Aulakh et al., 1990). For instance, where wheat received 60 kg P_2O_5 ha^{-1}, the succeeding soybean required only 60 kg P_2O_5 ha^{-1} for optimum yield instead of 80 kg P_2O_5 ha^{-1} (Aulakh et al., 2003).

Intercropping

In the earlier years, intercropping was followed to ensure the successful production of at least one crop in the event of the failure of the other(s). In the recent past, this system has been practiced to obtain high productivity and/or net returns from each field each year (Chandel et al., 1989; Sekhon and Singh, 2005). As mentioned in the previous section, legumes contribute N to the soil and plants through BNF. Intercropping of forage or grain legumes with nonlegume crops such as cereals, oilseeds, or sugar crops enriches the soil with N, and some N from legume is transferred to nonlegumes. In this way, the associated nonlegumes require lower amounts of N compared to their sole cropping. Intercropping and incorporation of legumes such as French beans (*Phasealus vulgaris* L.) and sunnhemp (*Crotalaria juncea* L.) in sugarcane *(Saccharum officinarum),* have established the beneficial effect of improving the N use efficiency on cane yield, as well as improvement in various soil physico-chemical properties (Shankaraiah and Nagaraju, 1997). However, application of higher doses of N to meet the N demand of cereal crops may reduce BNF in the legume crop. It is, therefore, imperative to use a moderate dose of N in legume/cereal intercropping systems (Reddy et al., 1980; Kumar and Ahlawat, 1986). Shading by associated cereal crops also reduces the growth and BNF of the pulse crop. Intercropping of climbing type of legumes such as cowpea could overcome this problem and improve N use efficiency.

Organic Manures and Composts

Organic manures such as FYM, swine and poultry manure, and composts contain not only macronutrients but also micronutrients. Their application improves the chemical, physical, and biological properties of the soil. Upon their application, mineralization of N in these manures usually starts quickly due to the relatively narrow C:N ratio, resulting in the accumulation of mineral N in soil. For example, Aulakh et al. (2000) observed that 12, 20, and 44 percent of FYM (N = 6 g kg^{-1}; C:N = 35), pressmud (N = 17.4 g kg^{-1}; C:N = 22), and poultry manure (N = 19.5 g kg^{-1}; C:N = 12) N mineralized in only sixteen days.

The combined use of synthetic fertilizers and organics helps reduce the dependence on fertilizers. Several field studies have shown that the combined use of FYM + FN (and P, and K in some cases) produced similar yields of rice, wheat, maize, barley (*Hordeum vulgare* L.), sorghum, sunflower (*Helianthus annuus* L.), linseed /flax (*Linum usitatissimum* L.), and soybean as a 50-100 percent higher use of FN (and P and K in some cases) alone (Table 7.6). Reduction in fertilizer usage by the combined use of chemical fertilizers and organics could help in reducing soil, water, and air pollution without sacrificing yield.

Green Manures

Green manure not only provides nutrients but also improves the physical and biological properties of the soil. Among frequently used GMs, sesbania/ dhaincha *(Sesbania aculeata),* sunnhemp *(Crotalaria juncea),* and cowpea *(Vigna unguiculata)* are excellent in terms of their characteristics for providing adequate nutrients and organic matter in a short growth period. A GM crop at forty-five to sixty days after sowing generally provides 100 kg N ha^{-1}, though the potential could exceed 200 kg N ha^{-1}. Sesbania GM, in sixty days, may produce 3-4 t ha^{-1} dry matter, accumulate 80-110 kg N ha^{-1}, and help save 60-90 kg N ha^{-1} in different crops and cropping systems (Singh et al., 1991; Aulakh et al., 2000, 2001a, 2004).

TABLE 7.6. Influence of INM using FYM and chemical fertilizers in different field crops

Crop	Treatment	Yield (kg ha^{-1})	Reference
Rice	120 kg N ha^{-1}	5,600	Maskina et al. (1988)
	80 kg N + 12 t FYM ha^{-1}	5,700	
	120 kg N + 26 kg P_2O_5 + 42 kg K_2O ha^{-1}	5,520	Yaduvanshi (2001)
	120 kg N + 26 kg P_2O_5 + 42 kg K_2O + 10 t FYMha^{-1}	6,420	
	60 kg N + 13 kg P_2O_5 + 21 kg K_2O ha^{-1}	4,270	
	60 kg N + 13 kg P_2O_5 + 21 kg K_2O + 10 t FYMha^{-1}	5,270	
	180 kg N + 39 kg P_2O_5 + 63 kg K_2O ha^{-1}	6,040	
Wheat	180 kg N ha^{-1}	3,399	Singh and Agarwal (2001)
	120 kg N + 10 t FYM ha^{-1}	3,295	
	120 kg N + 20 t FYM ha^{-1}	3,754	
	120 kg N + 30 t FYM ha^{1}	3,823	
	90 kg N ha^{-1}	3,730	Sushila and Giri (2000)
	45 kg N + 10 t FYM ha^{-1}	4,020	
	150 kg N ha^{-1}	3,927	Ranwa and Singh (1999)
	100 kg N + 10 t FYM ha^{-1}	4,072	
	100 kg N + 7.5 t vermicompost ha^{-1}	4,095	
	100 kg N + 10 t vermicompost ha^{-1}	4,120	
Maize	120 kg N + 26.2 kg P + 25 kg K ha^{-1}	2,900	Maskina et al. (1988)
	60 kg N + 13.1 kg P + 12.5 kg K + 6 t FYM ha^{-1}	3,000	
	60 kg N + 13.1 kg P + 12.5 kg K + 12 t FYM ha^{-1}	3,300	
Sorghum	80 kg N + 40 kg P_2O_5 ha^{-1}	4,680	Patidar and Mali (2004)
	60 kg N + 30 kg P_2O_5 + 10 t FYM ha^{-1}	4,900	
Barley	75 kg N ha^{-1}	1,659	Sharma et al. (2001)
	60 kg N + 10 t FYM ha^{-1}	1,668	

TABLE 7.6 *(continued)*

Crop	Treatment	Yield (kg ha^{-1})	Reference
Sunflower	80 kg N + 40 kg P$_2$O$_5$ ha^{-1}	1,861	Singh and Singh (1997)
	40 kg N 1 20 kg P2O5 1 10 t FYM ha^{-1}	1,842	
	120 kg N + 60 kg P$_2$O$_5$ ha^{-1}	1,966	
	80 kg N + 40 kg P$_2$O$_5$ + 10 t FYM ha^{-1}	1,951	
Linseed	50 kg N + 40 kg P$_2$O$_5$ + 20 kg K$_2$O ha^{-1}	1,117	Badiyala and Kumar (2003)
	37.5 kg N + 30 kg P$_2$O$_5$ + 15 kg K$_2$O + 5 t FYM ha^{-1}	1,382	
Soybean	50 kg N ha^{-1}	894	Saxena et al. (2001)
	25 kg N+5 t FYM ha^{-1}	1,567	

Most of the investigations have been done on the green manuring of the rice-wheat system, which is predominant in subtropical regions of South Asia. Integration of GM and chemical fertilizers could result in the saving of 50-75 percent of N fertilizers in rice (Singh et al., 1991; Meelu et al., 1994) and succeeding wheat (Aulakh et al., 2000). In a rice-wheat system, 50 percent of the NPK of rice can be substituted by GM without appreciable sacrifice in yield (Patra et al., 2000; Banwasi and Bajpai, 2001), and root length density, root volume, and root dry weight are also maximum (Banwasi and Bajpai, 2001). Further, green manuring with *Sesbania rostrata* increases the availability of micronutrients like Fe and Mn (Meelu et al., 1992). Studies from Pakistan also show beneficial effects of the green manuring of *Sesbania aculeata* on the productivity of rice and fertility status of soil (Zia et al., 1992). A recent study from Pakistan using ^{15}N-labeled GM (Ashraf et al., 2004) revealed that, in addition to the direct role of GM in rice production as a source of nutrients, its additional benefit is due to GM-induced increase in the uptake of native soil N.

In addition to the rice-wheat system, GM has also been used in other cropping systems. Application of *Sesbania aculeata* and *Crotalaria juncea* GMs at 10 t ha^{-1} to rice, compared to no GM application, significantly increased grain yield of rice by 1,600 and 1,100 kg ha^{-1}, and pod yield of the succeeding groundnut crop by 250 and 160 kg ha^{-1}, respectively (Prasad et al., 2002). In subtropical regions, rice is grown during the monsoon season until early October and rapeseed in autumn and winter (October-March).

However, with the advent of short-duration early growing rice cultivars, many farmers have adopted the advance transplanting of rice in mid-May, which does not leave enough time for raising a GM-crop prior to rice. Alternatively, a GM-crop could be raised in the rainy and mild season (September-October) after harvesting rice. Aulakh and Pasricha (1998) reported that growing a GM-crop for forty to forty-five days before the sowing of mustard around mid-October could provide 62 to 82 kg N ha^{-1}. Moreover, the irrigation needs of this mild and rainy season grown GM-crop are very low as compared to those for GM-crop grown in the hot-dry period prior to the transplanting of rice. A five-year study of Aulakh et al. (2004) showed that GM produced during the post-rice mild and rainy season with three irrigations, as compared to production during the pre-rice hot and dry season with seven irrigations, gave a 33 percent lower cost/benefit ratio, as well as higher N utilization by rice and rapeseed crops (70 percent versus 64 percent). The application of 20 t GM and 60 kg FN ha^{-1} to rice rather than 120 kg FN ha^{-1}, produced significantly greater yields of rice and rapeseed, while reducing the use of FN by >50 percent in rice and 17 percent in rapeseed. Combined application of GM and 100 kg FN ha^{-1} (which is the recommended N rate) to rapeseed improved the yield potential of rapeseed, illustrating the benefit of GM, which FN alone cannot achieve. While FN showed negligible residual effect on either of the succeeding crops, GM applied prior to rice produced 9 percent greater yield of the following rapeseed, and pre-rapeseed applied GM produced 35 percent greater yields of rice. The improved growth was related to N recovery of 67 percent as compared to only 52 percent with FN, presumably by synchronizing N supply with plant needs, and reducing N losses. Thus, the integrated use of GM and FN could attain crop yields higher than those produced with the recommended FN rates as well as sustain these production levels in semiarid-subtropical soils.

Besides increasing crop yields, improving N use efficiency, and reducing dependence on chemical fertilizers, integrated use of GM and FN reduces nitrate leaching in soils. For example, after four years of rice-wheat system, use of 120 kg FN ha^{-1} for both rice and wheat resulted in 35 kg of residual NO_3-N ha^{-1} in the soil profile (74 percent in the 90 to 150 cm soil depth), whereas only 19 kg NO_3-N ha^{-1} remained with 20 t GM ha^{-1} (GM_{20}), plus 60 kg FN ha^{-1} (FN_{60}) treatment (Figure 7.3b). However, the excessive use of FN in conjunction with GM could reduce fertilizer-use efficiency and crop yields due to lodging from excessive vegetative growth (Aulakh et al., 2000, 2004). Therefore, excess applications must be avoided to restrict leaching of nitrate in the soil profile (Figure 7.3), denitrification losses, and emission of N_2O (Aulakh et al., 2001b).

Crop Residues

Crop residues are generally considered as waste materials. Despite the availability of farm-borne organic materials in abundance, farmers do not use them. For example, foodgrain production in India has increased from 51 Mt in 1950-1951 to 197 Mt in 2000-2001 (Table 7.2). Although more than half of dry matter produced annually in cereals, legumes, root, and tuber crops is the aboveground inedible phytomass, hardly any nation keeps statistics of the CR produced as a part of total crop production. Calculation of the residual phytomass from the harvest index of major food crops that are grown on nearly 50 percent of India's cultivable area revealed that N and P removed annually by the CR is equivalent to approximately 30 and 34 percent of the nutrients used in the chemical fertilizers (Aulakh, 2001). Potassium removal by CR represents approximately five times as much as is supplied by fertilizers. As a conservative estimate, in India alone more than 100 Mt of CR, including those of wheat and rice, are disposed of by burning each year. Burning of CR not only results in loss of organic matter and nutrients but also causes atmospheric pollution due to the emission of toxic and greenhouse gases like CO, CO_2, and CH_4 that pose a threat to human and ecosystem health.

In situ trash management improves soil physico-chemical and biological properties of sugarcane-growing fields (Shankaraiah and Nagaraju, 1997). Incorporation of CR increases organic C, total N, and available P and K contents in the soil (Beri et al., 1995; Meelu, 1996; Aulakh et al., 2001a). Further, CR also improve soil structure, reduce bulk density, and increase porosity and infiltration rate of soil (Pandey et al., 1985; Meelu, 1996; Aulakh et al., 2001a). Similar grain yields of rice were obtained with recommended fertilizer rate + 3 t rice residue ha^{-1} (Raju et al., 1993). Rice residue incorporation either to meet 50 percent or 25 percent N along with fertilizers increased hydraulic conductivity, water stable aggregates, porosity, soil water retention characteristics, and maximum water-holding capacity (Bellakki et al., 1998).

Most farmers burn their CR to facilitate quick seedbed preparation, and to avoid reductions in crop yields due to immobilization of nutrients by the incorporation of residues such as wheat and rice that have a wide C:N ratio. This problem could be addressed by integrating the incorporation of wheat residues (WR) and rice residues (RR) with FN and sesbania GM in irrigated rice-wheat system. Field studies have demonstrated that RR and WR (6 t ha^{-1}) did not depress grain yields of wheat and rice respectively, in the presence of 120 kg FN ha^{-1} (Table 7.7). Rice production was higher in the presence of WR when 88 kg N ha^{-1} of a prescribed 120 kg N ha^{-1} dose was

TABLE 7.7. Effect of integrated use of fertilizer urea N (FN), GM, WR, and RR on rice and wheat yields, nitrate leaching, soil bulk density, carbon sequestration, denitrification losses, and N_2O emissions in rice-wheat system

Treatment		Rice yield[a] (t ha^{-1})	Wheat yield[a] (t ha^{-1})	Nitrate leaching[b] (kg N ha^{-1})	Soil bulk density[b] (g cm^{-3})	SOC[b] (g kg^{-1})	Denitrification losses[c] (kg N ha^{-1})	N_2O emissions[c] (kg N ha^{-1})
Rice	Wheat							
Control		3.40	4.73	59	1.60	6.74	18	6.9
120 kg FN ha^{-1}	120 kg FN ha^{-1}	5.62	4.77	94	1.60	3.71	58	12.4
GM$_{20}$ + 32 kg FN ha^{-1} [d]	120 kg FN ha^{-1}	5.85	5.01	76	1.54	4.05	50	11.8
WR$_6$[e] + GM$_{20}$ + 32 kg FN ha^{-1}	120 kg FN ha^{-1}	5.92	5.01	–	1.50	4.92	52	11.8
RR$_6$[f]+120 kg FN ha^{-1}	120 kg FN ha^{-1}	5.63	5.00	–	1.54	4.33		
LSD$_{(0.05)}$		0.24	0.21	12	0.05	0.22	6	3.4

Source: Adapted from Aulakh, Khera, Doran, Kuldhip-Singh et al. (2000, 2001a,b), Aulakh (2001).

[a]Three-year (year two-four of the experiment) mean yields.

[b]Measured at the end of four-year experiment.

[c]Cumulative for pre-rice fallow and rice-growing period.

[d]Amount of 120 kg N ha^{-1} applied through 20 t GM ha^{-1} (88 kg N ha^{-1}) and the balance through fertilizer N (32 kg N ha^{-1}).

[e]WR applied at 6 t ha^{-1}.

[f]RR applied at 6 t ha^{-1}.

applied as GM-N, and the balance as FN, compared to 120 kg FN ha^{-1} alone. At the end of the four-year study, soil bulk density was significantly reduced with the application of WR and RR and GM. These studies have demonstrated the following.

1. The wide C:N ratio RR and WR can be incorporated shortly after crop harvest without affecting yields of the following crop, if the N supply is adequate, and 120 kg urea-N ha^{-1} to each crop of rice and wheat appeared to be sufficient to avoid adverse effects of CR.
2. After four years, use of 120 kg FN ha^{-1} for both crops resulted in 35 kg of residual NO_3-N ha^{-1} in the 150 cm soil profile, whereas only 17 kg NO_3-N ha^{-1} remained with GM_{20} plus FN treatment, decreasing potential for groundwater nitrate contamination.
3. Significant amounts of residue C could be sequestered in soil.
4. Denitrification is a major N loss process under wetland rice but integrated management of WR, GM, and FN significantly reduces gaseous N losses as compared with FN alone.
5. Integrated use of CR, GM, and FN did not alter the N_2O emissions from soil and loading of the atmosphere.

Thus, adopting such integrated management strategies of FN, GM, and CR can result in an environmentally sound sustainable crop production system by enhancing soil fertility and C sequestration, and reducing denitrification losses, nitrate leaching, and emission of greenhouse gases.

Biofertilizers

Biofertilizers consist of inoculants of (1) *Rhizobium, Bradyrhizobium,* or other associated N fixers; (2) P solubilizing microorganisms (PSM); (3) Arbuscular mycorrhizae fungi (AM); and (4) Azolla. The great potential for BFs to provide plants with enhanced nutrient supply, resulting in higher productivity has been extensively reviewed (Pareek and Chandra, 2005; Aulakh and Sharma, 2005).

In legumes, inoculation of seed with *Rhizobium* results in a saving of N fertilizer. For instance, a review of various studies with groundnut and soybean in India revealed that in areas where these crops are a new introduction, *Rhizobium/Bradyrhizobium* inoculation often results in large increase in crop yield and protein content (Aulakh and Patel, 1991; Aulakh and Pasricha, 1997). However, in soils where legume crops are regularly grown, the native population of *Rhizobium* builds up and inoculation may not be beneficial. In sugarcane, inoculation with *Azotobacter* and *Azospirillum* brings a 20-25 percent reduction in FN requirement of sugarcane besides improving the residual N content of the soil (Shankaraiah and Nagaraju, 1997).

Phosphorus solubilizing microorganisms comprise many heterotrophic soil bacteria and fungi, which secrete organic acids lowering soil pH, which in turn dissolves immobile forms of soil phosphate (Pasricha et al., 1996; Kapoor, 1996). Some of the hydroxyacids may chelate with Ca, Al, Fe, and Mg resulting in effective solubilization of soil P. The rhizosphere is an active site for P transformation because bacteria and fungi engaged in P transformations are 20 to 50 times more abundant than in bulk soil. Few studies have shown significant effects of seed/soil inoculation with PSM in increasing P uptake, protein content, and grain yield of chickpea (*Cicer arietinum* L.), pigeonpea, lentil (*Lens cultinaris* Medik.), and mungbean (Singh and Tilak, 2001; Acharya and Biswas, 2002). Yield of pea was significantly increased by 460-550 kg ha^{-1} due to inoculation of seed with *Pseudomonas striata* (Gaur, 1990). Similarly, a significant increase in the yield of urdbean *(Vigna mungo)* (Tomar et al., 1993) was observed due to PSM. Goel et al. (2002) observed the formation of 68 to 115 percent more nodules, 1.18 to 1.35 times shoot dry weight, and significantly higher plant N content in chickpea with the coinoculation of *Pseudomonas* spp. and *Rhizobium* as compared to single inoculation with *Rhizobium*. These benefits were shown to be due to the antagonistic effects of *Pseudomonas* spp. to fungal pathogens that cause various plant diseases. The phosphate solubilizing microorganisms *Agrobacterium radiobacter* and *Bacillus megaterium* bring about improvement in P use efficiency with pressmud, which is used as a cheaper source of P than synthetic fertilizer (Shankaraiah and Nagaraju, 1997). In rice and wheat, a saving of fertilizer P equivalent to 30-50 kg P_2O_5 ha^{-1} is possible with the integrated use of PSM and low grade rock phosphate (Hedge and Dwivedi, 1994).

The AM fungi are found to invade roots of many field crops and form a symbiotic relationship. They help the growing plant in absorption of P, certain micronutrients, and water from the soil and protect the plant from certain pathogens. Mycorrhizae perhaps do not utilize unavailable P sources but rather facilitate P uptake by increasing the absorptive surface area of the root system, root phosphatase activity, and production of organic acids (Sanders and Tinker, 1971; Singh and Tilak, 2001). Dual inoculation of AM fungi *(Glomus* spp.) and two PSM *(Bacillus circulans* and *Cladosporium herbarium*) with and without rock phosphate in a P deficient sandy soil showed synergistic effects by helping the plant to obtain P from even insoluble P sources such as rock phosphate (Singh and Kapoor, 1999; Ali and Mishra, 2000). Similarly, co-inoculation of AM and *Asperigillus niger* improved the nutrient and biochemical conditions of urdbean and mungbean

(Rao and Rao, 1996). Response to AM inoculation is governed by several factors including soil, host plant, inoculums, and environmental effects (Wani and Lee, 1996). Chhonkar (1994) reviewed the literature on the role of PSM and AM and concluded that *in vitro* microbes increase the availability of P to plants. However, the information available so far is inadequate to generalize the extent to which BFs can increase the availability of P to plants under field conditions.

Azolla *(Anabaena azollae)*, a small water fern, which has a worldwide distribution (Moore, 1969) is found commonly floating on water in ponds and shallow ditches. It fixes N_2 in the cavities of its upper lobes (Moore, 1969) and is used as a GM in rice-growing countries. The fern grows well in a temperature range of 20-30°C and soil pH of 5.0-7.0, and could produce 333 t GM equivalent to 840 kg N ha^{-1} year^{-1} under favorable conditions in the humid tropics. However, high temperatures during summer and low temperatures in winter in the subtropical regions are limiting factors for the growth of Azolla (Beri et al., 1984). Therefore, with optimum growth only during spring and autumn months, it may produce 55 t GM and 139 kg N ha^{-1} year^{-1}, which upon application to rice could be as efficient as FN in increasing yields (Figure 7.4). Applying 1 t Azolla + 40 kg N + 20 kg P_2O_5 + 20 kg K_2O ha^{-1} gave a rice yield of 2.12 t ha^{-1}, higher than 80 kg N + 20 kg P_2O_5 + 20 kg K_2O ha^{-1} (1.98 t ha^{-1}), indicating that combinations of organic and inorganic N could decrease the requirement for inorganic FN (Singh and Singh, 1993).

Pressmud

Pressmud cake (PMC)—a biosolid by-product of the sugar industry using sugarcane in South Asia is an excellent source of N and P for field crops as it contains 1.0-2.5 percent N, 0.25-0.65 percent P, and 0.40-0.85 percent K on a dry weight basis. Several studies have shown that PMC could supply significant amounts of N and P to crops, especially when used in conjunction with chemical fertilizers. It could increase the efficiency of applied P fertilizers (Narval et al., 1990; Trivedi et al., 1995; Singh et al., 1997; Singh et al., 2003), increase the solubility of native P (Khanna and Roy, 1956; Singh et al., 2005), and improve soil productivity by improving Soil Organic Carbon (SOC) and available P (Table 7.8). Application of 20 t pressmud ha^{-1} (fresh weight basis) results in a saving of 25 percent of recommended inorganic fertilizers in rice, maize, wheat, sugarcane, and soybean (Jurwarkar et al., 1995).

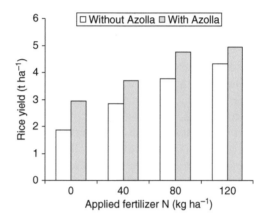

FIGURE 7.4. Impact of azolla manure alone and with different rates of fertilizer N on rice yield in a subtropical region. *Source:* Prepared from data of Beri et al. (1984).

TABLE 7.8. Effect of PMC and P application to rice on SOC and available P content of soil after one year of rice-wheat rotation

Treatment to rice[a]	Soil organic C (%)		Available P (kg ha^{-1})	
	After rice	After wheat	After rice	After wheat
PMC$_0$ P$_0$	0.58	0.50	7.0	6.8
PMC$_0$ P$_{13}$	0.59	0.50	8.0	7.7
PMC$_0$ P$_{26}$	0.61	0.50	9.5	8.5
PMC$_5$ P$_0$	0.62	0.54	12.2	8.9
PMC$_{10}$ P$_0$	0.64	0.57	18.5	12.4
LSD$_{0.05}$	0.03	0.03	2.3	1.6

Source: Adapted from Singh et al. (2005).

[a]Subscripts for PMC denote 0, 5, and 10 t ha^{-1} of PMC and for P denote 0, 13, and 26 kg of fertilizer P ha^{-1}.

Sewage Sludge and Industrial Effluents

Application of sewage sludge to agricultural soils, and irrigation of field crops with sewage water and untreated industrial effluents alone, or in combination with ground/canal water are common practices worldwide and are used in South Asia, especially in the vicinity of large cities, as these are considered reusable sources of essential nutrients and organic matter (Smith, 1996). However, some of the heavy metals present in sewage water and untreated industrial effluents could be toxic to plants and cause health hazards to animals and/or humans. It has been documented that irrigation with sewage water increases soil electrical conductivity and organic C, decreases soil pH, and could result in the accumulation of heavy metals (e.g., Al, Cr, Ni, and Pb) in the plough layer of agricultural soils in India and Pakistan (Singh and Verloo, 1996; Tirmizi et al., 1996; Brar and Arora, 1997), as well as in plants such as cauliflower (*Brassica oleracea* conniver. Botrytis), potatoes *(Solanum tuberosum),* spinach (*Spinacia oleracia* L.), and other leafy vegetables (Brar and Arora, 1997; Zayed et al., 1998; Brar et al., 2000; Sharma et al., 2005). Therefore, it is desirable to find out the critical toxic ranges of heavy metals and pollutants in such effluents, and the behavior of different plants for the uptake and accumulation of such elements, before their practical use alone or integrated with other organic and inorganic sources. These resources could become part of INM with the prior knowledge of nutrient dynamics in soils due to the nature of different sewage and industrial treatment works (if) performed in different catchments allowing the edaphic mineralization characteristics of biosolids.

Balanced Fertilization

Balanced fertilization is essential to obtain high yield while minimizing the cost of production. In potato, a maximum yield of 40.8 t ha^{-1} was obtained when 180 kg N + 100 kg P$_2$O$_5$ + 150 kg K$_2$O ha^{-1} were applied, compared to a tuber yield of only 14.6 t ha^{-1} in the control plots without NPK (Imas and Bansal, 2002). Furthermore, increasing K rates decreases the yield of small grade tubers and increases the proportion of large marketable tubers. Interestingly, K application also dramatically decreased the incidence of late blight *(Phytophthora infestans).* A study from Bangladesh showed the additive effect of S + Zn + B on rice; the combination of all three nutrients increased the rice yield by 41.8 percent (4.5 t ha^{-1}) over the control, while application of S, Zn, and B alone resulted in 23.3, 21.7, and 14.6 percent yield increase, respectively (Islam et al., 1997).

At adequate levels of 40 kg S and 80 kg P$_2$O$_5$ ha^{-1}, the seed and oil yield of soybean increased from 600 and 111 kg ha^{-1} in the control to 1,740 and 412 kg ha^{-1}, respectively, but application of 20 kg S and 120 kg P$_2$O$_5$ ha^{-1}

created an imbalance and reduced the seed and oil yield to 1,580 and 360 kg ha^{-1}, respectively (Aulakh et al., 1990).

TECHNICAL REQUIREMENTS FOR INM

The nutrient recommendations for various crops differ, both in terms of amount and nutrient balance. Recommended applications vary with many factors, including the inherent capacity of the soil to provide these nutrients, nutrient requirements of the crops, crop yield potential, and the climate/region they are grown in. Coarse-textured soils may require higher amounts of nutrients compared to the fine-textured soils. Moreover, coarse-textured soils are often deficient in many more nutrients (such as Zn, Fe, and Cu) as compared to fine-textured soils.

Nutrient recommendations should be based on soil testing. Generally soils are tested for N, P, and K. While soil testing for N, P, and K itself has not become a common practice in South Asia, testing for micronutrients is still limited (Dangarwala, 2001). Chemical soil tests have long been used to estimate the nutrient availability in soils, to predict the probability of obtaining a profitable response to applied nutrients, and to provide a basis for recommendation on the amount of fertilizers needed (Sekhon, 1992). However, soil testing has some analytical and interpretative problems. Plant analysis may also be used as both a diagnostic and recommendation tool, as it reflects the nutrient status of the plant, and thus is an indirect way of determining the ability of the soil to supply nutrients to the crop.

Fertilizers need to be applied at the appropriate time to meet the nutrient demands of the crop. In the case of wetland rice, N application in three equal splits (at transplanting, tillering, and panicle initiation) has been found to be more efficient in increasing rice yield than single or two split applications (Meelu and Morris, 1984). Not only is the rate of application important but also the source of nutrients. In the case of wetland rice, urea was significantly superior to calcium ammonium nitrate, whereas in wheat both sources of N were equally effective (Yadvinder-Singh et al., 1990). The slow release N fertilizers have also shown promising results in rice (Meelu et al., 1992).

The use of N fertilizers in South Asia is very low compared to many other parts of the world. However, it has been reported that in Punjab—an agriculturally advanced state in India—water in the wells located near village habitations contains higher nitrate content (on average 16.8 and 19.7 µg NO_3-N l^{-1} in June and September, respectively) than those in the vicinity of cultivated fields (4.4 and 4.5 µg NO_3-N l^{-1} in June and

September, respectively) (Singh and Sekhon, 1977). The elevated levels are due to disposal practices for animal wastes on the outskirts of the village. High values in September samplings were due to leaching of nitrates in rain water during the monsoon period. This suggests that not the chemical fertilizer use but the disposal of animal wastes in the open was the source of the contamination. Therefore, management of nutrients from animal wastes must be improved, not only to capture the value of the nutrients in the manure, but also to avoid nitrate content in well waters, and reduce the risk of disease transmission from pathogens in the manure.

High-yielding dwarf varieties of wheat were introduced in India and other South Asian countries in the mid-sixties. Since then P fertilizers have been regularly applied to wheat, resulting in a buildup of soil phosphorus levels to such an extent in many instances that one could either altogether forego the use of fertilizer P for sometime, or switch to a reduced maintenance dose to match the crop removal of P (Sekhon, 1992; Aulakh et al., 2003; Kumar, 2005). For this reason, guidelines need to be devised to help farmers determine the amount of P that is required to optimize crop production and avoid excess or unnecessary applications. Further, response of P is higher in *rabi* (winter season) crops than *kharif* (summer season) crops, as discussed in earlier sections.

Soluble P fertilizers are expensive; however, use of indigenous rock phosphates can help save foreign exchange in developing countries, especially in South Asia. For example, studies in Sri Lanka have shown that the partially acidulated phosphate rock, prepared from Eppawala rock phosphate, which is locally available, could become a commercial phosphatic fertilizer for use in annual crops in Sri Lanka (Abeykoon, 1992). Similarly, studies from India and other countries show a great potential in indigenous rock phosphates (Sharma and Sangrai, 1992; Prasad, 2000).

Recommendations for K seldom exceed 25-50 percent of the crop removal (Sekhon, 1992). However, the crop responses to K have been increasing with time both in rice and wheat (Bhargava et al., 1985; Aulakh and Malhi, 2005). Further, *kharif* crops (e.g., maize and rice), rather than *rabi* crops (e.g., wheat), respond more to K application (Meelu et al., 1992).

Sulfur deficiency in Indian soils is increasing (Tandon, 1995; Aulakh, 2003a), mainly due to the increased use of sulfur-free fertilizers during the last four decades, use of high-yielding crop varieties, and intensive cropping. Deficiency of micronutrients (Zn, Fe, Mn, Cu, Mo, and B) in the soil is also widespread (Prasad, 2000). There is a need to monitor regularly the status of secondary and micronutrients to make sure the supply of these nutrients is sufficient for obtaining high productivity and maintaining soil status.

CONSTRAINTS FOR IMPLEMENTATION OF INM

No doubt INM has many merits, however, this system has not been adopted as widely as it should be. This clearly shows that some constraints exist for the utilization of INM. These constraints may include:

Nutrient Recommendations

Current nutrient recommendations for crop production focus primarily on the application of different types of synthetic fertilizers. There is a requirement for research focusing on INM, incorporating locally available organic nutrient sources applied in combination with synthetic fertilizers, to form the basis for recommendations. Further, extensive efforts should be made to demonstrate the merits of INM to producers, using local examples, so that this system is adopted on a larger scale.

Availability of Fertilizers

Chemical fertilizers are available on the market in sufficient quantities, leaving little incentive for the farmers to consider other nutrient sources. Fertilizers containing not only single but two, three, or even more nutrients are commonly available on the market. So a single source can meet different types of nutrient requirements of the crop.

Availability of Organics

Organic sources of nutrients are not available in sufficient quantities. In India, it is estimated that only 25 percent of the total nutrient needs of agriculture can be met by utilizing various organic sources such as FYM, CR, urban and rural wastes, green manuring, and BFs (Prasad, 2000).

Yield Levels

With the use of organics alone, the yields are generally low (Singh and Sekhon, 2002), which are neither economical nor acceptable to farmers. Grain yields of rice were maintained with partial substitution of GM-N, but yields were lower with GM-N alone (Budhar and Palaniappan, 1997). Research and technology transfer addressing the benefits of combining the use of synthetic fertilizers with available organic sources to optimize the efficiency of both sources is needed.

Ignorance About the Long-Term Benefits of INM

Farmers of South Asia, in general, are not well aware of the long-term benefits of INM in sustaining soil productivity and quality as well as environmental benefits. They need to be educated regarding the reasons for the continuing deterioration of soil quality and declining crop yields, other than soil, water, and air pollution. Establishment of permanent long-term INM and conducting field days on such sites could be helpful in disseminating the benefits of INM.

RESEARCH GAPS AND FUTURE NEEDS

Dangarwala (2001) suggested that there is an urgent need to increase facilities for the analysis of available micronutrients in soil from farmers' fields. Furthermore, in many of the INM studies, soil analysis is limited to N, P, and K and the micronutrients are not included; this needs to be rectified. Fertilizer recommendations should be made based on soil testing and calibration with response curves generated from local research. Balanced and judicious fertilization can provide higher yields at optimum costs while minimizing environmental pollution.

Apart from the recommended rates of NPK, the crops may require an additional supply of FYM, or GM, or an extra dose of NPK, to produce the highest yields (Gill et al., 1994; Mathukia et al., 1999). There is a need to make recommendations for nutrient application based on a benefit to cost ratio. Further, nutrient management recommendations should be based on the cropping system rather than on an individual crop basis.

As discussed earlier, FYM has great impacts on crop yields and soil productivity, however, the availability of FYM for use on the farm is very limited, even though South Asia supports a large proportion of the world cattle population. This is mainly because animal dung is widely burnt as a source of domestic fuel in this region. There is a need to provide other sources of fuel, such as wood, cooking gas, or farm-borne–waste-based biogas so that FYM could be spared for use on the farm in as large quantities as possible.

Like FYM, green manuring also has beneficial effects in improving the chemical, physical, and biological properties of the soil. However, farmers show little interest in growing GM crops; rather they prefer to grow crops that provide them with some immediate income. In such cases, growing of grain legumes such as mungbean or urdbean can provide dual benefits, since the producer can harvest the seed, and then use the rest as GM. However, if the seed is removed, the N contribution from grain legumes will not be as high as from GM crops.

Biofertilizers also have great potential as a tool in INM. However, many times the response to BFs is not visible. This could be due to a variety of reasons although a common problem is ineffective rhizobia. So there is a need to ensure supply of effective rhizobia and other BFs that is easily accessible to farmers.

Extension and social workers could play a vital role in (1) educating the farmers about the long-term benefits of INM; and (2) convincing them about its merits for sustaining high crop yields, and improving soil productivity while minimizing environmental pollution.

SUMMARY AND CONCLUSIONS

Research trials in different South Asian countries have clearly demonstrated that balanced fertilization through integrated nutrient supply is the best method of sustaining crop yield and quality without adversely affecting the environment. Integrated nutrient management is an essential practice for sustainable crop production.

Long-term experiments have clearly shown that even with the application of NPK fertilizers there is a declining trend in productivity over time, possibly due to deficiency in secondary and micronutrients or deterioration in the physical properties of the soil. Further, neither organic manure nor chemical fertilizers alone can sustain high crop yields; however, the integrated use of organics and synthetic fertilizers based on soil testing can overcome this problem. The organic resources available in South Asia include FYM, GM, composts, CR, BFs, pressmud, sewage sludge, etc. In long-term multilocation field experiments in India it has been amply demonstrated that 50 percent of the N needs of rice could be supplied through FYM or *Sesbania* GM (Hedge, 1998; Katyal et al., 1998). There is a need to study INM strategies in other major crops. For this, strategies include (1) nutrient recommendations be made on a cropping systems basis rather than on an individual crop basis; (2) shallow-rooted crops be grown in rotation with deeper and extensive rooting crops; (3) fertilizer P be added to winter-grown crops as summer crops of subtropical regions can effectively utilize residual P.

Compared to many developed countries, fertilizer use in the countries of South Asia is very low. The available evidence does not suggest severe environmental pollution of soil or groundwater due to the current usage of chemical fertilizers (Sarkar, 1990) except in a few specific situations (Aulakh et al.,

2001a,b). However, there is a need to continuously monitor such effects critically in future.

Efficient resource use is required to maintain soil quality, optimize the economics of production, and reduce the risk of negative environmental impacts. In the future, there will still be an increasing population and ever-growing pressure on the land base to support the requirements of the population. Agriculture is likely to become increasingly intense. Meanwhile, the cost of nutrient inputs is likely to rise because of trends in energy costs. Soil quality must be maintained, as this is the base of agricultural production. So, nutrient management strategies must use nutrients effectively and ensure that the quality of the soil is maintained or improved. Organic resources cannot be wasted, both for their supply of nutrients, and for their potential benefits on soil quality. The key is to find a way to effectively utilize both organic and synthetic nutrient resources in a cost effective way.

REFERENCES

Abeykoon, V. (1992). Effectiveness of partially acidulated phosphate rock (PAPR) prepared from Eppawala rock phosphate. In Beri, V., M.R. Chaudhary, P.S. Sidhu, N.S. Pasricha, and M.S. Bajwa (eds.) *Proceedings of the International Symposium on Nutrient Management for Sustained Productivity.* Vol. 2. pp. 24-25. Ludhiana, India: Department of Soils, Punjab Agricultural University.

Acharya, C.L. and A.K. Biswas. (2002). Integrated nutrient management in pulse based cropping systems—Present status and future prospects. In Ali, M., S.K. Chaturvedi, and S.N. Gurha (eds.) *Pulses for Sustainable Agriculture and Nutritional Security.* pp. 1-12. Kanpur, India: Indian Society of Pulses Research and Development.

Ahmad, N. and T. Muhammad. (1998). Fertiliser, plant nutrient management, and self-reliance in agriculture. *Pakistan Development Review.* 37: 217-233.

Ahmed, S. and S.M. Rahman. (1992). Management guidelines for micronutrient fertilization in Bangladesh soils. In Beri, V., M.R. Chaudhary, P.S. Sidhu, N.S. Pasricha, and M.S. Bajwa (eds.) *Proceedings of the International Symposium on Nutrient Management for Sustained Productivity.* Vol. 2. pp. 59-60. Ludhiana, India: Department of Soils, Punjab Agricultural University.

Ali, M. and J.P. Mishra. (2000). Nutrient management in pulses and pulse-based cropping systems. *Fertiliser News.* 45(4): 57-69.

Antil, R.S., D. Singh, V. Kumar, and M. Singh. (1989). Effect of preceding crops on yield and nitrogen uptake by rice. *Indian Journal of Agronomy.* 34: 213-216.

Arora, R.P., M.S. Sachdev, Y.K. Sud, V.K. Luthra, and B.V. Subbiah. (1980). Fate of fertilizer N in a multiple cropping system. In *Soil Nitrogen as Fertilizer or Pollutant.* pp. 3-22. Vienna, Austria: International Atomic Energy Agency.

Ashraf, M., T. Mahmood, F. Azam, and R.M. Qureshi. (2004). Comparative effects of applying leguminous and non-leguminous green manure and inorganic N on the biomass yield and nitrogen uptake in flooded rice (*Oryza sativa* L.). *Biology and Fertility of Soils.* 40: 147-152.

Aulakh, M.S. (1994). Integrated nitrogen management and leaching of nitrates to groundwater under cropping systems followed in tropical soils of India. In *Transactions of 15th World Congress of Soil Science.* 5: 205-221.

Aulakh, M.S. (2001). Impacts of integrated management of crop residues, fertilizer N and green manure on productivity, nitrate leaching, carbon sequestration, denitrification and N_2O emissions in rice-wheat system. In Ji, L., G. Chen, E. Schnug, C. Hera, and S. Hanklaus (eds.) *Fertilization in Third Millennium Fertilizer Food Security and Environmental Protection; Proceedings of the 12th World Fertilizer Congress.* Vol. 3, pp. 1550-1556. Braunschweig, Germany: International Scientific Center of Fertilizers (CIEC), and Beijing, China: Chinese Academy of Sciences.

Aulakh, M.S. (2003a). Crop responses to sulphur nutrition. In Abrol, Y.P. and A. Ahmad (eds.) *Sulphur in Plants.* pp. 341-358. Boston/London: Kluwer Academic Publishers.

Aulakh, M.S. (2003b). Enhancing phosphatic fertilizer-use efficiency in sub-tropical regions. *Fertilizers & Agriculture (IFA, Paris).* (September 2003): 6-7.

Aulakh, M.S. and G.S. Bahl. (2001). Nutrient mining in agro-climatic zones of Punjab. *Fertiliser News.* 46(4): 47-61.

Aulakh, M.S., J.W. Doran, and A.R. Mosier. (1992). Soil denitrification—Significance, measurement, and effects of management. *Advances in Soil Science.* 18: 1-57.

Aulakh, M.S., B.S. Kabba, H.S. Baddesha, G.S. Bahl, and M.P.S. Gill. (2003). Crop yields and phosphorus fertilizer transformations after 25 years of applications to a subtropical soil under groundnut-based cropping systems. *Field Crops Research.* 83: 283-296.

Aulakh, M.S., T.S. Khera, and J.W. Doran. (2000). Mineralization and denitrification in upland, nearly-saturated and flooded subtropical soil. II. Effect of organic manures varying in N content and C:N ratio. *Biology and Fertility of Soils.* 31: 168-174.

Aulakh, M.S., T.S. Khera, J.W. Doran, and K.F. Bronson. (2001a). Managing crop residue with green manure, urea, and tillage in a rice-wheat rotation. *Soil Science Society of America Journal.* 65: 820-827.

Aulakh, M.S., T.S. Khera, J.W. Doran, and K.F. Bronson. (2001b). Denitrification, N_2O and CO_2 fluxes in rice—Wheat cropping system as affected by crop residues, fertilizer N and legume green manure. *Biology and Fertility of Soils.* 34: 375-389.

Aulakh, M.S., T.S. Khera, J.W. Doran, Kuldip-Singh, and Bijay-Singh. (2000). Yields and nitrogen dynamics in a rice—Wheat system using green manure and inorganic fertilizer. *Soil Science Society of America Journal.* 64: 1867-1876.

Aulakh, M.S. and S.S. Malhi. (2004). Fertilizer nitrogen use efficiency as influenced by interactions of N with other nutrients. In Mosier, A., J.K. Syers, and J.R. Freney (eds.) *Agriculture and the Nitrogen Cycle: Assessing the Impacts of*

Fertilizer Use on Food Production and the Environment. pp. 181-193. Covelo, CA: Island Press.

Aulakh, M.S. and S.S. Malhi. (2005). Interactions of nitrogen with other nutrients and water: Effect on crop yield and quality, nutrient use efficiency, carbon sequestration, and environmental pollution. *Advances in Agronomy.* 86: 341-409.

Aulakh, M.S. and N.S. Pasricha. (1997). Role of balanced fertilization in oilseed based cropping systems. *Fertiliser News.* 42(4): 101-111.

Aulakh, M.S. and N.S. Pasricha. (1998). The effect of green manuring and fertilizer N application on enhancing crop productivity in mustard-rice rotation in semiarid subtropical regions. *European Journal of Agronomy.* 8: 51-58.

Aulakh, M.S., N.S. Pasricha, and A.S. Azad. (1990). Phosphorus-sulphur interrelationship for soybean in P and S deficient soils. *Soil Science.* 150: 705-709.

Aulakh, M.S., N.S. Pasricha, and G.S. Bahl. (2003). Phosphorus fertilizer response in an irrigated soybean–wheat production system on a subtropical, semiarid soil. *Field Crops Research.* 80: 99-109.

Aulakh, M.S. and M.S. Patel. (1991). Soil-related constraints and management for oilseed and pulse production in India. In *Soil-Related Constraints in Crop Production.* pp. 129-144. New Delhi, India: Indian Society of Soil Science.

Aulakh, M.S. and P. Sharma. (2005). Nutrients dynamics and their efficient use in pulse crops. In Singh, G., H.S. Sekhon, and J.S. Kolar (eds.) *Pulses.* pp. 241-278. Udaipur, India: Agrotech Publishing Academy.

Aulakh, M.S. and B. Singh. (1997). Nitrogen losses and fertilizer N use efficiency in irrigated porous soils. *Nutrient Cycling in Agroecosystems.* 47: 197-212.

Aulakh, M.S., D. Singh, and U.S. Sadana. (2004). Direct and residual effects of green manure and fertilizer nitrogen in rice-rapeseed production system in the semiarid subtropics. *Journal of Sustainable Agriculture.* 25: 97-114.

Aulakh, M.S., R. Wassmann, and H. Rennenberg. (2001). Methane emissions from rice fields—Quantification, role of management, and mitigation options. *Advances in Agronomy.* 70: 193-260.

Badiyala, D. and S. Kumar. (2003). Effect of organic and inorganic fertilizers on growth and yield of linseed (*Linum usitatissimum*) under mid-hill conditions of Himachal Pradesh. *Indian Journal of Agronomy.* 48: 220-223.

Bahl, G.S. and M.S. Aulakh. (2003). Phosphorus availability dynamics in soils, and its requirements in field crops. *Fertiliser News.* 48(3): 19-29.

Bahl, G.S. and N.S. Pasricha. (1998). Efficiency of P utilization by pigeonpea and wheat grown in a rotation. *Nutrient Cycling in Agroecosystems.* 51: 225-229.

Bahl, G.S. and N.S. Pasricha. (2000). N-utilization by maize (*Zea mays* L.) as influenced by crop rotation and fieldpea (*Pisum sativum* L.) residue management. *Soil Use and Management.* 16: 230-231.

Bahl, G.S., N.S. Pasricha, M.S. Aulakh, and H.S. Baddesha. (1986). Effect of incorporation of crop straw residues on the yield of mustard and wheat. *Soil Use and Management.* 2: 10-12.

Bahl, G.S., N.S. Pasricha, M.S. Aulakh, and H.S. Baddesha. (1988). Influence of different grain legumes on nitrogen availability in soil, and fertilizer requirement of succeeding wheat crop. *Indian Journal of Ecology.* 15(1): 99-101.

Balasubramanian, V., B. Alves, M.S. Aulakh, M. Bekunda, Z. Cai, L. Drinkwater, D. Mugendi, C. van Kessel, and O. Oenema. (2004). Crop, environmental and management factors affecting fertilizer N use efficiency. In Mosier, A., J.K. Syers, and J.R. Freney (eds.) *Agriculture and the Nitrogen Cycle: Assessing the Impacts of Fertilizer Use on Food Production and the Environment.* pp. 19-33. Covelo, CA: Island Press.

Banwasi, R. and R.K. Bajpai. (2001). Effect of integrated nutrient management on root growth of wheat in a rice-wheat cropping system. *Agricultural Science Digest.* 21: 1-4.

Bellakki, M.A., V.P. Badanur, and R.A. Setty. (1998). Effect of long-term integrated nutrient management on some important properties of a Vertisol. *Journal of the Indian Society of Soil Science.* 46: 176-180.

Benbi, D.K. and S.P.S. Brar. (1992). Predicting micronutrient availability in alkaline soils of Punjab. In Beri, V., M.R. Chaudhary, P.S. Sidhu, N.S. Pasricha, and M.S. Bajwa (eds.) *Proceedings of the International Symposium on Nutrient Management for Sustained Productivity.* Vol. 2. pp. 56-58. Ludhiana, India: Department of Soils, Punjab Agricultural University.

Beri, V., O.P. Meelu, and B. Raj. (1984). Studies on multiplication, nitrogen fixation and utilization of *Azolla. Bulletin, Indian Society of Soil Science.* 13: 387-391.

Beri, V., B.S. Sidhu, G.S. Bahl, and A.K. Bhat. (1995). Nitrogen and phosphorus transformations as affected by crop residue management practices and their influence on crop yield. *Soil Use and Management.* 11: 51-54.

Bhandari, A.L., J.K. Ladha, H. Pathak, A.T. Padre, D. Dawe, and R.K. Gupta. (2002). Yield and soil nutrient changes in a long-term rice-wheat rotation in India. *Soil Science Society of America Journal.* 66: 162-170.

Bhandari, A.L., A. Sood, K.N. Sharma and, D.S. Rana. (1992). Integrated nutrient management in a rice-wheat system. *Journal of the Indian Society of Soil Science* 40: 742-747.

Bhargava, P.N., H.C. Jain, and A.K. Bhatia. (1985). Response of rice and wheat to potassium. *Journal of Potassium Research.* 1: 45-61.

Biswas, B.C., N. Prasad, and R.K. Tewatia. (1992). Fertiliser use and environmental quality in India. *Fertiliser News.* 37(11): 15-17.

Borlough, N.E. and C.R. Dowswell. (1994). Feeding to human population that increasingly crowds a fragile planet. Keynote lecture, *15th World Congress of Soil Science.* pp. 1-15. Acapulco, Mexico: International Soil Science Society.

Brar, M.S. and C.L. Arora. (1997). Concentration of microelements and pollutant elements in cauliflower (*Brassica oleracea* conniver. Botrytis var. botrytis). *Indian Journal of Agricultural Science.* 67: 141-143.

Brar, M.S., S.S. Malhi, A.P. Singh, C.L. Arora, and K.S. Gill. (2000). Sewage water irrigation effects on some potentially toxic trace elements in soil and potato plants in northwestern India. *Canadian Journal of Soil Science.* 80: 465-471.

Brown, S. and H. Schreier. (2000). Nutrient budgets: A sustainability index. In Allen, R., H. Schreier, S. Brown, and P.B. Shah (eds.) *The People and Resource Dynamics Project: The First Three Years (1996-1999). Proceedings of a Workshop.* pp. 291-299. Yunnan Province, China: Baoshan.

Budhar, M.N. and S. Palaniappan. (1997). Integrated nutrient management in lowland rice (*Oryza sativa*). *Indian Journal of Agronomy.* 42: 269-271.

Chandel, A.S., K.N. Pandey, and S.C. Saxena. (1989). Symbiotic nitrogen fixation and nitrogen benefits by nodulated soybean (*Glycine max* (L.) Merril) to interplanted crops in northern India. *Tropical Agriculture (Trinidad).* 66: 73-77.

Chhonkar, P.K. (1994). Mobilization of soil phosphorus through microbes: Indian experience. In Dev G. (ed.) *Phosphorus Researches in India.* pp. 120-125. Gurgaon, India: Potash and Phosphate Institute of Canada, India Programme.

Dangarwala, R.T. (2001). Need for sustaining balanced supply of micronutrients in soils rather than their correction. *Journal of the Indian Society of Soil Science.* 49: 647-652.

De, S.L.M., H.P.S. Jayasundara, D.N.S. Fernando, and M.T.N. Fernando. (1993). Integration of legume-based pasture and cattle into coconut farming systems in Sri Lanka. *Journal of the Asian Farming Systems Association.* 1: 579-588.

FAI. (2004). *Fertiliser Statistics 2003-2004.* New Delhi, India: Fertiliser Association of India.

Gardner, W.K., D.A. Barber, and D.G. Parbery. (1983). The acquisition of phosphorus by *Lupinus albus* L. III. The probable mechanism by which phosphorus movement in the soil/root interface is enhanced. *Plant and Soil.* 70: 107-124.

Gaur, A.C. (1990). *Phosphate Solubilizing Microorganisms as Biofertilizers.* New Delhi, India: Omega Scientific Publishers.

Gill, H.S. and O.P. Meelu. (1983). Studies on the utilization of phosphorus and causes for its differential response in rice-wheat rotation. *Plant and Soil.* 74: 211-222.

Gill, M.S., T. Singh, and D.S. Rana. (1994). Integrated nutrient management in rice (*Oryza sativa*)-wheat (*Triticum aestivum*) cropping sequence in semi-arid tropic. *Indian Journal of Agronomy.* 39: 606-608.

Goel, A.K., S.S. Sindhu, and K.R. Dadarwal. (2002). Stimulation of nodulation and plant growth of chickpea (*Cicer arietinum* L.) by *Pseudomonas* spp. antagonistic to fugal pathogens. *Biology and Fertility of Soils.* 36: 391-396.

Goswami, N.N. and N.S. Pasricha. (1992). Dynamics of phosphorus in soil. In Bajwa, M.S., N.S. Pasricha, P.S. Sidhu, M.R. Chaudhary, D.K. Benbi, and V. Beri (eds.) *Proceedings of the International Symposium on Nutrient Management for Sustained Productivity.* Vol. 1. pp. 30-42. Ludhiana, India: Department of Soils, Punjab Agricultural University.

Gurung, G.B., D.P. Sherchan, and R.K. Shrestha. (1996). *Integrated Nutrient Management Studies on Potato + Maize Mixed Cropping System.* Pakhribas Agricultural Centre Working Paper No. 137, Dhankuta, Kathmandu, Nepal.

Handa, B.K. (1987). Nitrate content in groundwater in India. *Fertiliser News.* 32(6): 11-29.

Hegde, D.M. (1998). Integrated nutrient management effect on rice (*Oryza sativa*)-wheat (*Triticum aestivum*) system productivity in sub-humid ecosystem. *Indian Journal of Agricultural Sciences.* 68: 144-148.

Hedge, D.M. and B.S. Dwivedi. (1994). Crop response to biofertilisers in irrigated areas. *Fertiliser News.* 39(4): 19-26.

Hundal, H.S. and G.S. Sekhon. (1976). Efficiency of Mussoorie rock phosphate as a source of fertilizer phosphorus to guar (*Cyamopsis tetragonaloba*) and groundnut (*Arachis hypogaea*). *Journal of Agricultural Science, Cambridge.* 87: 665-669.

Imas, P. and S.K. Bansal. (2002). Potassium and integrated nutrient management in potato. In Khurana, S.M.P., G.S. Shekhawat, S.K. Pandey, and B.P. Singh (eds.) *Potato, Global Research and Development. Proceedings of the Global Conference on Potato. Potash Research Institute of India, Gurgoan, and Indian Society of Agronomy.* pp. 744-754. New Delhi, India.

Islam, M.S. (2001). Soil fertility issues: the challenges, possible intervention and the IPNS concept. *Agro Chemicals Report.* 1(2): 32-35.

Islam, M.R., T.M. Riasat, and M. Jahiruddin. (1997). Direct and residual effects of S, Zn, and B on yield and nutrient uptake in a rice-mustard cropping system. *Journal of the Indian Society of Soil Science.* 45: 126-129.

Islam, M.S. and U.K. Saha. (1998). Integrated nutrient management of Gangetic floodplain soils of Bangladesh. In Rahman, M.A., M.S. Shah, M.G. Murtaza, and M.A. Matin (eds.) *Integrated Management of Gangas Floodplains and Sundarbans Ecosystem. Proceedings of the National Seminar.* pp. 173-187. Khulna, Bangladesh: Khulna University.

Jurwarkar, A.S., P.L. Thewale, U.H. Baitule, and M. Moghe. (1995). Sustainable crop production through integrated plant nutrition system—Indian experience. *RAPA Publication.* 12: 87-95.

Kabba, B.S. and M.S. Aulakh. (2004). Climatic conditions and crop residue quality differentially affect N, P and S mineralization in soils with contrasting P status. *Journal of Plant Nutrition and Soil Science.* 167: 596-601.

Kalra, Y.P. and Soper, R.J. (1968). Efficiency of rape, oat, soybean and flax in absorbing soil and fertilizer phosphorus at seven stages of growth. *Agronomy Journal.* 60: 209-212.

Kapoor, K.K. (1996). Phosphate mobilization through soil microorganisms. In R.K. Behl, A.L. Khurana, and R.C. Dogra (eds.) *Plant Microbe Interaction in Sustainable Agriculture.* pp. 46-61. New Delhi, India: Bio Science Publishers.

Katyal, V., S.K. Sharma, and K.S. Gangwar. (1998). Stability analysis of rice (*Oryza sativa*)-wheat (*Triticum aestivum*) cropping system in integrated nutrient management. *Indian Journal of Agricultural Sciences.* 68: 51-53.

Khanna, K.L. and P.K. Roy. (1956). Studies on factors affecting soil fertility in sugar belt of Bihar. VI. Influence of organic matter on phosphate availability under calcareous soil conditions. *Journal of Indian Society of Soil Science.* 4: 189-192.

Kumar, A. (2005). Effect of long-term fertilizer management and crop rotations on accumulation and movement of phosphorus in soil. MSc Thesis, Department of Soils, Punjab Agricultural University, Ludhiana, India.

Kumar, A. and I.P.S. Ahlawat. (1986). Effect of planting geometry and nitrogen fertilization in pigeonpea based cropping systems. *Indian Journal of Agronomy.* 31: 112-114.

Lal, S. and B.S. Mathur. (1988). Effect of long term manuring, fertilization and liming on crop yield and some physico-chemical properties of acid soil. *Journal of the Indian Society of Soil Science.* 36: 113-119.

Manandhar, R. (2001). Integrated plant nutrient system in Nepal. *Agro Chemicals Report.* 1(4): 29-39.

Mani, D. and V.P. Yadav. (2000). Mitigating soil pollution through integrated nutrient management system. *Bioved.* 11(1-2): 75-77.

Marschner, H. (1995). *Mineral Nutrition of Higher Plants, Second Edition.* San Diego, CA: Academic Press.

Maskina, M.S., Bijay Singh, Yadvinder Singh, H.S. Baddesha, and O.P. Meelu. (1988). Fertilizer requirement of rice-wheat and maize-wheat rotations on coarse textured soils amended with farmyard manure. *Fertilizer Research.* 17: 153-164.

Mathukia, R.K., V.B. Ramani, P.K. Chovatia, and R.L. Joshi. (1999). Effect of integrated nutrient management on yield and quality of sugarcane (*Saccharum officinarum* L.). *Indian Sugar.* 48: 839-842.

Meelu, O.P. (1996). Integrated nutrient management for ecologically sustainable agriculture. *Journal of the Indian Society of Soil Science.* 44: 582-592.

Meelu, O.P. and R.A. Morris. (1984). Integrated management of plant nutrients in rice and rice-based cropping systems. *Fertiliser News.* 29(12): 65-70.

Meelu, O.P., S.P. Palaniappan, Yadvinder-Singh, and Bijay-Singh. (1992). Integrated nutrient management in crops and cropping sequences for sustainable agriculture. In Bajwa, M.S., N.S. Pasricha, P.S. Sidhu, M.R. Chaudhary, D.K. Benbi, and V. Beri (eds.) *Proceedings of the International Symposium on Nutrient Management for Sustained Productivity.* Vol. 1. pp. 101-114. Ludhiana, India: Department of Soils, Punjab Agricultural University.

Meelu, O.P., Y. Singh, and B. Singh. (1994). Green manuring for soil productivity improvement. *World Soil Resources Reports No.76.* Rome, Italy: FAO.

Mishra, M.M., K.K. Kapoor, K. Chander, and R.D. Laura. (1991). Sustainable agriculture: The role of integrated nutrient management—A review. *International Journal of Tropical Agriculture.* 9: 153-173.

Moore, A.W. (1969). Azolla: Biology and agronomic significance. *The Botanical Review.* 35: 17-34.

Narval, P., A.P. Gupta, and R.S. Antil. (1990). Efficiency of triple super phosphate and Mussoorie rock phosphate mixture incubated with sulphitation process pressmud. *Journal of Indian Society of Soil Science.* 38: 51-55.

Olson, R.J., R.E. Hensler, O.J. Attoe, S.A. Witzwl, and L.A. Peterson. (1970). Fertilizer nitrogen and crop rotation in relation to movement of nitrate nitrogen through soil profiles. *Soil Science Society of America Proceedings.* 34: 448-452.

Panaullah, G.M., P.K. Saha, and M.A. Saleque. (2001). Integrated nutrient management: IPNS in rice-rice cropping patterns in Bangladesh. *Agro Chemicals Report.* 1(2): 36-38.

Pande, S., S.B. Sharma, and A. Ramakrishna. (2000). Biotic stresses affecting legumes production in the Indo-Gangetic Plain. In Johansen, C., J.M. Duxbury, S.M. Virmani, C.L.L. Gowda, S. Pande and P.K. Joshi (eds.) *Legumes in Rice* and Wheat Cropping Systems of the Indo-Gangetic Plain—Constraints and Opportunities. pp. 129-155. Patancheru, Andhra Pradesh, India: International Crop Research Institute for the Semi-Arid Tropics and Ithaca, NY: Cornell University.

Pandey, S.P., H. Shankar, and V.K. Sharma. (1985). Efficiency of some organic and inorganic residues in relation to crop yield and soil characteristics. *Journal of the Indian Society of Soil Science.* 33: 175-181.

Pareek, R.P. and R. Chandra. (2005). Biological nitrogen fixation in pulses. In Singh, G., H.S. Sekhon, and J.S. Kolar (eds.) *Pulses.* pp. 279-312. Udaipur, India: Agrotech Publishing Academy.

Pasricha, N.S., M.S. Aulakh, N.S. Sahota, and H.S. Baddesha. (1980). Comparative response of groundnut and wheat to phosphorus in groundnut-wheat rotation. *Journal of Agricultural Science, Cambridge.* 94: 691-696.

Pasricha, N.S., M.S. Aulakh, and R. Vempati. (2002). Evaluation of available phosphorus soil test methods for peanut in neutral and alkaline soils. *Communications in Soil Science and Plant Analysis.* 33: 3593-3601.

Pasricha, N.S., Y. Singh, B. Singh, and C.S. Khind. (1996). Integrated nutrient management for sustainable crop production. *Journal of Research Punjab Agricultural University.* 33: 101-117.

Pathak, H., D.R. Biswas, and R. Singh. (2002). Fertilizer use and environmental quality. *Fertiliser News.* 47(11): 13-20.

Patidar, M. and A.L. Mali. (2004). Effect of farmyard manure, fertility levels and bio-fertilizers on growth, yield and quality of sorghum (*Sorghum bicolor*). *Indian Journal of Agronomy.* 49: 117-120.

Patra, A.K., B.C. Nayak, and M.M. Mishra. (2000). Integrated nutrient management in rice (*Oryza sativa*)-wheat (*Triticum aestivum*) cropping system. *Indian Journal of Agronomy.* 45: 453-457.

Pilbeam, C.J., B.P. Tripathi, R.C. Munankarmy, and P.J. Gregory. (1999). Productivity and economic benefits of integrated nutrient management in three major cropping systems in the mid-hills of Nepal. *Mountain Research and Development.* 19: 333-344.

Prasad, P.V.V., V. Satyanarayana, V.R.K. Murthy, and K.J. Boote. (2002). Maximizing yields in rice-groundnut cropping sequence through integrated nutrient management. *Field Crops Research.* 75: 9-21.

Prasad, R. (2000). Nutrient management strategies for the next decades: Challenges ahead. *Fertiliser News.* 45(4): 13-18, 21-25, 27-28.

Prasad, R. and J.C. Katyal. (1992). Fertilizer use related environmental pollution. In Bajwa, M.S., N.S. Pasricha, P.S. Sidhu, M.R. Chaudhary, D.K. Benbi, and V. Beri (eds.) *Proceedings of the International Symposium on Nutrient Management for Sustained Productivity.* Vol. 1. pp. 216-226. Ludhiana, India: Department of Soils, Punjab Agricultural University.

Prasad, R., S.N. Sharma, and S. Singh (1999). *Summer Mung for Sustaining Rice-Wheat Cropping System.* IARI Bulletin No. 10. New Delhi, India: Division of Agronomy, Indian Agricultural Research Institute.

Raju, R.A., K.A. Reddy, and M.N. Reddy. (1993). Integrated nutrient management in wetland rice (*Oryza sativa*). *Indian Journal of Agricultural Sciences.* 63: 786-789.

Randhawa, N.S. (1992). Nutrient management for sustained productivity and food security. Key Note Address, National Seminar on Sustained Productivity and Food Security, February 10-12. Ludhiana: Punjab Agricultural University.

Ranwa, R.S. and K.P. Singh. (1999). Effect of integrated nutrient management with vermicompost on productivity of wheat (*Triticum aestivum*). *Indian Journal of Agronomy.* 44: 554-559.

Rao, U.V. and A.S. Rao. (1996) Interactive effects of VAM fungi and *Aspergillus niger* on nutrient and biochemical constituents in two legumes. *International Journal of Tropical Agriculture.* 14: 115-121.

Reddy, K.C.S., M.M. Hussain, and B.A. Krantz. (1980). Effect of nitrogen level and spacing on sorghum intercropped with pigeonpea and greengram in semi-arid lands. *Indian Journal of Agricultural Sciences.* 50: 17-22.

Reddy, M.N. and K. Surekha. (1999). Role of chickpea in enhancing available P in chickpea-upland rice system in Vertisol. *Journal of the Indian Society of Soil Science.* 47: 805-808.

Regmi, A.P., J.K. Ladha, H. Pathak, E. Pasuquin, C. Bueno, D. Dawa, P.R. Hobbs, D. Joshi, S.L. Maskey, and S.P. Pandey. (2002). Yield and soil fertility trends in a 20-year rice-rice-wheat experiment in Nepal. *Soil Science Society of America Journal.* 66: 857-867.

Rekhi, R.S. and O.P. Meelu. (1983). Effect of complementary use of mung straw and inorganic fertilizer N on the nitrogen availability and yield of rice. *Oryza.* 20: 125-129.

Saha, U.K., M.S. Islam, and R.R. Saha. (1998). Yield performance and soil nutrient balance under integrated fertilization in rice-wheat cropping system in a Calcareous floodplain soil of Bangladesh. *Japanese Journal of Tropical Agriculture.* 42: 7-17.

Sanders, F.E. and P.B. Tinker. (1971). Mechanism of absorption of phosphate from soil by endogone mycorryhizas. *Nature.* 233: 278-279.

Sarkar, M.C. (1990). Long term effect of fertilisers on soil eco-system. *Fertilisers News.* 35(12): 81-85.

Saxena, S.C., H.S. Manral, and A.S. Chandel. (2001). Effect of inorganic and organic sources of nutrients on soybean (*Glycine max*). *Indian Journal of Agronomy.* 46: 135-140.

Sekhon, G.S. (1992). Fertilization of crops and cropping systems. In Bajwa, M.S., N.S. Pasricha, P.S. Sidhu, M.R. Chaudhary, D.K. Benbi and V. Beri (eds.) *Proceedings of the International Symposium on Nutrient Management for Sustained Productivity.* Vol. 1. pp. 55-70. Ludhiana, India: Department of Soils, Punjab Agricultural University.

Sekhon, H.S. and G. Singh. (2005). Pulse-based cropping systems. In Singh, G., H.S. Sekhon, and J.S. Kolar (eds.) *Pulses.* pp. 203-221. Udaipur, India: Agrotech Publishing Academy.

Shankaraiah, C. and M.S. Nagaraju. (1997). Integrated nutrient management in sugarcane based cropping systems. *Fertiliser News.* 42(11): 57-60.

Sharma, A.D., M.S. Brar, and S.S. Malhi. (2005). Critical toxic ranges of chromium in spinach (*Spinacia oleracia* L.) plants and in soil. *Journal of Plant Nutrition.* 28: 1555-1568.

Sharma, R.N. and M. Pal. (1986). Fertilizer management in pigeonpea-wheat cropping system. *Annals of Agricultural Research.* 7: 52-56.

Sharma, C.M. and A.K. Sangrai. (1992). Dissolution and transformation of indigenous rock phosphates in an acid Alfisol. In Beri, V., M.R. Chaudhary, P.S. Sidhu, N.S. Pasricha, and M.S. Bajwa (eds.) *Proceedings of the International Symposium on Nutrient Management for Sustained Productivity.* Vol. 2. pp. 26-27. Ludhiana, India: Department of Soils, Punjab Agricultural University.

Sharma, R.P., V.K. Suri, and N. Datt. (2001). Integrated nutrient management in summer barley (*Hordeum vulgare*) in a cold desert of Himachal Pradesh. *Indian Journal of Agricultural Sciences.* 71: 752-755.

Sherchan, D.P., S.P. Chand, and P.G. Rood. (1995). Developing sustainable soil management technologies in the hills of Nepal. In Cook, H.F. and H.C. Lee (eds.) *Soil management in sustainable agriculture. Proceedings of Third International Conference on Sustainable Agriculture.* pp. 99-409. London: Wye College, University of London.

Shinde, V.S., F.R. Khan, and C.D. Mayee. (1996). *Annual Report. Cropping System and Verification Function.* Aurangabad, India: NARP.

Singh, B. and G.S. Kalra. (1989). Growth and yield of succeeding wheat crop as influenced by different dates of sowing, plant spacing and phosphorus doses of preceding arhar crop. *Indian Journal of Agricultural Research.* 23: 169-174.

Singh, B. and G.S. Sekhon. (1977). Impact of fertilizer use on environmental pollution in Punjab: Present status and future projections. *Fertiliser News.* 22(3): 7-11.

Singh, C.P., N. Singh, N.S. Dangi, and B. Singh. (1997). Effect of rock phosphate enriched pressmud on dry matter yield, phosphorus and sulphur nutrition of mungbean (*Vigna radiata* L. Wilczek). *Indian Journal of Plant Physiology.* 2: 262-266.

Singh, G. and O.P. Singh. (1993). Integrated nutrient management in irrigated rice. *Annals of Agricultural Research.* 14: 486-488.

Singh, G. and H.S. Sekhon. (2002). Organic farming. In Singh, G., J.S. Kolar and H.S. Sekhon (eds.) *Recent Advances in Agronomy.* pp. 167-189. New Delhi, India: Indian Society of Agronomy.

Singh, G. and K.V.B.R. Tilak. (2001). Phosphorus nutrition through combined inoculation with phosphorus solubilizing rhizobacteria and arbuscular mycorrhizae. *Fertiliser News.* 46(9): 33-36.

Singh, H., Y. Singh, and K.K. Vashist. (2005). Evaluation of pressmud cake as source of phosphorus for rice–wheat rotation. *Journal of Sustainable Agriculture.* 26: 5-21.

Singh, J. and K.P. Singh. (1997). Integrated nutrient management in sunflower (*Helianthus annus*). *Indian Journal of Agronomy.* 42: 370-374.

Singh, R. and S.K. Agarwal. (2001). Growth and yield of wheat (*Triticum aestivum*) as influenced by levels of farmyard manure and nitrogen. *Indian Journal of Agronomy.* 46: 462-467.

Singh, S. and K.K. Kapoor. (1999). Inoculation with phosphate-solubilizing micro-organisms and a vesicular-arbascular mycorrhiza fungus improves dry matter yield and nutrient uptake by wheat grown in a sandy soil. *Biology and Fertility of Soils.* 28: 139-144.

Singh, S.P. and M.G. Verloo. (1996). Accumulation and bioavailability of metals in semi-arid soils irrigated with sewage effluents. Meded. *Fac. Landbouwkd. Toegep. Biol. Wet. University Gent.* 61: 63-67.

Singh, Y., C.S. Khind, and B. Singh. (1991). Efficient management of leguminous green manures in wetland rice. *Advances in Agronomy.* 45: 135-189.

Singh, Y.B., Singh, R.K. Gupta, C.S. Khind, and J.K. Ladha. (2003). Managing pressmud cake for nitrogen and phosphorus nutrition of crops in a rice-wheat rotation. *International Rice Research Notes.* 28(1): 59-61.

Smith, S.R. (1996). *Agricultural Recycling of Sewage Sludge and the Environment.* London, England: CAB International.

Spalding, R.F. and L.A. Kitchen. (1988). Nitrate in the intermediate vadose zone beneath irrigated cropland. *Ground Water Monitor Review.* 8: 89-95.

Srinivasan, K., R. Shantha, and M. Ramasamy. (1991). Effect of summer pulses on the growth and productivity of succeeding *kharif* maize. *Indian Journal of Pulses Research.* 4: 51-55.

Sushila, R. and G. Giri. (2000). Influence of farmyard manure, nitrogen and biofertilizers on growth, yield attributes and yield of wheat (*Triticum aestivum*) under limited water supply. *Indian Journal of Agronomy.* 45: 590-595.

Tandon, H.L.S. (1995). Sulphur deficiencies in soils and crops: Their significance and management. In Tandon, H.L.S. (ed.) *Sulphur Fertilisers for Indian Agriculture—A Guidebook.* pp. 1-23. New Delhi, India: Fertiliser Development and Consultation Organisation.

Tandon, H.L.S. and P. Narayan. (1990). *Fertilizers in Indian agriculture—Past, Present and Future (1995-2000).* New Delhi, India: Fertiliser Development and Consultation Organisation.

Tilak, K.V.B.R. and G. Singh. (1996). Integrated nutrient management in sustainable agriculture. *Fertiliser News.* 41(3): 29-35.

Tirmizi, S.A., T. Javed, A. Saeed, and S. Samina. (1996). A study of the inorganic elements in vegetables and soil samples of the polluted and non-polluted areas of Bahawalpur city, Pakistan. *Hamdard Medicus.* 39: 90-95.

Tomar, S.S., M.A. Pathan, K.P. Gupta, and U.R. Khandkar. (1993). Effect of phosphate solubilizing bacteria at different levels of phosphate on black gram (*Phaseolus mungo*). *Indian Journal of Agronomy.* 38: 131-133.

Trivedi, B.S., P.M. Bhatt, J.M. Patel, and R.C. Gami. (1995). Increasing efficiency of fertilizer phosphorus through addition of organic amendments in groundnut. *Journal of Indian Society of Soil Science.* 43: 627-629.

USEPA. (1985). *Nitrate/nitrite health advisory (draft).* Washington, DC: U.S. Environmental Protection Agency, Office of Drinking Water.

Vig, A.C., G.S. Bahl, and M. Chand. (1999). Phosphorus—Its transformation and management under rice-wheat system. *Fertilizer News.* 44(11): 33-46.

Wani, S.P. and K.K. Lee. (1996). Role of microorganisms in sustainable agriculture. In Behl, R.K., A.L. Khurana, and R.C. Dogra (eds.) *Plant Microbe Interaction in Sustainable Agriculture*. pp. 62-88. New Delhi, India: Bio Science Publishers.

Wassmann, R., H.U. Neue, J.K. Ladha, and M.S. Aulakh. (2004). Mitigating greenhouse gas emissions from rice–wheat cropping systems in Asia. In Wassmann, R. and P.L.G. Vlek (eds.) *Tropical Agriculture in Transition—Opportunities for Mitigating Greenhouse Gas Emissions*. pp. 65-90. Dordrecht, the Netherlands: Kluwer Academic Publishers.

Williams, P.H. (1992). The role of fertilizers in environmental pollution. In Bajwa, M.S., N.S. Pasricha, P.S. Sidhu, M.R. Chaudhary, D.K. Benbi, and V. Beri (eds.) *Proceedings of the International Symposium on Nutrient Management for Sustained Productivity*. Vol. 1. pp. 195-215. Ludhiana, India: Department of Soils, Punjab Agricultural University.

Yadav, M.K., M. Raj, and R.P. Yadav. (2002). Effect of integrated nutrient management and legume based crop rotation on available nutrients balance under rainfed conditions. *Indian Journal of Dryland Agricultural Research and Development*. 17: 95-99.

Yaduvanshi, N.P.S. (2001). Effect of five years of rice-wheat cropping and NPK fertilizer use with and without organic and green manures on soil properties and crop yields in a reclaimed sodic soil. *Journal of the Indian Society of Soil Science*. 49: 714-719.

Yadvinder-Singh, O.P. Meelu, and Bijay-Singh. (1990). Relative efficiency of calcium ammonium nitrate under different agroclimatic conditions. *Fertiliser News*. 25(4): 41-45.

Zayed, A., C.M. Lytle, J.H. Qian, and N. Terry. (1998). Chromium accumulation, translocation and chemical speciation in vegetable crops. *Planta*. 306: 395-399.

Zia, M.S., M. Munsif, M. Aslam, and M.A. Gill. (1992). Integrated use of organic manures and inorganic fertilizers for the cultivation of low land rice in Pakistan. *Soil Science and Plant Nutrition*. 38: 331-338.

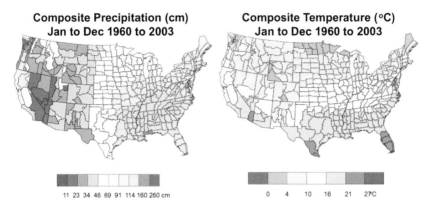

Composite Precipitation (cm)
Jan to Dec 1960 to 2003

11 23 34 46 69 91 114 160 260 cm

Composite Temperature (°C)
Jan to Dec 1960 to 2003

0   4   10   16   21   27°C

FIGURE 3.3. Composite average precipitation and temperature data for the continental United States from 1960 to 2003. *Source:* www.cdc.noaa.gov.

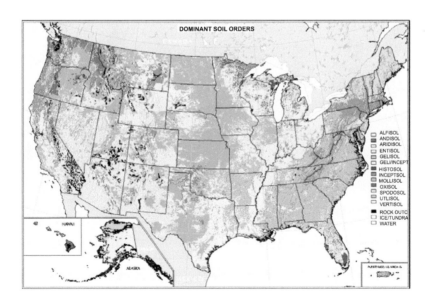

DOMINANT SOIL ORDERS

ALFISOL
ANDISOL
ARIDISOL
ENTISOL
GELISOL
GELI/INCEPT
HISTOSOL
INCEPTSOL
MOLLISOL
OXISOL
SPODOSOL
UTLISOL
VERTISOL

ROCK OUTC
ICE/TUNDRA
WATER

HAWAII

ALASKA

FIGURE 3.4. Dominant soil orders in the United States. *Source:* http://www.nrcs
.usda.gov/technical/land/meta/m4025.html.

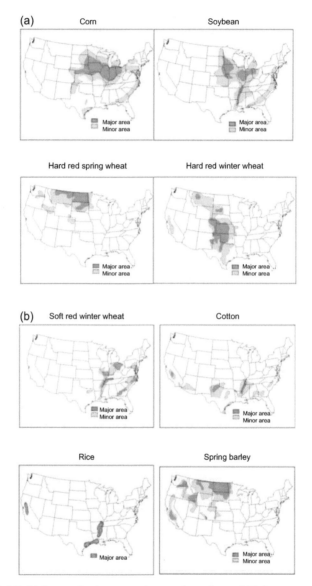

FIGURE 3.5. Major and minor growing regions (a and b) for corn, soybean, wheat, cotton, barley, and rice in the continental United States. *Source:* USDA-Office of the Chief Economist (2007). Major world crop areas and climatic profiles, North America. Available online: http://www.usda.gov/oce/weather/pubs/Other/MWCACP/world_crop_country.htm#northamerica (accessed April 2007).

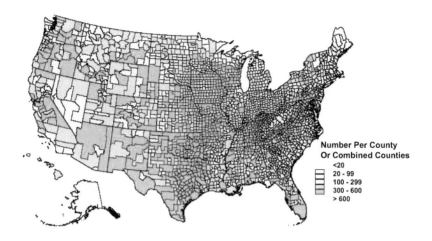

FIGURE 3.13. Confined animal feed operations in the United States in 1997.

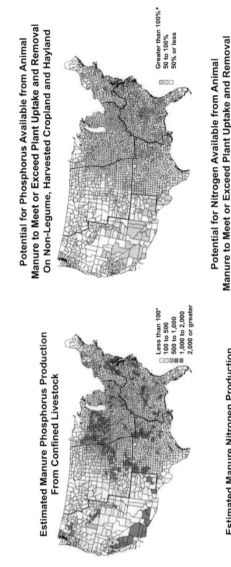

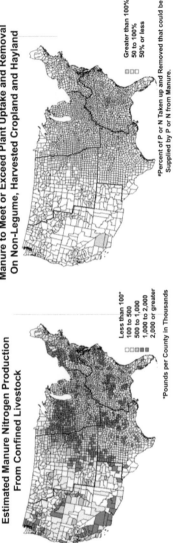

Estimated Manure Phosphorus Production From Confined Livestock

Less than 100*
100 to 500
500 to 1,000
1,000 to 2,000
2,000 or greater

Potential for Phosphorus Available from Animal Manure to Meet or Exceed Plant Uptake and Removal On Non-Legume, Harvested Cropland and Hayland

Greater than 100%*
50 to 100%
50% or less

Estimated Manure Nitrogen Production From Confined Livestock

Less than 100*
100 to 500
500 to 1,000
1,000 to 2,000
2,000 or greater

*Pounds per County in Thousands

Potential for Nitrogen Available from Animal Manure to Meet or Exceed Plant Uptake and Removal On Non-Legume, Harvested Cropland and Hayland

Greater than 100%*
50 to 100%
50% or less

*Percent of P or N Taken up and Removed that could be Supplied by P or N from Manure.

FIGURE 3.14. Estimated N and P production in animal manures and potential cropland available to assimilate these nutrients in the United States.

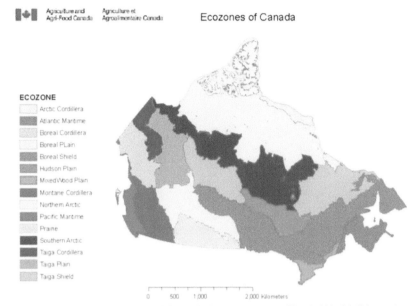

FIGURE 4.1. Agroecozones of Canada (map generated by A. Waddell from data on http://sis.agr.gc.ca/cansis/nsdb/ecostrat accessed April 27, 2007). *Source:* Adapted from AAFC (2005).

Chapter 8

Integrated Nutrient Management: Experience from China

Bao Lin
Jianchang Xie
Ronggui Wu
Guangxi Xing
Zhihong Li

INTRODUCTION

Nutrients from soil and organic manure have maintained Chinese agriculture for thousands of years. Trends in wheat and rice yields, over the past 2,000 years, evaluated based on historical records (Liu, 1992), show that organic manure application allowed for nutrient recycling in agricultural systems. Thus, soil fertility has been maintained and/or enhanced and crop yields have been sustained over time. Crop yields increased slowly when nutrients were supplied solely in organic forms. Over this period, wheat yield increased from 397 to 733 kg ha^{-1}, or increased by 85 percent. Rice yield increased from 302 to 1,465 kg ha^{-1}, or nearly three times over the past 2,117 years, taking 182 years for an increase of each 100 kg of rice yield. Improvements in crop variety and tillage systems also contributed to yield increases. Although crop yields increased substantially during this period, population grew even faster, from 50 to 450 million, or sevenfold. Therefore, the increased grain supply was insufficient to match the rise in population. To ensure a rapid increase in crop production, it is necessary to increase nutrient supply beyond that provided by recycling of organic nutrients within the agricultural system. The development of inorganic fertilizers provided a means of enhancing the nutrient supply for agricultural production. The introduction of chemical fertilizers into agricultural systems was a historical imperative for social development in China.

In this chapter, we highlight the discussion on fertilizer nutrients and their integrated management, and the link between fertilizers and environment. The historical Chinese agricultural development proves that the reasonable utilization of fertilizer resources is one of the important technical guarantees for agricultural sustainability. The use of organic manure is an essential ring within the matter and energy circulation in the agroecosystem. Chemical fertilizers increase the strength of the circulation. With 9 percent of the world's farmland, China feeds 22 percent of the world's population. China's success depends on integrated nutrient management (INM). However, the past thirty years saw a consistent increase in the use of fertilizer resources, but a gradual decline in their use efficiency, with a resultant increase in environmental risk. Therefore, the goal of INM should turn from seeking a high and stable yield to striving for yield, quality, and profit, as well as a safe environment for sustainable agriculture.

MAJOR AGRICULTURAL REGIONS
AND CROPPING SYSTEMS

China has a 9.6 million km^2 land area constituting 6.5 percent of the world total. China's land area is composed of 130 million hectares (Mha) of farmland, 162 Mha of forest (including shrubs, woods, and fruit orchards), and 290 Mha of usable grassland. Among the land used for agricultural purposes in 2002, arable land, orchard, forest, pasture, and fishery accounted for 43.2, 2.3, 32.2, 20.6, and 1.6 percent, respectively (GTCSES, 2003). Based on the fact that its population is 1.3 billion people, China has 0.1 ha of farmland per capita while the world average is 0.287 ha per capita. It is evident that the farmland in China is stressed by the high population density. However, the multi-cropping index (cropping intensity) has been about 158 percent, which effectively enlarges the cropping areas. The climate, landscape, soil, crop, and social/economic conditions are quite different from place to place. The nation can be divided into four agricultural regions (Figure 8.1), and each region may have several subregions (Liu, 1996; Wang and Li, 2003).

Southern Paddy Cropping Region

The southern paddy cropping region, located south of the Qinling Mountains and the Huaihe River and to the east of the Tibetan Plateau is the most

☐ Northwest Inland Grazing-Cropping Region
▨ Northern Dryland Cropping Region
☐ Qinghai-Tibetan Plateau Grazing-Cropping Region
▥ Southern Paddy Cropping Region

FIGURE 8.1. Four major agricultural regions in China. *Source:* Adapted from CCAR (1981).

important agricultural base in China. It accounts for 25 percent of the land, 58 percent of the rural population, and 38 percent of the farmland of the nation's total. The mean temperature in January is above 0°C. The annual precipitation is more than 800 mm and the crop-growing season lasts for more than eight months. The region spans from subtropical to tropical zones. The main soils consist of paddy soil, purple soil, red soil, and yellow soil. The water, heat, and soil resources are abundant and in balance with one another. The paddy fields account for two-thirds of the total farmland in this region. Approximately 90 percent of the paddy fields in China are in this region. The cropping patterns are two or three crops a year, resulting in a cropping index of 150-250 percent.

Middle and Low Reaches of the Yangtze River Region

The grain output from this region comprises one third of the national total while the rice makes up half of the nation's total. Up to the south of the Yangtze River, production is mainly double-cropping rice whereas in the north of the Yangtze, it is mainly mono-cropping rice plus winter wheat, rapeseed, or fallow. This region is prone to natural disasters which threaten agricultural production. It is necessary to readjust cropping patterns and enlarge the cropping index.

Southwest Region

The farmland in the southwest region accounts for only 5 percent of the nation's total. It is mainly mono-cropping rice plus winter rapeseed, broad bean, or wheat. Most of the paddy field is fallow in winter. Drought spells often prevail. The multi-cropping potential is great in this region.

South China Region

The South China region is the only region in China suitable for tropical crops. The triple-cropping includes double rice crops and a winter crop. Typhoons and crop damage from cold weather occur frequently. To improve agricultural potential, it is necessary to strengthen field infrastructure construction and readjust cropping systems, and to develop special practices in agricultural production, such as mini-tunnels in fields and greenhouses in the cities.

Northern Dryland Cropping Region

The northern dryland cropping region is one of the most important agriculture regions in China. It lies north of the Qinling Mountains and the Huaihe River, and east of 400 mm isohyet, accounting for 19 percent of land, 37 percent of the rural population, and 50 percent of the nation's total farmland. The mean temperature in January is below 0°C. The annual precipitation is less than 800 mm. The crop-growing season ranges between three and eight months. The region spans from warm temperate, to moderate and cold temperate regions. The main soil types in the region are calcareous soil, black soil, and saline-sodic soil. In terms of potential productivity, the natural conditions for agricultural production are next to those in the southern paddy cropping region. The cropping patterns include mono-cropping, double-cropping in a year, as well as triple-cropping in two years. The multi-cropping index is about 130 percent.

Northeast Region

The farmland per capita in the northeast is relatively higher than in the south agricultural regions in China. The area is comprised of a wide plain with fertile soil, but production is limited due to lack of heat. The only cropping pattern is one crop a year, such as corn, soybean, rice, and spring wheat. Improved productivity in this region would require adjusting the cropping systems, intensive cultivation, and increasing land use efficiency.

Huang-Huai-Hai Region

Three-quarters of the land in the Huang-Huai-Hai region is plain, where most of the wheat and cotton planted in China grow. An annual winter wheat-summer corn rotation predominates. Most cotton fields are fallow in winter and the rest are undercropped with winter wheat. Soil salinization is a serious concern. Much of the farmland produces moderate or low yields and water resources are limited. It is necessary to introduce water-saving irrigation systems in order to increase water use efficiency.

Loess Plateau Region

Most of the hilly plateau in this region is covered by loess soil. Dryland farming is predominant. This region is in urgent need of integrated management practices due to soil erosion problems.

Northwest Inland Grazing-Cropping Region

The northwest inland grazing-cropping region is located to the west of the 400 mm isohyet and north of the Tibetan Plateau, deep in the continent. It amounts to 32 percent of the land, 5 percent of the rural population, and 12 percent of the nation's total farmland. The small population is scattered across the vast area. The climate is dry and there is wide natural grassland and the Gobi desert. Grazing is the principal agriculture in this region. The main soils are chestnut soil (kastanozem), steppe sierozem, and desert soil. The farmland is rare and scattered, mainly relying on irrigation. Soil desertification and pasture degradation are the primary agricultural problems in this region.

Inner Mongolia and Great Wall Belt Region

The grazing-cropping region of the Inner Mongolia and the Great Wall belt is located in the transitional zone between the subhumid and semiarid regions. The cultivation is very extensive with a rotation of one-year cropping and one-year fallow.

Gansu-Xinjiang Region

In more than half of the area of the Gansu-Xinjiang region, the precipitation is less than 100 mm. The lands are characterized by desert pasture or desert. The main crops are spring wheat, corn, and cotton. Xinjiang is another cotton production base in China. Due to plentiful sunshine, the yield is quite high if irrigation is available. Emphasis should be given to take advantage of the light and heat resources for higher water use efficiency.

Qinghai-Tibetan Plateau Grazing-Cropping Region

The Qinghai-Tibetan plateau is an important grazing and forest region, including the whole Tibet Autonomous Region, most of Qinghai, western Sichuan, northwestern Yunnan, and southwestern Gansu. It accounts for 24 percent of the nation's total land. Two-thirds of the plateau is higher than 4,500 m above sea level. The main soils are alpine meadow soil, subalpine meadow soil, steppe soil, calcic brown soil, and sierozem. The region is suitable for livestock grazing. Cold tolerant crops such as hulless barley and oat can be planted in the area below 4,000 m above sea level.

AGRICULTURAL PRODUCTION, FERTILIZATION, AND NUTRIENT BALANCE

Fertilizer Consumption and Agricultural Production

China has used chemical fertilizers for 100 years. Prior to 1949, there were only two small-scale N fertilizer (ammonium sulfate) manufacturers and two N-recovery workshops in which ammonia was produced by the by-product of coking plant. The total production and importation were approximately 120,000 and 600,000 tonnes (t) of nitrogen, respectively. Chemical fertilizers were mainly used in coastal provinces. After the founding of the People's Republic of China, the development of chemical fertilizer production increased, and large tonnages were imported. Since the1970s, the production and consumption of chemical fertilizers in China have increased rapidly (Table 8.1). In 1980, the total consumption of chemical fertilizers in China was 12.7 million tonne (Mt) nutrients. It reached 25.9 Mt in 1990 and 35.9 Mt in 1995. The average annual increment was 1.55 Mt, or a rate of increase of 12 percent within fifteen years. This was the fastest period of

TABLE 8.1. Evolution of human population, arable land, fertilizer production/ consumption, and grain production in China

Year	1950	1960	1970	1980	1990	2000	2002
Population (m)	552	662	830	987	1,134	1,237	1,252
Arable land (Mha)	100	105	101	98.3	95.7	130[a]	130
Fert. production (Mt N + P_2O_5 + K_2O)	<0.12	0.41	2.43	12.3	18.8	31.9	36.9
Fert. Consumption (Mt N + P_2O_5 + K_2O)							
Inorganic	0.07	0.66	3.51	12.7	25.9	41.5	43.4
Organic	4.28	6.58	10.70	11.3	15.0	18.3	–
Grain production (Mt)	112	195	240	321	452	462	457
Grain yield (t ha^{-1})	1.04	1.47	2.01	2.75	3.98	4.26	4.34

Source: Adapted from China Agriculture Yearbook (1981-2003). China Chemical Industry Yearbook (1997-2003).

[a]Updated data after the agricultural survey in 1996.

increase in fertilizer use in the history of China. In 2002 the fertilizer output was 36.9 Mt, and consumption was 43.4 Mt, with both being the highest in the world.

In recent years, the shortage of chemical fertilizers in China has been alleviated. Although more potash and the high-analysis compound fertilizers are still imported, considerable amounts of urea are exported. The consumption of chemical fertilizers is still increasing. The small proportion of P and K relative to N in fertilizer output and consumption has been readjusted to some extent. The ratio of N:P has become more in balance with crop removal, but the ratio of K:N remains low. Low-analysis N and P fertilizers, such as ammonium bicarbonate and single superphosphate, have been tremendously reduced. In 2002, urea accounted for 58 percent of the total domestic produced N fertilizers; the high-analysis P fertilizers and compound fertilizers such as NPK, diammonium phosphate (DAP), nitrophosphate (NP), and triple superphosphate (TSP) made up 45 percent of P-containing fertilizers; compound fertilizers accounted for 24 percent of total fertilizers consumed.

China is the country with the largest population in the world, over 1.3 billion at present, and expected to grow to 1.6 billion by 2030. Increasing efficient application of chemical fertilizers seems to be the most effective way for China to meet the demand for food in the future. The total grain output of China went up from 100 Mt in 1949 to 457 Mt in 2002 (China Statistics Yearbook, 2003). The increase of NPK fertilizer consumption coincides with the increased grain output. Taking N fertilizer as an example, the regression analysis of chemical N fertilizer application to grain yield shows that during the fifty years from 1949 to 1998, there was a correlation coefficient of 0.98 (n = 50) and a regression coefficient of 14.5 between Chinese annual grain output and chemical N fertilizer consumption (Zhu and Chen, 2002).

During its long history, Chinese agriculture depended on input of organic manure. After the founding of the People's Republic of China, chemical fertilizer consumption grew consistently, with the development of the chemical industry and agricultural production. The chemical fertilizer consumption in China increased by almost forty times from 1960 to 1990, while organic manure input merely doubled. Up to 2000, about 70 percent of the nutrient input came from chemical fertilizers, and the remaining 30 percent from organic manure (Table 8.1). Nevertheless, N and P mainly come from chemical fertilizers, while K is primarily supplied by organic manure (Table 8.2).

TABLE 8.2. Overview of field nutrient balance between input and removal in China (Mt)

Year		1949	1957	1965	1975	1980	1985	1990	1995	2000
Input										
OM	N	1.62	2.49	2.93	4.10	4.16	5.03	5.26	6.11	6.52
	P_2O_5	0.79	1.23	1.38	1.94	2.06	2.56	2.80	3.30	3.44
	K_2O	1.87	2.86	3.06	4.62	5.09	6.21	6.93	7.60	8.32
IF	N	0.01	0.32	1.21	3.64	9.43	12.60	17.40	22.20	24.70
	P_2O_5	–	0.05	0.55	1.61	2.90	4.19	6.68	10.40	11.60
	K_2O	–	–	0.003	0.13	0.37	0.98	1.82	3.36	5.24
Removal	N	2.91	5.11	5.22	7.49	8.67	11.10	13.00	13.70	16.60
	P_2O_5	1.38	2.36	2.37	3.34	3.78	4.79	5.59	5.77	6.64
	K_2O	3.06	5.62	5.60	8.13	9.33	12.10	13.80	14.60	17.40

Source: Adapted from Li et al. (2000).

Note: OM = organic manure, IF = inorganic fertilizer. Data in 2000 were projected.

China is the largest chemical fertilizer consumer in the world at present. Chemical fertilizer plays an enormous part in Chinese agricultural production. The results from fifty-two long-term (over ten years) fertilizer experiments show that chemical fertilizers contribute 40.8 percent to grain production on a national average (Ma et al., 2002). In recent years, the structure of Chinese agriculture has been readjusted. The planting areas of vegetables and fruits have been increased and some fields are being reforested. Currently, the Chinese government is promoting measures for enlarging the planted acreage and reducing nonagricultural purpose fields for higher total grain output, in order to secure food safety.

Status of Nutrient Balance in Farmland

The nutrient budget and balance for agricultural production varies substantially from location to location and from year to year. Also, because different estimators use different parameters, especially for input and output of nutrients, the estimated results may be very different. Also, some factors used in calculations of regional balances have not been reliably measured, such as the N losses from leaching out of soil, denitrification, or ammonia volatilization.

Nutrient deficits have occurred in Chinese farmland for many years. Since the 1980s, with increasing chemical fertilizer input, the farmland nutrient balance has changed a great deal (Table 8.2). In 2000, the total input of organic and inorganic fertilizers was 31.2, 15.0, and 13.6 Mt N, P_2O_5, and K_2O, respectively, in the ratio of 1:0.48:0.43, while the removal of nutrients by crops was in the ratio of 1:0.40:1.05. The difference between the input and removal led to an excess of 2.24 Mt N and 4.88 Mt P_2O_5, and a deficit of 3.83 Mt K_2O (Li et al., 2000). The deficit of large amounts of K was the most important reason for the consequent expanding area of K deficiency since the 1970s.

Since 1997, Nanjing Institute of Soil Sciences, Chinese Academy of Sciences (CAS) has cooperated with International Potash Institute (IPI), to investigate the field nutrient balance in several cropping systems in the seven provinces (Heilongjiang, Shaanxi, Henan, Jiangsu, Hunan, Sichuan, and Guangxi) which represented five typical cropping systems such as one crop a year, two crops a year, three crops in two years, five crops in two years and three crops a year. The results in Table 8.3 roughly reflected the nutrient budget situation of the whole country, showing that N and P were in excess for all provinces except Guangxi where the N was in deficit. The N loss was estimated at 45 percent in the calculation. Therefore, overapplication of N seemed to be prevailing (Zhou et al., 2000).

TABLE 8.3. The field nutrient budget in representative provinces in China (kg ha^{-1})

Province	Cropping system	N		P_2O_5		K_2O	
		I	R	I	R	I	R
HLJ	A	108	102	107	24.7	28.5	56.8
SX	A, B	194	123	52.9	26.3	39.7	53.1
HN	B, C	136	123	173	115	84.1	141
JS	B, D	481	394	355	209	195	235
HuN	B, E	583	253	431	357	382	433
SC	E	323	249	278	155	90	236
GX	E	416	504	345	193	343	520

Source: Adapted from Zhou et al. (2000).

Note: I = input, R = removal, HLJ = Heilongjiang, SX = Shaanxi, HN = Henan, JS = Jiangsu, HuN = Hunan, SC = Sichuan, GX = Guangxi.

A, B, C, D, and E stand for one crop a year, two crops a year, three crops in two years, five crops in two years, and three crops a year, respectively.

The China National Agro-Technical Extension and Service Center (NATESC) established sixteen monitoring stations from 1988 to 1997 in the northeast region with a one-crop-a-year system, in order to study the crop yield, fertilizer rates, soil fertility changes and soil nutrient balance under conventional fertilization, and management practices with different cropping patterns in various places. The results showed that N was in excess except for corn, P was in excess, but K was deficit in all studied cropping systems (Wang et al., 2000). The Soil and Fertilizer Institute (SFI) of the Chinese Academy of Agricultural Sciences (CAAS) established eight long-term monitoring stations from 1998 to 1999 in Huang-Huai-Hai region with a rotation system of winter wheat–summer corn. The result revealed that the field nutrient budget of N and P was in excess, but K was in deficit at all times (Huang et al., 2000).

The above national and provincial statistics and the estimation of field nutrient balance from the various provinces indicate that N and P are oversupplied but K is undersupplied. Most of the excess N moves by various means into the atmosphere and water bodies. In spite of the past overfertilization of P, soil P is still consistently deficient for optimum crop production. Therefore, current rates of P application appear reasonable to correct deficiencies and enhance soil fertility, without excess environmental risk. The most urgent readjustments should be to optimize N application and provide a reasonable N:K ratio to improve soil and crop K status.

The Overview of Farmland Soil Nutrients and Their Trends

Nitrogen

The soil total N in farmland ranges from 0.4 to 3.8 g N kg^{-1}, with the black soil in northeast China being the highest and soils in Huang-Huai-Hai Plain and loess plateau in northwest China being the lowest. Normally, the N rich soil has total N > 2 g kg^{-1} with organic carbon between 18 and 23 g kg^{-1}. About 60.5 percent of farmland, which is mostly black soil, in Heilongjiang Province has soil total N > 2 g kg^{-1}. Most of the farmland in China has total N < 2 g kg^{-1}. For example, soil total N less than 2 g kg^{-1} accounts for as much as 99.5 percent in Henan Province in north China plain, and 99.2 percent in Shanxi Province on loess plateau. Therefore, N is a major element that needs to be applied to almost all fields in China (Lu, 1998).

With the increase in N consumption, the grain output has continuously increased and both the input and removal of N in soil have accelerated. Since the 1980s, the input of N in agricultural production has been greater than the removal. Thus, storage of N in farmlands has increased and the soil N level is gradually rising. However, it is unlikely that N consumption in China will decline due to the requirement of a high yield for feeding 1.3 billion people. In some developed areas, the N input should be controlled to prevent negative impacts on the environment.

Phosphorus

Total soil P generally varies between 200 and 1,100 mg kg^{-1}, being lower in the south due to weathering process. Soil available P level determines directly the capacity of soil P supply. The P-rich soil has a soil test over 20 mg P kg^{-1} and P deficit soil is <10 mg P kg^{-1}. In China, only a limited area of soils contains more than 20 mg P kg^{-1}, whereas approximately 4.67 Mha of farmland has soil available P <10 mg kg^{-1}. Phosphorus deficiency occurs mainly in the calcareous soil in the north, Baijiang soil (a kind of meadow podzolic soil) in the northeast, red soil, purple soil, and low-yield paddy soil in the south of China (Lu, 1998).

Excess soil P appears in every ecological region in China due to a long period of P input. Nowadays, the soil P level has been increased in areas with consistent P application. Therefore, the exploitation of the accumulated soil

P should be a basic consideration. Although fixed P does not readily move in soil, the consistent accumulation increases the risk that movement will occur, resulting in environmental problems.

Potassium

In northern China, the degree of weathering is low. Soil clays derived from native rocks contain a lot of illite (hydrous mica) and montmorillonite minerals. Unless the texture is very coarse, the soil K content is rather high. However, in the region up to the Yangtze River, with a stronger weathering process, the illite content in soil clay is diminished, while the kaolinite and gibbsite (hydrargillite) increased, thus resulting in a decline of the soil K supply potential from north to south. Soils with available K over 500 mg K kg^{-1} in the plow layer are 19.8 percent in northeastern China, but 2.6 percent in southern China; the area with <50 mg K kg^{-1} is 1.7 percent in the north, but 52.2 percent in the south (Xie et al., 2000).

The national field nutrient balance of K has been consistently negative over time, leading to mining of soil K reserves. Since the late 1970s, K fertilizers have been applied in southern China. But, except for a few districts, soil K fertility is still decreasing. In northern China, K fertilizers have not been applied to all soils. As a result, K deficiency occurs in the north.

Secondary and Micronutrients

The variations in soil Ca and Mg concentrations are tremendous, depending on parent materials, weathering conditions, leaching intensity, and tillage practices. In rainy, high-temperature, and humid areas, the soil Ca and Mg contents are much lower than those in arid and semiarid regions. The rainy areas with light textured, low organic carbon content soils are generally deficient in S. The soils deficient in secondary nutrients account for 46 percent of total farmland, among which both Ca and S deficient soils account for approximately 20 percent, and Mg deficient soils make up about 4 percent (Tang, 1996).

The distribution pattern of micronutrient contents is soil and climate dependent. The soil B content declines from north to south and from west to east. The western inland arid region is a B rich area, while the southeastern coastal region is a low or B deficient area. The low pH soils in southern China contain more Zn than in northern calcareous soils, while Mo deficient soils occur everywhere. The northern calcareous soils contain less Mn than acid southern soils. The area of various kinds of micronutrient deficiencies in China is about 160 Mha while only 17 Mha, or 11.1 percent of the required area, is replenished by micronutrients. Because the deficient area is large, it could be expected that the requirement of secondary and micronutrients should increase with the increase in crop yields and the expansion of cash crop growing areas.

INTEGRATED NUTRIENT MANAGEMENT (INM)

Chinese farmers emphasize INM to consistently increase crop yield, quality, and economic returns, as well as to protect the eco-environment and improve soil fertility. In the past fifty years, the NPK nutrient input has gradually turned from organic sources to a combination of organic and inorganic sources. The consumption of chemical fertilizer N and P has now well surpassed those in organic manure.

China has a long history in using organic manures. According to historical record, as early as 3,000 years ago, rotted or decomposed weeds were used as manures (Guo and Lin, 1985). The application of organic manure is a major component of traditional Chinese agriculture. Farmers have accumulated rich experience in the collection, preparation, preservation, and application of organic manure. The long-term addition of organic manure played an important role in maintaining soil fertility and providing a balanced supply of various nutrients for crops. In respect to organic nutrients and relevant managements, organic fertilizers are divided into three categories: organic manures, green manures, and biofertilizers.

Organic Manures

Availability

Organic manures are the materials taken from, prepared, and applied in local rural areas with organic matter in them that can supply multiple nutrients. According to a survey conducted by the Ministry of Agriculture, based on different organic sources, features, and preparation procedures, organic manure was also classified into ten types: excretions, composts, plant residues, green manures, fertile soils, seed residual cakes, seaweeds, urban and rural wastes, humic acid carriers, and marsh gas residues. Among the 433 varieties collected, some were original resources such as excretions and plant residues while some were derived resources such as composts (NATESC, 1999). Thus, only the original resource materials were calculated in Table 8.4 (Lin, 1998).

Among the organic manure resources, there are large quantities of pig dung, large livestock dung, and plant residues (including ash of plant residue), each accounting for about one-third of the total organic manure.

TABLE 8.4. The quantity and sources of organic manure in China in 1990

Sources	Quantity (million)	Nutrients (Mt)				
		N	P_2O_5	K_2O	Total	%
Human excretion	798	0.64	0.24	0.24	1.12	7.3
Pig dung	358	1.54	1.00	2.07	4.61	30.0
Sheep dung	211	0.30	0.21	0.16	0.67	4.4
Large livestock dung	129	1.82	0.94	1.31	4.07	26.5
Plant residue		0.71	0.21	1.74	2.66	17.3
Plant ash	–		0.21	1.40	1.61	10.5
Green manure	4.3 Mha	0.31	0.09	0.23	0.63	4.1
Total		5.32	2.90	7.15	15.37	100
	$N:P_2O_5:K_2O$			1:0.54:1.35		

Source: Adapted from Lin (1998).

Organic manure is relatively low in N, but rich in P and K and may therefore play an important part in regulating N:P:K ratio in balanced fertilization. The various kinds of organic carbon serve as both energy sources for microbial activity and favorable materials for soil structure stabilization.

Management

The management of organic manure is evolving with the development of agricultural production in China. First, in the epoch of organic manure as the predominant nutrient resource, organic manure played a role both in improving soil fertility, and supplying available nutrients. In order to reduce its volume for transportation, it was composted and fermented. The preparation process is microbial and results in compost with a narrower C/N ratio than raw manure. To accelerate the fermentation, several inoculants of fiber decomposing bacteria were isolated. Later on, liquid extracted from horse dung was used as inoculum. Both kinds of inoculants could increase fermentation temperature to as high as 70°C, shortening the fermentation period. The high temperature during fermentation also kills the pathogens and eggs of parasites harmful to human beings and animals, and kills microbes and weed seeds harmful to crops. The preparation can also be combined with nutrient preservation techniques such as mixture with soil and sealing of the pile surface (Liu and Jin, 1991). Second, China has abundant experience in multiple uses of recycled organic wastes. For example, there is a "mulberry-fishpond" system in southern China. The droppings of silkworms were fed to pigs, the pig dung fed to fish, and the mud deposits of the fishpond applied to mulberry orchards. On the basis of this model, several recycling rings were also developed, such as chick droppings feeding pigs, pig dung producing marsh gas, and deposits from the gas-pit used for cultivating mushrooms. Also, various livestock excretions, plant residues, cotton seed cake, and rapeseed cake are used as fish feed, as mushroom culture media, or as raw materials for producing marsh gas, before being applied as organic manure. The longer the recycling of the food chain, the more energy and nutrients can be exploited from organic materials. The combination of cropping and feeding lowers cost, increases income, and reduces pollution (Jin, 1989).

China produces more than 500 Mt of plant residues every year, of which about 20 to 40 percent returns to the field. Due to the wide C/N ratio, crop residues applied directly to the field often result in N deficiency because of microbial immobilization during decomposition. Therefore, plant residues are often used as bedding for animal stables or composted before being returned to the field. A small portion of the residue may be used as animal fodder before being returned to the soil after excretion. Because the application of chemical N fertilizers has increased, nutrient supply to crops primarily comes from chemical fertilizers. In addition, with the widespread use of machinery in agriculture, the area that receives direct return of plant residues has increased. Residues can be returned directly to the soil through either incorporation or mulching. The plant stalk must be chopped into shorter lengths before being incorporated into the 16 to 20 cm thick plow layer. The plant residue mulch can be made by manpower during the following crop growing season. For instance, in the north China plain wheat straw from the last crop can be placed onto the interrow area during summer corn growing, or stubble can be cut 10 to 15 cm high when harvesting and left untouched when planting corn. Soil microbial activity and the residue decomposition after direct plant residue return are not the primary point of this section and so will not be discussed in detail.

In recent years, because of the readjustment of agricultural structure and improvement of people's lifestyles, new problems are appearing in the management of organic manure resources. For example, large scale industrialized animal husbandry farms, or intensive livestock operations are beginning to emerge. The traditional practice of bedding with plant residue or soil is becoming out of date and manure produced may now consist of pure animal excretion. Large quantities of untreated animal waste become a source of serious pollution in some places. With dense urban populations, the amount of urban waste generated from human excretion and other activities increases.

In the past, farmers would buy wastes from cities as manure for land application. However, at present, producers will not accept them even free of charge. Due to increasing coal consumption by farmers, use of plant residue as a fuel is declining. Small-scale paper mills are closed due to restrictions on pollution. Little of the straw produced can be made into paper, thus excessive amounts of straw remain unused. These organic wastes could be fermented, deodorized, desiccated, and added with amounts of chemical fertilizers, to be made into organic-inorganic compound fertilizers, which would be more easily stored, transported, and sold than unmodified wastes (Zhang, 2000).

Crop Response

Crop response to organic manure application differs substantially due to variations in manure sources and their nutrient contents. The excretions and bedding material from one pig in a year can compost to approximately 2,000 to 2,500 kg of high quality farmyard manure that can increase grain yield by 50 to 75 kg. More than sixty field experiments were conducted in Shandong and Hebei Provinces by the SFI, CAAS. Results showed the control plot that received no organic manure had a wheat yield of 3,530 kg ha^{-1} for one crop. The plot treated with 30 t ha^{-1} of organic manure had a wheat yield of 4,050 kg ha^{-1}, or a yield increase of 14.7 percent (SFI, 1994).

According to the results of long-term fertilizer experiments, organic manures can increase yield of rice or upland crops, whether applied alone or with chemical fertilizers, with the crop response increasing year after year. The data in Table 8.5 show that applied organic manure alone produced obvious crop responses in a long-term experiment (Lin et al., 1996). The response of upland crop was higher than rice. The crop response to organic manure applied with N, P, K fertilizers was still higher. However, in most trials, organic and inorganic fertilizers did not have a positive interaction.

Most experimental results show that direct return of plant residue has an inconsistent yield increase. There were 129 direct residue return field trials conducted from 1975 to 1985, by the SFI, CAAS on winter wheat, corn, and soybean. Results showed that the proportion of trials with >5 percent yield increase, >5 percent yield loss, and no effect amounted to 70.4, 6.0, and 23.6 percent, respectively. With direct plant residue return for several continuous crops, the first and second crops may show no yield response or a yield decline, but the third and fourth crops will show a positive response. The optimal returning rates for wheat or rice straw are 2 to 3 t ha^{-1} and for corn stalk is 4.5 t ha^{-1} (Liu and Jin, 1991).

TABLE 8.5. Crop response to organic manure in ten years from 1981 to 1990 (t ha^{-1})

Treatment	Double rice (n = 14) Early rice	Late rice	Paddy + upland crops (n = 14) Rice	Upland crop	Double upland crops (n = 13) Wheat	Corn	Mono upland crops (n = 11)
Control	3.75	3.465	4.29	1.49	1.97	3.11	2.97
M	0	0	5.49	2.24	2.99	4.05	4.02
Increase (%)			28.00	50.50	51.90	30.40	35.40
NPK	5.51	5.30	6.60	3.12	4.94	5.79	4.25
MNPK	5.87	5.78	6.99	3.39	5.48	6.05	4.77
Increase (%)	6.50	9.10	5.90	8.70	10.90	4.40	12.40

Source: Adapted from Lin et al. (1996).

Note: M = organic manure. Rates = 15 to 22.5 t ha^{-1} of pig manure or 4.5 to 6 t ha^{-1} of straw in southern China for each crop; 30 to 60 t ha^{-1} compost in northern China for one year.

Green Manures

China has a long history in growing green manures. The book titled *Record of Nature,* written in the third century mentions that "the vetch plant, yellowish green with purple flowers, is planted under rice, growing vigorously. It can improve soil fertility and the leaf is edible." Today, it is still used as green manure. In 1950, the area seeded to green manure was 1.73 Mha, which then expanded six times to 12 Mha by the peak year of 1976. Since the 1980s, the acreage of green manure has decreased gradually. It decreased to 4.3 Mha in 1990. Based on the biomass yield and nutrient content, the total nutrients produced were 635,000 t, accounting for only 4.1 percent of total organic manure input. Currently, green manures are grown on less than 6 Mha. The main reason for the decline in planting area is that green manure occupies one crop season without immediate economic return and China is short in farmland per capita. Use of chemical fertilizers rather than green manures is viewed as a more viable option.

Use of green manure is concentrated in the rice-growing regions in the middle and lower reaches of the Yangtze River, including the Jiangsu, Zhejiang, Anhui, Jiangxi, Hubei, Hunan, and Guangdong Provinces; whereas planting of green manure in other provinces is sporadic (Jiao, 1986).

Green manures are divided into winter and summer green manures according to their planting and application season. Winter green manure is sown in autumn or winter and grows over the winter months, serving as the base fertilization for the next spring crop. Planting Chinese milkvetch (*Astragalus sinicus* L.) in rice fields is a predominant practice. Summer green manure is sown in spring or summer and grows in the summer months, serving as the base fertilization for fall crops. Common sesbania (*Sesbania Cannabina* Pers.) is the dominant summer green manure crop. Also, green manure crops are botanically classified into leguminous, grass, aquatic, and perennial green manures. There are 232 species of green manures in China, among which are 170 wild and cultivated leguminous green manures, forty-six grass and sixteen other green manures mainly belonging to the crucifer and composite families.

Green manures planted in China are mainly leguminous plants with higher nutrient contents and narrower C/N ratios than non-leguminous crops. Based on an average of eighteen leguminous green manure species, 1,000 kg of fresh biomass contains 5.4 kg N, 0.63 kg P, 3.9 kg K, and 80.4 kg C, with a C/N ratio between 10 and 20. After being incorporated into the soil, they decay readily (SFI, 1994). Green manure supplied approximately 0.75 Mt of N in 1980, based on an estimated 10 Mha of growing area, 15 t ha^{-1} fresh biomass, and 0.5 percent of N concentration. In addition, *Azolla* also provided approximately 97,500 t of N, estimated from 1.3 Mha of growing area, with 25 t ha^{-1} of biomass containing 0.3 percent of N, the estimated N supply was 97,500 t. The above two sources added 0.85 Mt of N (Chen and Hu, 1982). The root systems of leguminous green manure plants are deep. Crucifer plants such as rapeseed (*Brassica* sp.) and radish (*Raphanus sativus* L.), and the knotweed family such as common buckwheat (*Fagopyrum esculentum* Moench.) have a strong ability to absorb P and K from the deeper portions of the soil, thus biocycling the two nutrients into the plow layer after the green manure is produced. Planting and incorporating green manure can also improve saline and low-yielding soils. For example, if common sesbania and yellow sweetclover [*Melilotus officinalis* (L.) Desr] are sown on a saline soil with 6 to 7 mg kg^{-1} total salt content in 0 to 5 cm layer, the sesbania can be established but the sweetclover can only tolerate up to 3 mg kg^{-1} total salt content. Sesbania tolerates flood and sweetclover tolerates drought. While the green manure is growing, evaporation is reduced, thus inhibiting salt movement to the soil surface. After incorporation of the green manure, the soil salt content in both the 0 to 10 cm and 10 to 20 cm layers decreased (Chen, 1994). Whether or not the incorporated green manure increases soil organic matter remains controversial. In twenty-two long-term experiments conducted on various soils in seventeen provinces, five years of continuous incorporation of 22.5 to 30 t ha^{-1} fresh grass each year was compared with fallow fields. Soil organic carbon content increased from 0.6 to 1.2 g kg^{-1}, on an average, depending on application rates, C/N ratio of incorporated green manure, and the original soil fertility (Chen, 1994).

Based on an average of 588 experiments in southern China, the incorporation of each 1,000 kg of fresh biomass increased rice yield by 38 to 80 kg. Experiments conducted by the National Green Manure Experimental Network from 1981 to 1986 showed that incorporating winter green manure for four years at a rate of 2.7 to 4.5 t ha^{-1} year^{-1} in dry matter increased rice yield by 0.3 to 1.9 t ha^{-1}. In 1,500 experiments in southern China, incorporating 15 to 20 t ha^{-1} fresh *Azolla* into rice fields increased rice yield by 0.6 to 0.8 t ha^{-1} (Chen, 1988). The optimal period for incorporation of leguminous green manure is between peak flowering and early pod bearing and for grass green manure is early ear sprouting. At these times, the amount of succulent fresh biomass is high and the decomposition and nutrient release after incorporation is rapid (Jiao, 1986).

There are many ways to encourage use of green manure. It is more attractive if green manures are used as green fodder. For instance, the milkvetch, or milkvetch with rye (*Secale cereale* L.) can be planted in winter with the plant harvested as green feed, then the ratoon and stubble used as manure. Some green manure plants are also vegetables or grains. With broad bean (*Vicia faba* L.) or garden pea (*Pisum satium* Sens Ampl.), the green beans could be used as vegetables and the plant residue incorporated as green manure. Some green manure plants are also industrial materials, such as sesbania. Sesbania gel can be extracted from seeds and used in paper, textile, printing and dyeing, well-drilling industries, and substituting for guar extracted from cluster bean. Its stalk can be used as fuel, and fallen leaves and stubble as manure.

Biofertilizers

Biofertilizers are microbial fertilizers, inoculants, or bacterial manures. They are products that contain living microbes. The research work on microbial fertilizers began in the 1940s. The researches on bacteria for N-fixation, and dissolving P and silicates were carried out in the 1950s. The microbial fertilizers "5406," an *actinomyces* antibiotic, was investigated and applied in the 1960s. Vesicular arbuscular mycorrhizae were studied in the 1970s and the 1980s. Nevertheless, the development of biofertilizers was unstable due to inadequate product supervision, and lack of a criterion of product quality. On the other hand, the necessary conditions of use and application techniques have not been clear. Since the 1990s, the biofertilizer management and product standards have been developed after summarizing research, production, and application experience from past decades (Li et al., 2004).

Microbial fertilizers in China, especially root-nodule bacterium products, have been widely recommended and utilized. In the 1950s and the 1960s, legume crops such as peanut and soybean were commonly inoculated with root-nodule bacterium products. From 1950 to 1955, inoculated peanuts were sown on up to 600,000 ha in the Shandong, Hebei, and Henan Provinces. Based on the results from 1,392 experiments, the yield increase by peanut root-nodule bacterium product was 12 to 21 percent, or 200 to 285 kg ha^{-1} peanut pod. The value/cost ratio (VCR) was 40:1. From 1956 to 1985, the cumulative extension area was 2.28 Mha. In the 1960s, the inoculation of milkvetch root-nodule bacterium was the key measure for successfully expanding milkvetch from the Yangtze River Valley to both the southern and northern regions, where it had never been planted before. In the 1970s, the extension of root-nodule bacterium product added with 2 percent ammonium molybdate in Shandong and southern Henan demonstrated an evident yield increase. The 1980s saw the success of a study on root-nodule bacterium coated on the seed-pill of legume forage sown by airplane. The root-nodule bacterium inoculant exists in the seed coating to prevent adverse conditions from inhibiting the root-nodule bacterium after sowing. After seed germination, the bacterium can infect the root in time to form a nodule. The extension acreage was relatively large (Li et al., 1989).

At present, China has preliminarily established a microbial fertilizer industry structure, including more than 500 enterprises, with 3 Mt of annual output. The twelve types of products registered in the Ministry of Agriculture include those of root-nodule, N-fixation, silicate, P-dissolving, and organic manure decay bacteria. The Ministry of Agriculture has released the standard on microbial fertilizer application in agricultural production, and the whole standard system is under construction (Li et al., 2004).

Management of Chemical Fertilizer Nutrient

Chemical fertilizers significantly increased crop yields. Compared to a control, application of N and P fertilizers significantly increased wheat yield by 56.6 percent (1,260 trials) and corn yield by 46.1 percent (629 trials) in the 1980s. Nitrogen, P, K application when compared with no N, P, K, increased yield by 40.8 percent in 829 rice trials, 48.6 percent in 62 cotton trials, 64.5 percent in 64 rapeseed trials, and 17.8 percent in 115 soybean trials. It is evident that optimal fertilizer application increased crop yield by 50 percent or so compared with no fertilizer input. About one-third of the total grain production comes from fertilizer application (Lin, 1991).

Major Application Approaches and Experiences

According to half a century's practice, the major approaches and experiences in fertilizer application are as follows.

Balanced Fertilization

Balanced fertilizer applications are necessary to eliminate the gap between the nutrients required for an objective crop yield goal and the nutrients that can be supplied by the soil. Balanced nutrient application is the most effective fertilizer application technology. As chemical fertilizers contain concentrated nutrients, the application rates are less, and the supply of the proper rate is easier than with organic sources, where the composition and the release patterns are not as well defined. Therefore, application of chemical fertilizers is a major tool for balanced fertilization.

For instance, before the 1960s, the P and K in Chinese soils were rich, but crops needed more N than the soils could supply. Optimal use of fertilizer N achieved a balanced NPK supply and yield increases were evident. Up to the 1960s, with more and more N fertilizer consumed and higher crop yield produced, the supply of P from the soil was insufficient to support the yield potential and many regions experienced P deficiencies. Phosphorus became the yield-limiting factor and the effectiveness of N fertilizer alone diminished. However, balanced application of N and P in combination enhanced crop response significantly.

With time, the cropping systems changed and the multi-cropping index increased. One crop a year changed into two crops a year. Two crops a year changed into three crops a year. The high-yield and hybrid crop varieties, such as those of corn and rice became more prevalent. Increased crop production led to greater nutrient removal and higher nutrient demand. The N and P consumption was increasing constantly, but little K was applied, and shortages of soil K supply appeared. According to the summarized data from three southern provinces in 1980, K fertilizer combined with N or NP enhanced rice yield (Table 8.6). Thus, the history of fertilization in China is a history of the importance of balanced fertilization based on the variation of soil nutrient supply and crop requirement (Lin and Li, 1989).

TABLE 8.6. Effect of N, NP, and NPK fertilizers on rice yields in three provinces

Province	Treatment	Rice yield (Kg ha^{-1})	Yield increase by P or PK		Yield increment by K	
			Kg ha^{-1}	%	Kg ha^{-1}	%
Hunan	N	5,037	–	–	–	–
	NK	5,324	–	–	287	5.7
	NP	5,225	188	3.7	–	–
	NPK	5,679	642	12.7	454	8.7
Guangdong	N	3,768	–	–	–	–
	NK	4,196	–	–	428	11.4
	NP	3,861	93	2.5	–	–
	NPK	4,566	798	21.2	705	18.3
Zhejiang	N	4,248	–	–	–	–
	NK	4,575	–	–	327	7.7
	NP	4,419	171	4.0	–	–
	NPK	5,067	819	19.3	648	14.7

Source: Adapted from Lin and Li (1989).

Combined Use of Inorganic with Organic Fertilizers

The combined application of chemical fertilizers with organic manure has been the main approach and experience in China since the 1970s. This combination is beneficial for producing nonpolluted, high-quality, and nutritionally rich agricultural products, avoiding negative environmental impacts and enhancing long-term sustainability. There are three major advantages to INM.

First, numerous long-term experiments demonstrate that crop yields can be increased by the combination of N, P, K fertilizers with organic manure. Figures 8.2 and 8.3 are the results of fourteen rice-upland crop annual rotation experiments conducted for a ten-year period. They show that, with rice or upland crops, organic manure applied alone produced a significant yield increase. Nitrogen fertilizer alone could only maintain high yields in the first two years of the studies. Later on, as other soil nutrients, mainly P and K, were exhausted, yields dropped dramatically. Nitrogen fertilizer combined with organic manure led to higher yields and less yield variability than N fertilizer alone over the ten-year study, because the organic manure

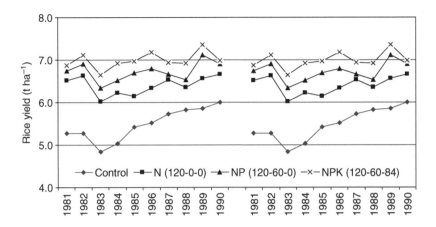

FIGURE 8.2. Effect of fertilizer combinations on rice yield from a rice-upland crop rotation in long-term trials (left-hand side = without organic manure; right-hand side = with organic manure at a rate of 15 to 22.5 t ha^{-1} fresh compost or 4.5 to 6 t ha^{-1} dry straw for each crop; number of trials = 14). *Source:* Adapted from Lin et al. (1996).

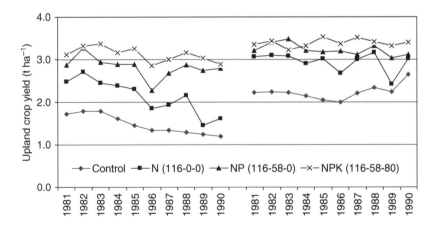

FIGURE 8.3. Effect of fertilizer combinations on upland crops (wheat or barley) from a rice-upland crop rotation in long-term trials (left-hand side = without organic manure; right-hand side = with organic manure at a rate of 15 to 22.5 t ha^{-1} fresh compost or 4.5 to 6 t ha^{-1} dry straw for each crop; number of trials =14). *Source:* Adapted from Lin et al. (1996).

replenished the nutrients. The NP fertilizers and N, P, K fertilizers both produced high yields, but when they were combined with organic manure, the yield was even higher. The same trend was also observed in other cropping regions (Lin et al., 1996).

Second, the combination of chemical fertilizers with organic manure improves the product quality of cash crops, vegetables, and fruits. Chinese tobacco growers have a tradition of applying organic manure, especially cakes. They even state "no manure no quality." However, disadvantages of organic manure may exist, including the unstable nutrient content, unpredictable nutrient release rate, and the possible presence of excess chloride salts. The combination of chemical fertilizers with organic manure produces tobacco leaves with lower total N and protein contents, higher total saccharide content, and optimal total N:total saccharide:nicotine ratio. The proportion of higher-quality leaves is high, thus leading to a higher production value. Evaluation of chemical fertilizer combined with barnyard manure or with cakes, on tea, showed that a series of quality indicators such as tea polyphenols, catechins, caffeine, and water extracts were better than chemical fertilizers alone and much better than no fertilizer. The combination of chemical fertilizers with organic manure on leafy vegetables lowered the nitrate content, increased the VCR, and increased the mineral element contents (Zhang, 2003).

Third, long-term fertilizer experiments demonstrated that at the end of ten years, soil organic carbon, one of the key indicators of soil fertility, declined where no fertilizer or N fertilizer alone was applied. In contrast, the soil organic carbon content was stable or climbed slightly with N, P fertilizers or N, P, K fertilizer treatment and increased with the combined application of chemical fertilizers with organic manure (Table 8.7). The trend of variation of total soil N was similar to that of organic carbon while the variation of soil P was greater. At the end of ten years, soil total P in the surface layer decreased by 0.03 to 0.07 g kg^{-1} for the no fertilizer control and for N fertilizer alone, whereas it increased by 0.09 g kg^{-1} for NP and NPK treatments, which had both received 45 to 75 kg P$_2$O$_5$ ha^{-1}. Furthermore, soil P increased by 0.31 to 0.56 g kg^{-1} with the combined application of chemical fertilizers with organic manure. The variation of soil Olsen P was also evident. At the end of ten years, Olsen P decreased with no fertilizer and N fertilizer alone treatments, increased by 10 mg kg^{-1} when P was added, and increased by 15 to 20 mg kg^{-1} for the combination treatment of NP plus organic manure. Soil total K content in the no K treatment was reduced by 1.7 g hg^{-1} (mean of seven trials), while soil available K in the surface layer remained unchanged when 60 to 120 kg K$_2$O ha^{-1} was applied, but in the K plus organic manure treatments soil available K increased slightly.

TABLE 8.7. Impacts of different treatments on soil organic carbon (g kg^{-1}) after ten years' experiments

Cropping pattern	Sampling time	Treatment							
		CK	N	NP	NPK	M	MN	MNP	MNPK
Double rice (n = 4)	Before trial	15.3	15.3	15.3	15.3	–	–	–	15.3
	Ten years after	14.5	13.6	14.4	15.3	–	–	–	18.0
Rice-upland crops (n = 7)	Before trial	13.1	13.1	13.1	13.1	13.1	13.1	13.1	13.1
	Ten years after	12.4	12.9	13.2	13.5	13.8	14.0	14.4	14.1
Double-upland crops (n = 7)	Before trial	8.2	8.0	8.0	8.0	7.3	7.1	7.3	7.4
	Ten years after	7.7	8.2	8.6	8.6	8.4	8.5	8.5	8.5
Mono-upland crop (n = 5)	Before trial	11.1	11.4	11.1	11.2	10.7	11.8	12.5	12.4
	Ten years after	10.7	10.7	11.4	11.5	11.8	13.6	13.6	12.4

Source: Adapted from Lin et al. (1996).

The combined use of chemical fertilizers with organic manure decreased soil bulk density, increased soil porosity and field capacity, and reduced leaching of nitrate into groundwater. It also increased soil respiratory strength (determined by CO_2 release) and earthworm population (Lin et al., 1996).

The Reasonable Distribution of Fertilizers in Cropping Systems

Use of INM must consider the effective distribution of nutrients over the growing season and from year to year. For example, N fertilizer has its major effect on crop growth in the year of application, with little carryover effect to the following crop. In contrast, P uptake by the crop in the year of application is often relatively low and there can be a significant carryover in the soil over time. The residual P can have a great effect on subsequent crops, and this residual benefit should be assessed in nutrient management. The P and K requirement characteristics of various crops and their potential responses are important considerations in nutrient planning through a cropping cycle.

A good example is the rice-upland cropping system. Phosphorus is applied to upland crops, with the following rice taking advantage of the carryover effect from the residual P. Some experiments showed that when P was applied to upland crops it produced yield increases, but when applied to rice there was no effect. However, after inundation, the soil P supply ability became stronger. One reason may be that flooding reduces ferric phosphate into ferrous phosphate, which is a more soluble and available form of P. In addition, the decomposition of organic matter in soils generates organic acids, which may chelate Ca ions in calcareous soils, increasing the solubility of Ca-P compounds, and the availability of P; in acid soils, they may chelate Fe or Al, decreasing P immobilization. Some experiments with the rotation system of wheat-rice produced in a year showed that the yield increased by 80 percent, and the P uptake increased by 61 percent when P was applied on wheat, with the subsequent rice crop receiving carryover effects as compared to P application on rice with the wheat receiving the carryover effect (Li et al., 1998).

A second example is that of legume-grass cropping systems, where P is applied to legume crops, especially on leguminous green manure, leading to yield increases. On rice fields in southern China, incorporated milkvetch treated with P increases the P uptake rate of the following rice as much as 66.2 percent, with 1 kg of P producing 5 kg of rice. If the same amount of P is applied to rice directly, the P uptake rate is merely 14.7 percent, or one kg of P produces 1.5 kg of rice. On fields in northern China, incorporating sesbania treated with P increases the yield of the next wheat crop by 67.9 percent, with one kg of P producing 4 kg of wheat. If the same amount of P is applied to wheat with incorporated sesbania without P application, the yield increase of wheat is only 38.9 percent, or one kg of P producing 2.3 kg of wheat (Chen, 1994).

A third example is that of double rice cropping where the K effect increases in the order: late rice > winter crops > early rice. When K fertilizer is limited, it is better to apply it to late rice. In double rice-milkvetch cropping system, or double rice-rapeseed system, K is better applied on milkvetch or rapeseed (Li et al., 1998).

In winter wheat-summer corn annual rotation, the effect of P applied on winter wheat is better than that of K. However, the effect of K on summer corn is better than that of P. Therefore, it is recommended that most P and less or no K be applied to winter wheat, and most of the K, and less or no P be applied to summer corn, to increase the output of the whole year (Lin et al., 1996). Therefore, consideration of nutrient management through crop rotation rather than individual crops in the rotation may improve the overall nutrient use efficiency and productivity of the cropping system.

Proper management of residues and tillage practices may improve soil nutrient status and productivity over time. Plant residue mulch, and tilling and planting along isohyets on slope land can reduce water runoff and soil erosion, improving soil fertility. No-till on rice fields in Taihu Valley, and changing constantly inundated fields into semi-drained rice fields in some districts in Sichuan Province might improve soil physical- and chemical-characteristics and nutrient regime, increasing rice yield (Gao and Wang, 1989).

The Implementation of Integrated Nutrient Resource Management

The use of organic manure, including green manure, is based on the individual farmer's household production and utilization. The government can encourage greater and more efficient use of organic manure resources by farmers through financial and technical support. For example, in some rural areas, marsh gas-pits were constructed with government loans and technical instruction. On wider areas, local governments organized agro-mechanical departments to develop stalk-cutting machinery and extended the technology for use of plant residue mulch (Li et al., 1998). Here we will mainly discuss chemical fertilizer resource utilization, or fertilizer recommendations, also called formula fertilization in China. To implement fertilizer recommendations, China has produced a great deal of data on soil nutrients and fertilizer responses. Nevertheless, since there are 240 million small farmer households in China, it is difficult to target a fertilizer recommendation for each household or each piece of the field.

The Nutrient Management Based on Nationwide Soil Survey, Fertilizer Trials, and Regionalization

Under the leadership of the Ministry of Agriculture, the second nationwide soil survey was initiated in 1979 and finished in 1994, taking sixteen years. More than 2,000 counties in thirty-one provinces (excluding Taiwan) completed their soil survey reports. In addition to soil maps, the twelve maps related to nutrient management and fertilization are: soil organic matter, total N, pH, calcium carbonate, available P, available K, and available micronutrients including Fe, Mn, Cu, Zn, B, and Mo. The national level maps are in 1/4,000,000; the provincial level maps are in 1/500,000 or 1/1,000,000; the county level maps are in 1/50,000. The *Data of Soil Survey in China* was published, serving as a valuable tool for integrated nationwide nutrient management (The National Soil Survey Office, 1998).

During the period from 1981 to 1983, the SFI, CAAS conducted more than 5,000 chemical fertilizer experiments in various climate-soil belts in China, as well as more than 100 long-term fertilizer experiments. Up to seventy of these experiments had complete data for more than ten years. These experiments were on grain, oil-bearing, fiber, sugar, and vegetable crops, and served to clarify the N, P, K optimal rates, ratios, and yield-increase effects for various crops on various soils (NCFEN, 1986). An international cooperative program on nutrient management focusing on K fertilizer is still going on.

On the basis of the above-mentioned works, national regionalization was carried out. According to N, P, K fertilizer consumption and crop responses, as well as the future chemical fertilizer requirement of cropping industry development, the regions were divided to provide the basis for suitable chemical fertilizer distribution (SFI, 1986).

The Major Organizations of Reasonable Fertilization Programs

Soil and Environment Division and Fertilizer Division, NATESC, and the Ministry of Agriculture are the major departments for instructing fertilization. Their branches are set up on a county level with each county establishing a Soil and Fertilizer Station. Technicians at the stations take part in the soil survey and fertilizer experiments, and they are relatively familiar with the local situation. They are the major extension personnel for delivering technical services to farmers. Meanwhile, with the development of the market economy, there are more and more chemical fertilizer dealers. At present, there are three major channels: the first includes the China National Agricultural Means of Production Group Corporation and Chemical Fertilizer International Trading Corporation of the Chinese Import and Export Corporation (SINOCHEM). They have established a domestic selling network, reaching directly to retailers at the grassroot level. The second is composed of fertilizer manufacturers who sell their fertilizer goods. The larger scale manufacturers set up their own agro-chemical service departments. The third includes agro-tech extension stations, plant protection stations, and soil and fertilizer stations. They trade fertilizers, pesticides, and other agricultural means of production. These three major kinds of dealers provide farmers with optimal fertilization programs. In the near future, foreign chemical fertilizer companies will also join this service. For the time being, they can only sell their goods through Chinese traders.

Research in Agricultural Institutions and Agricultural Universities

In the 1980s, the Ministry of Agriculture organized the Project of Formula Fertilization, which was conducted in nineteen provinces as a balanced fertilization study based on a soil test. The methodology of the soil test was routine: selecting optimal extracting solutions with correlation analysis for soil available nutrient test; calibrating available nutrient test results by biological experiments to determine the indices of soil nutrient abundance classification, which would be the basis for fertilizer recommendation. The methods selected were the Olsen and Bray methods for soil available P and 1 mol L^{-1} NH_4OAC as the extracting solution for soil available K. The indices of high, medium, and low for nutrients in various soils were set up, which were valuable references for P and K recommendation (SFI, 1994). At the same time, extracting solutions developed by Agro Services International were introduced for quick testing and evaluating of various soil nutrients (Beijing Office, 1992). However, because of variations in crop types (field crops/horticultural crops) and cropping practices (open field/protected field such as plastic mini-tunnels), the use of these indices and recommendation methods are still limited. The N recommendations based on soil inorganic N test for basal application and based on plant nitrate test for topdressing were studied (Li, 1999). The micronutrient recommendation is based on critical values indicated by soil test or plant test (Li and Zhu, 1991).

In the Chinese countryside, the farmer's household is still the unit of production and the scale is small. Each household has merely 0.54 ha farmland on average (based on the calculation of 130 Mha farmland divided by 240 million households). Furthermore, the half hectare of land might be divided into several smaller pieces. Therefore, it is difficult to provide specific fertilization programs to farmers based on plots. The available method was regionalized at a county or a township level according to factors including topography, soil types and texture, fertilizer application history, and cropping system. The recommendation for a certain region was based on nutrient variation and soil tests. The procedure is currently being studied (Huang et al., 2002). Meanwhile, some fertilizer manufacturers produced several formula universal grades for a region or specific mixtures for a crop based on local soil nutrient data, and may even take soil samples for testing, considering the main crops. The mixed/blended fertilizers produced crop response after local farmers used them (Sino-Arabic Chemical Fertilizer Corporation, 2004). In a word, the INM of the extensive small householder farming is a major issue that we must study and solve.

BENEFITS AND LIMITATIONS
OF CHEMICAL FERTILIZERS

There is no doubt that application of chemical fertilizer has been and will be the most effective measure to meet the increasing demand for food of the growing population. After thousands of years' cultivation, Chinese farmlands are relatively low in soil nutrients, especially soil N. Thus, addition of adequate amounts of fertilizers to those arable soils is vital to enhance crop production and provide enough food and fiber for the large population. Previous studies showed that more than 40 percent of increased crop production was attributable to application of chemical fertilizers (Ma et al., 2002).

Overfertilization should be prevented to avoid negative impacts on the environment. Excess nutrient applications can disturb the natural cycling of N and P, leading to the emission of gases such as N_2O and NO_X, and hydrogenous nitride such as NH_X; surplus P can migrate into water bodies in runoff. Thus excess fertilization can trigger and exacerbate global and regional environment problems, such as the greenhouse effect, water body eutrophication, and acid rain. Some environmental concerns of N fertilizers follow.

N_2O Emission from Farmland

Nitrous oxide in farmland is formed through nitrification and denitrification processes in the soil. Xing (1998) indicated that the N_2O emitted from the farmlands of China was as much as 0.398 Mt of N in 1995; and based on the method and some parameters offered by the Intergovernmental Panel on Climate Change (IPCC) for calculating emission of greenhouse gases (IPCC, 1996), the direct emission of N_2O from the farmland of China was estimated at 0.336 Mt of N (Xing and Yan, 1999). The results from the two sources are quite close.

Upland fields, the major contributor of N_2O in China, account for 78 percent of the total N_2O emission from farmlands. Researchers used to believe that N_2O emission from paddy fields was very limited (Freney et al., 1981) or might be ignored (Khalil et al., 1990). However, rice fields are also a source of N_2O because 86 percent of the paddy field in China is under rice-upland crop rotation, for example, one or two crops of rice in summer and upland wheat or rapeseed in winter. Moreover, the water management in rice is unique, since it undergoes alternation of flooding and draining at least once in each growing season, and during late maturity, the field will be drained dry to facilitate harvesting of the rice and sowing of the winter crops. The wet-dry cycling is most favorable for N_2O generation. Even in the growing season of rice, the N_2O emission from the rice field may account for 9 percent of the total from the field (Xing, 1998).

NH_3 Volatilization in Agriculture

Ammonia (NH_3) volatilization is one of the major paths for gaseous loss of N from farmland and agriculture. Approximately 11 percent of the chemical N fertilizer applied is lost by NH_3 volatilization (Xing and Zhu, 2000). In 1995, chemical N fertilizers and animal excretions contributed 2.71 and 3.35 Mt of NH_3-N volatilization, totaling 6.06 Mt in China (Xing and Zhu, 2002).

N Translocation from Farmland to Water Bodies

The amount of nitrogen migrating into water bodies through runoff and leaching is estimated at 1.24 and 0.5 Mt of N, or 5 and 2 percent, respectively, of the total amount of N fertilizers applied (Zhu and Chen, 2002). Nitrogen leaching from farmlands in China is a rather complicated problem, which varies greatly with different natural and social conditions, and with land use regimes.

Rice fields, accounting for one quarter of the country's total farmland, are distributed mainly in the regions up to the Yangtze River, where the annual precipitation ranges between 1,000 and 2,000 mm and the N fertilizer application rate is around 400 to 600 kg N ha^{-1} $year^{-1}$. A rice-upland crop rotation system with double or triple cropping each year prevails, that is, one or two crops of rice in summer and one upland crop (wheat, rapeseed, or beans, etc.) in winter. On rice fields, whether rice or upland wheat is grown, the leaching of fertilizer N is always low, ranging between 0.02 and 3.01 percent during the rice growing season and slightly higher during the wheat season. Leaching is closely related to N rates. The higher the N application rate, the more N is leached out (Table 8.8).

The leaching of fertilizer N from rice fields in south China is not severe either, ranging around 5 to 10 kg N ha^{-1} $year^{-1}$ at the N rate of 400 to 500 kg N ha^{-1} $year^{-1}$ (Lu et al., 1996). The reason for low N leaching from rice fields is that the plow layer of the field is kept submerged underwater during the rice-growing season. Denitrification takes place not only in the submerged plowing layer, but also in the deeper water-saturated subsoil layer during the rice- or wheat-growing seasons (Xing and Zhu, 2002). Nevertheless, observations at an experiment site of upland red soil in south China

TABLE 8.8. Nitrogen leaching from the rice field

Location	N rate (kg N ha^{-1} year^{-1})	Proportion of leaching to application (%)		Method used	Source of data
		Rice season	Wheat season		
Jiangning, Jiangsu	600[a]	1.80	3.40	Lysimeter	Zhu et al. (2000)
Changzhou, Jiangsu	254	0.56	–	Lysimeter	Lu (1999)
	388	3.01	–		
Hangzhou, Zhejiang	100	0.02	–	Lysimeter	Xiong (2002)
	150	0.03	–	Lysimeter	
	100	–	8.61	^{15}N	
	1,000	–	35.10		
Shenyang, Liaoning	150	0.27	–	Lysimeter	Zhang et al. (1984)
	250	0.46	–	Lysimeter	

[a]The total N application for two crops, rice and wheat.

with an annual precipitation of 2,128 mm showed that the N leaching there could account for 15-27 percent of the N applied (Lu et al., 1996). As the data collected in the observations were quite limited, much uncertainty existed in the calculation of N leaching from the upland fields in south China.

The Huang-Huai-Hai Plain, with a mean annual precipitation of 640 mm, is cultivated with two crops a year or three crops in two years. Nitrogen leaching from the farmland accounts for 2 to 17.2 percent of the total N applied. Although N leaching has a definite relationship with N rate, it varies dramatically (Table 8.9). It is evident that N leaching in the north China plain is not high, possibly due to a number of factors. About 60 to 70 percent of the precipitation occurs during June, July, and August, coinciding with the season when crops grow vigorously and absorb more NO_3^2 from the soil than in any of the other growing stages. Also, the over-wintering crops can absorb NO_3^2 in the plow layer during the winter months. Finally, precipitation occurs rarely in winter and spring in this region, so downward movement of NO_3^2 will be minimal until irrigation starts in spring.

TABLE 8.9. Nitrogen leaching in the North China plain

Location	N rate (kg N ha^{-1} year^{-1})	Ratio of leaching to added (%)	Method used	Source of data
Langfang, Hebei	173	3.6	Lysimeter	Dai and Zhao (1992)
Beijing	150	0.9	Lysimeter	Yuan et al. (1995)
	225	2.1		
Beijing[a]	75	1.8	Lysimeter	Sun et al. (1994)
	300	17.2		

[a]The data are means of seven years (1986-1992).

Nitrate Accumulation in North and Northwest Regions

The loess plateau, lying between 34 to 40° N and 102 to 114° E, is generally covered with a 50 to 150 m thick layer of loess. The region has a mean annual precipitation of 450 mm, concentrated mainly in July, August, and September. A large part of the region can have only one crop a year. As a result of the limited precipitation and thick loess layer, the groundwater table often stays at the depth of 50 to 60 m, or at 100 to 200 m in some places. Therefore, the NO_3^- leached out of the cultivated layer accumulates in the subsoil layer at a depth of 20 to 60 cm or 60 to 90 cm (Lu et al., 1996; Yi and Xie, 1993). Even when the field is left in summer fallow, the deepest NO_3^- could penetrate is only 1-1.5 m (Peng et al., 1981). In an apple orchard that has been treated with N fertilizer for eight years at a rate of 900 kg N ha^{-1}year^{-1}, the total accumulation of NO_3^- in the 4 m thick soil layer could be as high as 3,414 kg N ha^{-1}year^{-1} (Lu et al., 1998).

Accumulation of NO_3^- in the subsoil layers of farmland is also observed in the north China plain. In Beijing for example, most of the NO_3^- leached out of the plow layer stays in the 40 cm soil depth and little moves as far as the 140 cm depth (Sun et al., 1999). During the summer corn-growing season, NO_3^- could also move down only to 130 cm or slightly deeper; and during the wheat-growing season NO_3^- accumulates in the 40 to 60 cm depth. Currently, due to the overexploitation of groundwater, the water table has been falling drastically. For example, in the Beijing region it has dropped to 18-40 m and continues to fall at a rate of 1 to 2 m year^{-1} (Wu, 1998). The NO_3^- accumulated in the subsoil will not move sufficiently to reach the groundwater.

Reports from north China (Zhang et al., 1995) and northwest China (Lu et al., 1998) indicated that the NO_3^2 content in the groundwater was found to exceed the criteria. Whether the higher content of NO_3^2 in the groundwater comes from the downward movement of NO_3^2 from the upper soils driven by irrigation or very heavy rainfall, or from other sources needs to be verified.

Nitrogen Pollution of Water Bodies in the Economically Developed Regions

In the past few years, algae blooms have occurred frequently in some lakes in China. Eutrophication has now become an important environmental problem in the country. Sun and Zhang (2000) reported, for instance, that out of the twenty-eight lakes investigated, seventeen were moderately to extremely eutrophied. Although few studies were conducted on the impacts of farmland on N concentrations in the lakes, some individual reports stated that the value was around 7 to 35 percent (Jin, 1995).

The Taihu Lake region, located in the Yangtze River Delta is an economically developed, densely populated area, where the N and P tests have already exceeded the critical levels for eutrophication of water bodies (Table 8.10). The mean concentration of NO_3^2 in the 5 to 6 m deep shallow groundwater, the major source of drinkable water in the rural area has reached 7.98 $\pm$ 9.65 mg NO_3-N L^{-1}. Water in 28 percent of the monitored wells in the region has NO_3^2 concentration beyond the standard (10 mg NO_3-N L^{-1}) set by WHO for drinkable water. Currently, small-urban sewage, rural human and animal excretion, and feeds for aquatic production are found to be the major sources of N and P in the rivers and lakes. In the economically devel-

TABLE 8.10. Nitrogen and phosphorus concentrations of the surface/groundwater in the Taihu Lake Valley

Water body	NO_3^- (mg-L^{-1})	NH_4^+ (mg N L^{-1})	Total N (mg N L^{-1})	PO_4^{-3}-P (mg P L^{-1})
River	1.36 ± 0.68 (n = 23)	5.69 ± 4.26 (n = 23)	7.99 ± 4.44 (n = 23)	0.20 ± 0.24 (n = 13)
Lake	1.27 ± 0.98 (n = 7)	0.33 ± 0.43 (n = 7)	1.88 ± 1.46 (n = 7)	0.022 ± 0.085 (n = 7)
Well	7.89 ± 9.65 (n = 40)	<0.2		

Source: Adapted from Xing and Zhu (2001).

oped parts of the countryside, the night soil and animal dung are no longer used as manure, but are released directly into water systems.

FERTILIZER AND AGRICULTURE SUSTAINABILITY

Application of chemical fertilizers for higher agricultural production while minimizing their detrimental impacts on the environment is both a challenge we have to face and a target we must strive for. To answer the challenge, the following management options are available in China.

Optimizing N Application Rate

In fertilizing for high yield, maintaining N rates within the range for maximum economic yield is a key for linking high yield and environmental protection (Zhu and Chen, 2002).

Readjustment of N:K Ratio by Increasing Input of K Fertilizers

During the period from 1984 to 1995, the grain output stagnated at a level between 40 and 45 Mt, while the consumption of chemical N fertilizer increased by 77.3 percent, indicating that the crop response to N fertilizers was significantly lower as compared to that before 1984. Further increases in crop yield with increased N fertilization requires an adjustment of the N:P:K ratio to reflect crop demand. In 1995, the $N:P_2O_5:K_2O$ ratio was 1:0.45:0.16. It is quite obvious that the proportion of K fertilizer was too low. For that, Chinese scientists recommended that the $N:P_2O_5:K_2O$ ratio be readjusted to 1:(0.4-0.45):0.30 by 2010. The readjustment will not only improve the yield response to N and P fertilizers, particularly the former, but also help reduce N_2O emission from the farmland and migration of excess NO_3^2 into water bodies.

Reduction of Large Regional Differences in N Fertilizer Distribution

The difference between regions in consumption of chemical N fertilizer is huge in China. According to the 1995 provincial statistics data on chemical N fertilizer consumption, about one-third of the chemical N fertilizers consumed is concentrated in coastal provinces in southeast China, whose area of farmland accounts for only one-fifth of the country's total, whereas only 17.6 percent was used by eleven provinces or autonomous regions

(Heilongjiang, Jilin, Inner Mongolia, Shanxi, Ningxia, Gansu, Qinghai, Xinjiang, Yunnan, Guizhou, and Tibet), with an area of 36.6 percent of the nation's total. Such a difference in distribution is evidently related to a series of factors, such as development level, population density, climate, and irrigation conditions. If the enormous gap in the distribution of chemical N fertilizers can be diminished on the basis of the gradual improvement of other conditions, it would not only further increase the output of agricultural production, but also alleviate the impact of overfertilization of chemical N fertilizer on the environment in some regions.

Emphasis on Utilization of Organic Manure

With the increase in consumption of chemical fertilizers, the collection and utilization of organic manure is declining. It not only wastes resources, but brings about the potential for environmental pollution. In 1998, the agricultural usage of the total resources of organic manure was merely 32.7 percent (Table 8.11). The current scaled hog and chick farms produce approximately 300 Mt of excretion annually, of which around 20 percent is applied to the field. The traditional methods of handling organic manure with collection, preparation, and preservation are not suited for today's situation. In the future, the processing of human and animal excretion will need

TABLE 8.11. Nitrogen resources in organic manure and their agricultural utilization rates in China in 1998

Organic manure	Total (Mt N)	Used in agriculture (Mt)	Rate in agriculture (%)
Excreta from urban areas	0.686	0.091	13.3
Excreta from rural areas	2.051	0.503	24.5
Pig dung	3.790	1.310	34.6
Cattle dung	3.654	1.296	36.4
Large animal dung	0.909	0.212	23.3
Sheep dropping	0.916	0.147	16.0
Poultry dropping	0.703	0.240	34.1
Bean cake	0.879	0.703	80.0
Plant residues	3.053	0.916	30.0
Total	16.553	5.418	32.7

Source: Adapted from Zhu and Chen (2002).

to be industrialized and commercialized. Planting green manure crops can reduce chemical fertilizer application. The rice yield-increases in the past few years may open a door for planting rice in summer and green manure crops in winter.

Reduce the Proportion of Ammonium Bicarbonate in Chemical N Fertilizer and Extend the Use of Nitrification Inhibitors and Slow-Releasing Fertilizers

Ammonium bicarbonate has been the major type of chemical N fertilizer for a long period of time. Even in 2001, its output remained at 20 percent of the total chemical N fertilizers used in China. The NH_3 volatilization during its production, transportation, storage, and application is much higher than that of urea. Reduction or even elimination of the production and use of ammonium bicarbonate can reduce entry of NH_3 into the atmosphere. The addition of dicyanodiamide (DCD) to the ammonium bicarbonate can also improve its effectiveness and enhance crop yield (Zhang et al., 1997).

China is a country that mainly uses ammonium-based chemical N fertilizers, such as urea and ammonium bicarbonate. Nitrate forms of fertilizers occupy a very small proportion. In 2001, the ammonium nitrate output accounted for only 3.2 percent of the total output of chemical N fertilizer. Therefore, the use of nitrification inhibitors to reduce emission of N_2O from farmland might have great potential. The effect of slow-releasing fertilizers on reduction of N_2O emission has also received great attention (Shoji and Kauno, 1994). In recent experiments, the use of slow-releasing coated urea demonstrated the practical value in reducing emission of N_2O from rice fields in China.

The Chinese population is still growing and the area of farmland is declining. It is essential to continuously upgrade the yield and product quality. Fertilizer is one of the most important inputs to attain this goal. In order to avoid diminishing returns and growing environmental risks from increasing fertilizer input, effective integrated plant nutrient management is imperative.

Plant nutrient management in China is moving from a single purpose to a multipurpose orientation. For a long time, the primary issue of Chinese agriculture was food shortage. To feed 1.3 billion people, fertilization was targeted at increasing yield. Although the idea of High Yield, High Quality, and High Efficiency was presented earlier, the high yield portion of the equation dominated. After several decades' effort, China has attained self-sufficiency in its food supply since the mid 1990s. As Chinese agriculture moves forward, we must pay more attention to the quality of agricultural products, economic return and eco-environment, along with yield. There is still a long way to go. For example, we have established some long-term monitoring stations not only to monitor changes in the quality of soil fertility as in the past, but changes in the quality of the soil environment for the future. We paid close attention to the soil nutrient regime, by increasing fertilizer use to meet crop demands in the past. We must take advantage of biological technology to create genetic resources with highly efficient nutrient absorption to improve fertilizer use-efficiency. We have to increase the agricultural utilization rate of organic manure resources and develop new highly efficient fertilizers. We have to absorb experience from foreign countries when building up regional and national information-support nutrient management systems, to realize the optimization and integration of Chinese nutrient resource management.

REFERENCES

Beijing Office, Potash and Phosphate Institute of Canada. (1992). *The Systematic Approach to Soil Nutrient Study.* Beijing: China Agro-SciTech Press (in Chinese).

CCAR. (1981). China Comprehensive Agriculture Regionalization. Zhou Li-Shan (ed.) Comprehensive Agriculture Regionalization. Beijing: China Agricultural Press (in Chinese).

Chen, L. (1988). Green manure cultivation and use for rice in China. In IRRI (ed.) *Green Manure in Rice Farming.* pp. 63-71. Los Baños, Laguna, Philippines: International Rice Research Institute.

Chen, L. (1994). Use of green manure in China as agricultural systems commercialized. In IRRI (ed.) *Green Manure Production Systems for Asian Rice lands.* pp. 43-50. Los Baños, Laguna, Philippines: International Rice Research Institute.

Chen, T. and J. Hu. (1982). The situation and prospect of biotic N fixation resource in China. *Soil and Fertilizer* 2: 9-11 (in Chinese).

China Statistics Yearbook. (2003). Beijing: China Statistical Press (in Chinese).

Dai, T. and Z. Zhao. (1992). A case study on the chemical form conversion and leaching of nitrogen and phosphorous in the irrigated plain area of Haihe River. *Acta Science of Circumstantial.* 12: 497-501 (in Chinese).

Freney, J.R., O.T. Denemead, I. Watanabe, and E.T. Craswell. (1981). Ammonia and nitrous oxide losses following applications of ammonia sulfate to flooded rice. *Australian Journal of Soil Research.* 32: 37-45.

Gao, X. and D. Wang. (1989). The development and prospect of tillage science in China. *Soil and Fertilizer.* 4: 18-22 (in Chinese).

GTCSES (General Team of Countryside Society Economic Survey), National Bureau of Statistics. (2003). *China Rural Statistical Yearbook.* Beijing: China Statistical Press (in Chinese).

Guo, J. and B. Lin. (1985). The history of fertilizer research in China. *Soil Bulletin.* 5: 237-239 (in Chinese).

Huang, S., J. Jin, L. Yang, Y. Zuo, and M. Cheng. (2002). The effect of regionalized balanced fertilization on N fertilizer use efficiency and soil nutrient balance. *Soil and Fertilizer.* 6: 3-7 (in Chinese).

Huang, S., J. Jin, Y. Zuo, L. Yang, and M. Cheng. (2000). The site study on farmland nutrient balance status and its evaluation. *Soil and Fertilizer.* 6: 14-18 (in Chinese).

IPCC. (1996). Climate change 1995. In Houghton, J.T., L.G. Meira Filho, B.A. Callander, N. Harris, A. Kattenberg, and K. Maskell (eds.) *The Science of Climate Change.* p. 572. Cambridge, UK: Cambridge University Press.

Jiao, B. (1986). *Green Manure in China.* Beijing: China Agriculture Press (in Chinese).

Jin, W. (1989). The four decades of organic manure research. *Soil and Fertilizer.* 5: 35-40 (in Chinese).

Jin, X. (1995). Eutrophication of lakes in China. In Jin, X. (ed.) *Lacustrine Environment in China.* pp. 267-322. Beijing: Ocean Press (in Chinese).

Khalil, M.A.K., R.A. Rasmussen, M.X. Wang, and L Ren. (1990). Emission of trace gases from Chinese rice fields and biogas generators: CH_4, N_2O, CO, CO_2, Chlorocarbons and hydrocarbons. *Chemosphere.* 20(1-2): 207-226.

Li, J., B. Lin, G. Liang, and G. Shen. (2000). The assessment on the prospect of fertilizer use in China. In Zhou, J. (ed.) *The Balance and Management of Farmland Nutrients.* Nanjing, China: Publishing House of Hehai University (in Chinese).

Li, J., D. Shen, and X. Jiang. (2004). The situation, management and standard establishment of microbial fertilizers in China. *Sulfur Compound Fertilizer and Sulfuric Acid Information.* (1-2): 2-4 (in Chinese).

Li, Q., Z. Zhu, and T. Yu, eds. (1998). *The Fertilizer Issue in Chinese Agricultural Sustainability.* Nanchang, China: Jiangxi SciTech Press (in Chinese).

Li, Y., J. Hu, and C. Ge. (1989). A review of leguminous root-nodule bacterium in 40 years. *Soil and Fertilizer.* 4: 45-48 (in Chinese).

Li, Z. (1999). The study and application of N recommendation on the basis of soil and plant tests. A PhD dissertation of Post-Graduate College, China Agriculture University (in Chinese).

Li, Z. and Q. Zhu. (1991). *Micronutrients.* Nanjing, China: Jiangsu SciTech Press (in Chinese), pp. 100-108.

Lin, B. (1991). Bring the fertilizer yield-increase effect into full play. In Chinese Society of Soil Science (ed.) *The Situation and Prospect of Chinese Pedology.* pp. 29-36. Nanjing, China: Jiangsu SciTech Press (in Chinese).

Lin, B. (1998). China fertilizer structure, evolution of effectiveness, problems and related strategies. In Li, Q., Z. Zhu, and T. Yu (eds.) *The Fertilizer Iissue in Chinese Agricultural Sustainability.* pp. 12-17. Nanchang, China: Jiangxi SciTech Press (in Chinese).

Lin, B. and J. Li. (1989). The evolution of fertilizer effect in 50 years in China and balanced fertilization. In The Soil and Fertilizer Institute, CAAS (ed.) *The Proceedings of International Balanced Fertilization Symposium.* pp. 43-51. Beijing, China: Agro-SciTech Press (in Chinese).

Lin, B., J. Lin, and J. Li, eds. (1996). *The Variation of Crop Yield and Soil and Fertilizer in Long-Term Fertilization Condition.* Beijing: China Agro-SciTech Press (in Chinese).

Liu, G. (1992). The recycling of nutrient elements and agriculture sustainability. *Journal of Soil Science.* 29(3): 251-256 (in Chinese).

Liu, G. and W. Jin, eds. (1991). *The Organic Manure in China.* Beijing: China Agriculture Press (in Chinese).

Liu, S. (1996). *China Geography.* Beijing: Higher Education Press (in Chinese).

Lu, D., Y. Dong, and B. Sun. (1998). A study on effect of nitrogen fertilizer use on environment pollution. *Plant Nutrient and Fertilizer Science.* 4: 8-15 (in Chinese).

Lu, D., X. Yang, H. Zhang, W. Dai, and W. Zhang. (1996). A study on the characteristics of movement and leaching loss of NO_3-N in Lou Soil in Shaanxi and its influencing factors. *Plant Nutrient and Fertilizer Science.* 2: 289-296 (in Chinese).

Lu, R., ed. (1998). *The Fundamentals of Pedo-Plant Nutrition Science and Fertilization.* Beijing: Chemical Industry Press (in Chinese).

Lu, Y. (1999). In situ study of NO_3-N leaching in paddy soil of Taihu Valley. *Journal of Soil Science.* 30: 113-114 (in Chinese).

Ma, C., F. Yang, X. Gao, and L. Chen, eds. (2002). *A Historical Review of China Chemical Fertilizers in Last 100 Years.* Beijing: China Agro-SciTech Press (in Chinese).

NATESC (China National Agro-Technical Extension and Service Center), ed. (1999). *The Organic Manure Resources in China. Beijing*: China Agriculture Press (in Chinese).

The National Soil Survey Office, ed. (1998). *China Soil.* Beijing: China Agriculture Press (in Chinese).

NCFEN (National Chemical Fertilizer Experimental Network). (1986). Evolution of effectiveness of NPK and major practices for increasing crop yield in China. *Soil and Fertilizers.* 1: 1-8; and 2: 1-8 (in Chinese).

Peng, L., X. Peng, and Z. Lu. (1981). The seasonal variation of soil NO_3-N and the effect of summer fallow on the fertility of manured loessial soil. *Acta Pedological Sinica.* 18: 212-222 (in Chinese).

SFI (The Soil and Fertilizer Institute, CAAS), ed. (1986). *Chemical Fertilizer Regionalization in China.* Beijing: China Agriculture Press (in Chinese).

SFI (The Soil and Fertilizer Institute, CAAS), ed. (1994). *The Fertilizer in China* pp. 117-119, 132, 390-430. Shanghai: Shanghai SciTech Press (in Chinese).

Shoji and H. Kanno. (1994). Use of poloylefin-coated fertilizer increasing fertilizer efficiency and reducing nitrate leaching and nitrous oxide emissions. *Fertilizer Research.* 39: 147-152.

Sino-Arabic Chemical Fertilizer Corporation. (2004). Patent of agricultural means of production: The fertilizer of SACF compound formula. *China Cooperation Newspaper,* January 29, p. 2 (in Chinese).

Sun, K., G. Zhang, J. Yao, Y. Wang, and W. Qiao. (1999). Effect of continuous application of fertilizer on the crop yields and the accumulated NO_3-N in soil profiles. *Journal* of *Soil Science.* 30: 262-264 (in Chinese).

Sun, S. and C. Zhang. (2000). Nitrogen distribution in the lakes and lacustrine of China. *Nutrient Cycling in Agroecosystems.* 57: 23-31.

Sun, Z., X. Liu, and S. Yang. (1994). A study on nitrogen in precipitation and percolation of soil in Beijing region. *Soil and Fertilizers.* 2: 8-10 (in Chinese).

Tang, J. (1996). The second nationwide soil survey and scientific fertilization. *Proceedings of International Fertilizer and Agricultural Development Conference.* Beijing, China. pp. 38-44. Beijing: China Agro-SciTech Press (in Chinese).

Wang, L. and J. Li, eds. (2003). *Cropping Science.* Beijing: Science Press (in Chinese).

Wang, R., D. Huang, and Y. Cui. (2000). Causes of soil fertility changes in Northeast China—A report from soil fertility monitoring project. *Soil and Fertilizer.* 6: 8-13 (in Chinese).

Wu, C. (1998). *China Economic Geography.* pp. 6-91. Beijing: Science Press (in Chinese).

Xie, J., J. Zhou, and R. Hardter, eds. (2000). *Potash and Agriculture in China.* Nanjing, China: Publishing House of Hehai University (in Chinese).

Xing, G. (1998). N_2O emission from cropland in China. *Nutrient Cycling in Agroecosystems.* 52: 249-254.

Xing, G. and X. Yan. (1999). Direct nitrous oxide emission from agricultural fields in China estimated by the revised 1996, IPPC guidelines for national greenhouse gases. *Environmental Science and Policy.* 2: 355-361.

Xing, G. and Z. Zhu. (2000). An assessment of N loss from agricultural fields to the environment in China. *Nutrient Cycling in Agroecosystems.* 57: 67-73.

Xing, G. and Z. Zhu. (2001). The environmental consequences of altered nitrogen cycling resulting from industrial activity, agricultural production, and population growth in China. *The Scientific World.* 1(S2): 70-80.

Xing G. and Z. Zhu. (2002). Regional nitrogen budgets for China and its major watersheds. *Biogeochemistry.* 57/58: 405-427.

Xiong, Z. (2002). Quantitative evaluation and environmental impact of fate of fertilizer N in farmland under rice-wheat rotation system. A PhD Dissertation of Post-Graduate College, Chinese Academy of Sciences. pp. 20-49 (in Chinese).

Yi, X. and C. Xie. (1993). Pollution caused by the leaching N application rate in Lou soil. *Agro-Environmental Protection.* 12: 250-253 (in Chinese).

Yuan, F., Z. Chen, Z. Yao, C. Zhou, Y. Song, and X. Li. (1995). NO_3-N transformation accumulation and leaching loss in surface layer of Chao Soil in Beijing. *Acta Pedologica Sinica.* 32: 388-399 (in Chinese).

Zhang, F. (2000). *Industrialization is a Trend of Organic Manure Development.* (unpublished) (in Chinese).

Zhang, F. (2003). The function of organic manure in non-polluted agricultural production. In Lin, B. (ed.) *Chemical Fertilizer and Non-polluted Agriculture.* pp. 141-158. Beijing: China Agriculture Press (in Chinese).

Zhang, F., Z. Cao, and X. Xiong. (1984). The dynamic process of leaching N in soil-plant systems studies with [15]N. *Environmental Science.* 1: 2-22 (in Chinese).

Zhang, W., Z. Tian, N. Zhang, and X. Li. (1995). A report on nitrate pollution of groundwater caused by application of N fertilizers in North China. *Plant Nutrition and Fertilizer Sciences.* 1(2): 80-87 (in Chinese).

Zhang, Z., S. Bi, J. Li, Y. Feng, and W. Wu. (1997). Characteristics and application benefit of long-term effective ammonium carbonate. *Science Bulletin.* 42(8): 874-878 (in Chinese).

Zhou, J., X. Chen, J. Xie, and R. Hartder. (2000). The nutrient balance status and management countermeasures in ecological systems in Chinese farmland. In Zhou, J. (ed.) *The Balance and Management of Nutrients in Farmland.* pp. 42-56. Nanjing, China: Publishing House of Hehai University (in Chinese).

Zhu, J., Y. Han, G. Liu, Y. Zhang, and X. Shao. (2000). Nitrogen in percolation water in paddy fields with a rice/wheat rotation. *Nitrogen Cycling in Agroecosystems.* 57: 75-82.

Zhu, Z. and D. Chen. (2002). Nitrogen fertilizer use in China—Contributions to food production, impacts on the environment and best management strategies. *Nutrient Cycling in Agroecosystems.* 63: 117-127.

Chapter 9

Integrated Nutrient Management: Experience from Rice-Based Systems in Southeast Asia

Dan C. Olk*
Mathias Becker
Bruce A. Linquist
Sushil Pandey
Christian Witt

INTRODUCTION

This chapter will focus on rice (*Oryza sativa* L.)-based systems, as this single crop dominates agriculture in Southeast Asia in a manner that has few parallels worldwide. When defined as including Cambodia, Indonesia, Lao PDR, Malaysia, Myanmar, Philippines, Thailand, and Vietnam, Southeast Asia accounts for about 8 percent of the world's population but a quarter of the global rice and grain production area (Maclean et al., 2002). Sixty-nine percent of its arable land is planted to rice when multiple annual crops are counted separately. Southeast Asian nations have seven of the eight highest means worldwide for percent of total caloric uptake that is derived from rice; the overall mean for the seven nations (all except Malaysia) is 60 percent, compared to a global mean of 21 percent.

Rice is grown in Southeast Asia primarily in rainfed fields or irrigated lowland fields. Rainfed rice accounts for 57 to 85 percent of total rice area

---

*This chapter was written as part of Dan Olk's official duties as an employee of the U.S. government.

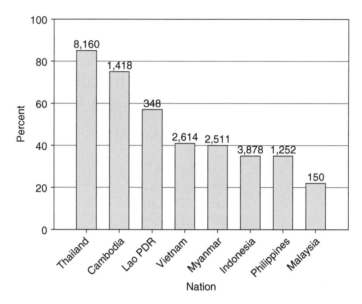

FIGURE 9.1. Rainfed lowland rice area as a percentage of total rice area for each Southeast Asian nation. Value above each bar is the area (thousands of hectares). *Source:* Adapted from Maclean et al. (2002).

in Cambodia, Lao PDR, and Thailand, and 2,241 percent in the other five nations (Figure 9.1; Huke and Huke, 1997). Corresponding proportions of irrigated lowland rice are 7 to 16 percent and 51 to 65 percent, respectively. Upland rice and deepwater rice each accounts for 10 percent or less of total rice area in any country except for Lao PDR. Average grain yields are highest in the irrigated (4.7 t ha^{-1}) and rainfed systems (2.9 t ha^{-1}), while yields in the other systems are less than 2 t ha^{-1}. The irrigated rice systems account for nearly 60 percent of the rice production in Southeast Asia, followed by rainfed systems with about 30 percent, whereas flood-prone and upland systems account for only 5 percent each. Fertilizer is generally not applied in upland rice cropping. Hence this discussion of integrated nutrient management (INM) will address only rainfed rice and irrigated lowland rice.

Management of nitrogen (N) fertilizer in rice production is complicated by the presence of floodwaters during some portion (rainfed rice) or all (irrigated lowland) of the growing season. Inorganic N is not stable in floodwaters for more than a few weeks, as nitrate will be denitrified as N gases, and ammonium will be volatilized as ammonia. Consequently, application of N fertilizer should be need-based and in synchrony with crop demand. In most rice soils, the soil indigenous supplies of phosphorus (P) and potassium (K) are insufficient to maintain high yields, particularly in the irrigated rice systems. Therefore fertilizer has to be applied at rates designed to overcome nutrient deficiencies and maintain adequate soil levels. Phosphorus, and in most cases K, can be applied at one time, early in the growing season, in contrast to N.

Integrated nutrient management is most commonly considered to be the simultaneous use of organic and inorganic fertilizers. Currently in Southeast Asia, this practice is more common in rainfed rice than in irrigated lowland rice; hence this chapter emphasizes rainfed rice. We define site-specific nutrient management as the calculation of fertilizer rates, to be supplementary to the estimated supplies of indigenous nutrients. Thus far this approach has been attempted primarily in irrigated lowland rice and for application of inorganic fertilizers. Both INM and site-specific nutrient management have primarily involved N, P, and K, so this chapter will not address management of secondary macronutrients or micronutrients.

The goals of rice farmers often differ between those in the rainfed rice and irrigated rice ecosystems. Rainfed rice farmers are generally subsistence-oriented and poorer than their lowland counterparts, having fewer resources and limited access to credit (Zeigler and Puckridge, 1995). The primary goals of the rainfed rice farmer are yield stability and risk (i.e., drought) avoidance, even if these goals preclude maximum yields (Dobermann and White, 1999). The irrigated lowland rice farmer, by comparison, is more likely to be a commercial producer, and so strives for high grain yields and increased profit. Implementation of INM in either ecosystem must be coordinated with the respective objectives. Yet in both ecosystems, nutrient management strategies must maintain or increase grain yields for at least the foreseeable future: the population of Southeast Asia is projected to increase by 54 percent from 1999 to 2050 (Maclean et al., 2002).

INTEGRATED NUTRIENT MANAGEMENT
IN RAINFED RICE

Rainfed lowland rice is grown on level to slightly sloping bunded fields (i.e., surrounded by small levees that retain floodwaters) with noncontinuous flooding of variable depth and duration (Zeigler and Puckridge, 1995). It occupies about 46 Mha, or 35 percent of the global rice area, mostly in South and Southeast Asia (Maclean et al., 2002). Rainfed lowland rice accounts for 82 percent of the total rice area in Thailand, Cambodia, and Laos, and these lands represent half of the total rainfed rice area in Southeast Asia

(Figure 9.1). Most of the published literature on rainfed lowland rice systems in Southeast Asia reports research conducted in these three countries, which thus forms the basis for this section. Reference will also be made on occasion to research conducted in smaller yet still significant areas of rainfed lowland rice in Vietnam, Malaysia, and Indonesia.

Surface hydrology strongly influences nutrient cycling and rice crop growth in rainfed lowland fields. It is most favorable for rice production in Laos and least favorable in Northeast Thailand. As discussed by Bell and Seng (2004), rainfall in the main rice growing areas is on average lowest in Northeast Thailand and highest in Laos (Table 9.1). The surface layers of

TABLE 9.1. Comparison of soils and rainfall in Northeast Thailand, Cambodia, and Laos

Country/region	Soils	Average annual rainfall
Northeast Thailand	Sandy textured soils are more common, and soils generally have lower clay contents than those of Laos and Cambodia. Also, coarse textured subsoils are more common. For example, in the Korat Plateau, which makes up most of Northeast Thailand, the Roi Et soil series makes up 56 percent of the lowland rice area. This soil has sandy loam surface and subsoils.	1,100-1,500 mm, except along the Lao border where it is similar to Laos
Cambodia	The dominant soil, which covers 25-30 percent of the lowland rice area, is a Prateah Lang (White et al., 1997)— a sandy surface soil over a loam or clay subsoil. Lowland rice soils (Prey Khmer) that have sandy surface soils and subsoils occupy 10-12 percent of the total area. The remaining soils have loam or clay surface and subsurface horizons.	1,250-1,750 mm in the main rice growing region (Nesbitt, 1997)
Laos	In the south and central regions where over 80 percent of the rice is grown, 68 percent of the surface soils of lowland rice are coarse textured (i.e., a clay content no greater than that of a sandy loam soil). More than 43 percent of the soils are classified as acrisols or alisols (FAO), which are soils that have a loam or clay B horizon (data from Soil Survey and Land Classification Center, Laos).	1,400-2,250 mm in the main rice growing area of central and southern Laos (Linquist and Sengxua, 2001)

lowland rice soils of Northeast Thailand are sandier and more permeable than those of Cambodia and Laos (Table 9.1). A significant number of soils in both Northeast Thailand and Cambodia also have sandy subsoils, further increasing water percolation. Rice soils of Northeast Thailand are likely to undergo flooding and drying cycles much more frequently than those of Laos and Cambodia (Fukai et al., 1998). The rainfed lowland rice soils of Vietnam, Malaysia, and Indonesia are generally relatively fine-textured and receive adequate rainfall, resulting in favorable surface hydrology.

General Considerations for INM in Rainfed Lowland Rice

Soil and Water Limitations

As in irrigated systems, N is the most limiting nutrient in rainfed lowland rice systems (Linquist et al., 1998; Wade et al., 1999). The second most limiting nutrient is usually P (Linquist et al., 1998), although in many areas P has been overapplied to the extent that currently there is no response to applied P. Potassium deficiency is less widespread, but responses to K are more common on sandy soils. In Cambodia, for example, K applications are recommended on all sandy soils (Dobermann and White, 1999). Sulfur deficiencies have also been reported in Laos, Cambodia, and Northeast Thailand (Blair et al., 1991; CIAP, 1996, 1997; Linquist et al., 1998); however, the extent of these deficiencies is not well-known. Little is known regarding micronutrient deficiencies in these soils. Rainfed lowland rice is grown at lower yield levels than is irrigated rice, therefore micronutrient deficiencies may not be as apparent. Zinc, which is deficient in some Asian rice soils (Randhawa et al., 1978), is a more likely problem for high pH soils than acidic soils.

Rice bronzing has been reported in Cambodia (Seng et al., 2001) and was observed by the authors in Laos. Although the loss of grain yields caused by this problem has not been confirmed, the disorder appears to be similar to that described for rice in Japan and Nigeria (Yamauchi, 1989). In Cambodia, anecdotal evidence suggests that the problem is more prevalent and more severe with the increased use of inorganic N and P fertilizers without any K fertilizer. Removal of rice straw from the field limits the return of K in soils receiving no or low rates of applied K.

Rainfed lowland rice soils in Thailand, Cambodia, and Laos are characterized as acid, sandy, and drought-prone, with a high degree of spatial and temporal variability in water availability (Zeigler and Puckridge, 1995). At the farm level, soil clay and soil organic matter (SOM) levels are associated with small changes in elevation (1 to 3 m) with mini water catchments.

Positions high in the toposequence tend to have lower levels of clay and SOM and to be more drought-prone (Homma et al., 2001, 2003). Farmers manage such variation by using different crop varieties and crop management strategies.

Periods of flooding and drying are common in rainfed soils. Flooding (anaerobic) and drying (aerobic) cycles affect the soil oxidation-reduction state and have large implications for the availability of several nutrients, as reviewed by Ponnamperuma (1972) and Kirk (2004). The resulting fluctuations in nutrient availability are heightened in sandy soils, which often have low contents of SOM and cation exchange capacity, and hence poor buffering capacities against pH changes that are caused by changes in the redox potential. As soils become oxidized during a drying period, the pH drops rapidly due to poor buffering capacity, which can trigger Al toxicity and therefore poor crop responses to fertilizers. Management of organic matter in these soils is essential to maintain cation exchange capacity (Ragland and Boonpuckdee, 1988; Willet, 1995). Wetting/drying fluctuations also have major effect on the accumulation and dissipation of soil mineral N (Buresh and De Datta, 1991), potentially increasing N losses compared to the loss of other nutrients (Schnier, 1995).

Wetting and drying also significantly affect P availability, which increases during the flooded phase and decreases during the aerobic phase. These changes correspond to the release of ferrous iron when the soil is waterlogged and the oxidation and precipitation of iron when the soil dries. Under repeated alternate flooding and drying cycles, soluble P is converted to unavailable iron phosphate forms at the expense of the soluble and aluminum phosphate forms (Patrick and Mikkelsen, 1971), resulting in inhibited crop uptake of P (Seng et al., 2004).

Availability of Organic Materials

In rainfed lowland rice systems, potential organic amendments differ widely in their availability and practicality of application. Rice straw and rice husks from the previous rice crops are both readily available. Animal manure may be available if the farmer or nearby farmers raise livestock. Off-farm sources of organic matter include leaf litter (Whitbread et al., 1999). Farmers can also grow their own organic fertilizers, such as the green manures that are discussed in the subsequent section.

Straw accounts for approximately 60 percent of the aboveground biomass and is probably the most available on-farm residue. Rainfed rice fields are usually left fallow during the dry season. Depending on variety and farmer practices, about half of the rice straw remains in the field following harvest. Livestock commonly graze this stubble straw during the dry season, but occasionally it is burned. The panicle straw, which is removed with the grain at harvest, is moved to a central location for threshing. Subsequently it is commonly burned or fed to livestock, despite the potential benefits of its return to the soil. While the benefits of straw applications have been noted, farmers find it bulky to transport and difficult to incorporate.

Rice husks account for about 20 percent of unmilled rice (Juliano and Bechtel, 1985) or about 10 percent of aboveground biomass. While they are bulky, rice husks are much easier to incorporate than straw. Yet they are usually left at the rice mills.

The number of livestock owned by farmers varies widely between and within countries. Given the increased use of tractors, many farmers no longer own buffaloes that have traditionally been used for field preparation. Therefore, farmyard manure (FYM) may not be an option for many farmers. The labor requirement for manure collection and application depends on whether the animals are free grazing, penned, or tethered.

Developing Appropriate INM Approaches
for Rainfed Lowland Rice

The following environmental and socioeconomic limitations should be considered when developing INM approaches for rainfed lowland rice. First, farmers need suitable management practices to account for the spatial variability in soil-water conditions that exist at the farm level. Second, nutrient management strategies must allow adjustments during the season as conditions change (Dobermann and White, 1999). Third, strategies need to improve fertilizer-use efficiency at suboptimal fertilizer rates. Finally, available organic amendments should be efficiently used and recycled.

The need for INM approaches that are specific to rainfed rice can be illustrated by examining management strategies for fertilizer N. Strategies for rainfed lowland rice are commonly derived from those for irrigated rice, which raises some concerns regarding their suitability. First, the objective of N management in irrigated rice is to optimize farm profits through increased yields. In the rainfed environment, N is typically applied at rates well below the yield-maximizing rates, due to risk and limited capital. Second, N fertilizer recommendations for irrigated rice can be based solely on the crop growth stage and not on precipitation patterns, as the fields are continually flooded, which enables crop uptake of applied N and minimizes denitrification losses regardless of application time. In the rainfed environment, water is not controlled and soil water conditions vary depending on rainfall. Thus, it may not be possible to apply N at the recommended growth stage.

Hence windows of opportunity are needed for optimal N application in rainfed rice. Based on research in Laos et al. (2003), split applications at transplanting, active tillering, and panicle initiation are recommended. This practice increased yields by 12 percent compared to a single application at transplanting. The benefit of fertilizer application was further increased when a higher proportion of the N was applied in synchrony with peak crop N demand at active tillering and panicle initiation. The authors recommended that the first N application be within the first thirty days after transplanting, and the final N application be from two weeks before, to one week after panicle initiation. Such "windows" provide the needed flexibility so that the farmer can meet crop N demand while applying N under flooded conditions to reduce denitrification losses.

Crop Response to Inorganic Fertilizers and Their Management

Yield response of rainfed rice to inorganic fertilizers varies considerably among countries. Different studies used widely varying fertilizer rates, complicating a synthesis of their crop response data, and recommendations for optimal fertilizer management. A basis for comparison can be obtained from the agronomic efficiency (AE) of fertilizer use, which is a unifying parameter across studies. This term is defined as the ratio of grain yield (kg) increase with N fertilizer application to the amount (kg) of N fertilizer applied.

In Northeast Thailand, Willet (1995), Ragland and Boonpuckdee (1988), and Wade et al. (1999) reported limited and erratic yield responses to inorganic fertilizers (Tables 9.2 and 9.3). Wade et al. (1999) compared fertilizer

TABLE 9.2. Rice yields from different treatments on a sandy Typic Paleaquult in Northeast Thailand

Treatment	Rice yield (kg ha^{-1})
(A) Control	1,775
(B) 156 kg ha^{-1} 16-16-8 + 95 kg ha^{-1} ammonium sulphate	1,888
(C) 18.8 t ha^{-1} compost	1,650
(B) + (C)	2,519

Source: Willet (1995). Role of organic matter in controlling chemical properties and fertility of sandy soils used for lowland rice in northeast Thailand. In *Soil organic matter management for sustainable agriculture,* R.D.B. Lefroy, G.J. Blair, and E.T. Craswell (eds.). ACIAR Proceedings No. 56, 109-114. Reprinted with permission.

TABLE 9.3. Average rice grain yield responses to NPK fertilizer and FYM

Environment grouping[a]	No. sites	NPK applied (kg ha^{-1})	FYM applied	Rice yields (t ha^{-1})			AE from NPK (kg kg^{-1})
				No fertilizer	FYM	NPK fertilizer	
Thailand-1	9	50-22-42	10 t ha^{-1} cattle manure + lime	1.94	3.05	2.69	6.6
Thailand-2	13			1.56	1.77	1.92	3.2
Philippines-1	5	140-17-33	8 t ha^{-1} chicken manure	1.99	4.69	5.01	15.9
Philippines-2	1			3.43	3.85	4.27	4.4
Indonesia-1	2	120-19-50	5 t ha^{-1} FYM	1.51	2.03	4.57	16.2
Indonesia-2	2			3.49	5.49	6.75	17.2

Source: Based on data from Wade et al. (1999).

[a]Groupings are the result of pattern analysis that involved a joint application of cluster analysis and ordination to a standard matrix (see Wade et al. (1999)).

responses across a number of countries and environments and found the poorest response was in Northeast Thailand (Table 9.3). At thirteen sites (Thailand-2) the AE of applied fertilizer was only 3.2 kg kg^{-1} and on average yields increased with fertilizer application by 0.36 t ha^{-1}. In the Thailand-1 group, the AE was 6.6 kg kg^{-1} and the yield increase was 0.75 t ha^{-1}. In the same study, the AE in the Philippines and Indonesia averaged 16 to 17 kg kg^{-1}, respectively, at most sites. Mean values in the Wade et al. (1999) study mask erratic responses across years for some sites. Similarly in Cambodia, the AE ranged from 4 to 10 kg kg^{-1} in 124 on-farm trials on four soil types (CIAP, 1996, 1997; Table 9.4).

In Laos, inorganic fertilizer use is generally more efficient than in Thailand and Cambodia. For example, one experiment was conducted in Laos at two sites with sandy soils to evaluate the rice yield response to inorganic fertilizers and on-farm residues (2 t ha^{-1} of manure, rice straw, or rice husks) (Linquist and Sengxua, 2001, 2003). At the Champassak site, yields increased by 1.5 t ha^{-1} (126 percent increase) in the first two years in response to inorganic fertilizer, and by 0.7 t ha^{-1} (43 percent increase) in the third year (Figure 9.2). At the Saravane site, yields increased by 1.7 t ha^{-1}

TABLE 9.4. Rice yields with and without the recommended fertilizer rate and the agronomic efficiency (AE, ratio of yield increase with fertilizer application to kg fertilizer applied) of all fertilizer nutrients (N-P-K) applied

Soil type	No. of sites	Recommended fertilizer rate kg ha^{-1} of N-P-K	Yield with no fertilizer t ha^{-1}	Yield with recommended rate t ha^{-1}	Yield increase t ha^{-1}	AE kg grain kg^{-1} nutrient
Toul Samrong	22	86-30-10	2.78	3.61	0.83	6.62
Prey Khmer	26	19-12-26	1.95	2.51	0.56	9.84
Bakan	37	73-64-20	2.78	3.37	0.59	3.76
Prateah Lang	39	67-24-16	2.37	3.13	0.77	7.18

Source: Calculated from CIAP (1996, 1997).

Note: Data represent a total of 124 on-farm experiments conducted in Cambodia between 1996 and 1997. Varieties were all modern and recommended for that site.

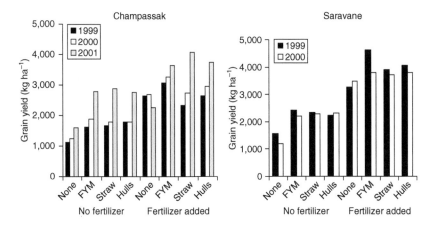

FIGURE 9.2. Rice yield response to annual applications of fertilizers (60, 13, and 18 kg ha⁻¹ of N, P, and K, respectively) and on-farm residues (FYM, rice straw, or rice hulls, all 2 t ha⁻¹ dry weight) at Champassak and Saravane, Laos, during the 1999, 2000, and 2001 wet seasons. *Source:* Adapted from Linquist and Sengxua (2001, 2003).

(107 percent increase) and 2.3 t ha⁻¹ (192 percent increase) in the first two years, respectively. The AE ranged from 8 to 25 kg kg⁻¹, similar to other studies of irrigated rice and rainfed rice (Linquist and Sengxua, 2003).

The cause of such variable responses to inorganic fertilizers is not fully understood but may be due to any of several reasons. First, as discussed above, the relatively low rainfall and broad extent of sandy soils in Thailand promote the occurrence of wetting and drying cycles, which can have major effects on the oxidation-reduction state of these low organic matter soils, and can reduce fertilizer use efficiency. Also, Wade et al. (1999) suggested that the poor crop response to nutrients found in their study of Northeast Thailand was due to the widespread use of an improved traditional variety (KDML 105); this type of variety often fails with exposure to drought, submergence, or prolonged waterlogging.

Interactions Between Inorganic and Organic Fertilizers

Mechanisms for Improved Nutrient Availability with Organic Fertilizers

Application of organic fertilizers can improve nutrient management through at least two mechanisms, especially for the coarse-textured, droughty soils that are common in rainfed lowland rice systems. First, SOM is a primary source of the soil cation exchange capacity, and regular incorporation of organic fertilizers will normally increase the level of SOM. For example, Naklang et al. (1999) reported that annual applications of leaf litter in Northeast Thailand increased total soil carbon from 24 to 37 percent (from 3.51 to 4.78 mg g^{-1}) and more than doubled labile soil carbon during a five-year period. Hence addition of organic fertilizers will normally increase the soil cation exchange capacity, thereby strengthening the soil buffering capacity against pH changes caused by oxidation-reduction reactions (Willet, 1995; Ragland and Boonpuckdee, 1988). Such changes in soil chemistry are likely to result in more efficient utilization of applied inorganic nutrients and synergistic benefits to the crop. In support of this idea, Seng et al. (2004) reported that the addition of straw to soils with alternating wet and dry periods decreased soil Al concentrations during the aerobic period and increased P availability by inhibiting soil reoxidation. Consequently, the P uptake by the rice crop increased by an amount that was due not merely to the P content of the added straw.

A second mechanism for improving nutrient management, through organic fertilizer amendment, is the addition of nutrients in the amended material. Most organic fertilizers are low in N and P but high in other nutrients. The amounts and percentages of some plant nutrients in straw and grain fractions at harvest are shown in Table 9.5. Straw contains more than 50 percent

TABLE 9.5. Macro- and micronutrients in rice grain and straw at harvest

	N	P	K	S	Ca	Mg	Mn	Zn	Cu
	Nutrient concentration								
	g kg^{-1}						mg kg^{-1}		
Grain	7.9	1.9	2.8	1.0	0.4	1.0	103	23	39
Straw	3.2	0.4	7.9	1.0	3.9	1.7	884	25	25
	Nutrient per tonne of grain yield (kg t^{-1})[a]								
Grain	7.9	1.9	2.8	0.9	0.4	1.0	0.1	0.02	0.04
Straw	4.8	0.6	11.8	1.4	5.9	2.6	1.3	0.04	0.04
Total	12.6	2.4	14.7	2.4	6.3	3.6	1.4	0.06	0.08
	Percent of nutrient in grain or straw at harvest								
Grain	62	76	19	41	7	28	7	38	51
Straw	38	24	81	59	93	72	93	62	49

Source: Data adapted from Linquist and Sengxua (2001).

[a]Assumes a harvest index of 0.4. Therefore, if rice grain yield is 1 t ha^{-1}, the straw yield would be 1.5 t ha^{-1}.

of the nutrients in aboveground plant biomass for all the nutrients measured except N, P, and Cu. Furthermore, 80 percent or more of crop K, Ca, and Mn remains in the straw at harvest. Annual removal of rice straw can, therefore, rapidly deplete soil nutrient reserves. This is especially a concern for sandy soils where nutrient reserves are inherently low. The exclusive use of N and P fertilizers, as is the common practice by many farmers, will accelerate the decline in soil fertility, especially in sandy soils. Organic fertilizer additions can add macro- and micronutrients not commonly found in inorganic fertilizers (Table 9.6). In the long term, application of either organic fertilizers or a comprehensive set of inorganic fertilizers may be required to maintain soil nutrient balances for sustainable crop production.

Field Observations of Integrated Application with Inorganic and Organic Fertilizers

When inorganic and organic fertilizers are applied simultaneously, the crop response can reflect any of three possible interactions: (1) no response to one fertilizer type if the other has already been applied; (2) an additive benefit; and (3) a synergistic benefit that exceeds the sum of individual benefits. At locations where rainfall is abundant, and favorable soil conditions mitigate against flooding and drying cycles, there is commonly a good response to either organic or inorganic fertilizer applied alone. When applied together, there is either no additional benefit or the benefits are additive. For example, in the Lao study by Linquist and Sengxua (2001, 2003) that found healthy yield responses to inorganic fertilizers, when on-farm organic resi-

TABLE 9.6. Nutrient concentration of some on-farm residues

Residue	N (g kg^{-1})	P (g kg^{-1})	K (g kg^{-1})	S (g kg^{-1})	Ca (g kg^{-1})	Mg (g kg^{-1})	Mn (mg kg^{-1})
Rice straw	3.2	0.4	7.9	1.0	3.9	1.7	884
Rice husks[a]	4.3-5.5	0.3-0.8	1.7-8.7	0.5	0.7-1.5	0.3	116-337
FYM[b]	5-10	1.2-1.7	2.2-2.6	na	na	na	na
Cattle dung[b]	3.5	1.1	0.9	na	na	na	na
Cattle urine[b]	8	0.2	2.6	na	na	na	na

Note: na = not available.

[a]Juliano and Bechtel (1985).

[b]Uexkull and Mutert (1992).

dues were applied alone, rice yields increased on average by 0.6 t ha^{-1} (48 percent increase) at Champassak in the first two years and by 1.2 t ha^{-1} (78 percent increase) in the third year (Figure 9.2). At Saravane, yields increased by 0.8 t ha^{-1} (47 percent increase) in the first year and 1.1 t ha^{-1} (90 percent increase) in the second year. Joint application of organic residues with inorganic fertilizers, though, did not significantly increase yields compared with inorganic fertilizers alone. In the third year, the benefits of fertilizer and on-farm residues were additive at the Champassak site, suggesting an evolving benefit of adding on-farm organic residues.

Results from Cambodia and Indonesia showed additive benefits when organic materials and inorganic fertilizers were both applied. On a coarse-textured soil in Java, Indonesia, Wihardjaka et al. (1998) reported that the addition of 5 t ha^{-1} of FYM alone or inorganic NPK fertilizer alone increased yields on average by 1.7 and 2.1 t ha^{-1}, respectively, during three years (Figure 9.3). When both fertilizer types were applied simultaneously, the yield response was additive, with yields increasing on average by 3.1 t ha^{-1}. On a fine-textured soil in Cambodia, rice yields increased by 0.9 t ha^{-1}

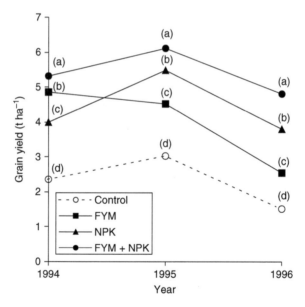

FIGURE 9.3. Rice yields from Java, Indonesia for three years where FYM (at 5 t ha^{-1}) and NPK (120-18-75 kg ha^{-1}, respectively) were applied in each season. Common letters in a given year indicate means are not significantly different by LSD$_{0.05}$. *Source:* Adapted from Wihardjaka et al. (1998).

(from 2.9 to 3.8 t ha^{-1}) with the addition of 10 t ha^{-1} of FYM and by 1.8 t ha^{-1} with the addition of NPK (Figure 9.4) (CIAP, 1997). When FYM and NPK were applied together, plus sulfur fertilizer, a presumably additive benefit resulted with yields increasing by 2.5 t ha^{-1}. The applied sulfur may have also contributed to the benefit.

In Northeast Thailand, by contrast, Ragland and Boonpuckdee (1988) and Willet (1995) reported failed attempts to increase rice yields on sandy soils through the conventional application of inorganic fertilizers alone. Yet strong synergistic benefits occurred with the application of both organic and inorganic amendments (Table 9.2). Little additional field evidence exists for such positive interactions. A number of other studies in Northeast Thailand showed generally poor responses to organic fertilizers applied alone. As discussed above, Wade et al. (1999) reported poor responses to inorganic and organic fertilizers applied separately (Table 9.3). Wonprasaid et al. (1996) compared several varieties in their response to FYM (6.25 t ha^{-1}) applied to very infertile, droughty rainfed soils; yield responses were only 0.3 to 0.5 t ha^{-1}. Supapoj et al. (1998) reported that rice straw (applied at rates of 6.25 to 18 t ha^{-1}) and rice husk (applied at 3.13 t ha^{-1}) applications increased rice yields by 10 percent to 15 percent in Northeast Thailand, much less than observed in Laos for similar amendments of on-farm residues (Figure 9.2).

Also in Northeast Thailand, Whitbread et al. (1999) evaluated the annual application (1.5 t ha^{-1}) of tree leaf litter to rice fields. All leaf litters had a positive effect on rice yields in each year, and rice yields increased on average by 43 percent in five years (Figure 9.5). Addition of *Cajanus cajan* leaf

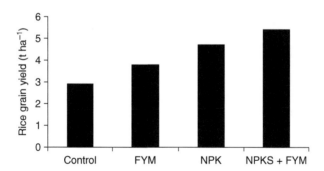

FIGURE 9.4. Rice yields from an experiment in Cambodia following application of farm yard manure (FYM, at 10 t ha^{-1}) and N, P, K, and S fertilizer (115-50-50-30 kg ha^{-1}). *Source:* Data calculated from CIAP (1997).

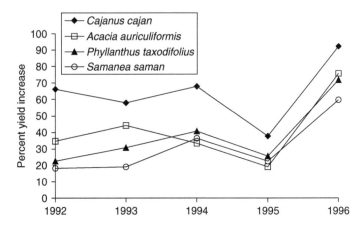

FIGURE 9.5. Percentage rice yield increase (compared to a control with no leaf litter) due to addition of tree leaf litter (1.5 t ha^{-1} annually). *Source:* Data calculated from Whitbread et al. (1999).

litter resulted in the highest rice yields, which was attributed to high nutrient concentrations of N, P, K, and S in the leaf biomass.

Managing Variability at the Farm Scale

These regional differences in response to organic amendments and inorganic fertilizers may provide insights into better nutrient management at the farm level. Specifically, variations in surface hydrology at the farm level can override the effect of rainfall patterns (Bell and Seng, 2004). Upper terrace soils have lower clay contents and fewer days with standing floodwaters than lower terrace soils (Fukai et al., 2000; Oberthür and Kam, 2000; Homma et al., 2003). Accordingly, upper terrace soils seem less likely to respond to either organic materials or inorganic fertilizers applied alone, but if both fertilizer types are applied synergistic interactions are possible. In the middle to lower toposequence positions, additive benefits seem more likely than synergistic interactions when organic fertilizers and inorganic fertilizers are applied together. These hypothesized extrapolations are, however, untested.

GREEN MANURES

Green manures have historically played an important role in lowland rice of Southeast Asia, and a wide range of soil-improving plant species has been cultivated throughout the region. Until the 1970s, the use of green manures was often the sole means for low-input farmers to improve the N nutrition of rainfed lowland rice. While no direct area estimates are available, extrapolation of several reports (compiled by IRRI, 1988) indicates that at least 10-15 Mha were once planted to green manure legumes in Southeast Asia alone. The fern *Azolla* was traditionally used for symbiotic N fixation with the cyanobacterium *Anabaena azollae* in the seasonally wet tropics and the subtropical regions of China and Vietnam, covering an estimated 7 Mha (Watanabe, 1991). Numerous flood- and drought-tolerant legumes were cultivated throughout the region for grain, fodder, or solely for soil improvement purposes. With the advent and rapid spread of relatively cheap mineral fertilizers, the area under green manures declined (Roger and Watanabe, 1986) and was estimated to be less than 5 Mha in the late 1980s (Rosegrant and Roumassett, 1988) and less than 3 Mha by 1995 (Garrity and Becker, 1995). With the recent opening of market economies in Vietnam and China, traditionally green manure countries, consumption of mineral N fertilizers has increased dramatically while the use of green manures in Southeast Asia has declined to possibly less than 1 Mha today (Becker, 2003).

Concerns mount over agricultural sustainability as most Asian countries move into a post–Green Revolution phase (Cassman and Pingali, 1994). During the past three decades, cropping systems have evolved from a single low-input crop of rainfed lowland rice to irrigated multiple rice cropping with high external input use, or intensified rainfed systems with lowland rice during the rainy season, and high value cash crops during the dry season (Garrity and Flinn, 1988). A widely observed decline of partial factor productivity, and in some instances even declining yield levels, have been associated with intensified lowland cultivation (Pingali et al., 1990; Cassman, Gines et al., 1996) and poor management of external inputs (Dawe et al., 2000, 2003). When faced with declining crop yield, farmers traditionally opted for the addition of green manures into the farming system (King, 1911). Today, farmers compensate for stagnating or declining yields with even higher use of external inputs (Cassman, Gines et al., 1996; Roger and Watanabe, 1986). Have Asian farmers forgotten an important part of their (agri)cultural heritage, or are green manures indeed a thing of the past?

The Diversity of Green Manure Systems

Green manure systems are highly diverse and can be differentiated by the site of production, the cropping niche, and the ecological requirements of the species. Thus, green manures can be grown either *ex situ* or *in situ*. The use of *ex situ* grown biomass (cut-and-carry system) for lowland rice is limited to a few site-specific production systems and cropping niches. For example, in the framework of the System of Rice Intensification, one recommendation calls for the collection of a *Graminaeae*-dominated biomass from surrounding fallow land for application to the lowland rice fields (Stoop et al., 2002; Uphoff, 2002). In this case, soil nutrients from a large area would be concentrated on a limited land surface to benefit a single rice crop. Obviously *ex situ* production is highly labor-intensive.

In the lowlands of Southeast Asia, green manure systems are predominantly based on *in situ* production and use of biomass, whereby green manures can be fitted into rice-based rotations in either the pre-rice or post-rice phase. Post-rice green manures may still occasionally be encountered in production systems with a single crop of rainfed rice, and they occupy the land during much of the dry season (i.e., *Indigofera tintoria* in the northern Philippines and forage legumes such as *Stylosanthes* and *Desmodium* in central Vietnam). However, much of the green manure use in the rainfed lowlands of Southeast Asia is concentrated in the relatively short pre-rice niche, where species are cultivated before the establishment of the rice crop and are incorporated into the soil during land preparation (Garrity and Flinn, 1988) or weeding operations (Watanabe, 1991). Prototype species include the floating fern *Azolla* (Watanabe and Roger, 1984) and a range of legumes (Becker, Ladha et al., 1995; Ladha et al., 1992). The stem-nodulating *S. rostrata* was cultivated on about 500,000 hectares in Myanmar during the early 1990s (Mar et al., 1995). The flood-tolerant *Sesbania cannabina* is occasionally encountered in lowland production systems in Thailand and Indonesia. *Sesbania sesban* and *S. speciosa* can be locally important and are grown in parts of Thailand, where they provide edible flowers in addition to an N-rich biomass (Arunin et al., 1988; Evans and Rotar, 1987). In unfavorable rainfed lowlands with an erratic onset of monsoon rains, namely in Laos and Northeast Thailand, an occasionally encountered risk-avoidance strategy is the use of the grain legume *Vigna radiata* as a green manure. It is dry broadcast together with rice at the onset of the rainy season. In case of a dry year or a late start of the monsoon rains, *Vigna* outcompetes the rice and provides edible food grains. If the rains start early, the *Vigna* biomass vanishes under the flooded soil conditions and provides an N-rich green manure to the lowland rice. This diversity of systems indicates a potentially large adaptability of green manures to the diverse cropping situations in the Southeast Asian lowlands.

The Performance of Green Manures

Green manure use in lowland rice-based systems has numerous reported benefits to either rice yield or other performance parameters of the cropping system. The quantification of such benefits was a primary focus of green manure literature until the mid-1990s, with special emphasis on either the addition of atmospheric N by symbiotic systems or on the saving of native soil N from losses by temporary immobilization in the growing biomass of the green manure vegetation (*Azolla,* legumes).

While few papers report N accumulation and N_2 fixation data from post-rice green manures (Yost and Evans, 1988), some 280 observations are published on the N yield, and N_2 fixation by flood-tolerant pre-rice legumes and *Azolla* (Figure 9.6). Nitrogen accumulation has varied with species, season, and site, and is influenced by water regime, soil fertility, photoperiod, inoculation, and crop growth duration (Becker et al., 1990; Buresh and De Datta, 1991; Becker, 2001). Accumulation values range from as low as 7 to more than 300 kg N ha^{-1}, with mean values of 88 kg N ha^{-1} for

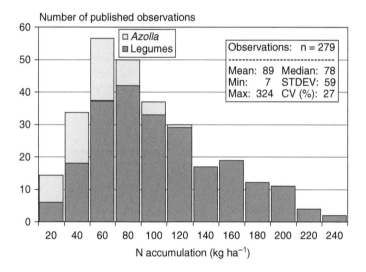

FIGURE 9.6. Frequency distribution of N accumulation by leguminous pre-rice green manures and *Azolla* in lowland rice-based systems of Southeast Asia. *Source:* Compiled from Becker, Ladha et al. (1995), Becker (2003), Buresh and De Datta (1991), Ladha et al. (1992), Peoples and Herridge (1990), Peoples et al. (1995), Rahman et al. (1996), Roger and Watanabe (1986), Van Hove (1989), Ventura et al. (1987), Vlek et al. (1995), Watanabe (1991), and Watanabe et al. (1989).

legumes, and 52 kg N ha^{-1} for *Azolla* (reviewed by Peoples and Herridge, 1990; Watanabe, 1991; Becker, Ladha et al., 1995; Peoples et al., 1995). The mean N accumulation from green manures in Southeast Asia (some 80 kg N ha^{-1}) corresponds to the average amount of mineral fertilizer N applied to irrigated lowland rice (84 kg N ha^{-1}) and exceeds that applied to rainfed lowland rice (63 kg N ha^{-1}) (Maclean et al., 2002). The major share of N accumulated in the biomass of N_2-fixing green manures is derived from atmospheric N_2 via biological fixation. Flood-tolerant legumes derive on average 78 percent of their N from fixation, while this value can exceed 85 percent for *Azolla* (Becker, Ladha et al., 1995; Roger and Watanabe, 1986; Watanabe, 1991). After incorporation, these materials can act as a slow release N source, which often, more closely match the temporal pattern of rice crop N demand than does a split application of mineral N fertilizer (Becker et al., 1994; Becker and Ladha, 1996). Consequently, average N losses from applied green manure are generally lower (14 percent) than those from split-applied urea (35 percent), resulting in rice yield increases from 0.3 to 3.6 Mg ha^{-1} (Becker, Ali et al., 1995).

Yet investment in a green manure crop is risky, given the large reported variability both in terms of N accumulation and yield responses. One cause of performance variability can be linked to available soil moisture during the pre-rice niche, and competition for water between the green manure and the rice crop, particularly in drought-prone and variable rainfall environments. Another main cause for variability appears to be the availability of soil P (Engels et al., 1995; Somado et al., 2003). The significance of P is highlighted in Figure 9.7, based on fifty-seven observations. Nitrogen accumulation by legumes never exceeded 50 kg ha^{-1} when the available soil P was less than 5-8 mg kg^{-1} Bray-I P. The available P requirement is even higher in the case of *Azolla*, where the critical value was determined to be 12-15 ppm Olsen-P (Watanabe and Ramirez, 1984). With P requirements of N-fixing crops exceeding that of rice by a factor of 1.7-2.3 (legumes), or 2.5-3.4 (*Azolla*), it becomes apparent that the use of N-fixing green manures is limited to environments with a relatively high soil P content.

Apart from N additions by biological N_2 fixation, green manures may also contribute to conserving both native soil N and mineral fertilizer N from gaseous or leaching losses (Buresh and De Datta, 1991). Ammonia volatilization is an important loss mechanism of mineral fertilizer N in flooded rice soils (De Datta and Buresh, 1989), and it is promoted by high floodwater concentrations of ammoniacal N, high temperatures, and an elevated floodwater pH (Vlek and Craswell, 1981; Roger and Watanabe, 1986). Recent evidence from field experiments in the Philippines underlines the potential of *Azolla* to not only reduce N losses and increase fertilizer N use efficiency

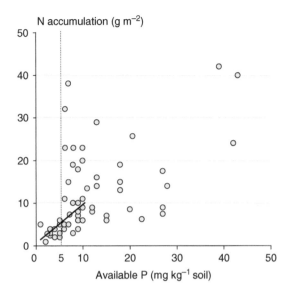

FIGURE 9.7. Relationships between the nitrogen accumulation by green manure legumes and available soil N (Bray-I). *Source:* Compiled from Becker et al. (1991), Becker and Ladha (1996), Becker (2003), Carsky et al. (2001), Engels et al. (1995), Somado et al. (2003), Ventura et al. (1987), Watanabe and Ramirez (1994), and Yadvinder-Singh et al. (1994).

but also to significantly increase the grain yield of lowland rice, particularly in soils of relatively high pH (de Macale, 2002). Dense covering of the floodwater by *Azolla* provides cooling shade and curbs the photosynthetic activity of floodwater organisms, which would dampen diurnal pH fluctuations. This muting of variations in temperature and diurnal pH, combined with ammoniacal-N interception by *Azolla*, can reduce N losses and save N in the green manure biomass for later use by rice (Vlek et al., 1995). The relevance of such strategies at farm level remains questionable, not least because of the large P requirements of *Azolla*.

Another type of N-saving by green manures is the "nitrate-catching effect" (George et al., 1992). In much of Southeast Asia, the rainfed lowland field lies idle between the harvest of a dry season crop and transplanting of lowland rice during the wet season. At the beginning of this dry-to-wet season transition period, the repeated drying and wetting cycles of the soil stimulate N mineralization, leading under aerobic soil conditions to the formation of nitrate-N. With the onset of the monsoon rains, nitrate losses of up to

90 kg N ha^{-1} can occur from the saturated soil (George et al., 1994), primarily through leaching in well-drained sandy soils and denitrification in saturated heavy-textured soils (De Datta and Buresh, 1989; George et al., 1995; Pande and Becker, 2003). Growing a short-cycled green manure as "nitrate-catch crops" during the transition season can assimilate soil N and protect substantial quantities from loss. The effectiveness of such nitrate-catching green manures in conserving soil N, and increasing the grain yield of subsequent rainfed lowland rice has been shown for both tropical and subtropical environments and for a range of soil types (Clément et al., 1995, 1998; George et al., 1994, 1998; Ladha et al., 1996). Published amounts of "saved" nitrate-N range from 12 to more than 90 kg N ha^{-1}, and yield increases in rainfed lowland rice range from 0.2 to 2.3 Mg ha^{-1} (George et al., 1995).

Green manure application can also result in diverse non-N benefits. (Becker, Ali et al., 1995; Kröck et al., 1988). These may include the mobilization of P, Si, Zn, Cu, and Mn as a result of increased microbial activity and a decrease in redox potential. Green manure can increase the cation exchange and water-holding capacities of the soil, and help reclaim alkaline and acid sulfate soils. Increased soil biological activity as a result of green manuring can accelerate the mineralization of persistent pesticides, reduce the incidence of bacterial diseases, insect pests, and nematodes in rice, and help control weeds (reviewed by Becker, 2003). Experiences from other regions in the world indicate that such additional benefits can be crucial for green manure adoption (Ladha and Garrity, 1994; Becker and Johnson, 1998; Tian et al., 2001).

Niches for Green Manure Use

Despite a large diversity of systems and species, and the numerous reported benefits, the use of green manure legumes in lowland rice production systems has dramatically declined over the past thirty years. Several reviews and regional surveys identified a high price (or low availability) of land and labor, high demands on the quality of crop management, and a relatively low price of mineral fertilizer as the major constraints to green manure use (Garrity and Becker, 1995; Becker, Ladha et al., 1995; Becker, 2001). Consequently, likely future niches for green manure use are situations where the availability of land and labor suffices for investment in a green manure, and where mineral fertilizer is either unavailable or costly, or where green manures far out-compete mineral N sources with regard to N use efficiency.

At an average application rate of 80 kg N ha^{-1}, green manure and mineral fertilizer N show similar agronomic N use efficiencies (Figure 9.8,

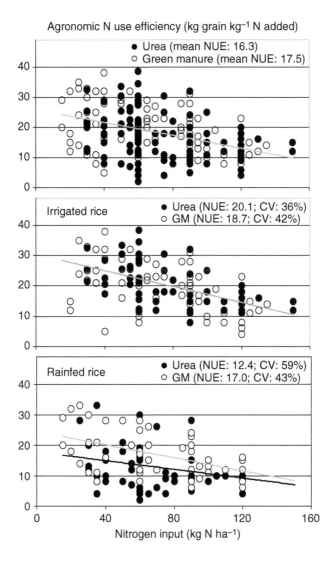

FIGURE 9.8. Fertilizer use efficiency of urea and green manure N by lowland rice in irrigated and rainfed production systems of Southeast Asia. *Source:* Compiled from Buresh and De Datta (1991); Becker et al. (1994, 1995); Becker (2003); De Datta and Buresh (1989); de Macale (2002); Garrity and Becker (1995); Tuladhar (2003); and Watanabe et al. (1989).

top). However, a differentiation of the 194 considered data points (ninety-seven paired comparisons of green manure with mineral N fertilizer) into irrigated (Figure 9.8, middle) and rainfed (Figure 9.8, bottom) production environments shows that urea-N tends to out-compete green manure-N in irrigated systems, while green manure-N appears more efficient than urea-N in rainfed lowland rice. A further differentiation of the same data set by soil textural class indicates that N use efficiency is similar for both N sources on clay soils, but in sandy soils it tends to be substantially greater for green manures (Figure 9.9). Thus, green manure use can compete with mineral N fertilizers in some unfavorable (sandy-textured) rainfed situations. Such conditions occur in Northeast Thailand, Laos, Central Myanmar, and Kalimantan. However, in these environments green manure growth can be severely limited by the widespread deficiency of soil P and possible water shortages, leading to competition for water between the green manure and the rice crop.

Besides these biophysical factors, growing land and labor shortages in most regions (Pingali et al., 1990) and the typical risk aversion behavior of Asian farmers (Ali, 1999) limit the adoption of such technologies that are management intensive, yet variable in their outcome. Consequently, green manures are not seen to gain any importance in the future and are likely to remain a strategy for isolated niches in Southeast Asian lowlands, such as:

1. sandy-textured rainfed lowlands with sufficient soil P in high-rainfall environments;
2. cropping systems with an extended dry-to-wet season transition period in environments with poor access to input markets; and
3. proximity to specialty markets, where high prices for organically grown commodities justify additional risks or investments associated with green manures.

INTEGRATED NUTRIENT MANAGEMENT IN IRRIGATED LOWLAND RICE

Soil Organic Matter and Soil Indigenous Nitrogen Supplies

If the amount of nutrients supplied by native soil fertility to a rice crop could be accurately estimated, organic and inorganic fertilizers could be applied more efficiently, and grain yields could be optimized. SOM provides half or more of the N taken up by an irrigated lowland rice crop (Broadbent, 1979; Cassman, Gines et al., 1996). This amount varies considerably among

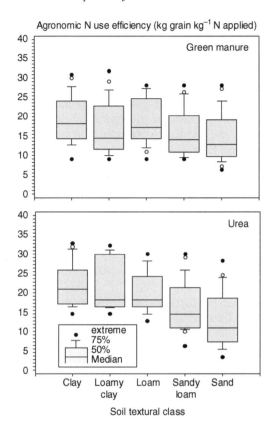

FIGURE 9.9. Fertilizer use efficiency of urea and green manure N by lowland rice in rainfed production systems of Southeast Asia as affected by soil texture. *Source:* Compiled from Becker et al. (1995); Becker (2003); Cassman De Datta et al. (1996); and Garrity and Becker (1995).

lowland rice fields and in the same field over time (Cassman, Gines et al., 1996; Dobermann et al., 2003a).

Unfortunately, a reliable means to estimate the native soil N supply from soil properties remains unidentified. Total soil N content is not a valid indicator of native soil N supply (Cassman, Dobermann et al., 1996; Olk et al., 1999). Thus increasing the SOM content of rice soils through crop management practices does not necessarily increase the crop N supply. Other laboratory analyses have been evaluated as measures of available soil N, but none has worked across an adequately broad range of rice soils and agronomic conditions (Thach, 1999).

Separate from the quantity of soil organic N, the quality, or chemical nature, of soil organic N may also influence the soil N supply. Although little work has been done in rice soils, one example is the accumulation of phenolic compounds under long-term cropping of continuous lowland rice (Olk et al., 1996, 1998). The phenols are incorporated into soil as part of woody tissues of crop residues, and their decomposition is slowed greatly by the submerged conditions. Phenol accumulation in a young humic fraction was associated with an inhibition of soil N mineralization from that fraction during the rice-growing season (Olk et al., 2007). Decreased N mineralization may be caused by covalent binding of soil organic N with the phenols, forming an N compound that is more recalcitrant under flooded conditions than are other organic N compounds. An agronomically significant quantity of this bound N form was found in a triple-cropped rice soil (Schmidt-Rohr et al., 2004), and this chemical stabilization has been shown to occur under laboratory conditions (Flaig et al., 1975). Further work is needed to demonstrate whether this chemical bonding inhibits soil N supply across a significant area of rice production and whether it contributes to the lack of correlation between quantity and availability of soil organic N.

Basing Fertilizer Rates on Native Soil Fertility

Conventional soil tests provide limited opportunities to improve current fertilizer recommendations in irrigated rice in Asia. Basic soil properties and rapid chemical extractions of soil showed little correlations with indigenous N, P, and K supply, measured as plant nutrient uptake in nutrient omission plots, across a wide range of on-farm environments in South and Southeast Asia (Dobermann et al., 2003b). In recent years, an alternative approach to estimating native soil fertility has been developed as part of a novel site-specific nutrient management strategy (Dobermann and White, 1999; Dobermann et al., 2002). In this approach, soil nutrient supplies are indirectly estimated as plant nutrient uptake or grain yield in nutrient omission plots (Dobermann et al., 2003a,b). An omission plot receives all nutrients except for the nutrient of interest. Plant-based estimates of soil nutrient supply integrate the supply of all indigenous sources under field conditions and also offer the capability to estimate the nutrient supplying power of straw, organic manures, irrigation water, biological N_2 fixation, and sediment input.

In the site-specific nutrient management approach, fertilizer N, P, and K requirements are calculated based on the difference between yield goal and nutrient-limited yield estimated in the nutrient omission plots (Witt et al., 2002; Dobermann et al., 2004). Major steps in the calculation of fertilizer requirements include:

1. selection of a suitable yield goal of not more than 70-80 percent of the potential yield;
2. use of the omission plots to estimate soil nutrient supplies based on yield;
3. estimation of fertilizer requirements based on the expected yield deficit between yield goal and the respective nutrient-limited yield in the omission plots (yield gain approach);
4. use of empirically derived standard values for fertilizer requirements per unit yield deficit; and
5. integration of a simple nutrient balance for calculating fertilizer P and K maintenance rates (yield gain plus nutrient balance approach).

Fertilizer requirements in the yield gain approach are estimated following the principles of the QUEFTS model (Janssen et al., 1990; Witt et al., 1999). Based on plant nutrient requirements and expected fertilizer recovery efficiencies, about 40-50 kg fertilizer N ha^{-1}, 7-12 kg fertilizer P ha^{-1}, or 22-41 kg fertilizer K ha^{-1} would be needed to raise yield by 1 t over the nutrient-limited yield in the respective omission plot (Witt et al., 2002; Witt and Dobermann, 2004). Nitrogen applications are scheduled based on crop need. In recent years, major progress has been made in the development, on-farm evaluation, and promotion of leaf color charts for real-time N management (Balasubramanian et al., 1999; Yang et al., 2003; Witt et al., 2003). Two approaches based on leaf color charts are currently promoted in Southeast Asia: (1) a full real-time N management during the growing season where the application of yield gain-specific fertilizer N rates is signaled once rice leaves have reached a critical color (e.g., in the Philippines); and (2) a location-specific schedule of fertilizer N splitting including the use of the leaf color chart (e.g., in south Vietnam) (Witt et al., 2002). In both approaches, less emphasis is given to the correct estimation of indigenous soil N supply in favor of corrective measures. These approaches allow N management to be based on the application of organic nutrient sources without the need to accurately determine their nutrient supplying power.

Fertilizer P and K rates are further adjusted using a newly developed nutrient balance approach to compensate for nutrient removal with grain and straw (Witt et al., 2002; Witt and Dobermann, 2004). Integrating a nutrient balance in the yield gain approach appears essential for developing meaningful short- and long-term fertilizer P and K strategies to avoid nutrient depletion. The few additional input parameters that are needed include the average amount of straw returned and additional rates of organic nutrient sources such as animal manure.

Site-specific nutrient management advocates the economic use of organic nutrient sources, including straw and manure, prior to the estimation of inorganic fertilizer needs. An example is in the Red River Delta near Hanoi, Vietnam, where FYM application is still widespread, in contrast to other major rice growing areas in Southeast Asia. Apart from a long tradition in the use of organic manure, farmers follow intensive crop management with high labor input because population density is high, farm sizes are small, and family labor is sufficiently available (Son et al., 2004). To test the hypothesis that inorganic fertilizer can fully substitute the nutrients supplied with FYM, on-farm experiments were conducted during two cropping seasons in 1999 (Table 9.7). Inorganic nutrient inputs were 13-26 percent greater in treatments without, than with FYM. Total nutrient inputs were adjusted slightly higher with FYM than without to compensate for the presumed lower availability of nutrients in FYM. Differences in nutrient uptake between treatments were relatively small, but the continuous supply of nutrients from FYM resulted in greater nutrient use efficiency. Furthermore, FYM application can improve other soil properties that were not taken into account, including cation exchange capacity, microbial activity, and soil structure. Yields were about 0.5 t ha^{-1} greater with FYM than without, and profit increased by about US\$ 60 ha^{-1} when additional labor costs for FYM application were considered. In conclusion, the approach described here is sufficiently robust to adjust inorganic fertilizer rates for the amount of FYM nutrients applied.

Straw Management and the Use of Organic Fertilizers

In Southeast Asia, rice straw is the sole organic amendment that is routinely applied to a significant acreage of lowland rice. It is normally cut at least 20 cm above the ground, leaving large amounts of straw biomass in the field particularly where combine harvesting is used, such as in Thailand. The remaining straw is not removed for feed purposes, as draft animals are no longer widely used. Hence it is generally burned, but significant portions are also incorporated. Management of this biomass offers significant potential benefits given its high content of N, K, carbon, and silicon. Optimal management is not straightforward, though, and agronomic objectives often clash with economic objectives.

A large number of studies concluded that straw incorporation can contribute substantially to nutrient cycling and improved soil health. Ponnamperuma

TABLE 9.7. Effect of FYM application on fertilizer use, nutrient uptake, grain yield, and the economics of site-specific nutrient management in rice (means of twenty-four farms × two seasons) in the Red River Delta, Hanoi, 1999

	+FYM	−FYM
Fertilizer applied (kg ha^{-1})		
N	92	104
P	14	19
K	60	76
N-P-K input from FYM (kg ha^{-1})		
N	21	0
P	17	0
K	34	0
Total nutrient input (kg ha^{-1})		
N	113	104
P	31	19
K	94	76
Total plant nutrient uptake (kg ha^{-1})		
N	104	98
P	23	22
K	121	111
Grain yield (t ha^{-1})	6.9	6.4
Increase in profit due to FYM (US$ ha^{-1})[a]		
Without labor cost for FYM	92	
With labor cost for FYM[b]	60	

Source: Adapted from Son et al. (2004).

[a]Δ = difference in gross return over fertilizer costs with and without FYM.

[b]Labor cost for FYM application estimated as US$1.54 (1 labor day) per 400 kg FYM.

(1984) summarized field studies of straw management in lowland rice fields by concluding that straw incorporation generally resulted in higher grain yield than did straw removal or burning, if incorporation was continued for a sufficient number of crop seasons. The benefit was greater in warmer climates like that of Southeast Asia, where toxic compounds released by incorporated straw decompose quickly before the transplanting of the next crop (Cho and Ponnamperuma, 1971) and in soils with adequate drainage.

Chanchareonsook et al. (1991) reported that straw compost incorporation markedly improved the nutrient status of several Thai soils. Grain yield increased in fine sandy soil but not in clay soil. In a Philippines trial of double-cropped rice, N supplied half as rice straw and half as urea for twenty consecutive crops performed significantly better than did treatments of green manure N sources, such as *Sesbania rostrata* and *Azolla microphylla* (Cassman, De Datta et al., 1996). In comparison with the green manure treatments, the straw-urea combination provided for greater extractable inorganic soil N, a larger increase in soil organic C and total N, and a larger residual benefit to rice N uptake in crops following cessation of treatments. The highest grain yields during and after treatment imposition, though, were achieved with urea.

The benefit of regular straw incorporation to rice crop performance might occur in several ways, as has been reported by studies in Southeast Asia and elsewhere. Both immobilization and mineralization of N can be enhanced (Yoneyama and Yoshida, 1977a; Shibahara et al., 1998), which in some studies has caused a decreased level of soil inorganic N and an increased level of soil organic N (Chou et al., 1982). These conditions would buffer soil inorganic N at levels conducive to adequate plant growth, while reducing excessive vegetative growth and N losses, such as through denitrification (Bacon et al., 1986), early in the season (Tanaka, 1978). Low, buffered soil N levels also appear to increase the amounts of N that is biologically fixed by soil microorganisms (Yoneyama et al., 1977).

Studies in Southeast Asia and East Asia indicate the need to coordinate the timing of N release from incorporated rice crop residues with the availability of fertilizer N and mineralization of native soil N. Kumazawa (1984) concluded that stable and high rice yields were possible in Japan only with application of both mineral N fertilizer and crop residue compost, which would best ensure a steady N supply at later crop stages. The majority of incorporated straw N appears to be mineralized early in the cropping period, as found in the Philippines (Yoneyama and Yoshida, 1977b) and Thailand (Masayna et al., 1985), whereas some mineralization of straw N continued throughout the cropping period. By comparison, the peak release of native soil N was generally slightly later than that of straw N. Working in a double-cropped rice rotation with a two-month fallow after each cropping period, Witt et al. (2000) found that incorporation of crop residues into the soil shortly after harvest improved the congruence between soil N supply and crop N demand for the next crop, compared to the conventional incorporation at the end of the fallow. This earlier incorporation resulted in greater remineralization of the fertilizer ^{15}N that had become immobilized into young humic fractions (Olk et al., 2007). Decomposition of crop residues with the earlier incorporation was aerobic, which may have led to the improved chemical nature of the humic fractions, and increased N mineralization potential as compared to the anaerobic decomposition with the later incorporation.

Straw incorporation can also improve other soil properties in addition to N availability. Because rice straw has high contents of K, silicon, and carbon (Ponnamperuma, 1984), its incorporation into the soil can increase availability of these nutrients (Dei, 1975). Soil physical properties can also be improved, including water percolation and ease of puddling (Dei, 1975), porosity, water retention, and bulk density (Beaton et al., 1992).

Principal nutritional issues with straw incorporation are early season biological N immobilization and production of organic acids in anaerobic soil conditions, including volatile fatty acids and phenolic acids (Tsutsuki, 1984). At sufficient levels in the soil, organic acids can impair the growth of the subsequent crop, probably through toxin actions. The deleterious effects of organic acids and N immobilization on crop growth have often been demonstrated in laboratory or greenhouse studies, but few if any studies have attempted to demonstrate their significance under realistic field conditions, that is, when residue incorporation is followed by the customary delay of at least two weeks before the next rice crop is planted. A rice crop planted one week after crop residue incorporation showed no ill effect in an Indonesian trial (Ismunadji, 1978). Crop residue incorporation in a Philippines study further benefited the subsequent rice crop if done four to six weeks before the flooding of the field compared to incorporation at the time of flooding (Witt et al., 2000), in part due to avoidance of N immobilization.

Yet straw incorporation has not always proven advantageous. Improved soil N availability and other soil benefits associated with straw incorporation have not always translated into yield increases. Incorporation of straw or FYM did not improve grain yield trends in seven long-term trials of continuous rice and eighteen trials of rice-wheat in Asia (Dawe et al., 2003). Alberto et al. (1996) found that straw addition resulted in a grain yield increase of only 0.4 t ha^{-1} but also significantly higher methane emissions. In practice, economic constraints often promote straw removal from the field. Alternatively, in regions where combine-harvesting is becoming more common, rice straw is burned. Reviewing Asian rice production, Tanaka (1974) concluded that the economics of that time did not encourage regular straw incorporation. Potential nutrition benefits appeared slight in relation to the expenses associated with incorporation, and little evidence existed at that time that mineral fertilizers cannot provide most of the benefits of straw. Earlier studies had indicated that straw burning might provide nearly all the nutrient input as incorporation of fresh straw while providing substantial labor savings and avoiding several pest problems. These conclusions would encourage straw incorporation in favorable lowland soils only when burning is not allowed due to pollution problems, and when no off-farm opportunities exist for the use of the straw, such as livestock feed or as an energy source for households. More recently, Dawe et al. (2003) concluded that straw or FYM incorporation may be profitable on-farm when used in conjunction with mineral fertilizers. They further concluded that the use of FYM is profitable where wages are low and FYM is available. The purchase of manure is generally not profitable because of high price, low nutrient content compared to mineral fertilizers, and high labor costs.

SOCIOECONOMIC CONSTRAINTS TO ADOPTION OF IMPROVED NUTRIENT MANAGEMENT TECHNOLOGIES

Patterns of Growth in Chemical Fertilizer Use

Starting in the mid-1960s, the Green Revolution was characterized in Asia by a rapid increase in the use of chemical fertilizers. In 1965, 4.9 million tonnes of elemental nutrients (NPK) was applied in the form of chemical fertilizers in major rice growing countries of Asia, and by 2002 this amount increased twelvefold to 59.5 million tonnes (Table 9.8). The mean application rate of NPK fertilizers per hectare grew from a negligible amount prior to the Green Revolution to 131 kg ha^{-1} in 2002 (Table 9.9). The growth rate in the use of chemical fertilizers was high at 7.9 percent per year during the advent of the Green Revolution (Table 9.8). Although the pace declined somewhat in subsequent decades, only in more recent years has the growth rate decreased to almost negligible levels.

This rapid increase in the use of chemical fertilizers was driven mainly by two factors: high yield response to fertilizers and declining real price of chemical fertilizers. The use of chemical fertilizers increased only as farmers intensified rice production and adopted modern rice varieties that are fertilizer-responsive. At the same time, technological innovations in the fertilizer industry led to a long-term decline in the real price of fertilizers, although fertilizer prices have increased somewhat again in recent years.

Economics of Adoption of Chemical Fertilizers

Economic factors that determine the adoption of a technology such as chemical fertilizers can be classified into those that are related to the specific conditions of a farm/farmer and those that are related to the wider

TABLE 9.8. Trends in fertilizer consumption in major rice-growing countries in Asia

Region	Nutrient consumption (Mt)[a]						Rate of increase (%/year)			
	1965	1978	1988	1992	1996	2002	1967-1976	1977-1986	1987-1996	1997-2002
Bangladesh	0.1	0.3	0.7	0.8	1.2	1.3	8.4	7.1	6.5	4.7
China	2.2	10.1	22.2	25.3	31.5	35.5	10.6	6.8	4.0	1.1
India	0.7	4.4	9.4	10.4	12.5	14.5	8.9	7.8	4.5	−0.3
Indonesia	0.1	0.7	2.0	2.2	2.4	2.7	14.4	12.9	2.3	3.1
Japan	1.4	1.7	1.4	1.3	1.2	1.0	−0.8	−0.7	−2.6	−2.4
Korea (Republic)	0.3	0.7	0.7	0.8	0.8	0.6	5.9	−0.1	1.2	−6.9
Myanmar	0.0	0.1	0.1	0.1	0.2	0.1	11.2	11.7	6.1	−3.3
Philippines	0.1	0.3	0.5	0.5	0.6	0.6	8.8	1.4	2.1	0.1
Thailand	0.0	0.2	0.6	0.8	1.3	1.5	10.5	8.0	9.9	2.5
Vietnam	0.0	0.3	0.5	0.8	1.3	1.7	8.3	7.6	13.9	3.8
Total	4.9	18.6	38.0	43.0	52.7	59.5	7.9	6.5	4.1	0.8

Source: Data for 1965-1996: FAO AGROSTAT Database, data for 1997-2002: FAOSTAT data (last updated April 2005).
[a]Nutrients expressed as the sum of N, P, and K. Conversion factor used for P is 0.83 and for K is 0.44.

TABLE 9.9. Fertilizer use in rice in the major rice-producing countries of Asia

Country	Share of rice in total fertilizer use	Total fertilizer use in 2002 NPK	Fertilizer use in rice in 2002 NPK[a]	Rice area 2004	NPK use[b]
		(Tg)	(Tg)	(Mha)	(kg ha^{-1})
	Col 1	Col 2	Col 3	Col 4	Col 5
Bangladesh	0.79	1.30	1.0	11.0	93
China	0.23	35.54	8.2	28.3	289
India	0.20	14.52	2.9	42.3	69
Indonesia	0.52	2.68	1.4	11.9	117
Japan	0.31	1.01	0.3	1.7	184
Korea (Republic)	0.40	0.56	0.2	1.0	225
Myanmar	0.89	0.13	0.1	6.0	19
Philippines	0.41	0.64	0.3	4.1	63
Thailand	0.61	1.48	0.9	9.8	92
Vietnam	0.69	1.66	1.1	7.4	154
Total		59.53	16.5	123.6	131

Source: Data for columns 1 and 2 are taken from IFA Web site (http://www.fertilizer .org/ifa, updated October 2004) and FAOSTAT data (last updated April 2005), respectively. Data for column 4 are taken from FAOSTAT data (last updated July 2005).

[a]By multiplying Columns 1 and 2.

[b]By dividing Column 3 by Column 4.

socioeconomic milieu within which farmers operate. The first type of factors includes the biophysical characteristics of the farm, the farmer's resource constraints, technological options available to the farmer, and the farmer's primary objective. This is assuming the conventional objective of increasing the economic profit associated with the use of a technology (or an input), the characteristics of the response function, the price of input under consideration, and price of outputs are the major factors that determine adoption. Risk considerations may also be important if farmers are risk-averse or if they misperceive the likely benefits.

Due to the high responsiveness of modern varieties to fertilizers in irrigated areas relative to rainfed areas, much of the expansion in fertilizer use occurred in irrigated areas. The expansion in fertilizer use during the past three decades can be described by the simplified conceptual model of Byerlee (1994). In the initial stages of the Green Revolution, farmers switched from traditional to modern varieties, and the production function shifted upward without greatly increasing fertilizer use. Both the yield gains and fertilizer use were modest. From this initial stage, farmers increased fertilizer use rapidly as they learned more about fertilizers. This phase of "input expansion" during the Green Revolution led to a rapid increase in fertilizer use. Active promotion of fertilizers through credit and subsidy schemes played a critical role during this phase in input expansion. This phase was best expressed in irrigated areas. In the "post–Green Revolution" phase, by which time most of the gains from input expansion had been achieved, the growth in fertilizer use has slowed, and farmers and researchers now look for opportunities to more efficiently use fertilizers.

Today fertilizer use in irrigated rice production is high in many countries and concerns are being expressed about the low input use efficiency. For example, fertilizer use is in excess of 180 kg of NPK ha^{-1} in China, Japan, and Korea (Table 9.9). In contrast, countries with larger shares of rainfed area have much lower application rates (Figure 9.10).

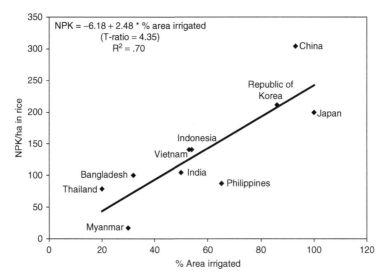

FIGURE 9.10. Effect of irrigated area (expressed as percent of total area) on fertilizer use (kg NPK ha^{-1}) in rice production. *Source:* Adapted from Irrigated area: IRRI (2002). Fertilizer NPK in rice: FAOSTAT data, 2004 (last updated February 2004).

Given these differences in the level of fertilizer application, nutrient management issues in irrigated and rainfed environments are different. Most rainfed areas have not yet gone through the input expansion phase. Hence, the key issue is how to make fertilizer use more profitable to farmers. In most of the irrigated areas, by contrast, the input expansion phase is past, and the concern is now about how to improve fertilizer use efficiency and the balance of different nutrients, so that the production costs and potential negative impacts on the environment can be reduced. Meeting this objective will require new technologies that use knowledge regarding crop and soil conditions in order to match the temporal pattern of nutrient supply with crop nutrient demand.

Three factors provide the economic incentive for adoption of such knowledge-intensive technologies (KITs) for efficient fertilizer management (Pingali et al., 1998). They are: (1) potential yield change; (2) the value of fertilizer saved; and (3) the cost of acquiring and using the technology. The economic benefit consists of any yield increase plus the value of fertilizer saved resulting from adoption of KITs. This benefit needs to exceed the costs associated with the use of KITs, which arise from the purchase of the KIT, plant or soil monitoring, and training in the use of KITs.

When the costs of using KITs are considered, simple and inexpensive devices such as leaf color charts tend to have an edge over more complex devices such as the chlorophyll meter. Soil-based indicators such as the indigenous nutrient supplying capacity may be more expensive to monitor and their usefulness less transparent to farmers compared to the leaf color chart.

In rainfed areas, the main issue is the current low rate of fertilization, which is due to both demand and supply factors. As described above, grain yields are low and variable and the AE of N fertilizer in rainfed areas is low and varies widely due to droughty soils and highly fluctuating water regimes. Low AE estimates were also obtained in a survey of Laotian data by Pandey and Sanamongkhoun (1998). Low and variable yields prompt farmers to reduce fertilizer use in order to minimize economic losses. Thus rainfed rice production needs improved rice varieties and crop management practices that stabilize the AE and increase yield levels. The adoption of modern varieties in rainfed areas is currently limited, with farmers growing mainly the traditional varieties that do not exhibit sufficient responsiveness to fertilizers.

Farmers' capacity to purchase (or supply) fertilizers is also constrained due to a number of socioeconomic factors. In rainfed areas, the production system is mainly subsistence-oriented and farmers' incomes are low. Rainfed areas tend to have poorly developed infrastructure such as roads for transporting fertilizers. This obstacle increases the effective cost of fertilizers, as transportation costs are added to the fertilizer price. In Laos, fertilizer use was found to be higher in areas with better access to markets than in areas with poor access (Pandey and Sanamoungkhoun, 1998). In addition, rainfed rice farmers typically have limited access to the institutional credit that was critical in encouraging the adoption of fertilizers by irrigated lowland rice farmers in the wake of the Green Revolution.

Factors Determining the Adoption of Organic Fertilizers

Farmers have traditionally relied on organic sources such as FYM, compost, straw, and green manure crops to maintain and enhance soil fertility. Close integration of livestock with rice cropping, and the low opportunity costs of land and labor in traditional systems, made the use of organic nutrients economically viable. However, the use of organic fertilizers for rice production has decreased over time as cheaper chemical fertilizers have taken their place, as described above for green manures.

Although organic amendments are generally considered to improve the long-term soil fertility, the empirical evidence for improved grain yields is mixed, especially for irrigated lowland rice as described above. Further, low nutrient content and high bulk result in higher effective prices of nutrients from organic sources than those from chemical fertilizers (Rosegrant and Roumasset, 1988; Ali and Narciso, 1994).

Although organic fertilizers are largely produced on-farm and no cash costs may be incurred in their production, the land and labor used for their production and incorporation into the soil have opportunity costs. Hence, whenever land can be used for the production of cash crops, legumes that are grown strictly for soil improvement cannot compete for land use. In addition, farm residues such as straw and compost may have alternative, more valuable uses, so using them on rice fields carries additional opportunity costs. When all opportunity costs were considered at current energy prices, the use of organic fertilizers turned out to be generally not economically viable (Ali and Narciso, 1994; Garrity and Flinn, 1988). Use of organic fertilizers tends to be viable only when the opportunity costs of land and labor are low and the production system is subsistence-oriented.

Characteristics of Production Systems and Nutrient Management Technologies

It has been suggested that the nutrient management practices of farmers are dependent on the nature of production systems and that these practices change as production systems evolve over time. Pandey (1998) developed a simplified typology of production systems (Figure 9.11) that is based on the evolutionary model of Boserup (1965, 1981). This typology uses population density and the level of economic development (proxied by income) as the major determinants of production systems to describe their linear evolution. In practice, diversity exists within these systems, and the transition from one system type to the next is not necessarily linear as described here.

In the initial stages of economic development, when the population density and income levels are low, the dominant features of the production system tend to be low-intensity land use and subsistence-oriented production (Type I). Farmers mostly adopt an extensive land use strategy, and rice is grown on only a portion of the land, often in rotation with natural fallow. The household supplies most of the labor, and the challenge for each household is how best to allocate labor between different enterprises. In fact, labor is the major constraint to production under this situation, and nutrient management technologies that improve labor productivity are suitable. Application of green manures or other organic fertilizers is not attractive, for it would compound the labor shortage. A substitution of labor by capital is

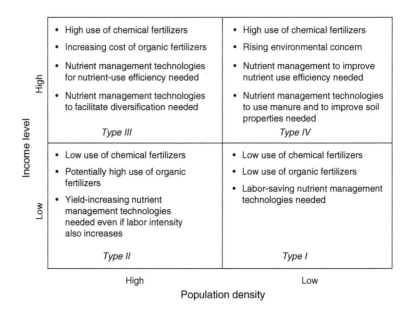

FIGURE 9.11. Nature of nutrient management technologies that are suitable to different types of rice systems. *Source:* Adapted from Pandey (1998).

restricted by the unavailability of cash in the subsistence orientation. Demand tends to be low for nutrient management technologies that increase yield trends. If yields are limited by unfavorable growing conditions or nutrient deficiencies, the likely response is to increase the area of production. Native soil N is the dominant N supply to crops. The domain for this type of production system in Southeast Asia is currently limited mainly to the lowlands of Laos and Cambodia.

As population pressure increases, the Type I production system evolves toward the Type II system where agricultural land use is intensified. Land is relatively scarce compared to labor, and farmers attempt to maximize land productivity through practices such as fertilization, bunding, and the replacement of dry seeding with transplanting. The production system is still subsistence-oriented, and incomes are low, although producers will take advantage of the few opportunities for market-based production. The use of purchased inputs such as mineral fertilizers tends to be limited. Instead, farmers use their abundant labor to augment the nutrient supply through composting, collecting and concentrating rice crop residues, green manures (post-rice, and increasingly pre-rice), and animal manures, making this domain perhaps the most attractive for the integrated use of organic and inorganic fertilizers. A closer integration of livestock with crop production in this system also enables economic use of animal manures and residues. The domain for the Type II system in Southeast Asia is currently restricted to unfavorable rainfed rice locations, such as remote and infrastructurally deficient regions.

Given further economic growth, the Type II production system transforms into the Type III system. Pressure on land continues to rise, not only for agricultural production but also for settlement and other nonagricultural activities as can be seen today across Southeast Asia. Production is once again limited by labor as growing urban centers and nonfarm activities compete for labor. Land and labor are increasingly replaced by capital, such as external inputs (fertilizers, pesticides) and mechanization. Mineral fertilizers become the main nutrient source, as the use of green manures and other organic sources decreases with the rising opportunity costs of land and labor. The production system moves from the stage of input expansion to input efficiency, and demand rises for nutrient management technologies that increase input use efficiency. Simultaneously, the developing infrastructure improves access to markets, and to agricultural extension and other means of knowledge acquisition. Agricultural production systems become increasingly commercialized and diversified. The income levels of farmers are high, for example, as a result of the rapid growth of the nonfarm sector (e.g., Thailand) or of rapid growth in farm productivity using new crop varieties (e.g., Punjab in India). This system is the domain for knowledge-based nutrient management, and currently it characterizes the intensive irrigated lowland areas and favorable rainfed areas of Southeast Asia.

The final phase is the evolution of highly commercialized and specialized rice production systems (Type IV). It characterizes the relatively advanced agricultural systems of developed countries. The rural population density declines as labor moves from agricultural to nonagricultural activities during economic growth. As a result, agricultural operations tend to be mechanized and farm size tends to expand. Emerging concerns are the rising environmental costs associated with the use of chemicals in agriculture, and maintenance of farm income at par with nonfarm incomes. Again, the demand for technologies that increase input use efficiency tends to be high, as input costs increase with rising labor costs. Achieving higher efficiency of fertilizer use becomes an important consideration. Concerns regarding safe disposal of wastes from specialized livestock industry may lead to the use of processed waste on crops. The use of organic fertilizers may therefore make a comeback, mainly for environmental reasons. Much of Southeast Asia will require considerably more time to reach this stage.

Currently the ideal niches for green manures and other organic fertilizers in Southeast Asia are limited to systems where both land and labor are available but capital for external inputs is lacking. Such a situation dominated much of the rural areas of Southeast Asia throughout the 1970s and 1980s but is now limited to a few remote areas that are Type II systems.

Implications for Research and Policies

Slow growth in the productivity of rainfed environments raises the issue of research resource allocation to rainfed environments. Is the agricultural research system investing enough in rainfed areas? Although opinions vary depending on the perspective, a consensus is emerging that the marginal productivity of research in rainfed areas may now be higher than that in irrigated areas (Fan and Hazell, 1999). If this is the case, a reallocation of research resources at the margin in favor of rainfed environments seems desirable. Further, adaptive research is essential to develop new technologies that are suitable to the heterogeneous rainfed environments. Availability of technologies that enable higher and more stable yields will provide incentives to farmers to apply more fertilizers and at greater efficiency.

Another potential area of policy reform for both rainfed rice and irrigated lowland rice is the extension system, which has traditionally been geared toward providing blanket recommendations. For higher input use efficiency and more integrated management, nutrient recommendations must be location specific. The current extension systems are not suitable for delivering knowledge to this level of detail. Needed are (1) upgraded capacity of extension systems; and (2) new approaches to extension that will utilize private sector and nongovernment organizations for efficient flow of information.

CONCLUSIONS

In recent decades, the nutrient supply for rice production in Southeast Asia has changed from completely organic fertilizers to a dominance in most places of inorganic fertilizers. For example, green manure use in the rainfed lowlands is largely limited to the short pre-rice niche and in isolated regions. Availability of animal manure has plummeted in both rainfed and irrigated lowland systems. While integrated use of organic and inorganic fertilizers will often improve soil properties and can provide similar or better yields as pure reliance on inorganic fertilizers, the associated opportunity costs in labor and land restricts the use of organic fertilizers to specific situations. These trends have occurred elsewhere in the world, due to the same mix of economic and population forces described above, but they have been dramatic in Southeast Asia as both the regional economy and population have grown rapidly.

This situation may change in the future if the economics of inorganic fertilizers are altered by declining global oil supplies, if growing environmental concerns, and further economic development reintroduce an emphasis on organic amendments, or if rice consumers begin to demonstrate significant interest in organically grown rice. All of these potential changes, however, appear to be several years in the future for Southeast Asia. INM will remain more attractive in coarse-textured rainfed rice soils than elsewhere. Perhaps greater research efforts in this region would enhance the frequency and effectiveness of INM.

Rice straw is the organic amendment that is most likely to grow in use across Southeast Asia, especially if its burning becomes prohibited across wide regions, and as farmers become better educated in favorable modes of incorporation. Increased use may also be made of waste materials that are generated by advancing economic activities, such as industry, wood production, or human waste from urban centers. Considerable knowledge gaps remain in the efficient disposal of these materials in ways that optimize agronomic benefits, especially on a site-specific basis.

A bright future awaits the development of fertilizer recommendations based on either plant-based diagnostic tools or native soil fertility. Farmers have little knowledge of their options for more efficient application of fertilizers while attaining improved yield targets. Their training will require either significant upgrading of the public extension services in Southeast Asia or more substantial involvement by the private sector.

REFERENCES

Alberto, M.C.R., H.U. Neue, A.B. Capati, R.U. Castro, L.M. Bernardo, J.B. Aduna, and R.S. Lantin. (1996). Effects of different straw management practices on soil fertility, rice yields, and the environment. In Attanandana, T., I. Kheoruenromne, P. Pongsakul, and T. Vearasilp (eds.) *Maximizing Sustainable Rice Yields through Improved Soil and Environmental Management.* pp. 721-731. Bangkok: Thailand Department of Agriculture.

Ali, M. and J.H. Narciso. (1994). Economic evaluation and farmers' perception of green manure use in rice-based farming systems. In Ladha, J.K. and D.P. Garrity. (eds.) *Green Manure Production Systems for Asian Ricelands. Selected papers from the International Rice Research Conference.* Manila, Philippines: International Rice Research Institute.

Arunin, S., C. Dissataporn, Y. Anuluxtipan, and D. Nana. (1988). Potential of *Sesbania* as green manure in saline rice soils in Thailand. In *Sustainable Agriculture: Green Manure in Rice Farming.* pp. 83-95. Manila, Philippines: International Rice Research Institute.

Bacon, P.E., J.W. McGarity, E.H. Hoult, and D. Alter. (1986). Soil mineral nitrogen concentration within cycles of flood irrigation: Effect of rice stubble and fertilization management. *Soil Biology and Biochemistry.* 18: 173-178.

Balasubramanian, V., A.C. Morales, R.T. Cruz, and S. Abdulrachman. (1999). On-farm adaptation of knowledge-intensive nitrogen management technologies for rice systems. *Nutrient Cycling in Agroecosystems.* 53: 59-69.

Beaton, J.D., H. Hasegawa, J.-C. Xie, J.C.W. Keng, and E.H. Halstead. (1992). Influence of intensive long-term fertilization on properties of paddy soils and sustainable yields. *Proceedings, International Symposium on Paddy Soils.* pp. 252-273. September 15-19, Nanjing, China.

Becker, M. (2001). Potential and limitations of green manure technology in lowland rice. *Journal of Agriculture in the Tropics and Subtropics.* 102: 91-108.

Becker, M. (2003). *Potential von Leguminosen zur Gründüngung und Einsatzmöglichkeiten im tropischen Reisanbau.* Norderstedt, Germany: BOD Publishers GmbH.

Becker, M., M. Ali, J.K. Ladha, and J.C.G. Ottow. (1995). Agronomic and economic evaluation of *Sesbania rostrata* green manure establishment in irrigated rice. *Field Crops Research.* 40: 135-141.

Becker, M., K.H. Diekmann, J.K. Ladha, S.K. De Datta, and J.C.G. Ottow. (1991). Effect of NPK on growth and nitrogen fixation of *Sesbania rostrata* as a green manure for lowland rice (*Oryza sativa* L.). *Plant and Soil.* 132: 149-158.

Becker, M. and D.E. Johnson. (1998). Legumes as dry season fallow in upland rice-based systems of West Africa. *Biology and Fertility of Soils.* 27: 358-367.

Becker, M. and J.K. Ladha. (1996a). Adaptation of green manure legumes to adverse conditions in rice lowlands. *Biology and Fertility of Soils.* 23: 243-248.

Becker, M. and J.K. Ladha. (1996b). Synchronizing residue N mineralization with rice N demand in flooded conditions. In Giller, K. and G. Cadisch (eds.) *Driven by Nature: Plant Litter Quality and Decomposition.* pp. 231-238. London: CAB International.

Becker, M., J.K. Ladha, and M. Ali. (1995). Green manure technology: Potential usage, limitations. A case study for lowland rice. *Plant and Soil.* 174: 181-194.

Becker, M., J.K. Ladha, and J.C.G. Ottow. (1990). Growth and nitrogen fixation of two stem-nodulating legumes and their effect as green manure on lowland rice. *Soil Biology and Biochemistry.* 22: 1109-1119.

Becker, M., J.K. Ladha, and J.C.G. Ottow. (1994). Nitrogen losses and lowland rice yield as affected by residue N release. *Soil Science Society of America Journal.* 58: 1660-1665.

Bell, R.W. and V. Seng. (2004). Rainfed lowland rice-growing soils of Cambodia, Laos and Northeast Thailand. In Seng, V., E. Craswell, S. Fukai, and K. Fisher (eds.) *Water in Agriculture.* pp. 161-173. ACIAR Proceedings 116. Phnom Penh, Cambodia, Nov 25-28, 2003.

Blair, G.J., N. Chinoim, R.D.B. Lefroy, G.C. Anderson, and G.J. Crocker. (1991). A sulfur soil test for pastures and crops. *Australian Journal of Soil Research.* 29: 619-626.

Boserup, E. (1965). *The Conditions of Agricultural Growth.* London: Allen and Unwin.

Boserup, E. (1981). *Population and Technological Change: A Study of Long-Term Trends.* Chicago: University of Chicago Press.

Broadbent, F.E. (1979). Mineralization of organic nitrogen in paddy soils. In *Nitrogen and Rice.* pp. 105-118. Manila, Philippines: International Rice Research Institute.

Buresh, R.J. and S.K. De Datta. (1991). Nitrogen dynamics and management of rice-legume cropping systems. *Advances in Agronomy.* 45: 1-59.

Byerlee, D. (1994). Technology transfer systems for improved crop management: Lessons for the future. In Anderson, J.R. (ed.) *Agricultural Technology: Policy Issues for the International Community.* pp. 208-230. CABI: Wallingford, United Kingdom.

Carsky, R.J., M. Becker, and S. Hauser. (2001). Mucuna cover crop fallow systems: Potential and limitations. In Dick, W.A. and J.L. Hatfield (eds.) *Sustaining Soil Fertility in West Africa.* SSSA Special Publication. pp. 111-135. Madison, WI: Soil Science Society of America.

Cassman, K.G., S.K. De Datta, S.T. Amarante, S.P. Liboon, M.I. Samson, and M.A. Dizon. (1996). Long-term comparison of the agronomic efficiency and residual benefits of organic and inorganic nitrogen sources for tropical lowland rice. *Experimental Agriculture.* 32: 427-444.

Cassman, K.G., A. Dobermann, P.C. Sta Cruz, G.C. Gines, M.I. Samson, J.P. Descalsota, J.M. Alcantara, M.A. Dizon, and D.C. Olk. (1996). Soil organic matter and the indigenous nitrogen supply of intensive irrigated rice systems in the tropics. *Plant and Soil.* 182: 267-278.

Cassman, K.G., G.C. Gines, M.A. Dizon, M.I. Samson, and J.M. Alcantara. (1996). Nitrogen-use efficiency in tropical lowland rice systems: Contributions from indigenous and applied nitrogen. *Field Crops Research.* 47: 1-12.

Cassman, K.G. and P.L. Pingali. (1994). Extrapolating trends from long-term experiments to farmers' fields: The case of irrigated rice systems in Asia. In Varnett, V., P. Payne, and R. Steiner (eds.) *Agricultural Sustainability: Economic, Environmental and Statistical Considerations.* pp. 63-86. London: John Wiley and Sons, Ltd.

Chanchareonsook, J., S. Vacharotayan, H. Wada, C. Suwannarat, S. Panichsakpatana, and T. Yoshida. (1991). Utilization of organic wastes and paddy soil fertility. In M. Ekasingh et al. (eds.) *International Seminar on the Impact of Agricultural Production on the Environment.* pp. 290-304. Chiang Mai, Thailand: Sang Silp Printing.

Cho, D.Y. and F. Ponnamperuma. (1971). Influence of soil temperature on the chemical kinetics of flooded soils and the growth of rice. *Soil Science.* 112: 184-194.

Chou, C.-H., Y.-C. Chiang, H.H. Cheng, and F.O. Farrow. (1982). Transformation of ^{15}N-enriched fertilizer nitrogen during rice straw decomposition in submerged soil. *Botanical Bulletin of Academia Sinica.* 23: 119-133.

CIAP (Cambodia-IRRI-Australia Project). (1996). *Annual Research Report.* Phnom Penh, Cambodia. p. 177.

CIAP (Cambodia-IRRI-Australia Project). (1997). *Annual Research Report.* Phnom Penh, Cambodia. p. 183.

Clément A., J.K. Ladha, and F.P. Chalifour. (1995). Crop residue effects on nitrogen mineralization, microbial biomass and rice yield in submerged soils. *Soil Science Society of America Journal.* 59: 1595-1603.

Clément A., J.K. Ladha, and F.P. Chalifour. (1998). Nitrogen dynamics of various green manure species and the relationship to lowland rice production. *Agronomy Journal.* 90: 149-154.

Dawe, D., A. Dobermann, J.K. Ladha, R.L. Yadav, L. Bao, R.K. Gupta, P. Lal, et al. (2003). Do organic amendments improve yield trends and profitability in intensive rice systems? *Field Crops Research.* 83: 191-213.

Dawe, D., A. Dobermann, P. Moya, S. Abdulrachman, B. Singh, P. Lal, S.Y. Li, et al. (2000). How widespread are yield declines in long-term rice experiments in Asia? *Field Crops Research.* 66: 175-193.

De Datta, S.K. and R.J. Buresh. (1989). Integrated nitrogen management in irrigated rice. *Advances in Soil Science.* 10: 143-169.

De Macale, M.A.R. (2002) *The Role of Azolla Cover in Improving the Nitrogen Use Efficiency of Lowland Rice.* Center for Development Reserach (ZEF), Ecology and Development Series, No 2. Göttingen, Germany: Cuvillier Verlag.

Dei, Y. (1975). The effects of cereal crop residues on paddy soils. *ASPAC Extension Bulletin No. 49.* Taipei, Taiwan: ASPAC Food & Fertilizer Technology Center.

Dobermann, A. and P.F. White. (1999). Strategies for nutrient management in irrigated and rainfed lowland rice systems. *Nutrient Cycling in Agroecosystems.* 53: 1-18.

Dobermann, A., C. Witt, S. Abdulrachman, H.C. Gines, R. Nagarajan, T.T. Son, P.S. Tan, et al. (2003a). Soil fertility and indigenous nutrient supply in irrigated rice domains of Asia. *Agronomy Journal.* 95: 913-923.

Dobermann, A., C. Witt, S. Abdulrachman, H.C. Gines, R. Nagarajan, T.T. Son, P.S. Tan, et al. (2003b). Estimating indigenous nutrient supplies for site-specific nutrient management in irrigated rice. *Agronomy Journal.* 95: 924-935.

Dobermann, A., C. Witt, and D. Dawe, eds. (2004). *Increasing the Productivity of Intensive Rice Systems through Site-Specific Nutrient Management.* Science Publishers, Inc., and International Rice Research Institute (IRRI), Enfield, NH (USA) and Los Baños (Philippines).

Dobermann, A., C. Witt, D. Dawe, S. Abdulrachman, H.C. Gines, R. Nagarajan, S. Satawatananont, et al. (2002). Site-specific nutrient management for intensive rice cropping systems in Asia. *Field Crops Research.* 74: 37-66.

Engels, K.A., M. Becker, J.C.G. Ottow, and J.K. Ladha. (1995). Influence of P and PK fertilization on N_2 fixation of the stem-nodulating green manure legume *Sesbania rostrata* in different marginally productive wetland rice soils. *Biology and Fertility of Soils.* 20: 107-111.

Evans, D. and P.P. Rotar. (1987). *Sesbania in Agriculture.* Westview Tropical Agriculture Series, No. 8. Boulder, CO: Westview Press.

Fan, S. and P. Hazell. (1999). *Should Developing Countries Invest More in Less-Favored Areas? An Empirical Analysis of Rural India.* Washington, DC: International Food Policy Research Institute.

FAOSTAT (2005). http://www.faostat.org.

Flaig, W., H. Beutelspacher, and E. Rietz. (1975). Chemical composition and physical properties of humic substances. In Gieseking, J.E. (ed.) *Soil Components,* Vol. 1. *Organic Components.* pp. 1-211. New York: Springer Verlag.

Fukai, S., J. Basnayake, and M. Cooper. (2000). Modelling water availability, crop growth, and yield of rainfed lowland rice genotypes in Northeast Thailand. In Tuong, T.P., S.P. Kam, L. Wade, S. Pandey, B.A.M. Bouman, and B. Hardy (eds.) *Characterising and Understanding Rainfed Environments.* pp. 111-130. Manila, Philippines: International Rice Research Institute.

Fukai, S., P. Sittisuang, and M. Chanphengsay. (1998). Increasing production of rainfed lowland rice in drought prone environments: A case study in Thailand and Laos. *Plant Production Science.* 1: 75-82.

Garrity, D.P. and M. Becker. (1995). Where do green manures fit in Asian rice farming systems? In Ladha, J.K and D.P. Garrity (eds.) *Green Manure Production Systems for Asian Ricelands.* pp. 2-12. Manila, Philippines: International Rice Research Institute.

Garrity, D.P. and J.C. Flinn. (1988). Farm-level management systems for green manure crops in Asian rice environments. In *Sustainable Agriculture: Green Manure in Rice Farming.* pp. 111-129. Manila, Philippines: International Rice Research Institute.

George, T., J.K. Ladha, R.J. Buresh, and D.P. Garrity. (1992). Managing native and legume-fixed nitrogen in lowland rice-based cropping systems. *Plant and Soil.* 141: 69-91.

George, T., J.K. Ladha, R.J. Buresh, and D.P. Garrity. (1994). Nitrate dynamics during the aerobic soil phase in lowland rice-based cropping systems. *Soil Science Society of America Journal.* 58: 269-273.

George, T., J.K. Ladha, D.P. Garrity, and R.O. Torres. (1995). Nitrogen dynamics of grain legume-weedy fallow-flooded rice sequences in the tropics. *Agronomy Journal.* 87: 1-6.

Homma, K., T. Horie, M. Ohnishi, T. Shiraiwa, N. Supapoj, N. Matsumoto, and N. Kabaki. (2001). Quantifying the toposequential distribution of environmental resources and its relationship with rice productivity. In Fukai, S. and J. Basnayake (eds.) *Increased Lowland Rice Production in the Mekong Region.* pp. 179-190. *Proceedings of an International Workshop,* Vientiane, Laos, Oct 3-Nov 2, 2000. ACIAR Proceedings No. 101.

Homma, K., T. Horie, T. Shiraiwa, N. Supapoj, N. Matsumoto, and N. Kabaki. (2003). Toposequential variation in soil fertility and rice productivity of rainfed lowland paddy fields in mini-watershed (*Nong*) in Northeast Thailand. *Plant Production Science.* 6: 147-153.

Huke, R.E. and E.H. Huke. (1997). *Rice Area by Type of Culture: South, Southeast, and East Asia. A Revised and Updated Data Base.* Manila, Philippines: International Rice Research Institute.

IRRI. (1988). *Sustainable Agriculture: Green Manure in Rice Farming.* Manila, Philippines: International Rice Research Institute.

IRRI. (2002). *World Rice Statistics 1993-94.* Manila, Philippines: International Rice Research Institute.

Ismunadji, M. (1978). Utilization of cereal crop residues and its agricultural significance in Indonesia. *Contributions, Central Research Institute for Agriculture (No. 37).* Bogor, Indonesia.

Janssen, B.H., F.C.T. Guiking, D. van der Eijk, E.M.A. Smaling, J. Wolf, and H. van Reuler. (1990). A system for quantitative evaluation of the fertility of tropical soils (QUEFTS). *Geoderma.* 46: 299-318.

Juliano, B.O. and D.B. Bechtel. (1985). The rice grain and its gross composition. In Juliano, B.O. (ed.) *Rice: Chemistry and Technology.* 2nd edition. AACC monograph series. St. Paul, MN: American Association of Cereal Chemists.

King, F.H. (1911). *Farmers of Forty Centuries; or, Permanent Agriculture in China. Korea and Japan.* Emmaus, PA: Rodale Press.

Kirk, G.J.D. (2004). *The Biogeochemistry of Submerged Soils.* Chichester, UK: Wiley Science.

Kröck, T.J., T. Alkämper, and I. Watanabe. (1988). Effect of an *Azolla* cover on the conditions in floodwater. *Journal of Agronomy and Crop Science.* 161: 185-189.

Kumazawa K. (1984). Beneficial effects of organic matter on rice growth and yield in Japan. In *Organic Matter and Rice.* pp. 431-444. Manila, Philippines: International Rice Research Institute.

Ladha, J.K., T. George, and B.B. Bohlool. (1992). *Biological Nitrogen Fixation for Sustainable Agriculture.* Dordrecht, The Netherlands: Kluwer Academic Publishers.

Ladha, J.K., D.K. Kundu, M.G. Angelo-Van Coppenolle, M.B. Peoples, V.R. Carangal, and P.J. Dart. (1996). Legume productivity and soil nitrogen dynamics in low-land rice-based cropping systems. *Soil Science Society of America Journal.* 60: 183-192.

Ladha, J.K., R.P. Pareek, and M. Becker. (1992). Stem-nodulating legume-Rhizobium symbiosis and its agronomic use in lowland rice. *Advances in Soil Science.* 20: 147-192.

Linquist, B and P. Sengxua. (2001). *Nutrient Management in Rainfed Lowland Rice in the Lao PDR.* Manila, Philippines: International Rice Research Institute.

Linquist, B. and P. Sengxua. (2003). Efficient and flexible management of nitrogen for rainfed lowland rice. *Nutrient Cycling in Agroecosystems.* 67:107-115

Linquist, B.A., P. Sengxua, A. Whitbread, J. Schiller, and P. Lathvilayvong. (1998). Evaluating nutrient deficiencies and management strategies for lowland rice in Lao PDR. In Ladha, J.K., L.J. Wade, A. Dobermann, W. Reichardt, G.J.D. Kirk, and C. Piggin (eds.) *Rainfed Lowland Rice: Advances in Nutrient Management Research.* pp. 59-73. Manila, Philippines: International Rice Research Institute.

Maclean, J.L., D.C. Dawe, B. Hardy, and G.P. Hettel, eds. (2002). *Rice almanac: Source Book for the Most Important Economic Activity on Earth.* 3rd edition. Manila, Philippines: International Rice Research Institute.

Mar, M., T. Saing, T. Thein, and R.K. Palis. (1995). *Sesbania* green manure program for rice farming in Myanmar. In *Proceedings, International Rice Research Conference* April 21-25, 1992. pp. 119-131. Manila, Philippines: International Rice Research Institute.

Masayna, W., H. Kai, and S. Kawaguchi. (1985). Nitrogen behavior in tropical wet-land rice soils. 2. The efficiency of fertilizer nitrogen, priming effect and A-values. *Fertilizer Research.* 6: 37-47.

Naklang, K., A. Whitbread, R. Lefroy, G. Blair, S. Wonprasaid, Y. Konboon, and D. Suriya-arunroj. (1999). The management of rice straw, fertilizers and leaf lit-ters in rice cropping systems in Northeast Thailand. 1. Soil carbon dynamics. *Plant and Soil.* 209: 21-28.

Nesbitt, H.J. (1997). Topography, climate and rice production. In Nesbitt, H.J. (ed.) *Rice Production in Cambodia.* Manila (Philippines): International Rice Re-search Institute.

Oberthür, T. and S.P. Kam. (2000). Perception, understanding, and mapping of soil variability in the rainfed lowlands of Northeast Thailand. In Tuong, T.P., S.P. Kam, L. Wade, S. Pandey, B.A.M. Bouman, and B. Hardy (eds.) *Character-ising and Understanding Rainfed Environments.* pp. 75-95. Manila, Philippines: International Rice Research Institute.

Olk, D.C., K.G. Cassman, N. Mahieu, and E.W. Randall. (1998). Conserved chemi-cal properties of young humic acid fractions in tropical lowland soil under inten-sive irrigated rice cropping. *European Journal of Soil Science.* 49: 337-349.

Olk, D.C., K.G. Cassman, E.W. Randall, P. Kinchesh, L.J. Sanger, and J.M. Anderson. (1996). Changes in chemical properties of organic matter with intensified rice cropping in tropical lowland soil. *European Journal of Soil Science.* 47: 293-303.

Olk, D.C., K.G. Cassman, G. Simbahan, P.C. Sta Cruz, S. Adulrachman, R. Nagarajan, Pham Sy Tan, and S. Satawathananont. (1999). Interpreting fertilizer-use efficiency in relation to soil nutrient-supplying capacity, factor productivity, and agronomic efficiency. *Nutrient Cycling in Agroecosystems.* 53: 35-41.

Olk, D.C., M.I. Samson, and P. Gapas. (2007). Inhibition of nitrogen mineralization in young humic fractions by anaerobic decomposition of rice crop residues. *European Journal of Soil Science.* 58: 270-281.

Pande, K.R. and M. Becker. (2003). Seasonal soil nitrogen dynamics in rice-wheat cropping systems of Nepal. *Journal of Plant Nutrition and Soil Science.* 66: 499-506.

Pandey, S. (1998). Nutrient management technologies for rainfed rice in tomorrow's Asia: economic and institutional considerations. In Ladha, J.K., L. Wade, A. Dobermann, W. Reichardt, G.J.D. Kirk, and C. Piggin (eds.) *Rainfed Lowland Rice: Advances in Nutrient Management Research.* pp. 3-28. Manila, Philippines: International Rice Research Institute.

Pandey, S. and M. Sanamongkhoun. (1998). *Rainfed Lowland Rice in Laos: A Socio-Economic Benchmark Study.* Manila, Philippines: International Rice Research Institute.

Patrick, W.H. and D.S. Mikkelsen. (1971). Plant nutrient behavior in flooded soils. In Olsen, R.A. (ed.) *Fertilizer Technology and Use.* 2nd edition. pp. 187-215. Madison WI: Soil Science Society of America.

Peoples, M.B. and D.F. Herridge. (1990). Nitrogen fixation by legumes in topical and sub-tropical agriculture. *Advances in Agronomy.* 44: 152-223.

Peoples, M.B., D.F. Herridge, and J.K. Ladha. (1995). Biological nitrogen fixation: an efficient source of nitrogen for sustainable agricultural production. *Plant and Soil.* 174: 3-28.

Peoples, M.B., J.K. Ladha, and D.F. Herridge. (1995). Enhancing legume N_2 fixation through plant and soil management. *Plant and Soil.* 174: 83-102.

Pingali, P.L., M. Hossain, S. Pandey, and L. Price. (1998). Economics of nutrient management in Asian rice systems: Towards increasing knowledge intensity. *Field Crops Research.* 56: 157-176.

Pingali, P.L., P.F. Moya, and L.E. Velasco. (1990). *The Post-Green Revolution Blues in Asian Rice Production.* IRRI Social Science Division Paper Series No 90-01. Manila, Philippines: International Rice Research Institute.

Ponnamperuma, F. (1972). The chemistry of submerged soils. *Advances in Agronomy.* 24: 29-97.

Ponnamperuma, F. (1984). Straw as a source of nutrients for wetland rice. In *Organic Matter and Rice.* pp. 117-136. Manila, Philippines: International Rice Research Institute.

Ragland, J. and L. Boonpuckdee. (1988). Fertilizer responses in NE Thailand. 3. Nitrogen use and soil acidity. *Thailand Journal of Soil Fertility.* 10: 67-76.

Rahman, M., A.K. Podder, C. van Hove, Z.N.T. Begum, T. Heulin, and A. Hartmann. (1996). *Biological Nitrogen Fixation Associated with Rice Production.* Dordrecht, The Netherlands: Kluwer Academic Publishers.

Randhawa, N.S., M.K. Sinha, and P.N. Takkar. (1978). Micronutrients. In *Soils and Rice.* pp. 581-603. Manila, Philippines: International Rice Research Institute.

Roger, P.A. and I. Watanabe. (1986). Technologies for utilizing biological nitrogen fixation in wetland rice: Potentialities, current usage and limiting factors. *Fertilizer Research.* 9: 39-77.

Rosegrant, M.W. and I.H. Roumassett. (1988). Economic feasibility of green manure in rice-based cropping systems. In *Sustainable Agriculture: Green Manure in Rice Farming.* pp. 13-27. Manila, Philippines: International Rice Research Institute.

Schmidt-Rohr, K., J-D. Mao, and D.C. Olk. (2004). Nitrogen-bonded aromatics in soil organic matter and their implications for a yield decline in intensive rice cropping. *Proceedings, National Academy of Sciences USA.* 101: 6351-6354.

Schnier, H.F. (1995). Significance of timing and method of N fertilizer application for the N-use efficiency in flooded tropical rice. *Fertilizer Research.* 42: 129-138.

Seng, V., R.W. Bell, and I.R. Willet. (2004). Amelioration of growth reduction of lowland rice caused by a temporary loss of soil-water saturation. *Plant and Soil* 265: 1-16.

Seng, V., C. Ros, R.W. Bell, P.F. White, and S. Hin. (2001). Nutrient requirements for lowland rice in Cambodia. In S. Fukai and J. Basnayake (eds.) *Increased Lowland Rice Production in the Mekong Region.* pp. 169-178. *Proceedings of an International Workshop, Vientiane, Laos.* Oct 30-Nov 1. ACIAR Proceedings 101.

Shibahara, F., S. Yamamuro, and K. Inubushi. (1998). Dynamics of microbial biomass nitrogen as influenced by organic matter application in paddy fields. I. Fate of fertilizer and soil organic N determined by ^{15}N tracer technique. *Soil Science and Plant Nutrition.* 44: 167-178.

Somado, E.A., M. Becker, R.F. Kuehne, K.L. Sawrawat, and P.L.G. Vlek. (2003). Combined effects of legumes with rock phosphorus on rice in West Africa. *Agronomy Journal.* 95: 1172-1178.

Son, T.T., N.V. Chien, V.T.K. Thoa, A. Dobermann, and C. Witt. (2004). Site-specific nutrient management in irrigated rice systems of the Red River Delta of Vietnam. In Dobermann, A. C. Witt, and D. Dawe (eds.) *Increasing Productivity of Intensive Rice Systems through Site-Specific Nutrient Management.* pp. 223-242. Enfield, NH. USA and Los Baños, Philippines: Science Publishers, Inc., International Rice Research Institute (IRRI).

Stoop, W.A., N. Uphoff, and A. Kassam. (2002). A review of agricultural research issues raised by the system of rice intensification (SRI) from Madagascar: Opportunities for improving farming systems for resource-poor farmers. *Agricultural Systems.* 71: 249-274.

Supapoj, N., K. Naklang, and Y. Konboon. (1998). Using organic material to improve soil productivity in rainfed lowland rice in Northeast Thailand. In Ladha, J.K., L.J. Wade, A. Dobermann, W. Reichardt, G.J.D. Kirk, and C. Piggin (eds.) *Rainfed Lowland Rice: Advances in Nutrient Management Research.* pp. 161-168. Manila, Philippines: International Rice Research Institute.

Tanaka, A. (1974). *Methods of Handling of Cereal Crop Residues in Asian countries and Related Problems.* ASPAC Extension Bulletin No. 43. Taipei, Taiwan: ASPAC Food & Fertilizer Technology Center.

Tanaka, A. (1978). Role of organic matter. In *Soils and Rice.* pp. 605-620. Manila, Philippines: International Rice Research Institute.

Thach, T.N.A. (1999). Prediction of crop nitrogen uptake and grain yield by soil nitrogen availability tests for irrigated lowland rice (*Oryza sativa L.*) in Nueva Ecija, Philippines. MS thesis, University of Philippines, Los Baños.

Tsutsuki, K. (1984). Volatile products and low-molecular-weight phenolic products of the anaerobic decomposition of organic matter. In *Organic Matter and Rice.* pp. 329-343. Manila, Philippines: International Rice Research Institute.

Uexkull, H.R. and E. Mutert. (1992). Principles of balanced fertilization. In: Proceedings of the Regional FADINAP Seminar of Fertilization and the Environment, Chang Mai, Thailand, September 1992.

Uphoff, N. (2002). Opportunities for raising yields by changing management practices: The system of rice intensification in Madagascar. In Uphoff, N. (ed.) *Agroecological Innovation: Increasing Food Production with Participatory Development.* pp. 145-161. Sterling, VA: Earthscan Publications.

Van Hove, C. (1989). *Azolla and Its Multiple Uses with Emphasis on Africa.* pp. 207-218. Rome: Food and Agriculture Organization.

Ventura, W., G.B. Mascanina, R.E. Furoc, and I. Watanabe. (1987). *Azolla* and *Sesbania* as biofertilizers for lowland nice. *Philippine Journal of Crop Science.* 12: 61-69.

Vlek, P.L.G. and E.T. Craswell. (1981). Ammonia volatilization from flooded soils. *Fertilizer Research.* 2: 227-245.

Vlek, P.L.G., M.Y. Diakite, and H. Mueller. (1995). The role of *Azolla* in curbing ammonia volatilization from flooded rice systems. *Fertilizer Research.* 42: 165-174.

Wade, L.J., S.T. Amarante, A. Olea, D. Harnpichitvitaya, K. Naklang, A. Wihardjaka, S.S. Sengar, M.A. Mazid, G. Singh, and C.G. McLaren. (1999). Nutrient requirements in rainfed lowland rice. *Field Crops Research.* 64: 91-107.

Watanabe, I. (1991). Biological nitrogen fixation in sustainable rice farming. In Dutta, S.K. and C. Sloger (eds.) *Biological N_2 Fixation Associated with Rice.* pp. 280-289. New Delhi, India: Oxford and IBH Publishers.

Watanabe, I. and C. Ramirez. (1984). Relationship between soil P availability and *Azolla* growth. *Soil Science and Plant Nutrition.* 30: 595-598.

Watanabe, I. and R.A. Roger. (1984). Nitrogen fixation in wetland rice field. In Subba Rao, N.S. (ed.) *Current Developments in Biological Nitrogen Fixation.* pp. 237-274. New Delhi, India: Oxford and IBH Publishing Co.

Watanabe, I. W. Ventura, G. Mascarina, and D.L. Eskew. (1989). Fate of *Azolla* and urea nitrogen applied to wetland rice. *Biology and Fertility of Soils.* 8: 102-110.

Whitbread, A., G. Blair, K. Naklang, R. Lefroy, S. Wonprasaid, Y. Konboon, and D. Suriya-arunroj. (1999). The management of rice straw, fertilizers and leaf litters in rice cropping systems in Northeast Thailand. 2. Rice yields and nutrient balances. *Plant and Soil.* 209: 29-36.

White, P.F., T. Oberthür, and P. Sovuthy, eds. (1997). *The Soils Used for Rice Production in Cambodia: A Manual for Their Identification and Management.* Manila, Philippines: International Rice research Institute.

Wihardjaka, A., G.J.D. Kirk, S. Abdulrachman, and C.P. Mamaril. (1998). Potassium balances in rainfed lowland rice on a light-textured soil. In Ladha, J.K., L.J. Wade, A. Dobermann, W. Reichardt, G.J.D. Kirk, and C. Piggin (eds.) *Rainfed Lowland Rice: Advances in Nutrient Management Research.* pp. 127-137. Manila, Philippines: International Rice Research Institute.

Willet, I.R. (1995). Role of organic matter in controlling chemical properties and fertility of sandy soils used for lowland rice in Northeast Thailand. In Lefroy, R.D.B., G.J. Blair, and E.T. Craswell (eds.) *Soil Organic Matter for Sustainable Agriculture in Asia.* pp. 109-114. Proceedings 56. Canberra, Australia: Australian Centre for International Agricultural Research.

Witt, C., V. Balasubramanian, A. Dobermann, and R.J. Buresh. (2002). Nutrient management. In Fairhurst, T. and C. Witt (eds.) *Rice: A Practical Guide for Nutrient Management.* pp. 1-45. Singapore and Manila, Philippines: Potash and Phosphate Institute & Potash and Phosphate Institute of Canada and International Rice Research Institute.

Witt, C., J.M.C.A. Cabrera-Pasuquin, R. Mutters, and S. Peng. (2003). Nitrogen content of rice leaves as predicted by SPAD, NIR, spectral reflectance and leaf color charts. *Agronomy Abstracts,* November 2-6, 2003, Denver Colorado [CD-ROM].

Witt, C., K.G. Cassman, D.C. Olk, U. Biker, S.P. Liboon, M.I. Samson, and J.C.G. Ottow. (2000). Crop rotation and residue management effects on carbon sequestration, nitrogen cycling and productivity of irrigated rice systems. *Plant and Soil.* 225: 263-278.

Witt, C. and A. Dobermann. (2004). Towards a decision support system for site-specific nutrient management. In Dobermann, A., C. Witt, and D. Dawe (eds.) *Increasing Productivity of Intensive Rice Systems through Site-Specific Nutrient Management.* pp. 359-395. Enfield, NH (USA) and Manila, Philippines: Science Publishers, Inc., and International Rice Research Institute (IRRI).

Witt, C., A. Dobermann, S. Abdulrachman, H.C. Gines, G.H. Wang, R. Nagarajan, S. Satawatananont, T.T. Son, P.S. Tan, Le Van Tiem, G.C. Simbahan, and D.C. Olk. (1999). Internal nutrient efficiencies in irrigated lowland rice of tropical and subtropical Asia. *Field Crops Research.* 63: 113-138.

Wonprasaid, S., S. Khunthasuvon, P. Sittisuang, and S. Fukai. (1996). Performance of contrasting rice cultivars selected for rainfed lowland conditions in relation to soil fertility and water availability. *Field Crops Research.* 47: 267-275.

Yamauchi, M. (1989). Rice bronzing in Nigeria caused by nutrient imbalances and its control by potassium sulfate application. *Plant and Soil.* 117: 275-286.

Yang, W.-H., S. Peng, A.L. Sanico, R.J. Buresh, and C. Witt. (2003). Estimation of leaf nitrogen status using leaf color charts in rice. *Agronomy Journal.* 95: 212-217.

Yoneyama, T., K.-K. Lee, and T. Yoshida. (1977). Decomposition of rice residues in tropical soils. IV. The effect of rice straw on nitrogen fixation by heterotrophic bacteria in some Philippine soils. *Soil Science and Plant Nutrition.* 23: 287-295.

Yoneyama, T. and T. Yoshida. (1977a). Decomposition of rice residue in tropical soils. III. Nitrogen mineralization and immobilization of rice residue during its decomposition in soil. *Soil Science and Plant Nutrition.* 23: 175-183.

Yoneyama, T. and T. Yoshida. (1977b). Decomposition of rice residue in tropical soils. I. Nitrogen uptake by rice plants from straw incorporated, fertilizer (ammonium sulfate) and soil. *Soil Science and Plant Nutrition.* 23: 33-40.

Yost, R. and D. Evans. (1988). *Green Manure and Legume Covers in the Tropics.* HITAHR, University of Hawaii. Research Series 055.

Zeigler, R.S. and D.W. Puckridge. (1995). Improving sustainable productivity in rice-based rainfed lowland systems of South and Southeast Asia. Feeding 4 billion people. The challenge for rice research in the 21st century. *GeoJournal.* 35: 307-324.

Chapter 10

Integrated Nutrient Management: Experience from South America

Bernardo van Raij
Alfredo Scheid Lopes
Eduardo Casanova
Martín Díaz-Zorita

INTRODUCTION

South America is a region of large contrasts, with diversity of climate and soils imposed by geology and the enormous differences in latitude and altitude. Several ecosystems are found in the region, including the Atlantic forest, Andean Altiplanos, and Pampas (Britannica, 2001). Each of these have specific characteristics that affect the agriculture that can be practiced.

Presently two opposing forces are acting in a global way. On the one hand, it is expected that there will be a high demand for food to supply the growing population of the world in the coming decades. South America is a continent that can substantially increase its cultivated area and also its crop productivity. On the other hand, there is an increasing awareness of the need to protect the world's biodiversity, avoiding excessive destruction of natural ecosystems. Better fertilizer use can help in both aspects, by increasing agricultural outputs, and by reducing the need for new areas through the increase in productivity. It can also reduce the negative impacts of fertilization in the environment.

South America is also a continent of large social contrasts, with a broad spectrum of agricultural technology in use, from the most primitive to the most advanced. In many places agricultural production is insufficient for local necessities, whereas in others large exportable surpluses are produced.

Very important advances were possible in some regions by increasing fertilizer and limestone use, transforming large areas of unfertile soil in very productive land. This also helped to reverse soil fertility decline in many areas. Soil testing is used in several parts of the continent as a basic tool for diagnosis of soil reaction and nutrient deficiencies.

Although the importance of organic matter is recognized, fertilization is mostly based on mineral fertilizers. Organic sources are used when available, but in most cases the nutrient concentrations in these materials are not considered. In some places, with intensive animal production, manure is treated as a waste to be disposed of rather than a nutrient source. On the other hand, increasing environmental concern encourages the disposal of much of the available organic residues in agriculture.

One of the major breakthroughs in agriculture is the practice of conservation agriculture, ensuring that the soil is always covered with organic materials, which results in the control of soil erosion, higher organic matter contents, increase in the biodiversity of soil flora and fauna, improvement of physical properties, and better recycling of water and nutrients.

This chapter will describe how South America manages plant nutrition and how things can be improved for the general benefit of the farmer, the environment, and society.

MAIN ECOSYSTEMS AND CROPPING SYSTEMS

In Table 10.1 information is given on the twelve South American countries. Brazil is a Portuguese speaking country and all the other countries except Guyana and Surinam speak Spanish. This means that culturally the countries of the continent are close and the similarity in languages helps communication.

South America extends over an area of 17,819,100 km^2 and is the fourth largest continent, covering about 12 percent of the World's land surface. The Andean Cordillera, localized in the west of South America, stretches continually for 7,500 km, from Tierra del Fuego in the extreme south, to Venezuela in the north. The Andes mountain range is the only volcanically active region of the continent. On the west of the Andes there is only a narrow strip of land and rivers are short. The main areas of the continent, on the eastern side of the Andes, have extensive rivers and large land areas suitable for agriculture.

The South American relief is highly diversified and the most important geographic regions are: the Andes, the second highest mountain range in the world, the Patagonian plateau, the Brazilian and Guyana plateaus, the

TABLE 10.1. Information on the countries of South America

Country	Total area km²	Habitants, millions	Hectares per habitant	Rural population, %	Major crops
Argentina	2,780,092	37.9	7.4	11	Wheat, maize, soybean, sorghum
Bolivia	1,098,581	8.7	12.6	39	Soybean, sugarcane
Brazil	8,514,204	169.6	4.8	19	Soybean, maize, sugarcane, coffee, rice, oranges, cocoa, pasture, forestry
Chile	756,626	15.6	4.9	15	Wheat, oats, barley, maize
Colombia	1,141,748	43.5	2.6	27	Coffee, cocoa, sugarcane, banana
Ecuador	283,561	13.1	2.2	37	Sugarcane, rice
Guyana	214,970	0.765	28.1	63	Banana, coffee, cocoa
Paraguay	406,752	5.8	7.0	45	Soybean, cotton, sugarcane
Peru	1,285,215	26.5	4.8	28	Coffee, rice, maize, potato
Surinam	163,829	0.421	38.9	27	Rice, banana
Uruguay	176,215	3.4	5.2	9	Rice, sugarcane, wheat
Venezuela	912,050	25.1	3.6	14	Sugarcane, banana, maize

Source: Adapted from Almanaque Abril (2001, 2003).

coastal plain on the Atlantic coast, and the Amazon basin. The highest point of the continent is the Aconcágua peak, 6,959 m, located in Argentina. The central plains contain the large hydrographic basins of the Amazon, the Plata, and the Orinoco rivers.

The continent presents a great variety of ecosystems, resulting from diverse combinations of latitude, altitude, climate, geology, and other geographic features, the most important being the Amazon region, the Cerrado, the Llanos, Pantanal, Chaco, Caatinga, the Atlantic forest, the Andean Altiplanos, and the Pampas. Soils, of course, also follow the pattern of diversity, influenced by the variety of combinations of the soil formation factors: parent material, climate, organisms, topography and time (Jenny, 1941), which determine land occupation by crops. Currently, South America still has 70 percent of its original vegetation (Abril, 2003a), but the occupation of the different ecosystems is not uniform.

The Atlantic Forest. Considered the ecosystem with the largest biodiversity in the world, the Atlantic forest extends from southeastern Brazil to the south, into Argentina and Paraguay. The original area was estimated as 130 Mha, of which only 9.5 Mha remain with the original vegetation. Presently, around 70 percent of the population and of the economy of Brazil are found in this region (Abril, 2003). One of the first regions occupied by agriculture, it is probably the most important agricultural area of the continent, presenting a large diversity of crops, including grains, coffee, sugarcane, cotton, fruits, vegetables, pasture, forestry, and others. It is also associated with poultry, swine, dairy cattle, and beef cattle production. Climatic conditions are tropical and subtropical and most agriculture is rainfed, although irrigation is increasing rapidly. In most cases, two sequential crops of grains are possible in a calendar year. Highly fertile soils derived from basic rocks were responsible for the high agricultural outputs in the past. Sugarcane in the colonial era, and coffee in Brazil since the nineteenth century, were greatly responsible for the development of the economy of the country.

The Pampas. This grassland covers approximately 52 Mha along the central part of Argentina, extending into Uruguay and the south of Brazil. It is an extension of a bigger plain wedged between the Andes in the west and the Guyana Brazilian shield in the east. Originally it was covered with grasses and with extensive areas of originally very fertile mollisols (Hall et al., 1992). It is extensively used for grain and cattle production, and also for flooded rice.

The Cerrado Region. Another important region for agriculture and animal production is the Cerrado region, covered with neotropical savanna vegetation. It covers 267 Mha, with about 76 percent of it mostly in central Brazil and the rest in Bolivia, Colombia, Guyana, and Venezuela (Borlaug and Dowswell, 1994). The soils are mainly low fertility oxisols and ultisols and agriculture is only possible with the intensive use of limestone and fertilizers. Agriculture, mainly rain fed, is dedicated to grain, coffee, cotton, cattle production, etc. Tens of millions of hectares of virgin Cerrado land is potentially arable and can be converted to agriculture in the future.

The Amazon Region. The ecosystem of South America that occupies the most extensive area is the Amazon region, which comprises the largest hydrographic system in the world. The Amazon basin is the most extensive plain area in the continent. The total length of the river—measured from the headwaters of the Ucayali-Apurimac river system in Peru—is about 6,400 km, which is slightly shorter than the Nile River, the World's longest. It is estimated that about one-fifth of all the water that runs off the Earth's surface is carried by the Amazon. The flood-stage discharge at the river's mouth is about 175,000 cubic meters per second. (Britannica, 2001). Stretching some 2,760 km from north to south at its widest point, the basin includes the greater part of Brazil and Peru, significant parts of Colombia, Ecuador, and Bolivia, and a small area of Venezuela. The climate of the Amazon is warm, rainy, and humid, which poses a challenge for conventional agricultural production. In spite of the extremely lush forest vegetation, soils are generally of very low fertility. As in the Cerrado region, oxisols and ultisols are the predominant soils, although extensive areas of lowlands close to rivers are often of high fertility, but in these soils agriculture is restricted due to the seasonal flooding of the rivers. The main economic activity is cattle production and a variety of products of subsistence farming. However, there is potential for several crops, such as oil palm, cocoa, brazil nuts, and a variety of native plants.

The Caatinga. In Brazil, the Caatinga is a semiarid region, with an important irrigated fruit production area. It also has production of cotton, beans, corn, cassava, banana, beef cattle, and goats.

The Llanos. These are wide native grasslands stretching across northern South America, bordered by the Andes in the north and the west, occupying western Venezuela and northeastern Colombia. The Venezuelan Llanos have been for a long time one of South America's major livestock-raising areas, with cattle production being predominant over other agricultural activities. In addition, sugarcane, cotton, and rice are grown on a commercial scale on the plains and coffee has become important on the northwestern and northern highland fringes of the river basin.

The Chaco. An alluvial plain in the interior of south central South America, the Chaco is a vast, arid, lowland bounded by the Andes mountains on the west and by the Paraguay and Paraná rivers on the east.

The Pantanal. Considered as a great ecotone, the Pantanal represents the union of four great ecosystems: the Amazon Forest, the Savanna, the Chaco, and the Atlantic Forest. The Pantanal is located in the midwest of the states of Mato Grosso and Mato Grosso do Sul in Brazil and also in Paraguay. This enormous catchment area, distinguished for its peculiar environment, is linked to a regime of periodic floods that affect both the biotic components and the processes of soil formation. Floods occasionally cover about

70 percent of the Pantanal area, and depending on the location and altitude, flooding lasts from three to nine months (Wade, 1999).

Although the other ecosystems of South America present less activity in agriculture, they need to develop agriculture and increase its productivity since they shelter a considerable part of the rural population of the continent.

AGRICULTURAL PRODUCTION
AND NUTRIENT BALANCES

Fertilizer use in Brazil, for example, is directly associated with crop production (Table 10.2). The data shown for the sixteen main crops of the coun-

TABLE 10.2. Brazilian agricultural production performance—average for three year periods for harvest years 1982-1984 to 2000-2002

Indicators	1982-1984	1985-1987	1988-1990	1991-1993	1994-1996	1997-1999	2000-2002
Agricultural production[a] (1 Mt dry matter)	90.2	107.2	112.8	112.7	129.3	137.5	155.0
Grain production[b] (1 Mt grains)	50.3	59.5	65.5	64.1	74.2	76.8	92.9
Harvested area, sixteen crops (1 Mha)	44.1	47.8	50.2	46.0	46.3	44.6	47.7
Fertilizer consumption[c] (1 Mt N + P_2O_5 + K_2O)	2.54	3.37	3.60	3.31	4.40	5.39	6.28
Agricultural yield per area (kg/ha harvested area)	2,047	2,241	2,245	2,451	2,793	3,089	3,252
NPK consumption (kg/ha of harvested area)	58	71	72	72	95	121	132
Population[d] (1 million inhabitants)	129.8	137.7	145.2	152.2	159.0	165.7	172.4
Per capita production (kg/inhabitant)	695	779	777	740	813	830	899

Source: Adapted from ANDA (2003).

[a]Considering the production of the sixteen main exportation and internal consumption products, adjusted to dry weight basis as: sugar cane, cassava, and potato (15 percent), and orange and tomato (10 percent).

[b]Cereals and oil crops.

[c]Considering the consumption in the year of planting.

[d]Considering the population in the year of harvesting.

try indicate an increase in agricultural production and grain production over a period of twenty years, but with a nearly unchanged cultivated area, indicating that higher productivity was responsible for the major part of the increase in agricultural output. It should be emphasized that, for many years and until the beginning of the 1980s, the average consumption of fertilizers ($N + P_2O_5 + K_2O$) in Brazil was around 50 kg ha^{-1}. It was only during the 1990s that a substantial increase occurred for the average yearly consumption, reaching 132 kg ha^{-1} in the three-year period 2000-2002. Unofficial data for the harvest of 2003 indicate a record production of grains of around 120 million metric tonnes (Mt), with an average consumption of fertilizers ($N + P_2O_5 + K_2O$) of 138 kg ha^{-1} for the sixteen major crops.

In Table 10.3 information is given on the cultivated area, production, average yield, and fertilizer use for the main crops in Brazil. The calculations were based on statistical information data of ANDA and FAO. The data show that fertilizer use is now an important factor for soil productivity in

TABLE 10.3. Area, yield, productivity, and fertilizer consumption for the main crops in Brazil in the year 2002

		Total		Fertilizer use				
					Nutrients			
	Area	yield	Productivity	Product	N	P$_2$O$_5$	K$_2$O	Total
Crop	1,000 ha	1,000 t	kg ha^{-1}		kg ha^{-1}			
Bean	4,371	3,431	785	122	8	13	7	28
Coffee	2,380	2,511	1,055	542	114	24	92	230
Corn	12,615	42,576	3,375	262	40	35	33	108
Cotton	725	2,212	3,051	960	83	130	122	335
Orange	827	18,490	22,358	438	55	24	45	124
Potato	148	2,913	19,681	2,873	121	362	195	678
Rice	3,169	10,271	3,241	193	27	35	20	82
Soybean	18,445	51,609	2,798	365	8	66	62	136
Sugarcane	5,215	372,231	71,377	447	55	51	110	216
Wheat	2,064	2,928	1,431	276	12	50	47	109
Other crops	6,685			246	43	45	39	127
Total	56,646			327	31	48	52	131

Source: Calculated with data of ANDA (2003).

Brazil. But, there are some important discrepancies that do not show up in Table 10.3. For example, common beans present an average yield of 785 kg ha^{-1}. Yet, the average for the northeast, with more than half of the cultivated area, is only 408 kg ha^{-1}, whereas it is 1,752 kg ha^{-1} in the center west. The same tendency occurs with corn, with a yield of 1,229 kg ha^{-1} in the northeast and 4,208 kg ha^{-1} in the south. Such large discrepancies, where low yields are associated with low fertilizer use, indicate the need to optimize fertilizer use to increase agricultural output.

With respect to the individual nutrients, some patterns can be recognized. The highest rates of nitrogen are applied for potato, coffee, sugarcane, cotton, and orange. High rates of nitrogen are also used for vegetables and fruits for export, which are considered under "other crops" in Table 10.3. The smallest rates are used in common beans, rice, wheat, corn, and in soybeans. This last crop usually does not require mineral nitrogen fertilization, since this nutrient is supplied by biological fixation. An important segment that does not appear isolated in Table 10.3 is improved pastures, that in the last few years have shown a tendency for increasing consumption of fertilizers, including nitrogen. One fact that calls for attention is the extremely low rate of nitrogen for practically all crops in the north and northeast regions (data not shown). An exception in the northeast region is the area used for the production of irrigated fruits for export.

For phosphorus, the highest rates are applied in cotton and potato. Medium rates are used in sugarcane, soybeans, and wheat, and low rates in rice, coffee, common beans, orange, and corn. For this nutrient the lowest rates are also applied for all crops in the north and northeast regions.

The rates of potassium are highest for cotton, potato, coffee, and sugarcane. Medium rates are applied for orange, soybeans, and wheat and low rates in rice, common beans, and corn. Among the regions, the north and northeast apply the lowest rates of potassium, considering all crops.

In Argentina, the situation of fertilizer use is different, as can be seen in Table 10.4. Although fertilizer use is low, productivities of corn and wheat are higher than in Brazil and are comparable for soybean. The reason is, of course, the high natural fertility of the Argentine soils. Potassium is not used in that country due to the high supply by the soil.

The pattern of fertilizer use for Venezuela is more comparable to Brazil, with the use of all the three nutrients (Table 10.5) reflecting the lower natural fertility of tropical soils.

Another relevant aspect to be discussed for mineral fertilization is the evolution of the $N/P_2O_5/K_2O$ consumption ratio in agriculture as a whole, as shown in Figure 10.1 for Brazil. During the 1950s that ratio was 0.33/1.00/0.50, which changed to 0.50/1.00/0.65 in the 1960s. In 1970, the ratio

TABLE 10.4. Area, yield, productivity, and nutrient use for the main crops in Argentina in the year 2002

				Nutrients applied		
	Cultivated area	Total yield	Productivity	N	P_2O_5	K_2O
Crop	(1,000 ha)	(1,000 t)	(t ha^{-1})		1,000 t	
Corn	3,047	18,400	6,039	264	110	0
Soybean	9,098	23,900	2,627	0	145	0
Sunflower	2,027	3,800	2,027	8	26	0
Wheat	7,108	15,900	2,237	407	319	0

Source: Calculated with data of Argentine Secretary of Agriculture, Livestock, Fisheries and Food (2002).

TABLE 10.5. Area, yield, productivity and nutrient use for the main crops in Venezuela in 2003

				Nutrients applied		
	Cultivated area	Total yield	Productivity	N	P_2O_5	K_2O
Crop	(1,000 ha)	(1,000 t)	(t ha^{-1})		1,000 t	
Corn	516	1,800	3,487	51.7	46.5	46.5
Rice	154	787	5,104	4.6	12.3	12.3
Sorghum	265	553	2,083	15.9	15.9	8.0
Sugar cane	138	8,874	64,318	24.1	17.9	17.9
Banana	41	735	17,886	12.3	2.0	24.7
Coffee	224	92	393	7.0	1.9	3.5

Source: Calculated with data of the Statistic Department of Ministry of Agriculture and Land (2003).

was 0.67/1.00/0.80, and in 1976 it changed to 0.37/1.00/0.52. That relative increase in the consumption of P_2O_5, in relation to N was a consequence of the increasing occupation of the soils of the "Cerrado" area in the center west region of the country. As those soils are extremely deficient in phosphorus, the relatively greater increase in the consumption of this nutrient was fully justifiable. Another important factor is the large amount of nitrogen added

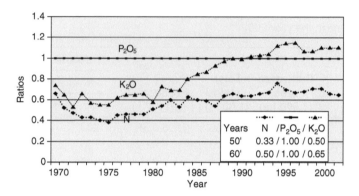

FIGURE 10.1. Evolution in nutrient consumption ratios in Brazilian agriculture ($N/P_2O_5/K_2O$) from 1970 to 2002. *Source:* Lopes et al. (2003).

to agricultural systems by biological nitrogen fixation due to soybean inclusion in crop rotations.

In 2002, the nutrient ratio was 0.79/1.00/1.14 for all $N/P_2O_5/K_2O$ consumed in Brazilian agriculture, which includes all the fertilizer consumed by the soybean crop. Excluding the soybean crop from the calculations, since that crop does not consume significant amounts of nitrogen fertilizers, this ratio would change to 1.18/1.00/1.21. By comparing this ratio with that of highly industrialized countries, which is 2.82/1.00/1.00 (Figure 10.2), one can infer that low rates of nitrogen might be an important factor limiting yield increases in a great number of crops in Brazilian agriculture (Figure 10.2).

The establishment of mineral balances (input minus removals) is a useful tool to compare the nutrients used in a country, region, state or province, or even a farm, with the amounts removed by the crop.

The relatively low consumption of nitrogen, in relation to phosphorus and potassium in Brazil was confirmed by a study of Yamada and Lopes (1999), involving calculations of nutrient exports (removal from the cultivated areas by the vegetal production removed from the field) concerning the sixteen main crops in Brazil. This study assumed an average efficiency of 60 percent for nitrogen, 30 percent for phosphorus, and 70 percent for potassium fertilizers. Taking into account data for the period of 1993-1996, these authors estimated an average annual deficit of 888,000 metric tonnes (t) of nitrogen, even when all the nitrogen of the soybean and common bean crops was considered as originating from biological N fixation. The estimated deficit for phosphorus was 414,000 t of P_2O_5, and that for potassium was 413,000 t of K_2O. These data reveal that in spite of the substantial increases observed in the consumption of fertilizers in Brazil in the last decades, the Brazilian agriculture is still "mining" from the soil a quantity of nutrients, especially nitrogen that should be replenished by fertilization. This situation can affect the long-term productivity and consequently sustainability of Brazilian agriculture. The total balance for Brazil (Table 10.6) indicates the need for higher application of fertilizers to fulfill the total needs of the crops.

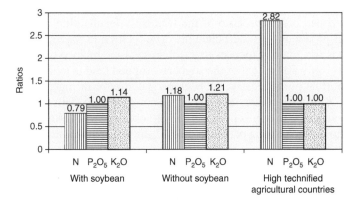

FIGURE 10.2. Relationship of consumption of $N/P_2O_5/K_2O$ in Brazilian agriculture in 2002 for the total area (with and without soybean) in comparison to countries with advanced production technology. *Source:* Adapted by Lopes et al. (2003) using data from ANDA (2003).

TABLE 10.6. Nutrient balance in Brazilian agriculture in 2002

Input as fertilizer	Removal by crops	Balance in total amounts	Balance per area
	1,000 t of the nutrient		(kg ha^{-1})
Nitrogen-N			
1,041	1,901	−859	−16.2
Phosphorus-P_2O_5			
805	1,319	−514	−9.7
Potassium-K_2O			
2,048	2,373	−324	−6.1

Source: Adapted from IBGE (2003) and ANDA (2003).

For Argentina, potassium is not used, due to the high supplying capacity of the soils. Disregarding potassium, nutrient balances for corn and wheat are adequate only for nitrogen and phosphorus, as shown in Table 10.7. In the Pampas, even with the increase of fertilizer use during the last decade, the overall nutrient balance in agricultural systems is still negative. In general, the use of P fertilizers contributes less than 40 percent of the extraction in the grains, and it is one of the major reasons for the depletion of soil phosphorus in agricultural soil from the Pampas (2001). Nitrogen in agricultural land is nearly balanced for cereal crops (i.e., maize or wheat) because of generally adequate N fertilization practices. However, it is strongly negative for oilseed crops, because of the lack of N fertilization practices for sunflower production, or inefficient nitrogen fixation in soybean crops. Although positive yield responses to soybean seed-inoculation are described frequently, only 50 percent of the soybean cropped land is inoculated. This is an important point because the bacteria specific for nitrogen fixation by soybean, *Bradyrhizobium japonicum,* is not native in soils from South America and inoculation is indispensable to obtain the benefits from nitrogen fixation. The bacteria persist in the soil, although inoculation can benefit the productivity in areas that were inoculated before.

The intensification of livestock production systems also contributes to soil P depletion in the Pampas, mostly because of the uneven distribution of manure in pasture soils, independent of meat or milk production levels. On the other hand, soils under alfalfa-based pastures provide a positive N balance related to biological fixation processes for N with native and introduced *Rhizobium* strains. If alfalfa is not adequately fertilized, it will accelerate P depletion. However, in 80 percent of the cases alfalfa is well fertilized.

TABLE 10.7. Average values for nutrient balances in Argentina

Crop	Nutrients used			Nutrients removed			Nutrient balance		
	N	P_2O_5	K_2O	N	P_2O_5	K_2O	N	P_2O_5	K_2O
	kg ha^{-1}								
Corn	87	36	0	88	18	5	−1	18	−57
Soybean	0	16	0	(158)	18	23	(158)	2	−230
Sunflower	4	13	0	70	11	8	−66	2	−8
Wheat	57	45	0	44	8	4	13	36	−4

Source: Adapted from FAO (2004).

Uruguay is also mainly in the Pampas region and fertilizer use for crops is similar to Argentina. From the 135,000 t of nutrients, 37 percent is used for pasture, 24 percent for paddy rice, and 23 percent for wheat (FAO, 2002).

Coincidence or not, the two enclosed countries in South America, that do not have an exit to an ocean, use small amounts of synthetic fertilizers (FAO, 2002). Paraguay consumes 69,000 t of nutrients divided for many crops, the most important being soybean (20 percent), maize (15 percent), wheat (14 percent), and sunflower (10 percent). This pattern indicates the influence of the agriculture of southern Brazil. Bolivia, a typical Andean country, uses only 17,000 t of nutrients, 45 percent for potato, 12 percent for rice, 12 percent for soybean, and 6 percent for each of maize, sugarcane, and wheat.

The other Andean countries have a pattern of fertilizer use for crops that varies with latitude (FAO, 2002). Chile applies 353,000 t of nutrients in a quite diversified agriculture, with main applications for wheat (24 percent), grassland (19 percent), maize (11 percent), sugar beet (8 percent), oat (6 percent), and grape (6 percent). Ecuador applies 172,000 t of nutrients mainly for banana (32 percent), maize (22 percent), rice (21 percent), and potato (11 percent). Colombia applies 568,000 t of nutrients mostly for coffee (33 percent), and also for paddy rice (16 percent), maize (13 percent), potato (8 percent), oil palm (6 percent), and banana (6 percent). Each of these countries has specific products for export: Chile-grapes; Ecuador-bananas; Colombia-coffee.

In Venezuela (Table 10.8), the average balance per area, for the six most important crops, is negative for N (-15 kg ha^{-1}), positive for P_2O_5 ($+192$ kg ha^{-1}), and negative for K_2O (475 kg ha^{-1}). For 2003, a year with low average productivity for the most important crops, the average N-P_2O_5-K_2O ratio was of 1.37/1.00/1.33 (Table 10.9), close to the data of Brazil with low rates of nitrogen used when compared to the highly developed countries.

In Argentina, the soils were originally of very high natural fertility, especially in the Pampas, and the use of mineral fertilizers has increased only recently. The predominant nutrient used is nitrogen, and the main sources are presented in Figure 10.3 on a time scale. Urea and urea derived fertilizer are the most important. MAP and DAP are also used and the quantities used indicate a limited and slightly decreasing use of phosphorus.

In Brazil, according to ANDA (2003), the most widely used fertilizers are single superphosphate, triple superphosphate, potassium chloride, urea, MAP, DAP, ammonium nitrate, ammonium sulfate, and thermophosphate. There is some limited direct use of reactive rock phosphates.

Limestone is widely used in the acid soils of South America. A unique amendment used for the amelioration of acid subsoils that present problems

TABLE 10.8. Average values for nutrient balances in Venezuela

	Nutrients used			Nutrients removed			Nutrient balance		
	N	P_2O_5	K_2O	N	P_2O_5	K_2O	N	P_2O_5	K_2O
Crop					kg ha^{-1}					
Corn	100	90	90	123	49	122	−23	41	−32
Sorghum	60	60	30	109	27	87	−49	33	−57
Rice	30	80	80	180	47	240	−150	33	−160
Sugarcane	175	130	130	64	37	154	111	93	−24
Banana	300	50	600	179	60	755	121	−10	−155
Coffee	30	8	15	55	6	62	−25	2	−47

Source: Adapted from Pequiven data for year (2003).

for root penetration due to a chemical barrier formed by excess exchangeable aluminum or low calcium content is phosphogypsum, a byproduct of the fabrication of phosphoric acid. Used in central Brazil, it is also an important source of sulfur, a nutrient widely deficient in that region.

Among organic products used as fertilizers and soil amendments, manure is the main source of nutrients. But two other sources are also important. One important source, especially of potassium, is the waste water of alcohol distillation, which is applied to the sugarcane crop.

Another source, far more important, is biologically fixed nitrogen. It is estimated that the amounts of nitrogen removed by the soybean crop represent more than two times the amount of nitrogen supplied by commercial fertilizers in Brazil. This is a highly important technical achievement. The inoculation of soybean crops with *Bradyrhizobium japonicum* is essential to guarantee adequate N supply to the crop by biological fixation of atmospheric N. The procedure takes into account the use of selected strains adapted for the different environments prevailing in the regions of production. Generalized inoculation practices are based on seed treatments, although the inoculants can also be used for liquid in-furrow treatments at planting. New liquid inoculant formulations provide not only effective microbial strains for N fixation but also other plant growth promoter effects that enhance soybean growth and yields. Other legume crops, annual or perennial, require different strain formulations, specific infection, and effective N biological fixation depending on plant growth requirements.

TABLE 10.9. Main crops per country, productions, productivities, mineral fertilizer used and N-P_2O_5-K_2O ratio in Venezuela in the year 2003

Crops	Area (1,000 ha)	Production (×1,000 t)	Productivity (kg ha^{-1})	Nutrients applied N	P_2O_5	K_2O	N-P-K ratio	N-P-K ratio P:1
				(×1,000 t)				
Maize	517	1,800.000	3.487	51.649	46.485	46.485	0.36-0.32-0.32	1.13-1.0-1.0
Rice	154	787.052	5.104	4.626	12.336	12.336	0.16-0.42-0.42	0.38-1.0-1.0
Sorghum	265	552.751	2.083	15.921	15.921	7.960	0.4-0.4-0.2	1.0-1.0-0.5
Sugarcane	138	8.874.000	64.318	24.145	17.940	17.940	0.4-0.3-0.3	1.33-1.0-1.0
Banana	41	735.060	17.886	12.329	2.054	24.658	0.32-0.05-0.63	6.40-1.0-12.6
Coffee	233	91.912	393.000	7.016	1.870	3.508	0.57-0.15-0.28	3.8-1.0-1.87
Average ratio for 2003							0.37-0.27-0.36	1.37-1.0-1.33

Source: Adapted from statistics of the Ministry of Agriculture and Land (2003).

435

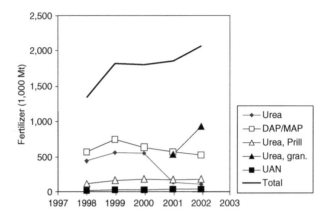

FIGURE 10.3. Use of fertilizers in the Pampas, Argentina. *Source:* FAO (2004).

For South America, statistics concerning the consumption of organic fertilizers, especially manure, are scarce. Brazil has the largest livestock herd in the world (176 million head), as well as large numbers of swine (33 million), equine (8.4 million), chicken and other birds (883 million), and sheep (24 million), according to IBGE (2003a,b). Animal production is a very good source for production of organic fertilizers. However, the commercial consumption of manure as organic fertilizers, or its use in farms where it is produced, is limited to special situations, such as horticultural and perennial crops and annual crops if located close to the intensive livestock production areas. In the grain crop production, the use of manure is important only in specific situations, such as in small subsistence or family farming systems or in great farms that integrate agriculture and animal production in confined or partially confined systems.

Even considering the low concentration of nutrients in animal manures, the 176 million head Brazilian herd could produce a total of 1.5 billion t of manure, containing, in million metric tonnes, about 8.5 of N, 3.9 of P_2O_5, and 9.3 of K_2O. These amounts of N, P, and K are at least one order of magnitude greater than the amounts of these nutrients supplied by the fertilizer used in Brazilian agriculture. However, intensive livestock production in Brazil represents just 5 percent of the total herd (Vitti and Malavolta, 1999), thus the amounts of nutrients available to agriculture are much smaller. This may change in the future because of increases in the confined cattle production and adoption of more stringent legislation requiring proper disposal of organic materials of all kinds of organic residues.

In Venezuela, Casanova (1996) indicated that sheep and chicken manure are used mainly in horticulture crops with the following results (kg t^{-1}, dry): 39 N; 3.5 P; 32 K; 2 Ca, and 2 Mg for sheep manure, and 20 N; 0.4 P; 25 K; 96 Ca, and 34 Mg for chicken manure. The rates of manure applied range from 3 to 5 t ha^{-1} in horticulture crops.

Some of the agronomic practices to improve nutrient reserves and soil characteristics such as structure recommended by integrated nutrient management (INM) are the use of cover crops, intercropping, and application of organic manures. A rotation of cereals and leguminous plants has been shown to reduce chemical fertilizer use by up to 30 percent, as cereals absorb the nitrates released from the decaying roots and nodules of leguminous plants. Soil management practice used by the farmers should increase the organic matter content of the soil. For example, assuming that the soil has 2 percent organic matter (low), 1 ha of that soil weighs 2×10^6 kg, it has 5 percent of N in the organic matter, the mineralization rate is 2 percent per year and if the corn crop has an efficiency of 50 percent, the amount of N produced by the organic matter to the soil will be:

2 percent of 2×10^6 kg ha^{-1}	=	40.000 kg ha^{-1} of organic matter
5 percent of 40.000	=	2.000 kg ha^{-1} of N
2 percent of 2.000	=	40 kg N ha^{-1} year
50 percent of 40	=	20 kg N ha^{-1} per year absorbed by the corn plants

If soil organic matter is increased, or more legume crops are included in rotations, then the supply of N by the soil might be increased, which shows the importance of one of the factors emphasized by INM—preserving organic matter in the soil, resulting in the saving of nitrogen applications.

Sewage sludge and municipal solid waste treatment plants are much more of an exception than a rule in Brazil, and the use of their by-products, after transformation into organic compost, is usually restricted to reforestation activities and lawns, with very little use in agriculture. Recent research focusing on the agricultural use of biosolids application has shown its viability for several crops in Brazil (Bettiol and Camargo, 2000). Estimated quantities and composition of municipal solid wastes in Brazil are presented in Table 10.10. Of the total municipal solid waste produced, 76 percent is deposited in open areas, the so-called lixões (open-air garbage pits), which create a great social discomfort with the landfill effluents, bad smell, and infestation of insects and animals that are vectors of diseases. Also of concern are the additional social and public health problems due to the great

TABLE 10.10. Composition of municipal solid wastes in Brazil

Garbage	Quantity (thousand tonnes/year)	Participation (%)
Organic material	23,725	50
Nonreclycled material	18,031	38
Recycled material	5,564	12
Total	47,450	100

Source: Adapted from AENDA (2001).

number of people that live reclaiming products from these pits. About 13 percent of the remaining municipal solid waste is deposited in controlled landfills and 10 percent in semicontrolled landfills, whereas 0.1 percent are incinerated, and only 0.9 percent are transformed into organic compost (AENDA, 2001).

FERTILIZERS AND SUSTAINABLE DEVELOPMENT

The worldwide use of synthetic fertilizers is associated with large increases in crop productivity. In Brazil, this can be inferred by the close relationship between the consumption of fertilizers and the production of the sixteen main crops (dry weight basis), for the periods from 1970/1971 to 2001/2002, as shown in Figure 10.4 (Lopes et al. 2003). These data demonstrate that the evolution of the Brazilian national production of these crops in the last three decades was more the result of an increase in the use of modern production technologies—with emphasis on the efficient use of synthetic fertilizers—than the simple expansion of the area planted with those sixteen crops. During the period, production of those sixteen crops increased 3.4 times and the increase in the consumption of fertilizers was 4.4 times, while the cropped area increased only 1.5 times, from 36.4 to 56.2 Mha.

These results offer a contrast to the view of critics of the Green Revolution, who often refer to fertilizers as one of the problems of modern agriculture and defend its reduction in agriculture. There is of course room for improvement of fertilizer management, as will be shown in many examples in this chapter and in this book. INM is one possibility. Yet, as was shown before, mineral fertilization often falls short of restoring the amounts of nutrients removed by crops of the agricultural systems. There is no magic.

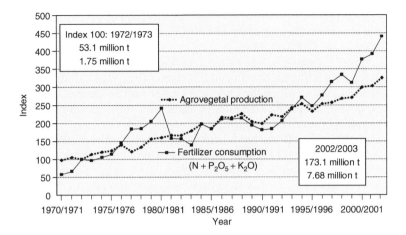

FIGURE 10.4. Relationship between agrovegetal production (dry-weight basis for sixteen crops) and fertilizer consumption (N + P_2O_5 + K_2O) in Brazil from 1970/1971 to 2002/2003. *Source:* Lopes et al. (2003).

If a plant contains 5 percent mineral nutrients in the dry matter, nutrients must have been taken up from the soil by the plants, independently of the type of management, conventional or organic.

Furthermore, the data shown also demonstrate clearly that properly managed synthetic fertilizers applied to deficient soils, guarantee an increase in productivity, resulting in less need to increase the area required to produce the same amounts of agricultural output. Lopes et al. (2003) demonstrated that, without an increase in productivity, to produce the same amount of agricultural products by the sixteen main crops, the cultivated area must be doubled. Recovery of nutrient depleted soils back to agriculture, and the cultivation of large areas of naturally infertile soils were made possible by the use of limestone and fertilizers, thus saving extensive areas that otherwise should be opened for agriculture.

It is also obvious that a productive agriculture is needed for food security in any country. In South America, this is not different, but average values are often misleading. For example, there are countries that produce exportable surplus of agricultural products, yet there is hunger, as part of the rural population is unable to produce to meet its own needs because of nutrient depleted soils. In agriculture, most fertilizer recommendations are for nutrients in mineral forms. Organic matter is recommended for horticultural crops and for the installation of some perennial crops, but much more for its beneficial effect as a soil conditioner than for the nutrient content. Only for sugarcane are the recommendations of residues from the sugar and alcohol industry included in the fertilizer tables.

Suggested average values of the composition of organic fertilizers are given together with the recommendation tables (Raij et al., 1996) to be considered in cases in which the chemical analysis of the materials is not obtained. As normal application rates of the organic fertilizers are rather high, the amounts of nutrients applied are substantial.

As stated before, Brazil has a herd of millions of cattle, pigs, and poultry but only a small part of the manure is used in agriculture, however, precise statistical values on nutrients applied to the soil in this way are not available. Some of the nutrients contained in corn and soybean production are used in animal feed and end up in manure, which is partially used in agriculture.

For sugarcane, vinesse (residual water from alcohol distillation) is applied in rates varying from 20 to 250 m^3 ha^{-1} and filter cake from 15 to 100 t ha^{-1}. The amounts of nutrients applied are high. Under these conditions, it is recommended that all the K and half of the P be discounted from the mineral fertilizer recommendations. Years ago, vinesse was dumped into rivers, but environmental rules make its use on soil compulsory, resulting not only in the reduction of pollution, but in profit for the production of sugarcane. This is so because vinesse is a good K fertilizer, contains small amounts of other nutrients, and improves soil structure because of the water soluble organic matter that it contains.

Certainly the most important nonfertilizer source of any nutrient is the supply of nitrogen through biological fixation by soybean. The total removal of nitrogen by soybeans is more than all the mineral nitrogen applied in agriculture. As one-third of the fixed nitrogen remains in plant residues left on the field, this is an important source of nitrogen for other crops in the rotation, especially corn and wheat.

The role of organic sources in agriculture is being reviewed in several regions of South America. Animal manure is an ancient organic fertilizer and will always remain important. However, it is questionable whether new facilities for bio-industries with confined animals, be it poultry, cattle, pigs, or others, should be made nowadays without a plan for the proper use of manure on agricultural land. Certainly for large installations, an environmental permit that includes a plan for adequate disposal of manure will be necessary. The same is basically true for several other types of industries that produce organic by-products or residues.

In South America, manure is an important source of nutrients not only for small farms, but also for larger farms that integrate animal production and agriculture. Sewage sludge or biossolids, and municipal composts are being produced and used increasingly. Yet, amounts available are still limited. In the case of sugarcane, bagasse is used to produce energy and also compost. Ashes and residual water from alcohol distillation are used as fertilizers. Several industries have residues that are or can be used in agriculture. Phosphogypsum finds a unique application as an acid subsoil amendment.

Thus, agriculture can play an important role in the recycling of organic and inorganic residues. From the point of view of the environmental authorities, in Brazil, the materials that can be recycled in agriculture can be classified as by-products, instead of residues.

Organic and synthetic fertilizers should be applied in adequate amounts to avoid contamination of subterranean water with nitrates, and surface water with nitrates and phosphates.

The Pampas region of Argentina has the potential for being a major soil C sink because most soils of the area have been degraded due to intensive tillage cultivation practices, with significant losses of organic matter occurring during the last century. Actually, with most of the area turning into no-tillage agricultural systems, the conditions are adequate for modifying this trend and increasing the amount of C stored in the soil. In order to achieve this goal, the implementation of recommended production practices including crop sequences based on cereals and intensive use of fertilizers is required. In other countries of the continent, C sequestration in agriculture in minimum tillage systems is likewise in Argentina an issue of the moment, but practical information on the subject is scarce.

INTEGRATED NUTRIENT MANAGEMENT (INM)

Modern agriculture is more and more dependent on synthetic fertilizers. However, in the process of using them, often alternative sources that could be used profitably, not only for their content of mineral nutrients but also for the very important effect of organic matter on the physical properties of soils, are neglected. The management of nutrients is presently subject to some changes in approach. There is a growing concern about the importance of organic matter in soil for sustainable agricultural production. Increasing environmental awareness is another important factor that should be taken into account with respect to the use of fertilizers.

There are various sources for the nutrient supply of plants in agricultural systems. Finck (1995), in his overview of the management of plant nutrient in crop rotations and farming systems, points out that even considering that the concept of fertilization in its modern comprehensive form, based on improved soil fertility, has proved to be a powerful tool for enormous yield increases, the "capital," or stock, of farm nutrient resources has been somewhat neglected. The author stresses the need to put more emphasis, beyond the crop nutrient requirements, on the farm internal and external nutrient cycles. This idea finds conceptual support in the FAO approach known as "integrated plant nutrition systems"—IPNS—a nutrient management system that proposes the maintenance or adjustment of soil fertility and of plant nutrient supply to an optimal level, through optimization of the benefits from all possible sources of plant nutrients in an integrated manner (Roy, 1995).

These aspects are not yet considered in a clear manner in the present fertilizer recommendations in Brazil, but are often present. Such is the case of specific recommendations of by-products of the sugar and alcohol industry as fertilizers and recommendations of organic fertilizers for several crops (Raij et al., 1996).

Before discussing INM, it might be worthwhile to discuss how this approach is in line with the principles of sustainable development, environmental management, and how it fits into a more general approach of farm management with a view on sustainable agriculture.

According to the Agenda 21, sustainable development should consider the economic result along with environmental preservation and social interest. This concept applies perfectly to agriculture and nutrient management and is a key aspect for maintaining adequate nutrient balances (Roy et al., 2003). In other words, it means that for survival, it will not be possible to continue to consider economy, social development, and environmental preservation as independent issues (EIF, 1998). In the case of agriculture, the need of global and local food security is another motive to seek all ways to pursue sustainable development.

Integrated Nutrient Management contributes to environmental management, as can be concluded from some of the objectives of the norms of the series 14000 (Harrington and Knight, 2001) reduction of the use of natural resources; reduction of energy use; process improvement; reduction of environmental pollution; reduction of residues generation, recycling of materials.

In relation to INM, several aspects described in the guidelines for integrated crop production or other environmentally friendly approaches apply. Thus rotation, control of soil erosion, and improvement of soil fertility are general recommendations.

Sustaining and improving soil fertility is achieved by (1) maintenance of optimum humus level according to the characteristics of the location; (2) maintenance of high species diversity of fauna and flora (earthworms, cellulose decompositors, predatory mites etc.); (3) maintenance of the longest possible protection of the soil by crops or plant cover; (4) the lowest possible soil disturbance (physical and chemical).

For nutrient management and fertilizer use, some of the recommendations are as follows (Boller et al., 1999):

1. plant nutrient management plan considering each crop on a plot level and over an entire rotation;
2. preference for organic fertilizers when possible;
3. application of off-farm fertilizer inputs only to compensate the real exports and technical losses;
4. soil and plant analysis to adjust nutrient rates;
5. consideration of hidden nutrient sources such as additions through polluted air and water, animal feed, and mineralization potential of soil organic matter; and
6. determination of maximum nitrogen input, expressed in kg ha^{-1} per year, and periods of application adjusted for each crop and soil type.

Yet another approach to organize information for soil fertility management has been proposed by the International Fertilizer Industry Association in their published best agricultural practices for several regions of the world, including Latin America (IFA, 1998). IFA discusses the main aspects of fertilizer use, but the publication fails to organize the information in a form that can be considered "best practices," as compared, for example, with best laboratory practices, best practices for environmental management, or best practices for integrated crop production, which have rules that should be followed and are subject to inspection by third parties.

Whatever system is used, organic matter and recycling of organic materials in agriculture must be considered. Organic matter is not essential for plants, but it is essential for good soils. The addition of organic matter to soils always stimulates intensive microbiological activity. The rules of sustainable development stress the necessity to recycle as much as possible all sorts of residues, transforming wastes into by-products that can be used as fertilizers. Organic matter, especially manure, is not easy to handle. If nutrient concentration is low, cost/benefit ratios may not encourage the use of the organic material from an economic point of view, considering transport and handling costs. It is no wonder that synthetic fertilizers are used much more than organic sources. In South America, labor is becoming scarce in rural areas, as can be concluded from Table 10.1, and little labor is available for the collection of organic manure. Careful planning of the farm operation as a whole for the adequate use of manure is done by some farmers, but the great majority could certainly benefit from better organization and planning.

Organic sources of nutrients also have the advantage that they contain most nutrients needed by crops, which makes sense considering that most of the available organic matter is derived from plants that must contain all required nutrients. Furthermore, there is the advantage that organic matter improves the chemical, physical, and biological properties of soils.

Many possibilities are available for better management of nutrients. Certainly one of the most important is conservation agriculture which requires the soil to be covered at all times with crop or plant residues. Crop rotation is an important aspect for annual crops. Maintaining the soil covered, undisturbed by tillage, can increase organic matter. Intercropping is important for small farmers, but it is often not practical for mechanized agriculture. For perennial crops, intercropping is seldom practiced, but weeds also contribute to improve the organic matter of soils and by recycling nutrients, reducing losses.

The main application of microbiology in the continent is the inoculation of soybeans with *Bradirhyzobium japonicum,* an essential procedure to guarantee adequate nitrogen fixation. Mycorrhizae treatment for forest seedlings is another microbial contribution to the nutrition of these crops and partially for extensive crop production systems. In general, mycorrhizae contributes to crop root volume enlargement and enhances phosphorus crop uptake under limiting soil available phosphorus levels, although the nutrient is not added to the soil and the possibility of depletion should be of concern.

TECHNICAL REQUIREMENTS FOR INM

The large variety of soils and cropping systems in South America requires different approaches, making nutrient management for crops difficult. The diversity of soil fertility problems, the restriction of credit, and the high cost of fertilizers makes the proper adjustment of nutrient rates an important task. Thus, organization of appropriate fertilizer and lime recommendation guidelines are important. For this, a large research database is essential, covering the evaluation of soil fertility through chemical analysis, plant analysis, greenhouse and field trials with different types of fertilizers and crops, economic evaluation, and several basic studies related to soils and plants. Such information is present only in a few regions of the continent, especially in Brazil. A more detailed assessment of soil nutrient balance has been suggested by Roy et al. (2003).

In temperate regions, this type of information has been gathered for more than half a century. However, in tropical regions the peculiarities of the soils and their management for crop production have only been consistently studied more recently and most data have been obtained in the past fifty years. In Brazil, remarkable progress has been made in the last twenty-five years.

An equilibrated plant nutrition system within the philosophy of INM will consider all nutrients available from different sources, with especial attention to nutrient and organic matter sources available at the farm that involve no expenditure of money. Furthermore, fertilizers or other products that carry a cost for the farmers should be carefully and judiciously applied, especially in those cases where financial resources are scarce and farmers cannot afford high risk. This requires as a first step a knowledge of nutrient contents of plants, both in the harvested part and in the aboveground residues, and also of the expected yield. The next step is knowing the nutrients already available in soils, for annual crops, and in soil and plants for perennial crops. The third step is the knowledge of the nutrient availability of the different sources available.

Nutrient contents of some crops are given by many authors and will not be presented here. Definition of expected yields is a guessing exercise, but it must be done as accurately as possible, considering the records of yields of former crops, and adding an expectation for higher yields (Raij et al., 1996).

Soil analysis is the best tool available for determining nutrient availability in soils. The quality of the diagnosis provided is, however, strongly dependent on adequate sampling, the quality of soil testing methods to access bioavailability of nutrients, and adequate field calibration. In South America, soil analysis is used for fertilizer recommendations with variable degrees of success. There are differences in methodology, as shown in Table 10.11 for Argentina, Brazil, and Venezuela. Most of these methods are well known and will not be discussed. However, there are some unique developments in Brazil that deserve attention and will be commented upon later.

Soil testing for farmers has been available in South America since the end of the nineteenth century, but the number of samples analyzed was small until the second half of the twentieth century, when a strong increase in soil samples analyzed and fertilizer use in several countries occurred (Raij, 1980).

The rather small diversity of soil testing methods used in Brazil is the result of a joint project conducted by the North Carolina State University and the Ministry of Agriculture of Brazil. Details are described by Raij et al. (1994). The main points were the use of the same methods of soil analyses by all laboratories when the program started, the development of time saving equipment that allowed the simultaneous determination of a large number of samples, and a national network of communication and annual meetings of the laboratories. The methods introduced at that time were pH in water,

TABLE 10.11. Criteria or methods of soil analysis used for lime and fertilizer recommendations by the soil testing laboratories of Argentina, Brazil, and Venezuela

Country	Organic Matter	Mineral N	pH	Lime requirement	K	P	B	Cu, Fe, Mn, Zn
Argentina	Walkley Black	KCl extracted NO_3^2	Water	Cation saturation, increase base saturation and buffer	Exch.	Bray Kurtz 1	Hot water	DTPA
Brazil Embrapa	Not used	Not used	Water	Reduce Al, increase Ca, Mg	Exch.	Melich 1	Mehlich 1	Mehlich 1
IAC	Not used	Not used	$CaCl_2$	Increase base saturation	Exch.	Resin	Hot water	DTPA
Rolas	Used for N	Not used	Water	Increase pH (SMP buffer)	Exch.	Mehlich 1	Melich 1	0.1 M HCl
Venezuela UCV	Used for N	Not used	Water 1:1	Reduce Al, increase Ca, Mg	Exch.	Mehlih 1	Hot water	Mehlich 1
INIA	Used for N	Not used	Water 1:2.5	Reduce Al, increase Ca, Mg	Exch.	Olsen	Hot water	Not used

Source: Raij et al. (2001) and Universidad Central de Venezuela (1993).

determination of P and exchangeable K by the Mehlich 1 extraction, the extraction of Ca, Mg, and Al by 1 M KCl, and the calculation of lime required to neutralize exchangeable aluminum and increase in calcium and magnesium. As seen in Table 10.11, most of the laboratories in the country still use these methods, considering that about two-thirds of the soil samples are analyzed by Embrapa (Silva, 1999), and ROLAS (Tedesco et al., 1995) methods, and about one-third by the IAC methods (Raij et al., 1994).

The soil analysis of micronutrients is currently very important in Brazil. In 1996, soil analysis of micronutrients with hot water for B and DTPA for Fe, Cu, Mn, and Zn was introduced in the soil testing laboratories of the IAC system of soil analysis. However, most of the other laboratories use the Mehlich 1 method for Fe, Cu, Mn, and Zn extraction.

In the 1970s, field experiments were already pointing to shortcomings of the lime recommendation procedure and the P extraction with the Mehlich 1 extraction. In 1983, there was a major breakthrough in soil testing with the introduction in the State of São Paulo, of ion exchange resins for the determination of P, Ca, Mg, and K (Raij et al., 1986), the pH in 0.01 molar $CaCl_2$ solution, and the determination of the base saturation of soils by the increase of the base saturation of the cation exchange capacity at pH 7 (Raij, 1991; Raij et al., 1996), at values defined for each crop. The cation exchange capacity is determined by the sum of the exchangeable cations and the total acidity at pH 7, determined by the SMP buffer solution (Shoemaker et al., 1961; Quaggio et al., 1985). SMP stands for the initials of the authors of the buffer method.

Statistical information on soil analysis in Brazil is available only for the periods from 1972 to 1982 (Cabala-Rosand and Raij, 1983) and from 1983 to 1989 (Raij et al., 1996). Some selected data is presented in Table 10.11. More than one million samples are analyzed per year in the country, representing about one sample for 50 ha of crop land. The remarkable increase of soil samples analyzed in the state of São Paulo in the period 1981-1989, increasing from 101,000 to 255,000 samples per year, is largely due to improvement in soil testing.

In Argentina, because of the high fertility of its soils, soil testing for farmers is not a widely used practice. Most of the soil testing for available P and mineral N are performed before wheat or pasture seeding and maize, respectively. Less than 15 percent of the annually cropped land is analyzed and most of the fertilizer recommendations do not take into account soil fertility testing results. Foliar analysis for crop fertility management practices is used only by selected farmers and mostly for adjusting previous fertilization strategies based on crop potential yields and soil testing. However, the number of samples is increasing, as shown in Figure 10.5.

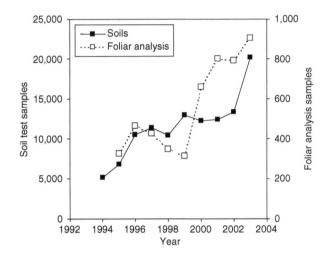

FIGURE 10.5. Evolution of soil and foliar analysis performed by "Tecnoagro," one of the largest private laboratories in the Pampas. *Source:* J. Lamelas, personal communication.

In Venezuela, it was estimated that about 50,000 soil samples should have been analyzed, but in practice only around 20,000 soil samples were analyzed by the national laboratories (Gilbert de Brito, 2003).

Besides soil tests, technical requirements for INM are based also on a large number of field experiments that have been conducted in the different countries. Most experiments deal with soil amendments and mineral and organic fertilizers. Experiments are essential to determine the optimum rates of amendments and nutrients and also for calibration of soil testing.

ACTUAL IMPLEMENTATION OF INM

The fertilizer recommendations for most of the crops in Brazil are quite well defined and published in the form of recommendation tables for the different parts of the country. There are three major publications that are "best sellers"; each of them has several tens of thousands of copies printed. They are known by popular designations: ROLAS, for the two southern states, Rio Grande do Sul and Santa Catarina (Comissão de Fertilidade do Solo—RS/SC, 1995), Bulletin 100, for São Paulo, prepared by IAC (Raij et al., 1996), and the 5th Approximation, which is the fifth edition of the recommendation tables for the state of Minas Gerais (Ribeiro et al., 1999). A more recent publication for the Cerrado region (Souza and Lobato, 2002) and two others, publications specific for the states of Pernambuco (Cavalcanti, 1998) and Espírito Santo (Dadalto and Fullin, 2001), are also available.

The recommendation tables define criteria for soil testing interpretation, and liming and fertilizer recommendations based on the best knowledge available at the time of publication. Recommendations take into account the yield goals for major crops. For cereals, previous soil use and application of manure are taken into account for fertilizer recommendation. Soybean in rotations, especially under minimum tillage, can supply a significant amount of nitrogen to subsequent crops. For some perennial crops (coffee, mango, and citrus), leaf nitrogen is important to establish nitrogen rates and this is built-in to some of the recommendation tables. Also, soil testing for micronutrients is built-in to some of the recommendation tables.

Thus a very adequate technical framework is available for fertilizer recommendations. Even in the case of organic fertilization, nitrogen information is taken care of by considering actual or previous applications, and the other nutrients are monitored by soil analysis.

Yet there is still a gap between what is recommended and what is used by producers. Fertilization based on standard NPK formulas for crops is still a rule. Furthermore, for large farms, soil samples are often analyzed for different plots, but the fertilization rates are not adjusted to the specific needs of each plot, considering instead the average for the farm. This contrasts with the procedure named "precision agriculture," for which the fertilization is very detailed, aiming to correct differences due to variability of soil fertility attributes.

In Argentina, the available information for crop fertilization and liming recommendations is not organized and published as it was described for Brazil. In general, public and private agricultural institutions provide continuous education training courses for updating the best recommended practices for consultants and also for farmers adapted for the different argentine production regions.

In Venezuela, fertilizer recommendation tables are used in all the soil and plant laboratories, which report results of soil analysis and amounts of fertilizer to apply considering the crop and the yield target. However, information is outdated, since the recommendation tables were produced more than thirty years ago by the Agronomic National Research Center (FONAIAP), using field calibration for the crops in different regions, and varieties and hybrids that are not grown anymore. Therefore, a national effort to update the fertilizer recommendation tables with the new hybrids and varieties of higher yield potentials is under way and will be published soon. Another problem is that the actual tables consider only N and P and therefore they should be completed considering also the nutrients Ca, Mg, S, and micronutrient requirements of crops.

Lime Requirement

Since the mid-1960s, lime requirement in Brazil was based mainly on a procedure based on the neutralization of exchangeable aluminum (Kamprath, 1970). The assumption was that, for highly weathered soils, exchangeable aluminum was the main component of soil acidity and that liming those soils to neutrality might cause detrimental effects to crop yields. Raij et al. (1983) provided evidence that even for highly weathered oxisols, liming rates much higher than the ones needed to neutralize aluminum produced yield increases in maize, soybeans, and cotton, provided fertilization with nutrients needed for the crops was adequate. The review of Raij and Quaggio (1997) includes several experiments that confirm those findings. Thus the criterion for liming was changed to another based on the increase of the base saturation of the cation exchange capacity at pH 7 to target values, specific for each crop.

Lime requirement (LR), in metric tonnes per hectare of $CaCO_3$ equivalent in the 0.20 cm soil layer, according to the base saturation concept, is calculated by:

$$LR = \frac{CEC(V_2 - V_1)}{1,000}$$

where CEC is given in $mmol_c \ dm^{-3}$, V_1 is the original percent base saturation of the soil, and V_2 the target value of the percent base saturation to be reached by liming (Quaggio et al., 1985). The CEC at pH 7 is calculated by the sum of exchangeable cations and total acidity. The determination of the latter was greatly facilitated by the procedure developed by Quaggio et al. (1985), which involves a simple measurement of pH of a soil suspension using the SMP buffer solution for the determination of the total acidity of the soils at pH 7. An advantage of the base saturation approach is the flexibility to adjust the calculations to the specific requirement of crops.

Subsoil Acidity

Subsoil acidity may prevent root penetration below the plow layer. Both aluminum toxicity and/or calcium deficiency are usually associated with the problem. Calcium deficiency is not so common but it was shown to impair root development of cereal crops in an acid oxisol of Cerrado in central Brazil (Ritchey et al., 1982). On the other hand, aluminum toxicity is widespread and it is the main constraint for root development in subsoils. In soils with low calcium and high aluminum in the subsoil, gypsum application might increase yields. The salt leaching into the subsoil increases the activity of calcium and reduces the activity of aluminum, enhancing root development (Ritchey et al., 1980; Raij, 1988). Gypsum is recommended for several crops when exchangeable Ca in the subsoil is below 4 $mmol_c\ dm^{-3}$ and/or aluminum saturation of the effective CEC (sum of Ca, Mg, K, and Al) is above 40 percent. The amount of gypsum to be applied depends on soil texture and is calculated according to Souza and Lobato (2002) or Raij et al. (1996) by the following or similar expressions:

$$GR = 6 \times \text{clay content}$$

where GR is the gypsum requirement, in $kg\ ha^{-1}$, and the clay content is expressed in $g\ kg^{-1}$. In soils where the values of exchangeable aluminum are high ($>30\ mmol_c/dm^3$), the rates recommended may not be effective and gypsum application not economically feasible.

Macronutrients

Soil testing plays an important role in fertilizer recommendation because it allows the assessment of nutrient availability to plants before the crop, thus allowing the prescription of fertilizer rates adequate to each case.

In most Brazilian soils, phosphorus deficiency restricts severely the growth of crops, therefore the determination of reliable indexes of P availability is a matter of great interest. Acid extractants have been widely used in Brazil, especially the Mehlich-1 solution containing 0.05 mol L^{-1} HCl in 0.0125 mol L^{-1} H_2SO_4. However, the extraction of phosphorus from acid soils, with high clay content, and rich in iron and aluminum oxides with dilute solutions of strong acids is often inadequate, because the results are too low for some soils well supplied with phosphorus. Furthermore, the acid extraction may dissolve rock phosphates in soils treated with these fertilizers, thus showing high results in situations in which the actual P availability is low (Raij and Diest, 1980).

For many years soil analysis for phosphorus in Brazil was somewhat unreliable, because of the problems mentioned. After considerable research, in which several alternatives were evaluated, Raij et al. (1986) introduced into routine soil testing a procedure to extract phosphorus, as well as calcium, magnesium, and potassium from soils, using an ion exchange resin method.

In a review of seventy papers from the international literature, in which P uptake in pot experiments was correlated with extractable values obtained by different procedures, the extraction with ion exchange resin provided the highest average coefficient of determination (Silva and Raij, 1999). The data indicate that the resin extractable soil P is superior to all other methods in the evaluation of the availability of soil P for crops.

The ion exchange resin procedure for phosphorus has been in use since 1983 and presently ninety laboratories in ten states in Brazil have adopted it for routine analysis. Different limits of interpretation are given for groups of crops according to their increasing P requirements: forestry, perennial, annual, and horticultural (Raij et al., 1996).

Soil testing for potassium is also important for making fertilizer recommendations. Exchangeable potassium is well suited as a criterion to assess soil K availability in the weathered Brazilian soils where the contribution of nonexchangeable forms is small.

Studies involving crop response to exchangeable Mg in soils have shown that little or no yield increases are expected when Mg contents exceed $9 \text{ mmol}_c/\text{dm}^3$. The recommendation is for minimum values of $5 \text{ mmol}_c/\text{dm}^3$ of Mg for crops such as maize, wheat, rice, soybean, beans, sunflower, and $9 \text{ mmol}_c/\text{dm}^3$ of Mg for cotton, citrus, banana, and most horticultural crops (Raij et al., 1996). Usually Mg is applied in the form of dolomitic limestone.

Soil testing has been of little use for N recommendation in Brazil. Presently only two states of southern Brazil make use of the soil organic matter content to estimate rates of N application (Comissão de Fertilidade do Solo RS/SC, 1995). This criterion was abandoned in the state of São Paulo due to inconsistent correlations with nitrogen response (Cantarella and Raij, 1986). The nitrogen reserve in the soil, mostly in organic forms, is of course affected by crop and soil management.

In view of the results summarized above, Raij et al. (1996) recommended nitrogen fertilizer based on the expected yield and soil and crop management history for annual crops. For perennial crops, expected yield is the most important criterion for most crops, but leaf content is also considered for coffee, citrus, and mangos (Raij et al., 1996).

To include the soil and crop management history in the recommendation tables, the soil is placed in one of three classes of response to nitrogen: high, medium, or low response. The high response class groups soils of high fertility after a few years of continuous maize or other nonleguminous crop, first, years of no-till crop, and sandy soils subject to leaching losses. The medium response class is for acid soils that will be limed, with occasional leguminous or moderate applications of manure. The low response class includes soils uncropped for two or more years, soils previously cropped to pasture, intensive cropping with leguminous crops or green manure, and soils that received large quantities of manure. Thus nitrogen from legumes or manure is accounted for in an indirect manner. In minimum tillage systems, with soybeans in rotation for some years, maize production even without applied nitrogen can reach productivity levels of 5 to 9 t ha^{-1} (Sá, 1996). Fertilizer recommendation for sulfur according to the publication of Raij et al. (1996) and Souza and Lobato (2002) does not follow a general pattern throughout Brazil. For most crops the rate recommended guarantees a minimum S supply to the crops. In other cases the recommended rates are proportional to recommended N rates or take into account the yield goals. Although preliminary interpretation limits for soil sulfate-S are available, they are seldom used for fertilizer recommendation.

Micronutrients

Fertilization with micronutrients in Brazil is gaining importance as the soils are more intensively cropped, yields increase, and low fertility soils, especially those of Cerrados (central Brazil), are converted to agriculture.

In São Paulo state, hot water (for boron) and DTPA-TEA at pH 7.3 (for copper, iron, manganese, and zinc) were chosen as standard extractions. For most of the other states the Mehlich1 extraction solution is used for copper, iron, manganese, and zinc. Since 1996, micronutrient recommendations for several crops are made based on soil analysis as well as on the sensitivity of particular crops to micronutrient deficiencies. Zinc and boron are recommended for the largest number of crops, followed by manganese and copper. Iron is seldom recommended, but molybdenum is required for some crops, although in this case soil analysis is not used. Micronutrients may not be recommended for plants that receive large quantities of manure.

DISCUSSION OF CASES

Soil testing for fertilizer recommendation requires a quite complex organization to be efficient, starting with the necessary research and ending with the practical implementation of the recommendations, as was described by Raij (1980). The adoption of soil analysis by farmers is a good index of acceptance of the technique (Table 10.12).

There are several reasons for the increase of the number of soil samples analyzed in São Paulo (Table 10.12). One fundamental reason was the innovations in methods of soil analysis developed by the Instituto Agronômico

TABLE 10.12. Number of soil samples analyzed in Brazilian regions in selected years

Região	1,000 soil samples analyzed				
	1972	1976	1981	1985	1989
South	117 (43%)	128 (40%)	141 (34%)	187 (31%)	204 (28%)
São Paulo state	95 (36%)	102 (32%)	101 (25%)	165 (27%)	255 (35%)
States of Minas Gerais, Rio de Janeiro, Espírito Santo, and Centro Oeste	19 (7%)	51 (16%)	85 (21%)	179 (30%)	198 (28%)
North and northeast	36 (14%)	40 (12%)	82 (20%)	71 (12%)	62 (9%)
Total	267	321	409	602	719

Source: Adapted from Raij et al. (1994) and Cantarella et al. (1995).

and their introduction in routine soil testing in 1983. The basic philosophy behind the fundamental changes introduced in soil testing was that, instead of being simple and fast, methods of soil analysis should be the best possible from the agronomic point of view. The introduction of much better methods for assessing phosphorus and lime requirements increased confidence in soil testing because of the improved diagnosis for P and lime requirement provided for different crops.

One important step in Brazil was the organization of sample exchange programs that resulted in the substantial improvement of the analytical quality of the results, as described by Quaggio et al. (1994) for different regions of the country. These programs started under the designation of "quality control programs," but a more correct name is "test of proficiency." Participating laboratories analyze soil samples, and statistical evaluation allows the classification of the laboratories, with the possibility of the exclusion of laboratories that do not produce results within defined limits of error. The oldest program in the country is that of Embrapa Soils, valid for the country as a whole, that started around the mid-1960s, but was discontinued for some time. The most consistent system was the ROLAS laboratories, which never interrupted its activities. Another important system, the most innovative, is that of São Paulo, that has different methods from the other laboratories, as already described and, although a program of the State of São Paulo, it actually operates with laboratories of ten states of Brazil. Other programs are from the states of Minas Gerais and Paraná and from central Brazil.

The organization of fertilizer recommendations on the basis of soil analysis is another very important aspect, because only by calibration of the results of soil analysis with field experiments, complemented by research on fertilizer rates, can the information be used to achieve the best possible soil correction and crop fertilization. The publication of Raij et al. (1996) presents recommendation tables for 160 plant species, based on the IAC (Agronomic Institute of Campinas) methods of soil analysis, including micronutrient analysis and leaf analysis for selected crops.

The information given to farmers by the laboratories or by agronomists using recommendation tables contains the results of soil analysis, the recommendations for limestone and nutrient rates, and the expected nutrient balance for nitrogen, phosphorus, potassium, and sulfur. Computer programs are available to transform the recommendations in quantities of commercial products or organic fertilizer to be applied.

The fertilizer recommendations are available for São Paulo, as already mentioned, and for the following regions: Rio Grande do Sul and Santa Catarina (Comissão de Fertilidade do Solo—RS/SC, 1995), Minas Gerais (Ribeiro et al., 1999), Pernambuco (Cavalcanti, 1998) and Espírito Santo (Dadalto and Fullin, 2001), and for the Cerrado region (Sousa and Lobato, 2002).

On a broader view, two very different patterns of fertilizer use can be identified in South America, especially because of the original soil fertility. In the extreme south, the Pampas, mainly in Argentina, but extending also to large areas of Uruguay and the state of Rio Grande do Sul and Brazil with its originally high fertility, will need increasing amounts of fertilizers to replenish deficient nutrients and to attend higher productivity levels.

In the Pampas, crop yields and livestock production per unit of area are increasing yearly. However, the replenishment of the nutrients extracted in the grain and other agricultural products by fertilizers is insufficient. On average, only the equivalent of 40 percent of the P removed in the grain of the major crops of the Pampas is applied yearly with the fertilizers. As a result, the area with P deficiency increased from less than 50 percent to almost 80 percent of the Pampas between the late 1970s and the end of the 1990s (García, 2001).

Consequently, in the Pampas, P fertilizers are required for the adequate establishment and production of alfalfa pastures for achieving high-yielding wheat crops and for improving corn establishment, because most of the soils contain intermediate to low extractable P levels. The results from a soybean contest in the soybean belt region of the Pampas concluded that the highest soybean grain yields were achieved by farmers that apply P fertilizers and use inoculants (Díaz-Zorita, 2005). However, less than 20 percent of soybean crops are fertilized and the areas with low soil available P levels are increasing. Thus, the former fertile soils of the Pampas are now requiring generalized use of nutrient supply to achieve high-yielding crops and provide efficient returns for agricultural investments.

Advanced producers are using nutrient management practices for sequences of crops, with rates of fertilizers for the current crop and for the crops to be cultivated in the following seasons. This practice, known as "fertilizing the crop rotation," is used primarily for P fertilization. After analyzing the soil before planting wheat, the P fertilizer recommendation is applied taking into account the potential requirements of the following double crop of soybean. A similar approach is performed for S fertilization practices, providing S in combination with N for wheat or corn crops, and leaving the following soybean crop without S application.

In Brazil, since the 1970s, with most of the fertile soil areas occupied, increasing levels of limestone and mineral fertilizers became necessary to sustain and increase crop production. This was demonstrated by Figure 10.4 and discussed by Lopes et al. (2003). The same trend can also be observed in countries such as Paraguay and the lower parts of Bolivia which are influenced by the Brazilian developments.

Under general agricultural management, two recent developments are of great importance. One is the cultivation of the very low fertility soils of the oxisol order that are acid and deficient in almost all nutrients, but presenting favorable physical properties, especially in the Cerrado regions of the central region, and more recently in the south of the Amazon region, and also in the eastern part of the northeast region (Lopes, 1996). These regions have high rainfall in the summer and dry winters. Soybean is a crop remarkable for its high nitrogen fixation and new varieties can be grown almost everywhere, even in very sandy soils, provided the rainfall is high and well distributed in the rainy season. However, because of the high rainfall in these areas, erosion is a very serious problem. So the second important development is the use of minimum tillage that is resulting in a very efficient erosion control technology and is being adopted increasingly in large areas of South America.

Minimum tillage or conservation agriculture is contributing to the sustainability of agriculture, by avoiding erosion, allowing better replenishment of underground water reservoirs, and increasing soil organic matter. A comprehensive publication on the basis of soil fertility management under such systems is given in the recent book of Lopes et al. (2004).

Related to INM, two cases of application of some of its principles from Venezuela will be illustrated next. The first is for corn production. Every year between 300 and 400,000 ha of field corn mostly for human consumption is planted in Venezuela. Corn yields from 1989 to 2003 increased by 50 percent, but fertilizer consumption had a drastic decrease (Figure 10.6). Between 1994 and 2003, higher yield hybrids were introduced and there was a reduction in area seeded to corn, which resulted in a sharp reduction in fertilizer use (Figure 10.7). The rates of fertilizer applications during the

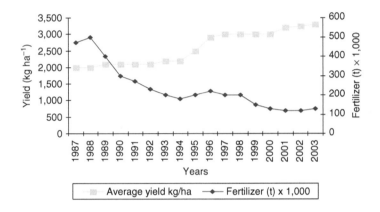

FIGURE 10.6. Corn yields and tonnes of fertilizers used in Venezuela from 1987 to 2003. *Source:* Pequiven (2006).

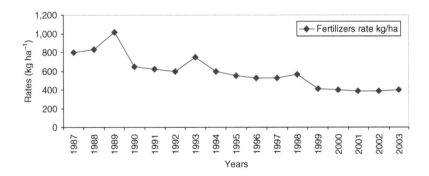

FIGURE 10.7. Rates of fertilizers applied in corn in Venezuela between 1987 and 2003. *Source:* Pequiven (2006).

period decreased from 1,014 kg ha^{-1} to just 400 kg ha^{-1} in 2004. The drop in fertilization rate was a consequence of the removal of fertilizer subsidies, in place in 1989. By 2004, when the subsidies were removed, the prices of fertilizers were affected by the open market (Casanova, 2005).

As already mentioned, the fertilization tables of Venezuela were developed thirty years ago and take only NPK into consideration. A comparison for corn fertilized by the existing recommendations (FONAIAP, 1989) and by an INM approach that considers the application of all needed nutrients shows that even without difference in the cost of fertilizers in the two cases, agronomic and economic efficiencies were much higher for a more balanced fertilization (Table 10.13).

Another case is related to recovery of degraded pasture. In Venezuela, after more than twenty years of research in the use of rock phosphate under different soil and climatic conditions, and annual and permanent crops (Barreiro et al., 2000; Casanova, 2002a,b; Casanova et al., 2002), the fertilizer industry decided to build a plant to produce partially acidulated phosphate rock (PAPR) with a capacity of 150,000 t per year. Depending on the acidulation grade (40 to 60 percent), the PAPR has around 26-30 percent P_2O_5, 50 percent of it water soluble, and the rest for residual effect. Potential use is important, since the country has 16 Mha of pastures, 11 million natural, and 5 Mha with introduced grasses like brachiarias and sylosanthes. One of the most recent commercial assessments of PAPR was made at Punta de Mata in the Monagas state program of recovery of degraded pasture for production of double-purpose livestock. The grasses on the thirty studied farms were *Brachiaria decumbens, Brachiaria humidicola,* and *Brachiaria brizantha.* They were established on well-drained savannah soils with P and K limitations and had a high degradation index due to the insufficient resting periods of the paddocks, and inadequate fertilizer application practices. The commercial area was established ten years prior to the study and currently had average values of 20 to 25 cm in height, a 30-40 percent coverage and a yield below 0.3 t ha^{-1} of dry matter by harvest. Crude protein levels are only 2-3 percent, which are considered quite low, in addition to low digestibility. Under these conditions, the zebu-type animals typically produced only three to five liters of milk per animal per day.

In an INM program, PARP was applied to these degraded pastures at a rate of 200 kg ha^{-1} at the beginning of the rainy season, and incorporated by plowing. The fields were seeded with three kg of *Stylosanthes capitata* seed, a tropical leguminous plant, along with *Brachiaria,* resulting in an excellent combination of gramineous and leguminous plants. The results were excellent (Table 10.14), the nutritive value of the pasture increased up to 6 percent of crude protein and with increase in P, Ca, Mg, Fe, Cu, Zn, and Mn levels

TABLE 10.13. Traditional and INM fertilizer recommendations for a corn farmer of Monagas State, Venezuela

	Nutritional Limitations	Nutrients (Kg ha^{-1})	Fertilizers (kg ha^{-1})	Fertilizer cost (US$ ha^{-1})	Average Yield (Kg ha^{-1})	Agronomic and economic fertilizer efficiency Kg grain/kg fertilizer and US$ grain/US$ fertilizer[a]
Method 1						
Traditional	−Mg −Ca Acidez −Zn −B −Mn	140 N (92 + 48) 96 P$_2$O$_5$ 108 K$_2$O (48 + 60)	200 Urea 400 12-24-12 100 KCl	68	3.200	4.6 kg/kg y 0.6 US$/US$
Total		344	700			
Method 2						
INM	None	125 N 80 P$_2$O$_5$ 70 K$_2$O 12 Mg 2 Zn 6 Mn 1.5 B	272 Urea 174 DAP 117 KCl Dolomite 6 ZnSO$_4$ 21 MnSO$_4$ 14 Borax 2 t Dolomite	69	5.000	8.28 kg/kg y 3.73 US$/US$
Total		310	604			

Source: Adapted from Casanova (2002b).

[a]The economic efficiency does not include the Dolomite cost.

TABLE 10.14. Pasture productivity parameters, nutritional level, milk productivity, and recuperation cost of a degraded pasture under PARP, and leguminous treatment compared with the traditional management of the farmer

	INM treatment	Degraded plot	Factor of increase
t Dry Matter/ha/harvest	1.80	0.30	6.0
% Crude Proteín	6.40	3.00	2.1
% P	0.15	0.08	1.9
% K	1.02	–	
% Ca	0.26	0.20	1.3
% Mg	0.46	0.25	1.8
mg/kg Fe	332	214	1.6
Mg/kg Cu	5	1	5
mg/kg Zn	45	37	1.2
mg/kg Mn	154	83	1.9
N of animals milking	100	100	
Liters milk/animal/day	8	5	1.6
Recuperación Cost ($/ha)	72	430	6

Source: Adapted from Casanova (2002b).

in the pasture. Milk production increased by 40 to 60 percent, averaging eight liters per animal per day for a herd of 100 animals. The gramineous and leguminous plant combination improved the grass nutritive quality with good adaptation, excellent consumption by the livestock, and easy recovery after shepherding. To recover the degraded pastures with the INM program cost US$72 ha^{-1} compared to the traditional way of recovery (tillage, seed, fertilizer, weed control, etc.) that costs US$430 ha^{-1}.

One of the premises of sustainable agriculture and particularly of INM is that the farmer-producer can live on his/her farm with a profitability level that is sufficient to provide his/her family and himself/herself a satisfactory standard of living and help contribute to producing foodstuffs for 25 million Venezuelans. The increase in milk production of about 3 l per animal per day, combined with a gain in animal weight of 120 kg per animal would allow the farmer to gain US$90 more per day in milk production and to gain US$15,600 more in meat production in a total of 100 animals (Table 10.15).

TABLE 10.15. Profitability in milk and meat with the application of PAPR to pasture in Monagas state (100 animals, liters/animal/day and weight (kg) of animal at the time of sale)

Site	Check plot	PARP + Urea	Price at the farm US$/l (milk) or US$/kg (meat)	Additional profit in US$/day (milk) or US$/100 animals (meat)
Monagas (milk)	5	8	0.30	90.00
Monagas (meat)	350	470	1.30	15,600.00

Source: Adapted from Casanova (2002b).

Note: Profit calculation: milk: 3 liters/animal/day × 0.30 US$/l × 100 animals; meat: 120 kg more of meat × 1.30 US$/kg at the farm × 100 animals.

RESEARCH GAPS AND FUTURE RESEARCH NEEDS

There are a tremendous number of topics related to INM that need research. However, in general more attention should be given to some far-reaching aspects.

The Pampas, a region with soils of originally high fertility, are becoming depleted and will need increasing additions of fertilizers. Research to determine adequate methods of soil analysis and to establish rates of nutrients to be applied should receive increasing attention.

Fertilization for conservation agriculture is another must. Research is needed on the dynamics of surface applied fertilizers and amendments. Cycling of nutrients and nutrient balances should be better characterized. Research and organization of information on organic sources of nutrients should be supported, as well as the role of organic matter in increasing yields.

Fertigation is another aspect that needs research, to avoid the waste of nutrients, and to ensure that the high yield potential is attained without environmental problems due to excessive use of fertilizers.

In a more comprehensive way, research is needed on nutrient application rates for new high-yielding hybrids and varieties, the effect of all organic sources in supplying nutrients to crops, the economic analysis of practical application of INM in the tropics, the environmental impacts of INM technology compared to traditional nutrient management, development of sustainability indicators (agronomic, economical, social, and environmental) for evaluation of nutrient management systems, and development of a communication network to facilitate the learning of new developments by farmers and agronomists.

SUMMARY AND CONCLUSION

South America is an important food producing area and has the potential to increase agricultural production by opening new areas to agriculture use, increasing productivity of currently cropped lands, and producing more than one crop per year in areas suitable for double cropping. For this, fertilizers are indispensable. For several decades in the past, increased agricultural production in Brazil was closely related to increased fertilizer use. Much of the increased agricultural output during the past decades was due to the increase in productivity. Organic sources of nutrients are available, but their use can be improved. Conservation agriculture is the new direction for the continent, for both annual and perennial crops. Synthetic fertilizers will play an important role in maintaining the sustainability of agriculture. Efforts must be made to adopt the principles around the world, for higher yields and profits and as an environmentally sound approach.

REFERENCES

Abril. (2001). *Almanaque Abril. Edição Mundo 2001.* São Paulo, Brasil: Editora Abril.

Abril. (2003). *Almanaque Abril. Edição Mundo 2003.* São Paulo, Brasil: Editora Abril.

AENDA. (2001). Revirando resíduos sólidos—editorial. *AENDA News,* v. 4, N° 40. São Paulo, Brasil: Associação das Empresas Nacionais de Defensivos Agrícolas.

ANDA. (2003). *Anuário estatístico do setor de fertilizantes—2002.* São Paulo, Brasil: Associação Nacional para Difusão de Adubos.

Andrigueto, J.R. and A.R. Kososky. (2002). *Legal Marks of Integrated Fruit Production in Brazil.* Brasilia, Brazil: MAPA/SARC, 2002.

Barreiro, I., E. Casanova, and J.R. Castillo. (2000). PAPR for sustainable agriculture. *Phosphorus and Potassium.* 228: 27-37.

Bettiol, W. and O. Camargo. (2000). *Impacto ambiental do uso agrícola do lodo de esgoto.* Jaguariúna, Brasil: Embrapa Meio Ambiente.

Boller, E.F., A. El Titi, J.P. Gendrier, J. Avilla, E. Jörg, and C. Malavolta. (1999). *Integrated Production: Principles and Technical Guidelines.* 2nd edition. Wädenswill, Switzerland.

Borlaug, N.E. and C.R. Doswell. (1994). *Feeding a human population that increasingly crowds a fragile planet.* Keynote address at the 15th World Congress of Soil Science. Acapulco, Mexico. Suppl. 1-15.

Britannica. (2001). South America. URL: http://www.britannica.com/bcom/original/article/0,5744,6548.html. Assessed August 7, 2001.

Cabala-Rosand, P. and B. van Raij. (1983). *A análise de solo no Brasil no período 1972-1981.* Campinas, Brasil, Sociedade Brasileira de Ciência do Solo.

Cantarella, H. and B. van Raij. (1986). Adubação nitrogenada no estado de São Paulo. In Santana, M.B.M. (ed.) *Simposio sobre adubação nitrogenada no Brasil.* (pp. 47-79) Ilhéus, Brasil: CEPLAC, SBCS.

Casanova, E. (1996). *Introducción a la ciencia del suelo.* Caracas, Venezuela: Universidad Central de Venezuela, Consejo de Desarrollo Cientifico y Humanistico.

Casanova, E. (2002a). Fertilización, nutrición y sustentabilidad de praderas. *Venesuelos,* Vol 7, Nº1 y 2 (1999) 33-37.

Casanova, E. (2002b). El uso de rocas fosfóricas y su efecto en la productividad de carne y leche en Venezuela. En: R. Tejos, W. Garcis, C. Zambrano, L. Mancilla y N. Valvuena (eds.). VIII Seminário Manejo Y Utilización de Pastos y Forrajes en Sistemas de Producción Animal. pp. 99-106.

Casanova, E. (2005). Políticas de distribución y uso de los fertilizantes para la producción agrícola en Venezuela: Pasado y Presente. Capítulo 9. En: Tierras Llaneras de Venezuela, J.M. Hetier y R. López F. (eds.), Serie Suelos y Clima SC-77, Institut de Recherche pour le Developpment, IRD, Francia, Centro Interamericano de Desarrollo e Investigación Ambiental y Territorial (CIDIAT), Mérida, Venezuela, y Universidad Nacional Experimental de los Llanos Occidentales Ezequiel Zamora. pp. 215-223.

Casanova, E., A. Salas, and M. Toro. (2002). Evaluating the effectiveness of phosphate fertilizers in some Venezuelan soils. *Nutrient Cycling in Agroecosystems.* 63: 13-20.

Cavalcanti, F.J. de A., ed. (1998). Recomendações de adubação para o Estado de Pernambuco (2ª Aproximação). Recife, Empresa Pernambucana de Pesquisa Agropecuária—IPA.

Comissão de fertilidade do Solo. (1995). Recomendações de adubação e de calagem para os estados do Rio Grande do Sul e de Santa Catarina. 3. ed. Passo Fundo: SBCS-Núcleo Regional Sul, 223p.

Dadalto, G.G. and F.A. Fullin. (2001). Manual de recomendação de calagem e adubação para o estado do Espírito Santo—4ª aproximação. Vitória, ES: SEEA/INCAPER, 2001. 266p.

Díaz-Zorita, M. (2005). El avance de la agricultura en Argentina: Cambios en el uso de pesticidas y fertilizantes. *Ciencia Hoy.* 15: 28-29.

EIF. (1998). *Integrated Crop Production: The New Direction for European Agriculture.* European Initiative for Integrated Farming. Bologne, France.

FAO. (2002). *Fertilizer Use by Crop.* 5th edition. Rome: FAO.

FAO. (2004). *Fertilizer Use by Crop in Argentina.* Rome: FAO.

Finck, A. (1995). From the fertilization of crops to the management of nutrients in crop rotations and farming systems. An overview. In *Integrated Plant Nutrition Systems. FAO Fertilizers and Plant Nutrition Bulletin* 12. pp. 67-82.

FONAIAP. (1989). *El cultivo del maíz,* Paquete Tecnológico, Prodetec. Turén, Portuguesa, Venezuela, 16p.

García, F.O. (2001). Phosphorus balance in the Argentine Pampas. *Better Crops Int.* 15: 22-24.

Gilabert de Brito, J. (2003). *Plan institucional para mejorar metodologiass analíticas com fines de fertilizacion de suelos.* Maracay, Aragua, Venezuela: Gerencia de Investigación y Programa de Tecnologías Agrícolas, INIA, 44p.

Hall, A.J., C.M. Rebella, C.M. Ghersa, and J.P. Culot. (1992). Field-crop systems of the Pampas. In Pearson, C.J. (ed.), *Field Crop Ecosystem.* pp. 413-450. Amsterdam: Elsevier.

Harrington, H.J. and H.J. Knight. (2001). *ISO 14000: Implementation.* New York: MaGrawHill.

IBGE. (2003a). http://www.ibge.gov.br/estatística/economia/agropecuaria/censoagro/brasil/tabela4brasil.shtm.

IBGE. (2003b). http://www.ibge.gov.br/estatística/economia/agropecuaria/censoagro/brasil/tabela1brasil.shtm.

IFA. (1998). *Best Agricultural Practices to Optimize Fertilizer Use in Latin America.* Paris: International Fertilizer Industry Association.

Isherwood, K.F. (1998). *Mineral Fertilizer Use and the Environment.* Paris: IFA.

Jenny, H. (1941). *Factors of Soil Formation.* New York: McGraw Hill.

Kamprath, E.J. (1970). Exchangeable Al as a criterion for liming leached mineral soils. *Soil Science Society of America Proceedings.* 34: 252-254.

Lopes, A.S. (1996). Soils under cerrado: A success story in soil management. *Better Crops International.* 10: 9-15.

Lopes, A.S., L.R.G. Guilherme, and C.A.P. Silva. (2003). *Vocação da terra.* São Paulo, Brazil: Associação Nacional para a Difusão de Adubos.

Lopes, A.S., S. Viethölter, L.R.G. Guilherme, and C.A. Silva. (2004). *Sistema de plantio direto: Bases para o manejo da fertilidade do solo.* São Paulo, Brasil: Associação Nacional para a Difusão de Adubos.

Pequiven. (2003). Base de datos sobre producción y importación de fertilizantes para la agricultura venezolana. Morón, Estado Caraboro, Venezuela.

Pequiven. (2006). Base de datos sobre producción e importación de fertilizantes para la agricultura Venezolana. Morón, Estado Carbobo, Venezuela.

Quaggio, J.A., H. Cantarella, and B. van Raij. (1994). Evolution of the analytical quality of soil testing laboratories integrated in a sample exchange program. *Communications in Soil Science and Plant Analysis.* 18: 1007-1014.

Quaggio, J.A. B. van Raij, and E. Malavolta. (1985). Alternative use of the SMP-buffer solution to determine lime requirement of soils. *Communications in Soil Science and Plant Analysis.* 16: 245-260.

Raij, B. van. (1980). Extension of soil testing results and fertilizer recommendation to the farmer. *Soils Bulletin.* pp. 217-222.

Raij, B. van. (1988). *Gesso na melhoria do ambiente radicular no subsolo.* São Paulo, Brasil: ANDA.

Raij, B. van. (1991). Fertilidade do solo e adubação. São Paulo/Piracicaba, Brasil: Ceres/Potafos.

Raij, B. van, J. C. Andrade, H. Cantarella, and J. A. Quaggio (eds.). (2001). Análise química para avaliação da fertilidade de solos tropicais. Campinas, Brazil: Instituto Agronômico.

Raij, B. van, A.P. de Camargo, H. Cantarella, and N.M. da Silva. (1983). Alumínio trocável e saturação por bases como critério para recomendação de calagem. *Bragantia.* 42: 149-156.

Raij, B. van, H. Cantarella, J.A. Quaggio, and A.M.C. Furlani. (1996). *Recomendações de adubação e calagem para o Estado de São Paulo.* 2nd edition. Campinas, Brasil, Instituto Agronômico. 285p. (Tec. Bull, 100).

Raij, B. van, H. Cantarella, J.A. Quaggio, L.I. Prochnow, G.C. Vitti, and H.S. Pereira. (1994). Soil testing and plant analysis en Brazil. *Communications in Soil Science and Plant Analysis.* 25: 739-751.

Raij, B. van and A. van Diest. (1980). Phosphate supplying power of rock phosphates in an oxisol. *Plant and Soil.* 55: 97-104.

Raij, B. van and J.A. Quaggio. (1997). Methods used for diagnosis and correction of soil acidity in Brazil: An overview. In Moniz, A.C. et al. (ed.) *Plant Soil Interactions at Low pH (205-214).* Campinas, Brasil: Sociedade Brasileira de Ciência do Solo.

Raij, B. van, J.A. Quaggio, and N.M. da Silva. (1986). Extraction of phosphorus, potassium, calcium and magnesium from soils by an ion-exchange resin procedure. *Communications in Soil Science and Plant Analysis.* 17: 544-566.

Resck, D.V.S., J.E. da Silva, A.S. Lopes, and L.M. da. Management systems in Northern South América. (2006). In: *Drylan Agriculture*, 2nd ed., Agronomy Monograph no. 23. (pp. 427-525). Madison, WI, USA: American Society of Agronomy.

Ribeiro, A.C., P.T.G. Guimarães, and V.H. Alvarez. (1999). *Recomendações para o uso de corretivos e fertilizantes em Minas Gerais.* 5ª aproximação. Viçosa, Brasil: Comissão de Fertilidade do Solo de Minas gerais.

Ritchey, K.D., J.E. Silva, and U.F. Costa. (1982). Calcium deficiency in clayey B horizons of Savannah Oxisols. *Soil Science.* 133: 378-382.

Ritchey K.D., D.H.G. Souza, E. Lobato, and O. Correa. (1980). Calcium leaching to increase rooting depth in a Brazilian Savannah Oxisol. *Agronomy Journal.* 72: 40-44.

Roy, R.N. (1995). Integrated plant nutrition systems basic concepts, development and results of trial network, initiation of project activities in AGLN and need for cooperation. pp. 49-66. In Dudal, R. and R.N. Roy (eds.) *Integrated Plant Nutrition Systems. FAO Fertilizer and Plant Nutrition Bulletin 12,* Rome: FAO.

Roy, R.N., R.V. Misra, J.P. Lesschen, and E.M. Smaling. (2003). *Assessment of Soil Nutrient Balance: Approaches and Methodologies.* Rome: FAO.

Sá, J.C.M. (1996). Manejo do nitrogênio na cultura do milho no sistema plantio direto. Passo Fundo, RS, Brasil: Aldeia Norte. 24p.

Shoemaker, H.E., E.O. McLean, and P.F. Pratt. (1961). Buffer methods for determining lime requirement of soils with appreciable amounts of extractable aluminum. *Soil Science Society of America Proceedings.* 25: 274-277.

Silva, F.C. da. (1999). organizer, *Manual de análises químicas de solos, plantas e fertilizantes.* Brasília, Brasil: Embrapa Comunicação para Transferência de Tecnologia.

Silva, F.C. da and B. van Raij. (1999). Disponibilidade de fósforo em solos avaliada por diferentes extratores. *Pesquisa Agropecuária Brasileira.* 34: 267-288.

Souza, D.M. G. de and E. Lobato. (2002) *Cerrado: correção do solo e adubação.* Planaltina: Embrapa Cerrados.

Tedesco, M.J., C. Gianello, C.A. Bissani, H. Bohnen, and S.J. Volkweiss. (1995). *Análise de solo, plantas e outros materiais.* Porto Alegre, Brasil: Departamento de Solos, UFRGS.

Universidad Central de Venezuela. (1993). Métodos de análisis de suelos y plantas utilizados en el Laboratorio General del Instituto de Edafología (1-89). Maracay, Vezuela: Facultad de Agronomia.

Vitti, G.C. and E. Malavolta. (1999). Atingir o patamar de produtividade alcançado com o uso de fertilizantes minerais via adubação orgânica: uma expectativa irreal? In Siqueira, J.O. et al. (eds.) *Inter-relação fertilidade, biologia do solo e nutrição de plantas.* pp.163-169. Viçosa, Brazil: Sociedade Brasileira de Ciência do Solo; Lavras, Brasil: Departamento de Ciência do Solo da Universidade Federal de Lavras.

Wade, J.S. (1999). The Brazilian pantanal and Florida everglades: A comparison of ecosystems, uses and management. In *Simpósio sobre recursos naturais e sócio-econômicos do Pantanal,* 2 pp. 29-37. 1996, Corumbá, MS. Manejo e conservação: Anais.Corumbá: Embrapa Pantanal.

Yamada, T. and A.S. Lopes. (1999) Balanço de nutrientes na agricultura brasileira. In Siqueira, J.O. et al. (eds.) *Inter-relação fertilidade, biologia do solo e nutrição de plantas.* pp. 143-161. Viçosa; Brasil: Sociedade Brasileira de Ciência do Solo; Lavras, Brasil: Departamento de Ciência do Solo da Universidade Federal de Lavras.

Chapter 11

Integrated Nutrient Management: Concepts and Experience from Sub-Saharan Africa

Andre Bationo
Job Kihara
Bernard Vanlauwe
Joseph Kimetu
Boaz S. Waswa
Kanwar L. Sahrawat

INTRODUCTION

In Africa, 28 percent of its landmass, 874 Mha, is potentially suitable for agricultural production. Of the potentially suitable agricultural land, 34 percent comprises arid and semiarid lands (ASAL) that are too dry for rainfed agriculture. The semiarid regions have a shortened length of growing period (75-129 days) compared to the subhumid (180-269 days), and the humid (>270 days) zones. The dynamics of agroecosystems show that the farming systems practiced have gone through diverse changes from traditional shifting farming systems to permanent and intensified arable and mixed farming systems. The changes are coping strategies to respond to the environment, and its changing biophysical and socioeconomic circumstances.

Africa has 340 million people, over half of its population living on less than US$1 per day, a mortality rate of children under five years of age, of 154 per 1,000, and a life expectancy of only forty-eight years (Benson, 2004). The average annual increase of cereal yield in Africa is about 10 kg ha^{-1}, the rate known as that for extensive agriculture neglecting external inputs like improved seeds and plant nutrients (Bationo et al., 2004). In 1996, fertilizer use in sub-Saharan Africa (SSA) was about 9 kg ha^{-1} compared to the global average use of 98 kg ha^{-1} (Gruhn et al., 2000). With 9 percent of the world's population, SSA accounts for less than 1.8 percent of global fertilizer use, and less than 0.1 percent of global fertilizer production (Bationo et al., 2004).

Twenty-eight percent of Africa's population is chronically hungry and heavily dependent on food imports (2.8 m tonnes of food aid in 2000 alone). Due to a high population growth rate (3 percent) compared to cereal grain yield (<1 percent) (Gruhn et al., 2000), per capita cereal production has decreased from 150 kg per person to 130 kg per person over the last thirty-five years, whereas Asia and Latin America realized per capita food increase from 200 kg per person to 250 kg per person during the same period. The increase in yields of the food crops has been more due to land expansion than realized from the crop productivity improvement potential (Table 11.1). The 7.6 percent yield increase of yam in west Africa for example, was mainly due to an area increase of 7.2 percent and only 0.4 percent due to improvement in crop productivity itself. Production of enough food for the growing population requires an increase from 1 billion tonnes per year at present to 2.5 billion t by 2030 (Walker et al., 1999). The Forum for Agricultural Research in Africa (FARA) and its member subregional organizations have developed a vision for agriculture in Africa, calling for a 6 percent annual growth in agricultural productivity (Bationo et al., 2004). In SSA, such increases in agricultural productivity are possible with judicious implementation of agricultural intensification programs to contribute 53 percent of the increase, with other increase coming from expansion in area (30 percent), and cultivation of cash crops (17 percent) (Nandwa, 2003).

TABLE 11.1. Percentage annual increase in crop yields of selected food crops due to land expansion and crop improvement potential in West Africa

Crops	Area (%)/year	Productivity (%)/year	Production (%)/year
Cassava	2.6	0.7	3.3
Maize	0.8	0.2	1.0
Yam	7.2	0.4	7.6
Cowpea	7.6	−1.1	6.5
Soybean	−0.1	4.8	4.7
Plantain	1.9	0.0	2.0

Source: Adapted from www.fao.org.

Macropolicy changes imposed externally in the last decade, such as structural adjustment and the removal of fertilizer subsidies, resulted in reduction in the use of external inputs (Gruhn et al., 2000), expansion in area of agriculture through the opening of new lands, and the reduction of the farmers' potential for investment in soil fertility restoration. Technological, environmental, sociocultural, economic, institutional, and policy constraints that hamper agricultural development in Africa have been identified. These constraints are:

1. low soil fertility;
2. fragile ecosystems;
3. overdependence on rainfall;
4. aging rural population and thus limited physical energies for production;
5. underdeveloped and degraded rural infrastructure;
6. insufficient research due to lack of motivation and inadequate facilities;
7. inadequate training and extension services;
8. high postharvest losses;
9. insufficient market;
10. lack of credit and insufficient agri-input delivery systems;
11. limited farmers' education and know-how;
12. brain-drain of African intellectuals;
13. policy instability; and
14. inconsistent agricultural policies and land tenure (Bationo et al., 2004).

Acquired immune deficiency syndrome (AIDS) has also emerged as a major constraint to agricultural production, especially in rural areas.

The problem of soil fertility degradation is not only a factor of biophysical aspects but also of socioeconomic factors (Figure 11.1). A wide range of socioeconomic factors such as macroeconomic policies, unfavorable exchange rates, poor producer prices, high inflation, poor infrastructure, and lack of markets diminish farmers' capacity to invest in soil fertility, as well as the returns obtained from such investments where value cost ratios are often less than two (2). These multiple causes of low soil fertility are strongly interrelated and the interactions between biophysical and socioeconomic factors call for a holistic approach in ameliorating the soil fertility constraints in SSA (Murwira, 2003).

There has been great concern that institutions of higher education are not making a significant contribution to the national agricultural research agenda. This is due in part to the limited funding of public agricultural research and development in SSA, which declined from 2.5 percent in the

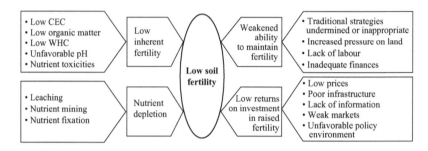

FIGURE 11.1. Biophysico-chemical and socioeconomic factors contributing to low soil fertility in Africa. *Source:* Adapted from Murwira (2003).

1970s to 0.7 percent in the 1980s. At the same time, agricultural investment in developed countries declined by 1 percent from 2.7 percent (Gruhn et al., 2000). From 1987 to 1997, World Bank global support to agricultural extension was 46.3 percent compared to 2.2 percent for agricultural higher education (Willett, 1998). The common trend in the African continent has been a decline in support for research in agricultural institutions.

Apart from water shortages, soil infertility is the major constraint to increased agricultural productivity in SSA. Soil fertility management is crucial for sustained production and requires inputs of nutrients to compensate those removed by crops and lost from the soil through various physical and biochemical processes. In the context of SSA, the approach based on supply of plant nutrients through organic and chemical sources, or integrated nutrient management (INM) needs to be followed. First, the use of chemical fertilizers is very low, and the use of manures and organic and crop residues (CRs) not only supply nutrients such as N, but also supply organic matter so essential for the soil's physical, chemical, and biological integrity. In this chapter, we will first provide an overview of soil fertility status in SSA, followed by the evolution of soil fertility paradigms before highlighting the experiences in the use of integrated soil fertility management (ISFM) in Africa.

OVERVIEW OF SOIL FERTILITY STATUS IN AFRICA

The fundamental biophysical root cause for declining per capita food production in smallholder farms in SSA (17 percent between 1980 and 1995) is soil fertility depletion. Although significant progress has been made in research, in developing principles, methodologies, and technologies for combating soil fertility depletion (Nandwa and Bekunda, 1998; Nandwa, 2001; Stoorvogel and Smaling, 1990; Smaling et al., 1997; Tian et al., 2001), soil infertility still remains the fundamental biophysical cause for the declining per capita food production in SSA over the past 3-5 decades (Sanchez et al., 1997). This is evident from the huge gaps between actual and potential crop yields (FAO, 1995).

During the past thirty years, soil fertility depletion has been estimated at an average of 660 kg N ha^{-1}, 75 kg P ha^{-1}, and 450 kg K ha^{-1} from about 200 Mha of cultivated land in thirty-seven African countries (Sanchez et al., 1997). Average depletion of major nutrients (N, P, and K) for the 1993-1995 period is shown in Figure 11.2. Stoorvogel et al. (1993) estimated annual net depletion of nutrients in excess of 30 kg N and 20 kg K per ha of arable

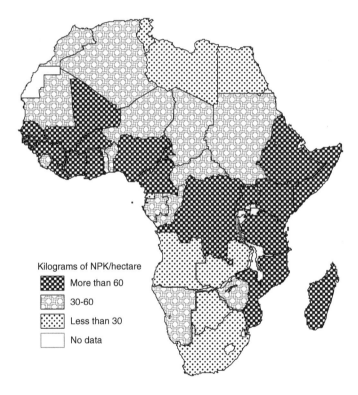

FIGURE 11.2. Fertilizer NPK depletion rates in Africa between 1993-1995.
Source: Adapted from Nandwa (2003).

land per year in Ethiopia, Kenya, Malawi, Nigeria, Rwanda, and Zimbabwe (Table 11.2).

Soil nutrient depletion is a major bottleneck in increasing land productivity in the region and has largely contributed to poverty and food insecurity. Soil nutrient depletion occurs when nutrient inflows are less than outflows. Nutrient balances are negative for many cropping systems indicating that farmers are mining their soils of nutrient reserves (Figure 11.3).

The inherent constraints in some soils have been exacerbated by their overexploitation for agricultural production. Large areas of soils of high production potential in SSA have been degraded due to continuous cropping without replacement of nutrients taken up in harvests (Murwira, 2003). Increasing population pressure of up to 1,200 persons per square kilometer (Shepherd et al., 2000) have necessitated the cultivation of marginal lands that are prone to erosion and other types of environmental degradation, and it is also no longer feasible to use extended fallow periods to restore soil fertility. The shortened fallow periods cannot regenerate soil productivity leading to nonsustainability of the production systems (Nandwa, 2001).

The negative effects of nutrient outputs exceeding inputs, manifested in negative nutrient balances, and the deficiencies of major nutrients are attributed primarily to nonuseful outflows such as burning/removal of biomass,

TABLE 11.2. Average nutrient balance of N, P, and K (kg ha^{-1} year^{-1}) for the arable land for some SSA countries (average of 1982-1984)

Country	N	P	K
Botswana	0	1	0
Mali	−8	−1	−7
Senegal	−12	−2	−10
Benin	−14	−1	−9
Cameroon	−20	−2	−12
Tanzania	−27	−4	−18
Zimbabwe	−31	−2	−22
Nigeria	−34	−4	−24
Ethiopia	−41	−6	−26
Kenya	−42	−3	−29
Rwanda	−54	−9	−47
Malawi	−68	−10	−44

Source: Adapted from Stoorvogel et al. (1993).

leaching, volatilization, erosion losses of nutrients, and the lack of water and waste recycling in agricultural systems (Figure 11.4). Considerable export of nutrients is via harvestable products, the goal and objective of agricultural production. Results from Nutrient Monitoring (NUTMON) (Gachimbi et al., 2002) studies demonstrate that for efficient return to increased agricultural production, enhanced nutrient availability will have to initially de-

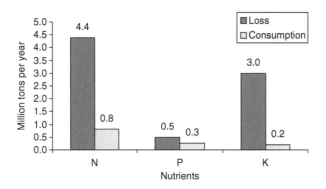

FIGURE 11.3. Major nutrient losses versus their application rates in Africa. *Source:* Adapted from Sanchez et al. (1997).

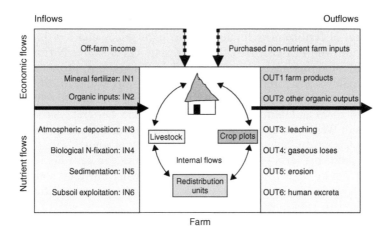

FIGURE 11.4. Nutrient inflows and outflows in a farm. *Source:* Adapted from Vlaming et al. (2001).

pend on the extent to which farmers minimize or eliminate nonuseful out-flows including residue burning, the loss of nutrients especially N via leaching, volatilization, and denitrification and loss of nutrients through erosion.

Among the plant nutrients, N and P are the most limiting for crop pro-duction in SSA. For example, studies undertaken at the West Africa Rice Development Association (WARDA) along a north to south transect in Cote d'Ivoire demonstrated that as one moved from north to south on a north-south transect, soil acidity increases and the deficiency of P becomes more important than that of N for cereals such as rice (*Oryza sativa* L.). These ob-servations are in accordance with the findings that the deficiency of P is more important than that of N in the humid forest zone of West Africa (Sahrawat et al., 2000, 2001). On the other hand, on soils in the savanna and savanna-forest transition zones, N deficiency is more important than P deficiency. As a general rule, P deficiency becomes increasingly important from north to south on a north-south transect in West Africa (Sahrawat et al., 2001).

Soil fertility, especially the organic matter status of soils also varies with ecology. For example, wetland soils located in the valley bottoms or lowest portion of the continuum or toposequence, generally cultivated to wetland rice, are higher in organic matter compared to soils in upland and transition zones of the toposequence (Sahrawat, 2004). Thus, the fertility of soils var-ies among ecosystems and ecologies and generalizations about the nutrient status of soils of general fertility may at times prove hazardous.

Nitrogen is commonly deficient and limits crop production in cultivated soils of the tropics (Sanchez, 1976). For most farmers in SSA, the use of mineral N fertilizers is limited due to high prices and low profitability (McIntire and Fussel, 1986). The only option is to source N from organic inputs, intercropping and rotations with N fixing crops, and through man-aged fallows using improved leguminous fallows (Tian et al., 2001).

High variability in total N content is observed both in different agroeco-logical zones and farm sections in the same zone. Different agroecological zones in West Africa for example recorded total soil N values between 0.5 and 1.6 g kg^{-1} (Table 11.3). Nutrient distribution studies within the same farm show that farm sections close to the homestead have high N content (0.9-1.8 g kg^{-1}) with distant bush fields showing low total soil N values (0.2-0.5 g kg^{-1}). The differences are attributed to farm-specific manage-ment practices (Table 11.4).

Nitrogen, phosphorus and organic matter reserves in soils in African agroecosystems also vary with resource endowment (Table 11.5). Recent studies in Kenya showed that land belonging to low resource endowed farm-ers can lose up to ten times more soil through soil erosion compared to land

TABLE 11.3. Total nitrogen, total P, and OC stocks of granitic soils in different agroecological zones in West Africa

AEZ	pH (H$_2$O)	OC (g kg^{-1})	Total N (g kg^{-1})	Total P (mg kg^{-1})
Equatorial forest	5.3	24.5	1.60	628
Guinea savanna	5.7	11.7	1.39	392
Sudan savanna	6.8	3.3	0.49	287

Source: Adapted from Windmeijer and Andriesse (1993).

TABLE 11.4. Total N, available P, and OC stocks of different subsystems in a typical upland farm in the Sudan-savanna zone

AEZ	pH (H$_2$O)	OC (g kg^{-1})	Total N (g kg^{-1})	Available P (mg kg^{-1})	Exchangeable K (mmol kg^{-1})
Home garden	6.7-8.3	11-22	0.9-1.8	20-220	4.0-24
Village field	5.7-7.0	5-10	0.5-0.9	13-16	4.0-11
Bush field	5.7-6.2	2-5	0.2-0.5	5-16	0.6-1

Source: Adapted from Prudencio (1993).

TABLE 11.5. Soil loss and soil fertility indicators (% of threshold level) as affected by farmers' wealth class in Machakos, Kenya

Group/Wealth class	Soil loss (t year^{-1})	SOC	Total N	Total P
High (wealthiest)	10	30	32	65
Medium (average)	40	29	30	63
Low (poorest)	100	28	29	62

Source: Adapted from Gachimbi et al. (2002).

belonging to high resource endowed farmers. The study showed that high- and medium-resource endowed farmers are better positioned to arrest or alleviate nutrient depletions in their farm holdings. They have better access to natural capital, manufactured or local resources, and can use external inputs, partly because of their financial status. Nutrient depletion studies at the farm level across Africa reveal a strong effect on socioeconomic conditions. Shepherd and Soule (1998) reported a negative carbon balance of 400 kg ha^{-1} year^{-1} for farmers with low resource endowment whereas a positive balance of 190 kg ha^{-1} year^{-1} was reported in fields of farmers with high resource endowment (Table 11.6).

Soil organic C (SOC) levels across farm fields show steep gradients resulting from long-term site-specific soil management by the farmer. According to Prudencio (1993), SOC status of various fields within a farm in Burkina Faso showed great variations with home gardens (located near the homestead) having 11-22 g kg^{-1} (Table 11.4), village field (at intermediate distance) 5-10 g kg^{-1}, and bush field (furthest) having only 2-5 g kg^{-1}. Usually, closer fields are supplied with more organic inputs as compared to distant fields due to the labor factor.

About 80 percent of the soils have inadequate supply of phosphorus, without which other inputs and technologies are not effective (Bationo et al., 2003; Gikonyo and Smithson, 2003). The importance of phosphorus in the integrated management of soil fertility is manifested by the fact that leguminous crops and cover crops in natural and managed fallows fail to take full advantage of biological nitrogen fixation (BNF) in the absence of adequate P levels in the soils (Sahrawat et al., 2001). Availability and total P levels of soil are low in SSA (Bache and Rogers, 1970; Mokwunye, 1974; Jones and Wild, 1975; Sahrawat et al., 2001). Despite its low levels of inherent soil P, the use of P in SSA is only 1.6 kg P ha^{-1} in cultivated land compared to 7.9 and 14.9 kg P ha^{-1} respectively for Latin America and Asia (Bationo et al., 2003).

The main factors contributing to soil fertility depletion in SSA are erosion by water and wind, especially in the semiarid and arid zone soils. The

TABLE 11.6. SOC balance, soil erosion, farm return, and household income at different farm resource endowment levels

Variable	Units	Farm resource endowment		
		Low	Medium	High
Soil C balance	Kg ha^{-1} year^{-1}	−400	−318	190
Soil erosion	T ha^{-1} year^{-1}	5.6	5.5	2.1
Farm returns	$ year^{-1}	3	70	545
Household income	$ year^{-1}	454	1,036	3,127

Source: Adapted from Shepherd and Soule (1998).

soil lost through erosion is about ten times greater than the rate of natural soil formation, while deforestation is thirty times greater than that in planned reforestation. Sterk et al. (1996) reported a total loss of 45.9 t ha^{-1} of soil by wind erosion during four consecutive storms. Buerkert et al. (1996) reported that in unprotected plots, up to 7 kg of available P and 180 kg ha^{-1} of organic carbon (OC) are lost from the soil profile within one year. The loss of the top soil, which can contain ten times more nutrients than the subsoil, is particularly worrying, since it potentially affects crop productivity in the long-term by removing the soil that is inherently rich in organic matter (Figure 11.5) and micronutrients. Runoff and soil loss will depend on soil types and their erodibility, and the implementation of landform configuration and soil and water conservation practices (Lal, 1980).

EVOLUTION OF SOIL FERTILITY PARADIGM IN AFRICA

During the past three decades, the paradigms underlying soil fertility management research and development efforts have undergone substantial change because of experiences gained with specific approaches, and changes in the overall social, economic, and political environment faced by various stakeholders. During the 1960s and 1970s, an external input paradigm was driving the agricultural research and development agenda. The appropriate use of external inputs, whether fertilizers, lime, or irrigation water, was believed to alleviate constraints to crop production (Vanlauwe, 2004). Following this paradigm together with the use of improved cereal germplasm, the "Green Revolution" boosted agricultural production in Asia and Latin

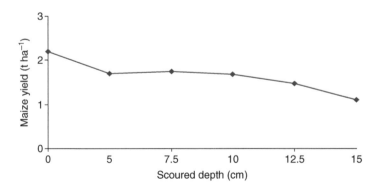

FIGURE 11.5. Effect of depth of soil mechanical desurfacing on maize grain yield at Mbissiri, north Cameroon. *Source:* Adapted from Roose and Barthes (2001).

America in ways not seen before. Organic resources were considered less essential. Sanchez (1976) stated that when mechanization is feasible and fertilizers are available at reasonable cost, there is no reason to consider the maintenance of soil organic matter (SOM) as a major management goal. However, application of the "Green Revolution" strategy in SSA resulted only in minor achievements for a variety of reasons (IITA, 1992). This, together with environmental degradation resulting from massive and injudicious applications of fertilizers and pesticides in Asia and Latin America between the mid-1980s and early 1990s (Theng, 1991), and the abolition of fertilizer subsidies in SSA (Smaling, 1993) imposed by structural adjustment programs, led to a renewed interest in organic resources in the early 1980s. The balance shifted from mineral inputs to low input sustainable agriculture (LISA) where organic resources were believed to enable sustainable agricultural production (Vanlauwe, 2004).

After a number of years of investment in research activities evaluating the potential of LISA technologies, such as alley cropping or live-mulch systems, several constraints were identified both at the technical (e.g., lack of sufficient organic resources) and the socioeconomic level (e.g., labor intensive technologies) (Vanlauwe, 2004). Sanchez (1994) also revised his earlier statement by formulating the Second Paradigm for tropical soil fertility research: "Rely more on biological processes by adapting germplasm to adverse soil conditions, enhancing soil biological activity and optimizing nutrient cycling to minimize external inputs and maximize the efficiency of their use." This paradigm recognized the need for both mineral and organic inputs to sustain crop production, giving way to the INM approach. The need for both organic and mineral inputs was advocated because (1) both resources fulfill different functions to maintain plant growth; (2) under most small-scale farming conditions, neither of them is available or affordable in sufficient quantities to be applied alone; and (3) several hypotheses could be formulated leading to added benefits when applying both inputs in combination (Vanlauwe, 2004).

INM is perceived as the judicious manipulation of nutrient inputs, outputs, and internal flows to achieve productive and sustainable agricultural systems (Smaling et al., 1996). It can be defined as a systematic, planned approach to soil fertility management on both a small and large scale in the context of both the farm and ecosystem as a whole, in which sound management principles and practices are followed throughout. This management approach involves the best possible combination of available nutrient management practices, in the context of biophysical resources, economic feasibility, and social acceptability. The INM concept is mostly applied to the use of organic and inorganic sources of nutrients in a judicious and efficient way. INM often comprises multidisciplinary teams of agronomists, soil scientists, livestock specialists, sociologists, anthropologists, and economists all working together, thus optimizing all aspects of nutrient cycling.

From the mid-1980s to the mid-1990s the shift in paradigm toward the combined use of organic and mineral inputs was accompanied by a shift in approaches toward involvement of the various stakeholders in the research and development process, mainly driven by the "participatory" movement. One of the important lessons learned was that the farmers' decision making process was not merely driven by the soil and climate but by a whole set of factors cutting across the biophysical, socioeconomic, and political domain. The Integrated Natural Resource Management (INRM) research approach was thus formulated and aimed at developing interventions that take all the above aspects into account (Figure 11.6) (Izac, 2000).

Currently, the ISFM paradigm, that forms an integral part of the INRM research approach with a focus on appropriate management of the soil resource, is adopted within the soil fertility research and development community (Figure 11.7). Like INRM, ISFM recognizes the important role of social, cultural, and economic processes regulating soil fertility management strategies. ISFM is also broader than INM as it recognizes the need of an appropriate physical and chemical environment for plants to grow optimally, besides a sufficient and timely supply of available nutrients (Vanlauwe, 2004).

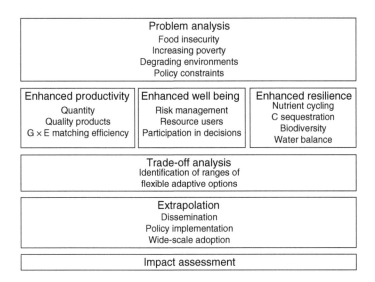

FIGURE 11.6. The INRM research approach. *Source:* Adapted from Izac (2000).

Integrated soil fertility management strategy

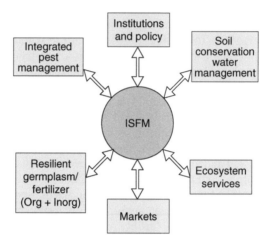

FIGURE 11.7. Integrated soil fertility management paradigm. *Source:* Bationo et al. (2003).

Within this new ISFM paradigm, considerable attention is given to the imbalances between nutrient inputs and outputs (nutrient(s) flows), and their agronomic, economic, and environmental consequences. Nutrient stocks, flows, and balances are increasingly being used for estimating nutrient depletion/accumulation. Nutrient depletion or enrichment of the system follows from the difference between total nutrient inputs (mineral fertilizer, organic inputs, deposition, BNF, sedimentation, and subsoil nutrient exploitation) and outputs (harvested products, (CR)s, leaching, gaseous losses, erosion, feces, and urine). Nutrient flows models are used at the plot and farm level to assist farmers and policymakers in evaluating the agronomic and environmental impact of their farm management practices. Nutrient balance may be characterized by a nutrient balance, made up of a number of nutrient inputs that exceed nutrient outputs ($\Sigma IN - \Sigma OUT >> 0_1$, termed surplus nutrient accumulation class); or where nutrient outputs exceed inputs ($\Sigma IN - \Sigma OUT << 0_1$, termed nutrient depletion class); or where $\Sigma IN - \Sigma OUT = 0_+$, which is termed balanced or equilibrium class.

The ISFM approach advocates careful management of soil fertility aspects that optimize production potential through the incorporation of a wide range of adoptable soil management principles, practices, and options for productive and sustainable agroecosystems. It entails the development of soil nutrient management technologies for adequate supply and feasible share of organic and inorganic inputs that meet the farmers' production goals and circumstances. The approach integrates the roles of soil and water conservation; land preparation and tillage; organic and inorganic nutrient sources; nutrient adding and saving practices; pests and diseases; livestock; rotation and intercropping; multipurpose role of legumes; and integrating the different research methods and knowledge systems (Kimani et al., 2003).

ISFM embraces multiplepurpose options (MPOs), which include INM (the technical backbone of the ISFM approach), biotic and abiotic factors and their relationships, livestock integration in crop production systems, use of local and indigenous knowledge together with science knowledge-based management system, and integration of policy and institutional framework. The major emphasis in the ISFM paradigm is on understanding and seeking to manage processes that contribute to change. This paradigm is closely related to the wider concepts of INRM, thereby representing a significant step beyond the earlier, narrower concept and approach of nutrient replenishment/recapitalization for soil fertility enhancement (Sanchez et al., 1997).

ISFM EXPERIENCES IN SSA

Semiarid Agroecological Zone

N and P Management

Nitrogen and P are the most limiting nutrients in the semiarid zones of Africa. Urea and calcium ammonium nitrate (CAN) are the most common sources of N used by farmers. Results of trials undertaken to evaluate these two sources and methods of application of nitrogen led to the conclusions that: (1) fertilizer N recovery by plant was low; (2) there is a higher loss of N with the surface point placement of urea (>50 percent) and the mechanism of N loss is believed to have been ammonia volatilization; (3) losses of N from CAN were less than from urea because one-half of the N in CAN is in the nonvolatile nitrate form; (4) although CAN is a lower N analysis fertilizer than urea, it is attractive as an N source because of its low potential for N loss via volatilization and its low soil acidifying properties (Table 11.7) (Bationo et al., 2003). Point placement of CAN outperformed urea point placed or broadcast, and ^{15}N data from similar trials indicate that uptake by plants was almost three times higher than that of urea applied in the same manner (Figure 11.8).

TABLE 11.7. Recovery ^{15}N fertilizer by pearl millet applied at Sadore, Niger, 1985

N source	Application method	Grain	Stover	Soil	Total
			(%)		
CAN	Point incorporated	21.3	16.8	30.0	68.1
CAN	Broadcast incorporated	10.9	10.9	42.9	64.7
Urea	Point incorporated	5.0	6.5	22.0	33.5
Urea	Broadcast incorporated	8.9	6.8	33.2	48.9
Urea	Point surface	5.3	8.6	18.0	31.9
SE		1.2	2.0	1.9	2.4

Source: Adapted from Christianson and Vlek (1991).

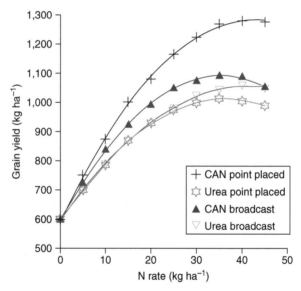

FIGURE 11.8. Effect of broadcast and point application methods for Urea and CAN on grain yield of pearl millet. *Source:* Adapted from Christianson and Vlek (1991).

Christianson and Vlek (1991) found that the optimum N rate for sorghum is 50 kg ha^{-1} and 30 kg ha^{-1} for pearl millet (*Pennisetum glaucum* L.). The N use efficiency can be increased through rotation of cereals with legumes and through the optimization of planting density. Bationo and

Vlek (1998) reported an N-use efficiency of 20 percent in the continuous cultivation of pearl millet, but its value increased to 28 percent when pearl millet was rotated with cowpea. Bationo et al. (1989) found a strong and positive correlation between planting density and response to N fertilizer.

Phosphorous deficiency is a major constraint to crop production and response to N is substantial only when both moisture and phosphorous are not limiting. Field trials were established to determine the relative importance of N, P, and K fertilizers in the Sahelian zone. The data in Table 11.8 indicates that from 1982 to 1986 the average control plot gave 190 kg grain ha^{-1} yield of pearl millet. The sole addition of 30 kg P_2O_5 ha^{-1} without N fertilizers increased the average pearl millet yield to 714 kg ha^{-1}. The addition of only 60 kg N ha^{-1} without P application did not increase the yield significantly over the control (average grain yield obtained was 283 kg ha^{-1}). Those data clearly indicate that P is the most limiting factor in those sandy Sahelian soils and there is no significant response to N without correcting first for P deficiency. Application of 120 kg N ha^{-1} resulted in pearl millet grain yield of 1,173 kg ha^{-1} compared to 714 kg ha^{-1} when only P fertilizers were applied. The addition of K did not increase significantly the yield of either grain or total dry matter of pearl millet in any year of the study.

Despite the fact that deficiency of P is acute on the soils of West Africa, local farmers use very low P fertilizers partly because of the high cost. The use of locally available phosphate rock (PR) could be an alternative to imported P fertilizers. For example, Bationo et al. (1987) showed that direct application of local PR may be more economical than imported water-soluble P fertilizers. Bationo et al. (1990) showed that Tahoua PR from Niger is suitable for direct application, but Parc-W from Burkina Faso has less potential for direct application. The effectiveness of local PR depends on its chemical and mineralogical composition (Lehr and McClellan, 1972; Chien and Hammond, 1978; Khasawneh and Doll, 1978), the most important feature being the ability of carbonate ions to substitute for phosphate in the apatite lattice which influences the solubility, and controls the amount of phosphorus available to crops (Smith and Lehr, 1966). The most reactive PR should have molar PO_4/CO_3 ratios less than 5. In Niger, Tahoua PR outperformed Kodjari PR (from Burkina Faso). The results are in agreement with the fact that the molar PO_4/CO_3 ratio is 23 for Kodjari PR and 4.88 for Tahoua PR, and Tahoua PR also has a higher solubility in Neutral Ammonium Citrate (NAC).

The solubility of PR in NAC is directly related to the level of carbonate substitution (Chien, 1977). Diamond (1979) classified African PR-based on citrate solubility as >5.4 percent high, 3.2-4.5 percent medium, and <2.7

TABLE 11.8. Effect of N, P, and K on pearl millet grain and total dry matter (kg ha^{-1}) at Sadoré and Gobery (Niger)

| Treatments | 1982 Sadoré | | 1983 | | 1984 Sadoré | | 1985 Sadoré | 1986 Sadoré | |
| | | | Sadoré | Gobery | | | | | |
	Grain	TDM	Grain	Grain	Grain	TDM	Grain	Grain	TDM
N0P0K0	217	1,595	146	264	173	1,280	180	180	1,300
N0P30K30	849	2,865	608	964	713	2,299	440	710	2,300
N30P30K30	1,119	3,597	906	1,211	892	3,071	720	930	3,000
N60P30K30	1,155	3,278	758	1,224	838	3,159	900	880	3,200
N90P30K30	1,244	3,731	980	1,323	859	3,423	1,320	900	3,400
N120P30K30	1,147	4,184	1,069	1,364	1,059	3,293	1,400	1,000	3,300
N60P0K30	274	2,372	262	366	279	1,434	290	230	1,500
N60P15K30	816	2,639	614	1,100	918	3,089	710	920	3,100
N60P45K30	1,135	3,719	1,073	1,568	991	3,481	1,200	980	3,500
N60P30K0	1,010	3,213	908	1,281	923	3,377	920	910	3,400
SE.	107	349	120	232	140	320	162	250	400
CV (%)	24	22	26	30	24	22	28	32	25

Source: Bationo, unpublished data.

Note: Nutrients applied are N, P$_2$O$_5$, and K$_2$O kg ha^{-1}, TDM = Total dry matter.

percent low. Based on this classification, only Tilemsi PR has a medium reactivity.

Phosphorous placement can drastically increase P use efficiency as shown with pearl millet and cowpea in an experiment involving broadcast and/or hill placed of different P sources. For pearl millet grain, P use efficiency for broadcasting SSP at 13 kg P ha^{-1} was 23 kg kg^{-1}, but hill placement of SSP at 4 kg P ha^{-1} gave a PUE of 83 kg kg^{-1} P. The PUE of 15-15-15 broadcast was 29 kg grain kg^{-1} P, whereas the value increased to 71 kg kg^{-1} P when additional SSP was applied as hill placed at 4 kg P ha^{-1}, and 102 when only hill placed of 4 kg P ha^{-1} of 15-15-15 was used. Hill placement of small quantities (4 kg ha^{-1}) of P attains the highest use efficiency with the efficiency decreasing with increasing quantity of P (Table 11.9).

TABLE 11.9. Effect of different sources of phosphorus (Single superphosphate-SSP, 15-15-15 NPK, and TPR-Tahoua rock phosphate) and the mode of application on pearl millet yield and phosphorus use efficiency (PUE in kg per kg P applied) in Karabedji, 1998 rainy season

P sources and method of application (selected treatments)	Grain		Total dry matter	
	Yield (kg ha^{-1})	PUE (kg kg^{-1})	Yield (kg ha^{-1})	PUE (kg kg^{-1})
Control	281		1,726	
SSP broadcast (13 kg P ha^{-1})	535	23	3,726	154
SSP broadcast + SSP hill placed (13 + 4 kg P ha^{-1})	743	27	5,563	226
SSP hill placed (4 kg P ha^{-1})	611	83	3,774	514
15-15-15 broadcast (13 kg P ha^{-1})	660	29	4,226	192
15-15-15 broadcast + hill placed (13 + 4 kg P ha^{-1})	1,493	71	7,677	350
15-15-15 (4 kg P ha^{-1})	690	102	4,767	760
TPR broadcast (13 kg P ha^{-1})	690	31	4,135	185
TPR broadcast + SSP HP (13 + 4 kg P ha^{-1})	663	22	4,365	155
TPR broadcast + 15-15-15 HP (13 + 4 kg P ha^{-1})	806	31	5,061	196
SE	84		194	

Source: Bationo, unpublished data.

Note: BC = broadcasting, HP = hill placement.

The efficiency of N fertilizers in the dryland soils depends on the level of land degradation. For example, whereas N efficiency was 8.6 kg grain per kg N, in degraded land, it increased to 15 kg grain per kg N in nondegraded land (Table 11.10).

Crop Residue Management

Crop residue management can play an important role in improving crop productivity in SSA. Numerous research reports show large crop yield increases as a consequence of organic amendments in the Sahelian zone of West Africa (Abdullahi and Lombin, 1978; Bationo et al., 1993; Bationo et al., 1998; Evéquoz et al., 1998; Pieri, 1986, 1989). Bationo et al. (1995) reported from an experiment carried out in 1985 on a sandy soil at Sadoré, Niger, that grain yield of pearl millet after a number of years had declined to only 160 kg ha^{-1} in unmulched and unfertilized control plots. However, grain yields could be increased to 770 kg ha^{-1} with a mulch application of 2 t CR ha^{-1} and to 1,030 kg ha^{-1} with 13 kg P as SSP plus 30 kg N ha^{-1}. The combination of CR and mineral fertilizers resulted in a grain yield of 1,940 kg ha^{-1}.

In different parts of SSA, crop or organic residue applications have been shown to increase soil P availability (Kretzschmar et al., 1991), enhance PR availability (Sahrawat et al., 2001), cause better root growth (Hafner et al., 1993), improve potassium (K) nutrition (Rebafka et al., 1994), protect young seedlings against soil coverage during sand storms (Michels et al., 1995), increase water availability (Buerkert et al., 1999), reduce soil surface resistance by 65 percent (Buerkert and Stern, 1995), and reduce topsoil temperature by over 4°C (Buerkert et al., 1999). These effects are stronger especially in the Sahelian zone, but weaker in other areas with lower temperatures, higher rainfall, and heavier soils (Buerkert et al., 1999). From incubation

TABLE 11.10. Use efficiency of N and P in degraded and nondegraded sites at Karabedji, Niger, 1998-2002

Fertilizer	Site condition	Efficiency (kg grain kg^{-1} N or P)
Nitrogen	Degraded	8.6
	Nondegraded	15.3
Phosphorus	Degraded	50
	Nondegraded	58

Source: Bationo, unpublished data.

studies under controlled conditions Kretzschmar et al. (1991) concluded that increases in P availability after CR application were due to a complexation of iron and aluminum by organic acids. The organic amendments have also been reported to reduce the capacity of the soil to fix P thereby increasing P availability for uptake and hence higher P use efficiency (Buresh et al., 1997; Sahrawat et al., 2001).

Availability of organic inputs in sufficient quantities and quality is one of the main challenges facing farmers and researchers today. In an inventory of CR availability in the Sudanian zone of central Burkina Faso, Sedga (1991) concluded that the production of cereal straw can meet the currently recommended optimum level of 5 t ha^{-1} every two years. However, McIntire and Fussell (1986) reported that on fields of unfertilized local cultivars, grain yield averaged only 236 kg ha^{-1} and mean residue yields barely reached 1,300 kg ha^{-1}. These results imply that unless stover production is increased through application of fertilizers and/or manure it is unlikely that the recommended levels of CR could be available for use as mulch. The availability of CR in smallholder farms in SSA is limited by the fact that there are many competing uses for biomass such as fodder and fuel for cooking (Figure 11.9).

In village level studies on CR, along a north-south transect in three different agroecological zones of Niger, surveys were conducted to assess

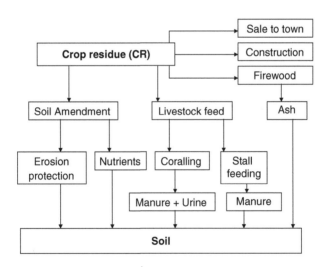

FIGURE 11.9. The competing uses of CRs in the West African semiarid tropics. *Source:* Bationo et al. (1995).

farm-level stover production, household requirements, and residual stover remaining on-farm. The results of these surveys showed that the average amounts of stover removed from the field by a household represented between 2 and 3.5 percent of the mean stover production (ICRISAT, 1993). At the onset of the rains the residual stover on-farm was between 21 and 39 percent of the mean stover production at harvest time. Although farmers require at least 2 T ha^{-1} of CR for mulch, only 250 kg ha^{-1} of CR is presently available on soil at planting time. Even if no data have been collected on the amount of CR lost by microbial decomposition and termites, cattle grazing is likely to be responsible for most of the disappearance of CRs. Similar losses were reported by Powell (1985) who found that up to 49 percent of sorghum and 57 percent of millet stover disappearance on the humid zone of Nigeria was due to livestock grazing. Sandford (1989) reported that in the mixed farming systems, cattle derive up to 45 percent of their total annual intake from CRs and up to 80 percent during periods of fodder shortage. Up to 50 percent of the total amount of CR and up to 100 percent of the leaves is eaten by livestock (van Raay and de Leeuw, 1971). Most of the nutrients are voided in the animal excreta, but when the animals are not stabled, nutrients contained in the droppings cannot be effectively utilized in the arable areas (Balasubramanian and Nnadi, 1980).

Manure Management

Manure, another farm-available soil amendment, is an important organic input in African agroecosystems. One of the earliest reported manure application studies in SSA was by Hartley (1937) in the Nigerian savannah. He observed that application of 2 t ha^{-1} farm yard manure (FYM) increased seed cotton yield by 100 percent, equivalent to fertilizers applied at the rate of 60 kg N and 20 kg P ha^{-1}. Palm (1995) has concluded that for a modest yield of 2 t ha^{-1} of maize the application of 5 t ha^{-1} of high quality manure can meet the N requirement, but this cannot meet the P requirements in areas where P is deficient. Bationo and Mokwunye (1991) found no difference between applying 5 t ha^{-1} of FYM compared to the application of 8.7 kg P ha^{-1} as single superphosphate (SSP) pointing to the role of manure in the availability of P through complexation of iron and aluminum (Kretzschmar et al., 1991). Other reports have shown that crop yields from the nutrient poor West African soils can be substantially enhanced through the use of manure (McIntire et al., 1992; Bationo and Buerkert, 2001; Sedogo, 1993). For Niger, McIntire et al. (1992) reported grain yield increases between 15 and 86 kg for millet and between 14 and 27 kg for groundnut per tonne of applied manure. Combined use of manure and mineral fertilizer show a long-term increase of sorghum yields over years (Figure 11.10).

The data in Table 11.11 (Panel A and B) summarizes the results of a number of trials on manure and manure + inorganic fertilizer conducted in research stations in some West African countries. The data shows that manure collected from stables and applied alone produces about 34 to 58 kg DM t^{-1} manure in cereal grain and 106 to 178 kg of DM t^{-1} manure in stover (Table 11.11 panel A). Application of manure with inorganic fertilizer gave yields of 80 to 90 kg of DM t^{-1} manure and 84 to 192 kg of DM t^{-1} manure grain and stover yields respectively (Table 11.11 panel B).

The quality of manure has been observed to vary with feeds, collection, and storage methods (Mueller-Samann and Kotschi, 1994; Mugwira, 1984; Ikombo, 1984; Probert et al., 1995; Kihanda, 1996). Current characterization studies indicate that manure quality is very variable, for example, %N 0.23-2.6; % P 0.08-1.0; %K 0.2-1.46; %Ca 0.2-1.3; and %Mg 0.1-0.5 (Table 11.12) (Williams et al., 1995). High quality manure has been defined as that with %N >1.6 or C:N ratios of <10; while low quality manure has N < 0.6 percent and C:N ratios of >17.

Several scientists have addressed the availability of manure for sustainable crop production. De Leeuw et al. (1995) reported that with the present livestock systems in West Africa the potential annual transfer of nutrients

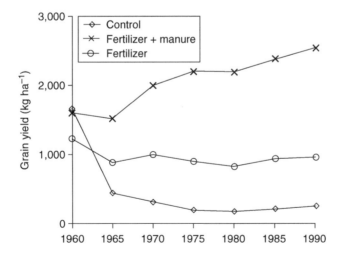

FIGURE 11.10. Sorghum grain yield as affected by mineral and organic fertilizers over time. *Source:* Adapted from Sedogo (1993).

TABLE 11.11. Results of manuring experiments at three sites in semiarid West Africa.

Panel A: Manure only

Location	Amount of manure applied (t ha^{-1})	Crop	Crop response[a] (kg of DM t^{-1} manure) Grain	Stover	Reference
M'Pesoba, Mali	10	Sorghum	35[b]	n.s.	1
Saria, Burkina Faso	10	Sorghum	58	n.s.	2
Sadore, Niger, 1987	5	Pearl millet	38	178	3
Sadore, Niger, 1987	20	Pearl millet	34	106	3

Panel B: Manure with inorganic fertilizer

Location	Amount of Manure (t ha^{-1})	Fertilizer (kg ha^{-1})	Crop	Crop response[a] (kg of DM t^{-1} manure) Grain	Stover
M'Pesoba, Mali	5	NPK: 8-20-0	Sorghum	90[c]	n.s.
Saria, Burkina Faso	10	Urea N: 60	Sorghum	80	n.s.
Sadore, Niger, 1987	5	SSP P: 8.7	Pearl millet	82	192
Sadore, Niger, 1987	20	SSP P: 17.5	Pearl millet	32	84

Source: Adapted from Williams et al. (1995).

Note: n.s. implies not specified. References: 1. Pieri (1989); 2. Pieri (1986); 3. Baidu-Forson and Bationo (1992).

[a] Responses were calculated at the reported treatment means for crop yields as: (treatment yield − control yield)/quantity of manure applied.

[b] Response of sorghum planted in the second year of a four-year rotation involving cotton-sorghum–groundnut-sorghum. Manure was applied in the first year.

[c] Estimated from visual interpolation of graph.

TABLE 11.12. Major nutrient (N, P, and K) composition of manure at selected sites in semiarid West Africa.

Location and type of manure	Nutrient composition (%)		
	N	P	K
Saria, Burkina Faso			
Farm yard manure	1.5-2.5	0.09-0.11	1.3-3.7
Northern Burkina Faso			
Cattle manure	1.28	0.11	0.46
Small ruminant manure	2.20	0.12	0.73
Senegal			
Fresh cattle dung	1.44	0.35	0.58
Dry cattle dung	0.89	0.13	0.25
Niger			
Cattle manure	1.2-1.7	0.15-0.21	–
Sheep manure	1.0-2.2	0.13-0.27	–

Source: Adapted from Williams et al. (1995).

from manure is 2.5 kg N and 0.6 kg P per hectare of cropland. Although the manure rates are between 5 and 20 t ha^{-1} in most of the on-station experiments, quantities used by farmers are very low and ranged from 1,300 to 3,800 kg ha^{-1} (Williams et al., 1995). This is due to insufficient number of animals to provide the manure needed; the problem becomes more pronounced especially in postdrought years (Williams et al., 1995). The amount of livestock feed and land resources available are also limited. Depending on rangeland productivity, it will require between 10-40 hectares of dry season grazing land and 3-10 hectares of wet season grazing rangeland to maintain yields on one hectare of cropland using animal manure (Fernandez et al., 1995).

The method of manure placement could have a significant effect on crop yields. Hill placement of manure performed better than broadcasting in on-farm trials in the Sahel. Broadcasting 3 t ha^{-1} of manure resulted in a pearl millet grain yield of 700 kg ha^{-1} whereas the point placement of the same quantity of manure gave about 1,000 kg ha^{-1} (Figure 11.11).

Cropping Systems

The most common cropping systems in this zone involve growing several crops in association as mixtures. This practice provides the farmer with

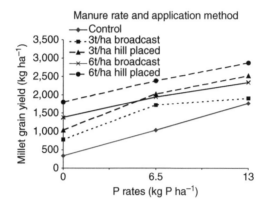

FIGURE 11.11. Millet grain yield response to P and manure rates and methods of application, Karabedji, Niger, 2002 rainy season. *Source:* Bationo, unpublished data.

several options for returns from land and labor, often increases efficiency with which scarce resources are used, and reduces dependence upon a single crop that is susceptible to environmental and economic fluctuations.

Intercropping. Traditional intercropping systems cover over 75 percent of the cultivated area in the semiarid Tropics (Steiner, 1984). In the Sudano-Sahelian zone, cereals such as millet and sorghum are traditionally intercropped with cowpea on small farms. In Niger, up to 87 percent of the millet area is intercropped (Swinton et al., 1984). A study by Norman (1974) in northern Nigeria has shown that only 8 percent of the area was planted to sole sorghum, while about 50 percent of the area was planted with sorghum in intercrop. Similar findings are reported by Fussell and Serafini (1985) for Nigeria, Niger, Burkina Faso, and Mali. The most common associations are cereal/cowpea, cereal/groundnut, and cereal/cereal such as millet/sorghum/maize, and millet/sorghum/cowpea. In these systems, cereals are normally sown first and act as the dominant crop. Norman (1974) concludes that mixed cropping is a strategy for the farmers' profit maximization and risk minimization. Production and income stability are important features of the systems, which also alleviate seasonal labor peaks (Abalu, 1976).

The yield advantages of intercrop systems vary from 10-100 percent in millet cropping systems (Fussell and Serafini, 1985). In western Niger, the combination of cowpea with millet has resulted in production advantages of 10-40 percent over four years and in Mali by 100 percent in maize/millet intercrop. For millet/groundnut systems there is a yield advantage of

28-53 percent (Table 11.13). There is a scarcity of data on nutrient use efficiency in intercropping systems as compared to monocropping systems.

Relay and sequential cropping systems: The performance of cultivars under relay and sequential systems has higher potential than traditional sole or mixed cropping (Shetty, 1984). In Mali, by introducing short season sorghum cultivars in relay cropping with other short duration cowpea and groundnut cultivars, substantial yields of legumes and sorghum were obtained as compared to traditional systems (IER, 1990; Sedogo and Shetty, 1991).

In the Sahelian zone, Sivakumar et al. (1990) analyzed the date of the onset and ending of the rains, and the length of the growing period. He found that an early onset of the rains offers the probability of a longer growing season while delayed onset results in a considerably shorter growing season. The analysis suggested that cropping management factors using relay cropping can increase soil productivity.

Crop rotation. Rotation of cereals and legumes can be used as a means of improving soil fertility and productivity. Several researchers (Bagayoko

TABLE 11.13. Yields of pearl millet and groundnut in sole crops and intercrops, and resultant land-equivalent ratio (LER) at Tara, Niger, rainy season 1989

Treatment[a]	Ground nut pods (t ha^{-1})	Millet grain (t ha^{-1})	LER[b]	Ground nut haulms (t ha^{-1})	Millet straw (t ha^{-1})	LER
Sole crop						
Groundnut (28-206)	1.29			2.62		
Groundnut (47-16)	0.99			3.13		
Groundnut [ICGS(E)11]	1.40			2.59		
Millet (CIVT)		1.29			3.70	
Intercrop						
CIVT and 28-206	0.71	1.20	1.48	1.28	2.95	1.29
CIVT and 47-16	0.66	1.04	1.46	1.44	3.05	1.28
CIVT and ICGS(E)11	0.71	1.31	1.53	1.22	3.32	1.37
SE	±0.08	±0.05		±0.17	±0.17	
CV (%)	16.6	16.2		16.8	19.0	

Source: Bationo, unpublished data.

[a]Randomized complete block design with four replications. Millet planted at 1 × 1 m and groundnut at 50 × 100 cm.

[b]LER = Sum of ratios of yield of each crop in mixture over yield of sole crop.

et al., 1996; Bationo et al., 1998; Klaij and Ntare, 1995; Stoop and Staveren, 1981; Bationo and Ntare, 2000) have reported cereal/legume rotation effects on cereal yields. Table 11.14 shows the effect of cowpea-millet rotation on millet grain and total biomass production. In a period of three years, there was an increase of about 3 t ha^{-1} of total dry matter production when millet was grown in rotation with cowpea.

Nitrogen use efficiency increased from 20 percent in continuous pearl millet cultivation to 28 percent when pearl millet was rotated with cowpea. Nitrogen derived from the soil is used better in rotation systems than with continuous millet (Bationo and Vlek, 1998). Nitrogen derived from the soil increased from 39 kg N ha^{-1} in continuous pearl millet cultivation to 62 kg N ha^{-1} when pearl millet was rotated with groundnut. Those data clearly indicate that although all the above biomass of the legume will be used to feed livestock and not returned to the soil, rotation will increase not only the yields of the succeeding cereal but also its nitrogen use efficiency (Bationo and Vlek, 1998).

The response of legumes to rotation was also significant and legume yields were consistently lower in monoculture than when rotated with millet (Figure 11.12). This suggests that factors other than N alone contributed to the yield increases in the cereal-legume rotations. It has been assumed by many workers that the positive effects of rotations arise from the added N from legumes in the cropping system. Some workers, however, have attributed the positive effects of rotations to the improvement of soil biological and physical properties, and the ability of some legumes to solubilize occluded P and highly insoluble calcium bounded phosphorus by legume root exudates (Gardner et al., 1981; Arhara and Ohwaki, 1989; Sahrawat et al., 2001). Other advantages of crop rotations include soil conservation (Stoop

TABLE 11.14. Millet grain and total dry matter yield at harvest as influenced by millet/cowpea cropping system at Sadore (Niger)

Cropping system	Grain yield (kg ha^{-1})			Total dry matter yield (kg ha^{-1})		
	1996	1997	1998	1996	1997	1998
Continuous millet	937	321	1,557	4,227	2,219	6,992
Millet after cowpea	1,255	340	1,904	5,785	2,832	8,613
P > F	<0.001	0.344	<0.001	<0.001	<0.001	<0.001

Source: Bationo and Ntare (2000).

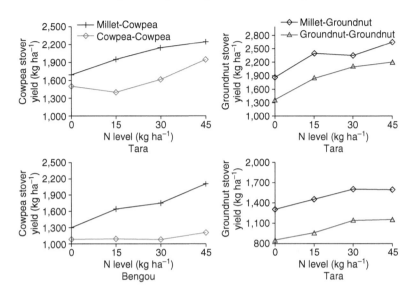

FIGURE 11.12. Effects of nitrogen and rotation on legume stover yield (kg ha^{-1}), average of four years (1989-1992) at Tara and Bengou, Niger. *Source:* Bationo and Ntare (2000).

and Staveren, 1981), organic matter restoration (Spurgeon and Grisson, 1965), and pest and disease control (Sunnadurai, 1973).

Changes in soil chemical properties from long-term cropping systems management trials, monitored in different agroecological zones of the Sudano-Sahelian region showed that rotations resulted in significantly higher soil pH, total N, and effective cation exchange capacity (ECEC) (Table 11.15). In the long-term cropping systems management studies in the Sahel, rotation systems were found to have higher levels of OC as compared to the continuous mono-cropping system (Figure 11.13). This could partially be due to the contribution made by the fallen leaves of the cowpea crop in the crop rotation.

Interaction Between Water and Nutrients

According to Rockstrom et al. (2003), the keys to improved water productivity and mitigating intraseason dry spells in rainfed agriculture are maximizing the amount of plant-available water and the plant water uptake capacity. This implies systems that partition more incident rainfall to soil storage and less to runoff, deep percolation, and evaporative loss, as well as

TABLE 11.15. Soil chemical properties after five years of experimentation at different sites in the Sudano-Sahelian region

Rotation	pH			Organic matter (%)			Total N (mg kg⁻¹)			ECEC (cmol kg⁻¹)		
	Sadore	Bengou	Tara	Sadore	Bengou	Tara	Sadore	Bengou	Tara	Sadore	Bengou	Tara
Initial	4.1	4.3	4.1	0.22	0.20	0.45	74	226	197	0.54	1.87	1.20
F-F	4.7	4.7	5.0	0.76	0.46	0.56	351	230	219	1.95	1.15	1.25
F-M	4.9	4.7	5.0	0.74	0.44	0.59	302	251	2.07	1.83	1.35	1.16
M-M	4.6	4.4	4.3	0.52	0.37	0.44	235	178	165	1.91	1.11	0.88
C-M	4.7	4.3	4.3	0.56	0.35	0.47	260	206	197	1.84	1.15	0.88
G-M	4.6	4.3	4.2	0.58	0.27	0.45	263	192	130	1.88	1.25	0.81
SE (DF 27)	0.11	0.09	0.10	0.03	0.04	0.033	12.0	10.67	21.3	0.123	0.107	0.077
CV (%)	4	4	4	9	20	13	8	10	23	13	17	15

Source: Bationo and Ntare (2000).

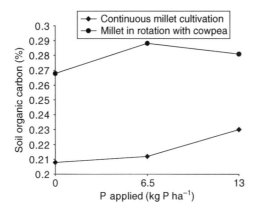

FIGURE 11.13. Effect of phosphorus and cropping system on SOC, Sadore, Niger, 1995. *Source:* Bationo and Buerkert (2001).

crops that provide more soil cover and root more deeply (Rockstrom et al., 2003). Loss of water and nutrients through runoff are major agriculture problems for inherently poor fertile soils in semiarid Africa. The intensification of crop production requires an integration of soil, water, and nutrient management that is locally acceptable and beneficial for smallholder farmers (Zougmore, 2003). In the central plateau of Burkina Faso, stone bounds alone doubled sorghum yield compared to plots without stone bounds and therefore, can reduce risks of crop failure in erratic rainfall years (Zougmore, 2003). Stone bounds consist of rows of stones constructed along the contours to check runoff and soil erosion (Rochette, 1989; Mando et al., 2001). Taonda et al. (2003) found that water harvesting alone with stone bounds did not improve yields, but the combination of water harvesting with stone bounds or zai plus manure increased sorghum yields over two times as compared to the control (Table 11.16).

A study by Zougmore and Zida (2000) has clearly shown the effects of the spacing between installations of stone bounds on decreasing runoff and erosion, and increasing yields (Table 11.17). A spacing of 50 m reduced runoff by 5 percent, a spacing of 33 m by 12 percent, and that of 25 m by 23 percent. Yield losses were reduced by 21 percent with a spacing of 50 m and 61 percent with a spacing of 25 m. With lower rainfall than normal, where yield increases were 58 percent with spacing of 50 m, it increased up to 343 percent with a spacing of 25 m. However, in a normal year, increases were

TABLE 11.16. Variation of the grain yields of sorghum during four years (kg ha^{-1})

Treatments	2000	2001	2002	Average
Control	353	393	215	331
Stone bounds	394	574	504	397
CP + Manure	1,026	1,168	1,072	789
Zaï + Manure	1,188	176	1,267	805

Source: Adapted from Taonda et al. (2003).

TABLE 11.17. Effects of spacing stone bounds on percent runoff, erosion, and grain yield

Spacing of stone bounds	% Decrease in runoff	% Decrease in erosion	Total grain yield		
			Poor rainfall year	Normal rainfall year	High rainfall years
50 m	5	21	58	1	
33 m	12	46	109	73	
25 m	23	61	343	56	

Source: Adapted from Zougmore and Zida (2000).

not as much and the trend was even negative in years where rainfall was higher than the long-term average.

Zougmore, Mando et al. (2003) found an average reduction in runoff of up to 59 percent in plots with barriers alone, but the reduction reached 67 percent in plots with barriers plus mineral N, and 84 percent in plots with barriers + organic N. On average, stone bounds reduced soil erosion more than grass strips (66 percent versus 51 percent). Integrated water and nutrient management may help alleviate poverty and may empower small-holder farmers to invest in soil management for better crop production (Zougmore, 2003).

Restoring favorable soil moisture conditions by breaking up the surface crust to improve water infiltration (half-moon technique) with appropriate nutrient management could be an effective method for the rehabilitation of degraded soil and improving productivity (Zougmore, Zida et al., 2003).

Animal drawn rippers and subsoilers could increase water productivity by increasing water infiltration and storage as well as root penetration (Rockstrom et al., 2003). Considerable yield increases above "farmers' practices" (i.e., flat cultivation and no fertilizer) could be realized by combining tied-ridged tillage with inputs of mineral N and P fertilizer, reaching maize grain yield levels of six times the prevailing yield under farmers' practices of approximate 1 Mg ha^{-1} (Jensen et al., 2003). Tied-ridging is a water harvesting method where cross ridges (ties) are constructed at specified intervals across ridges to form basins for storing water.

Humid and SubHumid Zone

Soil Fertility Problems

Soil constraints in the humid and subhumid tropical Africa can be divided into two major types: soil chemical and soil physical constraints. The chemical constraints include low nutrient reserves, low cation exchange capacity (CEC), aluminum toxicity, soil pH, and phosphorous fixation (Sanchez and Logan, 1992). The physical constraints to increased soil productivity are limited rooting depth, low water-holding capacity and susceptibility to soil erosion, soil crusting, and compaction.

Low nutrient reserves and cation exchange capacity are common in highly weathered soils such as Oxisols and Ultisols. These soils have limited capacity to retain and to supply cations and other major nutrients such as nitrogen (N), phosphorus (P), and sulfur (S) required by plants. Most of these soils are extremely nutrient-depleted and crop production can only be sustained by regular addition of external nutrients and by proper SOM management.

Nitrogen, a key nutrient for crop production, is the most mobile and also the most easily exhausted nutrient in the soil. Smallholder farmers rely on natural fallow periods and use of leguminous crops to restore soil nitrogen status (Nye and Greenland, 1960; Kwesiga and Coe, 1994). However, due to high population density and land pressure, long fallow periods are no longer sustainable. To sustain high crop yields in intensive and continuous crop production systems, nitrogen fertilizer input is required. Inclusion of legumes as green manure, cover crops, improved fallows, and use of agroforestry techniques in maize-based cropping systems are often employed to supply nitrogen and offer additional benefits such as weed suppression, soil erosion control, and soil structure amelioration.

The structural stability of many soils in subhumid zones, particularly the Alfisols, is very low and the aggregates are easily destroyed by rainfall. The breakdown of main aggregates is due to entrapped air and differential swelling, which results in the formation of microaggregates. Such crusting reduces aeration and infiltration of rainwater. Cultivation of such soils when wet leads to the formation of hard pans in the plough layer. Low organic matter content and low clay content are some of the conditions that lead to crusting and compaction.

N and P Management

In East Africa, evidence from a 23-year-old study at Kabete, Kenya, indicated that application of NP inorganic fertilizer, farmyard manure, and CRs is the best option to increase yield of crops such as maize (Figure 11.14) (Kapkiyai et al., 1999).

The data in Figure 11.15 clearly indicate the comparative advantage to combined organic and inorganic plant nutrients. The combination of both organic and inorganic P and N sources achieved more yield as compared to inorganic or organic sources alone. Many trials have been carried out on the usefulness of combining organic and inorganic plant nutrients using different proportions of each. The data in Figure 11.15 indicate for Mapira and Chinonda sites in Zimbabwe that there is advantage to combining the sources, whereas at Manjoro, the grain yield is the same when 100 percent of organic or inorganic plant nutrients are used.

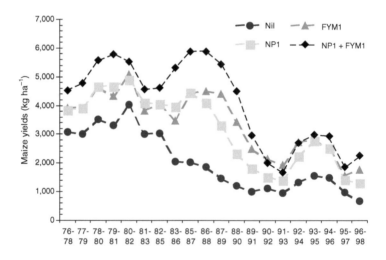

FIGURE 11.14. Effect of different soil fertility management strategies on average maize yields over a twenty-three-year period (1976-1998) of continuous cropping. *Source:* Adapted from Kapkiyai et al. (1999).

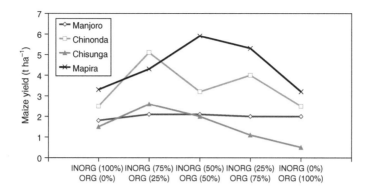

FIGURE 11.15. Maize grain yield obtained from 100 kg N applied in different proportions of manure and inorganic fertilizers, Murewa, Zimbabwe. *Source:* Nhamo and Murwira (2000).

The deficiency of P is found to be as important as or more important than that of N for crops such as upland rice grown in the humid forest or savanna zones in west Africa (Sahrawat et al., 2000, 2001). The direct application of PR to acidic pH soils (Uultisols and Oxisols) in the humid forest zone of SSA holds greater potential for boosting production of upland rice cultivars compared to production in acidic soils in the dry regions (Sahrawat et al., 2001). Somado et al. (2003) showed that in both pot and field experiments, significant responses of legume performance (*Aeschynomene afraspera*) including biomass production, nodulation, N accumulation and %N derived from atmosphere to PR application were observed in the lowland Ultisols of Cote d'Ivoire. The synergetic effects of PR and BNF on N and P cycling improved P nutrition and the total biomass of the subsequent lowland rice crop under pot conditions. However the impact of synergy between PR and BNF on the rice yield in the field was minimized due to asynchrony in legume manure nutrient release and demand. Nevertheless, the application of PR for driving the N and P cycles in legume-based rice production systems on acid Ultisols has potential and merits further evaluation under field conditions (Somado et al., 2003).

Agroforestry Systems

There is little doubt that under humid tropical conditions, agroforestry techniques are suited for the maintenance of soil fertility. This is because enhanced soil cover and root systems of the trees provide continuous soil protection, a favorable environment for soil biological processes and more efficient nutrient cycling.

Many years of research and experience have produced various highly successful and stable management practices for crop production. Similarly, through many generations of traditional wisdom and practical experience, farmers developed stable and viable multistory home garden production systems in the uplands. The particular methods that are most appropriate in any given locality will vary both within and among the humid forest regions. Local needs and opportunities, ecological circumstances, economic opportunities, and social and cultural status, as well as the status of land and water resources, determine which methods are most suitable.

Improved Fallows

Experiments on fallows date back to the 1930s with a well-known example of the "corridor system" which was tried and applied by the Belgians in the Democratic Republic of Congo (DRC) founded on the principles of shifting cultivation (Eckholm, 1976). Studies show that both the type and age of the fallow greatly influence the fertility of the soil at the end of the fallow period (Padwick, 1983). Soil productivity generally increases with increasing fallow length when the appropriate tree or shrub species are present. Fast-growing species can be expected to restore soil fertility more quickly and contribute to higher soil fertility for subsequent crops (Aweto, 1981). Planted legume species have been shown to be effective in improving soil fertility (Juo and Lal, 1977; Kang and Wilson, 1987; Gichuru and Kang, 1989).

Improved fallows are probably the most exciting developments for soil fertility improvement. Improved fallow experiments conducted in the eastern province of Zambia both on-farm and on-station showed that fallows with *Sesbania sesban* performed very well. *Sesbania sesban* planted fallows of one to two year rotation show potential in increasing maize yields with or without the application of inorganic fertilizers (Figure 11.16). Other multipurpose tree species (MPTs) such as *Tephrosia vogelii* and *Sesbania macrantha* have also shown promise for one to two year fallows.

In another study, maize grain yields were significantly increased by *Tephrosia candida* compared to the natural regrowth fallow at all lime levels following two years of continuous cropping with lime application (Table 11.18) (Gichuru, 1994). Overall, the yield increase due to the planted fallow was more than 200 percent. There was a strong interaction between the previous lime application and the fallow treatments indicating that the highest lime rate produced a significant effect on maize grain but only with *Tephrosia candida* fallow.

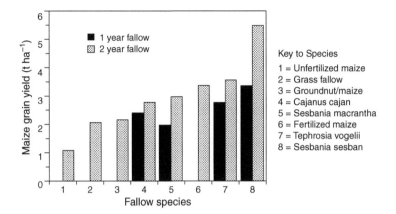

FIGURE 11.16. Effect of short rotation fallow on maize grain yield without inorganic fertilizer after one or two years fallow. *Source:* Mafongoya, unpublished data.

TABLE 11.18. Maize grain yields as influenced by fallow treatments and previous lime application

Lime rate (t ha^{-1})	Fallow (kg ha^{-1})		Means
	Natural regrowth	*Tephrosia candida*	
0	675	1,199	937
0.5	603	1,414	1,009
1.0	794	1,204	999
2.0	888	1,351	1,119
4.0	761	2,816	1,789
Means	744	1,597	
SE ±	133		194.1

Source: Mafongoya, unpublished data.

Alley Cropping (Hedgerow Intercropping)

Kang et al. (1984), described alley cropping as a stable alternative to shifting cultivation, which retains the basic features of bush fallow, such as providing shade and suppressing undergrowth, while fixing atmospheric nitrogen. The success of alley cropping depends on the use of suitable woody species, the successful establishment of the hedgerows and their appropriate management, and the design in terms of factors such as spacing and periodicity of pruning (Kang et al., 1990). The legume woody species used play a productive and/or protective role depending upon the dominant function(s) of the species (Nair et al., 1993). The productive role includes production of food, fodder, firewood, and various products. Alley cropping contributes to the maintenance of soil fertility under annual cropping through improved nutrient cycling.

The application of inorganic N fertilizers usually increases considerably the yields of alley crops, although the response varies with the hedgerow species used. Responses to other nutrients in alley cropping have also been reported. For instance, the application of 30 and 60 kg of P increased the pole bean *(Phaseolus vulgaris)* production in N-fertilized *Sesbania sesban* alley cropped systems (Yamoah and Burleigh, 1990). The beneficial effects of fertilization are sometimes delayed. Pruning alone, especially on infertile soils, cannot sustain productivity of continuous alley cropping (Palm et al., 1991; Szott, Palm et al., 1991; Szott, Fernandez et al., 1991).

The use of Gliricidia green manure can greatly increase maize yields over those obtained through continuous sole cropping. The beneficial effect of the mulch was significant in four of the five cropping seasons, when maize yields from intercropped treatments were substantially higher than those from control plots (both with and without the addition of mineral fertilizer), as exemplified by the results of the maize grain yields in both 1996 and 1997 (Table 11.19). Topsoil ammonium, nitrate, and total inorganic nitrogen (ammonium + nitrate) were significantly correlated with maize grain yields.

TABLE 11.19. Maize grain yield (t ha^{-1}) with and without gliricidia intercropping and fertilizer nitrogen in 1996 and 1997 at Makoka, Malawi

Fertilizer added (kg N ha^{-1})	1996		1997	
	No tree	Tree	No tree	Tree
0	1.0	4.8	0.4	3.5
24	3.5	6.1	2.0	3.6
48	4.2	6.7	2.1	4.3

Source: Adapted from ICRAF (1997).

Note: Least Significant Difference (LSD) (0.05): N rate = 0.45, tree biomass = 0.36 for 1996. LSD (0.05): N rate = 0.50, tree biomass = 0.41 for 1997.

Biomass Transfer System

Traditionally, farmers in Zimbabwe collected leaf litter from Miombo secondary forest as a source of nutrients to maize (Nyathi and Campbell, 1993). This practice is not sustainable in the long term since it mines nutrients from the forest ecosystems, in addition to the fact that Miombo litter collected is of low quality (Mafongoya and Nair, 1997). An alternative means of producing high-quality biomass is through establishment of on-farm biomass banks from which the biomass is cut and transferred to crop fields in different parts of the farm. In western Kenya, for example, the use of *Tithonia diversifolia, Senna spectabilis, Sesbania sesban,* and *Calliandra calothyrsus* planted as farm boundaries, woodlots and fodder banks, or found along the roads has proven beneficial in improving maize production (Maroko et al., 1998; Nziguheba et al., 1998; Palm, 1995; Palm et al., 2001). In a study by Gachengo (1996), tithonia green biomass grown outside a field and transferred into a field was found to be as effective in supplying N, P, and K to maize as an equivalent amount of commercial NPK fertilizer, and in some cases maize yields were higher with tithonia biomass than commercial inorganic fertilizer.

The effectiveness of biomass transfer as nutrient sources using organic inputs from MPT species depends on their chemical composition (Mafongoya and Nair, 1997). Leguminous trees can produce up to 20 t dry matter ha^{-1} year^{-1} prunings which contained enough nutrients to meet crop demand (Young, 1997; Szott, Palm et al., 1991). Biomass transfer using leguminous species is a far more sustainable means of maintaining nutrient balances in maize-based systems as these trees are able to fix atmospheric nitrogen. These systems can meet the N requirement of most crops in smallholder farming systems. Sometimes the biomass is first fed to livestock and then applied as manure to crops (Jama et al., 1997). However, these systems cannot meet the requirement of P and there is need to apply inorganic sources of P in addition to organic sources.

Integrated Nutrient Management trials in the continent have been used to establish the fertilizer equivalencies of locally available organic resources. Higher N content results in higher fertilizer equivalent values (Figure 11.17). For example *Tephrosia* (4 percent N), *Tithonia* (3.5 percent N), *Sesbania* (3.5 percent N), and Pigeon Pea (2.8 percent N) have been reported to have 93, 87, 36, and 33 percent fertilizer equivalencies, respectively (Palm et al., 2001). In a recent study in Kenya, *Tithonia diversifolia, Calliandra calothyrsus,* and *Senna spectabilis* had fertilizer equivalencies of 130, 72, and 68, respectively (Kimetu, 2002).

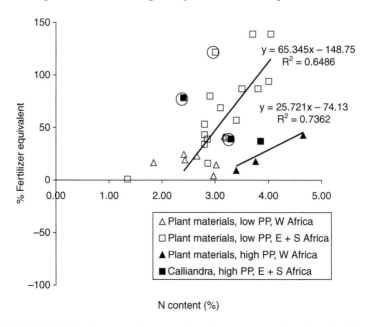

FIGURE 11.17. Fertilizer equivalency for different organic materials. *Source:* Adapted from Vanlauwe et al. (2002).

The varying quality of organic resources affects soil N balance through immobilization-mineralization processes. Research on characterization of organic materials has helped develop an organic resource database and translate the information into soil management practices relevant for, and targeted at, conditions experienced by farmers. A decision tree on the use of organic resources for INM, based on the amount of nitrogen, lignin, and polyphenol has been developed as a useful tool (Figure 11.18) (Palm et al., 1997, 2001; Singh et al., 2001). A farmer user-friendly decision tree provides indicators that farmers can use to predict the nutrient release potential of organic inputs (Figure 11.19) (Palm et al., 2001).

Cropping Systems

Intercropping and relay cropping. Intercropping and relay cropping of legume green manures have the advantage that crops are still produced while organic material is produced for soil amendment.

Mixed intercropping of cereals with legumes such as groundnuts (*Arachi hypogaea* L.), soybeans (*Glycine max* [L.] Merr), and *Phaseolus* beans or

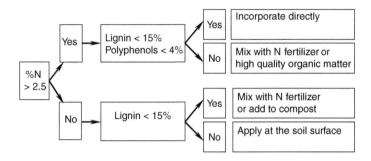

FIGURE 11.18. A decision support system tree for organic nutrient resources. *Source:* Adapted from Palm et al. (2001).

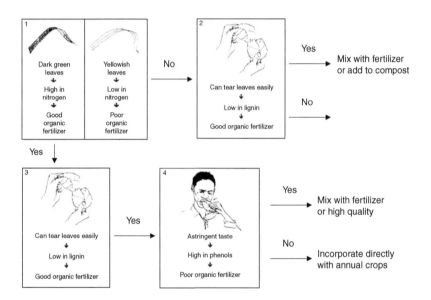

FIGURE 11.19. Farmer decision guide for selecting organic resource management options. *Source:* Adapted from Palm et al. (2001).

tree legumes such as pigeonpeas (*Cajanus cajan* L.) has been advocated (MacColl, 1989). Tropical grain legumes can certainly fix substantial amounts of N given favorable conditions, but the majority of this N is often harvested in the grain. Legumes such as soybean that have been subject to intense breeding efforts are very efficient at translocating their N into the grain, and, even when the residues are returned to the soil, there is generally a net removal of N from the field (Giller et al., 1994).

Relay planting can reduce the likelihood of competition with the crop where rainfall is limited, with the production of the green manure restricted by its ability to use residual water after the main cropping season (Mafongoya et al., 2003).

Rotation Systems. Many leguminous species are now grown as green manure and cover crops for erosion control, weed suppression, and for soil fertility restoration, often in rotation with food crops. Biological N contribution is probably the main reason why farmers include legume cover crops in cropping systems (Jeranyama et al., 1998). A legume cover crop may contribute N to a subsequent nonleguminous crop, reducing N fertilizer needs by 100 kg N ha^{-1} and more in some cases (Hesterman et al., 1992).

Many fast growing leguminous species such as mucuna (*Mucuna pruriens*), soya beans (*Glycine max*), *Lablab purpureus*, *Crotolaria ochroleuca*, and various species of the *Phaseolus* family can be especially useful as green manures and cover crops. In a study by Kamidi et al. (2000) at Matunda in western Kenya, Mucuna had the highest ground cover (72 percent) followed by *Crotalaria* (63 percent) and *Lablab* (54 percent). Soybeans and cowpeas gave the lowest ground cover (32 percent and 38 percent, respectively). The ground cover offered by these green manures greatly reduced soil erosion especially during the long rain season, as noted by Gachene et al. (2000).

A sole long-duration pigeonpea crop can provide up to 40 kg N ha^{-1} in fallen leaves and litter during its growth. A small harvest index, especially of the traditional, long-duration pigeonpea crop means that a relatively large proportion of the fixed N remains in the field which can give a substantial benefit to subsequent crops (Kumar-Rao et al., 1983; Mafongoya et al., 2003).

Leguminous plant materials provide higher quality organic inputs to meet N demands, but not P, but incorporating nonfood legumes in the farming systems require a sacrifice of space or time that is normally devoted to crop production. As such, legumes for soil fertility improvement have not been widely adopted by farmers (Jama et al., 1998).

CONCLUSIONS

Soil fertility depletion has been recognized as the major biophysical cause of declining food availability in smallholder farms in SSA. The latest figures show that some 200 million people or 28 percent of the SSA population are chronically hungry. Agricultural output should expand by at least 4 percent annually in order to ensure food security. Studies have clearly shown that the expansion of new farms cannot increase output by over 1 percent without accelerating environmental degradation. Consequently, productivity of land currently under cultivation should increase by at least 3 percent per annum. Any program aimed at reversing the declining trend in agricultural productivity, and preserving the environment for present and future generations must begin with soil fertility restoration and maintenance.

The low productivity of agriculture in SSA is strongly related to the low quality of the soil resource base. However, the general fertility and organic matter status of soils differ from upland to lowland ecologies. Wetland soils in the inland valley system of West and Central Africa are generally better endowed with fertility, especially SOM status, than their upland relatives. They have the potential for increasing agricultural productivity through an integrated use of crop genotypes adapted to the ecology and INM. Many of the arable soils are characterized by inherent or induced deficiencies of nutrients, particularly N and P. Low nutrient holding capacities, high acidity, and low organic matter are also constraints to soil productivity in SSA. The problem of low soil fertility is driven by a wide range of socioeconomic factors which diminish farmers' capacity to invest in soil fertility restoration.

Macroeconomic policies play a vital role in influencing the availability and accessibility of external inputs to replenish soil fertility. Unfavorable exchange rates, poor producer prices, high inflation, poor infrastructure, and lack of markets contribute to diminishing the capacity of farmers in SSA to adopt soil fertility enhancing technologies. This calls for a holistic approach on ISFM, and the need has been recognized for integration of socioeconomic and policy research besides technical research. Soil fertility restoration and maintenance, organic and inorganic sources, and management of plants and nutrients cannot be regarded as simple issues. ISFM embraces the full range of driving factors and consequences, namely biological, chemical, social, economic, and political aspects. The long-term and holistic approach of ISFM requires an evolutionary and knowledge intensive process, participatory research, and development focus rather than a purely technical focus.

REFERENCES

Abalu, G.O.I. (1976). A note on crop mixtures under indigenous conditions in northern Nigeria. *Journal of Development Studies.* 12: 11-20.

Abdullahi, A. and G. Lombin. (1978). *Long-term Fertility Studies at Samaru-Nigeria: Comparative Effectiveness of Separate and Combined Applications of Mineralizers and Farmyard Manure in Maintaining Soil Productivity under Continuous*

Cultivation in the Savanna, Samaru. Samaru Miscellaneous Publication. No. 75, Zaria, Nigeria: Ahmadu Bello University.

Arhara, J. and Y. Ohwaki. (1989). Estimation of available phosphorus in vertisol and alfisol in view of root effects on rhizosphere soil. In XI Colloquium Wageningen, Holland.

Aweto, P.M. (1981). Secondary succession and soil fertility restoration in southwestern Nigeria. III. Soil and vegetation interrelationships. *Journal of Ecology.* 69: 957-963.

Bache, B.W. and N.E. Rogers. (1970). Soil phosphate values in relation to phosphate supply to plants from some Nigerian soils. *Journal of Agricultural Science.* 74: 383-390.

Bagayoko, M., S.C. Mason, S. Traore, and K.M. Eskridge. (1996). Pearl millet/cowpea cropping systems yield and soil nutrient levels. *African Crop Science Journal.* 4: 453-462.

Baidu-Forson, J. and A. Bationo. (1992). An economic evaluation of a long-term experiment on phosphorus and manure amendments to sandy Sahelian soils: Using a stochastic dominance model. *Fertilizer Research.* 33: 193-202.

Balasubramanian, V. and C.A. Nnadi. (1980). Crop residue management and soil productivity in savanna areas of Nigeria. In: FAO (Food and Agriculture Organization of the United Nations). *Organic Recycling in Africa.* pp. 106-120. FAO soils bulletin 43. Rome, Italy.

Bationo, A. and A. Buerkert. (2001). Soil organic carbon management for sustainable land use in Sudano-Sahelian West African. *Nutrient Cycling in Agroecosystems.* 61: 131-142.

Bationo, A., A. Buerkert, M.P. Sedogo, B.C. Christianson, and A.U. Mokwunye. (1995). A critical review of crop residue use as soil amendment in the West African semi-arid tropics. In Powell, J.M., S. Fernandez Rivera, T.O. Williams, and C. Renard (eds.) *Livestock and Sustainable Nutrient Cycling in Mixed Farming Systems of Sub-Saharan Africa.* pp. 305-322. *Proceedings of the International Conference ILCA,* Addis Ababa.

Bationo, A., S.H. Chien, C.B. Christianson, J. Henao, and A.U. Mokwunye. (1990). A three year evaluation of two unacidulated and partially acidulated phosphate rocks indigenous to Niger. *Soil Science Society of America Journal.* 54: 1772-1777.

Bationo, A., C.B. Christianson, and W.E. Baethgen. (1989). Plant density and nitrogen fertilizer effects on pearl millet production in a sandy soil of Niger. *Agronomy Journal.* 82: 290-295.

Bationo, A., B.C. Christianson, and M.C. Klaij. (1993). The effect of crop residue and fertilizer use on pearl millet yields in Niger. *Fertilizer Research.* 34: 251-258.

Bationo, A., C.B. Christianson, and A.U. Mokwunye. (1987). Soil fertility management of the millet-producing sandy soils of Sahelian West Africa: The Niger experience. Paper presented at the Workshop on soil and crop management systems for rainfed agriculture in the Sudano-Sahelian zone, International Crops Research Institute for the Semi-Arid Tropics (ICRISAT), Niamey, Niger.

Bationo, A., J. Kimetu, S. Ikerra, S. Kimani, D. Mugendi, M. Odendo, M. Silver, M.J. Swift, and N. Sanginga. (2004). The African Network for Soil Biology and Fertility: New challenges and opportunities. In Bationo, A. (ed.) *Managing Nutrient Cycles to Sustain Soil Fertility in Sub-Saharan Africa.* pp. 1-23. Nairobi: Academy Science Publishers.

Bationo, A., F. Lompo, and S. Koala. (1998). Research on nutrient flows and balances in West Africa: State-of-the-art. Vol. 71: pp. 19-36. In Smaling, E.M.A. (Guest editor) Special Issue: *Nutrient Balances as Indicators of Production and Sustainability in Sub-Saharan African Agriculture. Agriculture Ecosystem and Environment.* Vol. 71. 346pp.

Bationo, A. and A.U. Mokwunye. (1991). Role of manures and crop residues in alleviating soil fertility constraints to crop production: With special reference to the Sahelian and Sudanian zones of West Africa. *Fertilizer Research.* 29: 117-125.

Bationo, A., U. Mokwunye, P.L.G. Vlek, S. Koala, B.I. Shapiro, and C. Yamoah. (2003). Soil fertility management for sustainable land use in the West African Sudano-Sahelian zone. In Gichuru, M.P., A. Bationo, M.A. Bekunda, P.C. Goma, P.L. Mafongoya, D.N. Mugendi, H.M. Murwira, S.M. Nandwa, P. Nyathi, and M.J. Swift (eds.) *Soil Fertility Management in Africa: A Regional Perspective.* pp. 253-292. Nairobi: Academy Science Publishers.

Bationo, A. and B.R. Ntare. (2000). Rotation and nitrogen fertilizer effects on pearl millet, cowpea and groundnut yield and soil chemical properties in a sandy soil in the semi-arid tropics, West Africa. *Journal of Agricultural Science.* 134(3): 277-284.

Bationo, A. and P.L.G. Vlek. (1998). The role of nitrogen fertilizers applied to food crops in the Sudano-Sahelian zone of West Africa. In Renard, G., A. Neef, K. Becker, and M. von Oppen (eds.) *Soil Fertility Management in West African Land Use Systems.* pp. 41-51. Weikersheim, Germany: Margraf Verlag.

Benson, T. (2004). Africa's food and nutrition security situation: Where are we and how did we get there? International Food Policy Research Institute 2020 Discussion paper 37. IFPRI: Washington.

Buerkert, A., K. Michels, J.P.A. Lamers, H. Marshner, and A. Bationo. (1996). Antierosive, soil physical and nutritional effects of crop residues. In Buerkert, A., B.E. Allison, and M. von Oppen (eds.) *Wind Erosion in Niger.* pp. 123-138. Dordrecht, The Netherlands: Kluwer Academic Publishers in cooperation with the University of Hohenheim.

Buerkert, A.K. and R.D. Stern. (1995). Crop residue and P application affect the spatial variability of non-destructively measured millet growth in the Sahel. *Experimental Agriculture.* 31: 429-449.

Buresh, R.J., P.C. Smithson, and D.T. Hellums. (1997). Building soil phosphorus capital in Africa. In Buresh, R.J. et al. (eds.) *Replenishing Soil Fertility in Africa.* pp. 111-149. Madison, WI: SSSA Spec. Publ. 51. SSSA.

Chien, S.H. (1977). Thermodynamic considerations on the solubility of phosphate rock. *Soil Science.* 123: 117-121.

Chien, S.H. and L.L. Hammond. (1978). A simple chemical method for evaluating the agronomic potential of granulated phosphate rock. *Soil Science Society of America Journal.* 42: 615-617.

Christianson, C.B. and P.L.G. Vlek. (1991). Alleviating soil fertility constraints to food production in West Africa: Efficiency of nitrogen fertilizers applied to food crops. In Mokwunye, A.U. (ed.) *Alleviating Soil Fertility Constraints to Increased Crop Production in West Africa.* pp. 21-33. Dordrecht, The Netherlands: Kluwer Academic Publishers.

De Leeuw, P.N., L. Reynolds, and B. Rey. (1995). Nutrient transfers from livestock in West African Agricultural Systems. In Powell, J.M., S. Fernandez-Rivera, T.O. Williams, and C. Renard (eds.) *Livestock and Sustainable Nutrient Cycling in Mixed Farming Systems of Sub-Saharan Africa.* Vol. 2. 569p. Technical papers. *Proceedings of an International Conference held in Addis-Ababa,* November 22-26, 1993. Addis Ababa, Ethiopia: ILCA (International Livestock Center for Africa).

Diamond, R.B. (1979). *Views on Marketing of Phosphate Rock for Direct Application.* Halifa, Israel. March 20-23, 1978. pp. 448-463. Special publication. IFDC-S1. Muscle Shoals, AL: International Fertilizer Development Center.

Eckholm, E.P. (1976). *Losing Ground: Myth and Reality in the Humid Tropics.* Norton: New York.

Evéquoz, M., D.K. Soumana, and G. Yadji. (1998). Minéralisation du fumier, nutrition, croissance et rendement du mil planté dans des ouvrages anti-érosifs. In Renard, G., A. Neef, K. Becker, and M. von Oppen (eds.) *Soil Fertility Management in West African Land Use Systems.* pp. 203-208. Niamey, Niger, March 4-8, 1997. Weikersheim, Germany: Margraf Verlag.

FAO. (1995). The State of Food and Agriculture. Rome, Italy: Food and Agriculture Organization of the United Nations.

FAO. FAO database: www.fao.org.

Fernandez-Rivera, S., T.O. Williams, P. Hiernaux, and J.M. Powell. (1995). Livestock, feed and manure availability for crop production in semi-arid West Africa. In Powell, J.M., S. Fernandez Rivera, T.O. Williams, and C. Renard (eds.) *Livestock and Sustainable Nutrient Cycling in Mixed Farming Systems of Sub-Saharan Africa. Proceedings of an International Conference.* pp. 149-170. Addis Ababa: ILCA.

Fussell, L.K. and P.G. Serafini. (1985). Associations de cultures dans les zones tropicales semi-arides d'Afrique de l'Ouest: stratégies de recherché antérieures et futures. (In Fr) In Ohm, H.W. and J.G. Nagy (eds.) *Technologies appropriées pour les paysans des zones semi-arides de l'Afrique de l'Ouest* pp. 254-278. West Lafayette, In: Purdue University.

Gachene, C.K.K., N.K. Karanja, and J.G. Mureithi. (2000). Effect of soil erosion on soil productivity and subsequent restoration of productivity using green manure cover crops: Data summary of a long-term field experiment. *Legume Research Network Project Newsletter.* 4: 8-10.

Gachengo, C.N. (1996) Phosphorus release and availability on addition of organic materials to phosphorus fixing soils. MPhil Thesis, Moi University, Kenya.

Gachimbi, L.N., A. de Jager, H. Van Keulen, E.G. Thuranira, and S.M. Nandwa. (2002). Participatory diagnosis of soil nutrient depletion in semi-arid areas of Kenya. Managing Africa's soils No. 26. 15p.

Gardner, M.K., D.G. Parbery, and D.A. Barker. (1981). Proteoid root morphology and function in legumes alnus. *Plant Soil.* 60: 143-147.

Gichuru, M.P. (1994). Regeneration of a degraded Ultisol with Tephrosia candida in the humid forest zone of South-Eastern Nigeria. *Interciencia.* 19(6): 382-386.

Gichuru, M.P. and B.T. Kang. (1989). *Calliandra calothyrsus* (Meissn.) in an alley cropping system with sequentially cropped maize and cowpea in South-Western Nigeria. *Agroforestry Systems.* 9: 191-203.

Gikonyo, E.W. and P.C. Smithson. (2003). Effects of Farmyard Manure, Potassium and their Combinations on Maize Yields in the High and Medium rainfall Areas of Kenya. In Bationo, A. (ed.) *Managing Nutrient Cycles to Sustain Soil Fertility in Sub-Sahara Africa.* pp. 137-149. Nairobi: Academy Science Publishers (ASP).

Giller, K.E., J.F. McDonagh, and G. Cadisch. (1994). Can biological nitrogen fixation sustain agriculture in the tropics? In: Syers, J.K. and D.L. Rimmar (eds.) *Soil Science and Sustainable Land Management in the Tropics.* pp. 173-191. Wallingford, England: CAB Int.

Gruhn, P., F. Goletti, and M. Yudelman. (2000). Integrated nutrient management, soil fertility, and sustainable agriculture: Current issues and future challenges. Food, agriculture and the environment discussion paper 32. Washington D.C., USA: IFPRI.

Hafner, H., J. Bley, A. Bationo, P. Martin, and H. Marschner. (1993). Long-term nitrogen balance of pearl millet (*Pennisetum glaucum* L.) in an acid sandy soil of Niger. *Journal of Plant Nutrition Soil Science.* 256: 264-176.

Hartley, K.T. (1937). An explanation of the effect of farmyard manure. *Empire Journal of Experimental Agriculture.* 19: 244-263.

Hesterman, O.B., T.S. Griffin, P.T. Williams, G.H. Harris, and D.R. Christenson. (1992). Forage legume-small grain intercrops: Nitrogen production and response of subsequent corn. *Journal of Production Agriculture.* 5: 340-348.

ICRAF. (1997). Annual report 1996. International Centre for Research in Agroforestry, Nairobi, Kenya.

ICRISAT. (International Crop Research Institute for the Semi-Arid Tropics). (1993). Annual Report 1992. ICRISAT Sahelian center, Niamey, Niger.

IER. (1990). Institut d'Economie Rurale. Rapport Annuel et Commissions Techniques Production Vivrières et Oléagineuses. Program coopératif ICRISAT/Mali, IER, Bamako, Mali.

IITA. (International Institute of Tropical Agriculture) (1992). *Sustainable Food Production in Sub-Saharan Africa: 1. IITA's Contributions.* Ibadan, Nigeria: IITA.

Ikombo, B.M. (1984). Effect of farmyard manure and fertilizer on maize in a semi-arid area of Eastern Kenya. *East African Agricultural and Forestry Journal* (Special issue: Dryland Farming Research in Kenya) 44: 266-274.

Izac, A-M.N. (2000). What Paradigm for Linking Poverty Alleviation to Natural Resources Management? *Proceedings of an International Workshop on Integrated*

Natural Resource Management in the CGIAR: Approaches and Lessons. August 21-25, 2000, Penang, Malaysia.

Jama, B., R.J. Buresh, J.K. Ndufa, and K.D. Shepherd. (1998). Vertical distribution of roots and soil nitrate: Tree species and phosphorus effects. *Soil Science of Society America Journal.* 62: 280-286.

Jama, B.A., R.A. Swinkels, and R.J. Buresh. (1997). Agronomic and economic evaluation of organic and inorganic sources of phosphorus in Western Kenya. *Agronomy Journal.* 89: 597-604.

Jensen, J.R., R.H. Bernhard, S. Hansen, J. McDonagh, J.P. Moberg, N.E. Nielsen, and E. Nordbo. (2003). Productivity in maize based cropping systems under various soil-water-nutrient management strategies in a semi-arid, alfisol environment in East Africa. *Agricultural Water Management.* 59(3): 217-237.

Jeranyama, P., B.O. Hesterman, and S.R. Waddington. (1998). The impact of legumes relay-intercropped into maize at Domboshava, Zimbabwe. In Waddington, S.R., H.K. Murwira, J.D.T. Kumwenda, D. Hikwa, and F. Tagwira (eds.) *Soil Fertility Research for Maize-Based Farming Systems in Malawi and Zimbabwe. Proceedings of the Soil Fert Net Results and Planning Workshop.* July 7-11, 1997, Africa University, Mutare, Zimbabwe. Soil Fert Net and CIMMYT-Zimbabwe, Harare.

Jones, M.J. and A. Wild. (1975). Soils of the West African savanna. Commonwealth Bur. Soils Tech. Comm. No. 55, Harpenden, England.

Juo, A.S.R. and R. Lal. (1977). The effect of fallow and continuous cultivation on the chemical and physical properties of an Alfisol in western Nigeria. *Plant Soil.* 47: 567-584.

Kamidi, M., F. Gitari, P. Osore, and D. Cheruiyot. (2000). Effect of green manure legume on the yield of maize and beans in Matunda Farm, Trans Nzoia district. *Legume Research Network Project Newsletter.* Issue No. 4: 8-10.

Kang, B.T., L. Reynolds, and A.N. Attra-Krah. (1990). Alley farming. *Advances in Agronomy.* 43: 316-359.

Kang, B.T. and G.F. Wilson. (1987). The development of alley cropping as a promising agroforestry technology. In Steppler, H.A. and P.K.R. Nair (eds.) *Agroforestry: A Decade of Development.* pp. 227-243. Nairobi, Kenya: International Council for Research in Agroforestry.

Kang, B.T., G.F. Wilson, and Lawson. (1984). *Alley Cropping: A Stable Alternative to Shifting Cultivation.* Ibadan, Nigeria: International Institute of Tropical Agriculture. 22p.

Kapkiyai, J.J., K.N. Karanja, J.N. Qureish, C.P. Smithson, and P.L. Woomer. (1999). Soil organic matter and nutrient dynamics in a Kenyan Nitisol under long-term fertilizer and organic input management. *Soil Biology and Biochemistry.* 31: 1773-1782.

Khasawneh, F.E. and E.C. Doll. (1978). The use of phosphate rock for direct application to soils. *Advances in Agronomy Journal.* 30: 155-206.

Kihanda, F.M. (1996). The role of farmyard manure in improving maize production in the sub-humid highlands of Central Kenya. PhD Thesis, The University of Reading, UK. 236p.

Kimani, S.K., S.M. Nandwa, D.N. Mugendi, S.N. Obanti, J. Ojiem, H.K. Murwira, and A. Bationo. (2003). Principles of integrated soil fertility management. In Gichuru, M.P., A. Bationo, M.A. Bekunda, P.C. Goma, P.L. Mafongoya, D.N. Mugendi, H.M. Murwira, S.M. Nandwa, P. Nyathi, and M.J. Swift (eds.) *Soil Fertility Management in Africa: A Regional Perspective.* pp. 51-72. Nairobi: Academy Science Publishers (ASP).

Kimetu, J.M. (2002). Determination of nitrogen fertilizer equivalent for organic materials based on quality and maize performance at Kabete, Kenya. MSc Thesis, Kenyatta. 64p.

Klaij, M.C. and B.R. Ntare. (1995). Rotation and tillage effects on yield of pearl millet (Pennisetum glaucum) and cowpea (Vigna unguiculata), and aspects of crop water balance and soil fertility in semi-arid tropical environment. *Journal of Agriculture Science.* 124: 39-44.

Kretzschmar, R.M., H. Hafner, A. Bationo, and H. Marschner. (1991). Long and short-term of crop residues on aluminium toxicity, phosphorus availability and growth of pearl millet in an acid sandy soil. *Plant Soil.* 136: 215-223.

Kumar-Rao, J.V.D.K., P.J. Dart, and P.V.S.S. Sastry. (1983). Residual effect of pigeonpea (Cajanus cajan) on yield and nitrogen response of maize. *Experimental Agriculture.* 19: 131-141.

Kwesiga, F. and R. Coe. (1994). The effect of short rotation *Sesbania sesban* planted fallows on maize yield. *Forestry Ecology Management.* 64: 199-208.

Lal, R. (1980). Soil erosion as a constraint to crop production. In Brady, N.C., L.D. Swindale, R. Dudal (eds.) *Priorities for Alleviating Soil Related Constraints to Food Production in the Tropics.* pp. 405-423. Los Baños, Laguna, Philippines: International Rice Research Institute.

Lehr, J.R. and G.H. McClellan. (1972). A revised laboratory reactivity scale for evaluating phosphate rocks for direct application, Bulletin 43, Tennessee Valley Authority, Muscle Shoals, AL, USA.

MacColl, D. (1989). Studies on maize (Zea mays L.) at Bunda, Malawi. II. Yield in short rotation with legumes. *Experimental Agriculture.* 25: 367-374.

Mafongoya, P.L., D.N. Mugendi, B. Jama, and B.S. Waswa. (2003). Maize based cropping systems in the subhumid zone of East and Southern Africa. In Gichuru, M.P., A. Bationo, M.A. Bekunda, P.C. Goma, P.L. Mafongoya, D.N. Mugendi, H.M. Murwira, S.M. Nandwa, P. Nyathi, and M.J. Swift (eds.) *Soil Fertility Management in Africa: A Regional Perspective.* pp. 73-122. Nairobi: Academy Science Publishers (ASP).

Mafongoya, P.L. and P.K.R. Nair. (1997). Multipurpose tree prunings as a source of nitrogen to maize under semiarid conditions in Zimbabwe. Nitrogen recovery rates in relation to pruning quality and method of application. *Agroforestry Systems.* 35: 47-56.

Mando, A., R. Zougmoré, N.P. Zombré, and V. Hien. (2001). Réhabilitation des sols dégradés dans les zones semi-arides de l'Afrique sub-saharienne. In Floret, C. and R. Pontanier (eds.) *La jachère en Afrique tropicale; de la jachère naturelle à la jachère améliorée: le point des connaissances.* pp. 311-339. Paris: John Libbey Eurotext.

Maroko, J.B., R.J. Buresh, and P.C. Smithson. (1998). Soil nitrogen availability as affected by fallow-maize systems in two soils in Kenya. *Biology and Fertility of Soils.* 26: 229-234.

McIntire, J., D. Bourzat, and P. Pingali. (1992). *Crop-Livestock Interactions In Sub Saharan Africa.* Washington, DC: The World Bank.

McIntire, J. and L.K. Fussel. (1986). *On-Farm Experiments with Millet in Niger. III. Yields and Economic Analyses.* Niamey, Niger: ISC (ICRISAT Sahelian Center).

Michels, K., M.V.K. Sivakumar, and B.E. Allison. (1995). Wind erosion control using crop residue: II. Effects on millet establishment and yield. *Field Crops Research.* 40: 111-118.

Mokwunye, A.U. (1974). Some reflections on the problem of "available" phosphorus in soils of tropical Africa, Samaru Conference Papers. 3: 1-11.

Mueller-Samann, K.M. and J. Kotschi. (1994). *Sustaining Growth: Soil Fertility Management in Tropical Smallholdings.* Germany: Margraf Verlag.

Mugwira, L.M. (1984). Relative effectiveness of fertilizer and communal area manures as plant nutrient sources. *Zimbabwe Agricultural Journal.* 81: 81-89.

Murwira, H.K. (2003). Managing Africa's soils: Approaches and challenges. In Gichuru, M.P., A. Bationo, M.A. Bekunda, P.C. Goma, P.L. Mafongoya, D.N. Mugendi, H.M. Murwira, S.M. Nandwa, P. Nyathi, and M.J. Swift (eds.) *Soil Fertility Management in Africa: A Regional Perspective.* pp. 293-306. Nairobi: Academy Science Publishers (ASP).

Nair, P.K.R. (1993). *An Introduction to Agroforestry.* Dordrecht, the Netherlands: Kluwer.

Nandwa, S.M. (2001). Soil organic carbon (SOC) management for sustainable productivity of cropping and agro-forestry systems in Eastern and Southern Africa. *Nutrient Cycling in Agroecosystems.* 61: 143-158.

Nandwa, S.M. (2003). Perspectives on soil fertility in Africa. In Gichuru, M.P., A. Bationo, M.A. Bekunda, P.C. Goma, P.L. Mafongoya, D.N. Mugendi, H.M. Murwira, S.M. Nandwa, P. Nyathi, and M.J. Swift (eds.) *Soil Fertility Management in Africa: A Regional Perspective.* pp. 1-50. Nairobi: Academy Science Publishers (ASP).

Nandwa, S.M. and M.A. Bekunda. (1998). Research on nutrient flows and balances in East and Southern Africa: State-of-the-art. *Agriculture Ecosystems and Environment.* 71: 5-18.

Nhamo, N. and H.K. Murwira. (2000). Use of manure in combination with inorganic N fertilizers on sandy soils in Murewa, Zimbabwe. TSBF Report 1997-1998. UNESCO: Nairobi.

Norman, D.W. (1974). Rationalizing mixed cropping under indigenous conditions: the example of northern Nigeria. *Journal of Development Studies.* 11: 3-21.

Nyathi, P. and B.M. Campbell. (1993). The acquisition and use of miombo litter by small-scale farmers in Masvingo, Zimbabwe. *Agroforestry Systems.* 22: 43-48.

Nye, P.H. and D.J. Greenland. (1960). *The Soil under Shifting Cultivation.* Technical Committee no. 51. Harpenden, UK: Commonwealth Bureau of Soils.

Nziguheba, G., C.A. Palm, R.J. Buresh, and P.C. Smithson. (1998). Soil phosphorus fractions and adsorption as affected by organic and inorganic sources. *Plant Soil.* 198: 159-168.

Padwick, G.W. (1983). Fifty years of experimental agriculture. II. The maintenance of soil fertility in tropical Africa: A review. *Experimental Agriculture.* 19: 293-310.

Palm, C.A. (1995). Contribution of agroforestry trees to nutrient requirements of intercropped plants. *Agroforestry Systems.* 30: 105-124.

Palm, C.A., C.N. Gachengo, R.J. Delve, G. Cadisch, and K.E. Giller. (2001). Organic inputs for soil fertility management in tropical agroecosystems: Application of an organic resource database. *Agriculture Ecosystems and Environment.* 83: 27-42.

Palm, C.A., A.J. McKerrow, K.M. Glasener, and L.T. Szott. (1991). Agroforestry in the lowland tropics: Is phosphorus important? In Tiessen, H., D. Lopez-Hernandez, and I. Salcedo (eds.) *Phosphorus Cycles in Terrestrial and Aquatic Ecosystems.* pp. 134-141. Regional Workshop No. 3. South and Central America. Scientific Committee on Problems of the Environment (SCOPE) and United Nations Environmental Programme, Saskatchewan Institute of Pedology, Saskatchewan. Canada.

Palm, C.A., R.J.K. Myers, and S.M. Nandwa. (1997). Combined use of inorganic and organic nutrient sources for soil fertility maintenance and replenishment. In Buresh, R.J., P.A. Sanchez, and F. Calhoun (eds.) *Replenishing Soil Fertility in Africa.* pp. 193-217. Madison, WI: SSSA Special Publication No. 51. Soil Science Society America.

Pieri, C. (1986). Fertilisation des cultures vivrières et fertilité des sols en agriculture paysanne subsaharienne. *Agronomie Tropicale.* 41: 1-20.

Pieri, C. (1989). *Fertilite de terres de savane. Bilan de trente ans de recherché et de developpmenet agricoles au sud du Sahara.* CIRAD. Paris, France: Ministere de la cooperation. 444p.

Powell, J.M. (1985). Yields of sorghum and millet and stover consumption of livestock in Nigeria. *Tropical Agriculture.* 62(1): 77-81.

Probert, M.E., J.R. Okalebo, and R.K. Jones. (1995). The use of manure on smallholder farms in semi-arid Eastern Kenya. *Experimental Agriculture.* 31: 371-381.

Prudencio, C.Y. (1993). Ring management of soils and crops in the West African semi-arid tropics: The case of the mossi farming system in Burkina Faso. *Agriculture, Ecosystems and Environment.* 47: 237-264.

Rebafka, F-P., A. Hebel, A. Bationo, K. Stahr, and H. Marschner. (1994). Short- and long-term effects of crop residues and of phosphorus fertilization on pearl millet yield on an acid sandy soil in Niger, West Africa. *Field Crops Research.* 36: 113-124.

Rochette, R.M. (1989). *Le Sahel en lutte contre la désertification: leçons d'expériences.* Weikersheim, Margraf.

Rockstrom, J., J. Barron, and P. Fox. (2003). Water productivity in rainfed agriculture: challenges and opportunities for small holder farmers in drought prone tropical agro-ecosystems. In J.W. Kijne., R. Barker, and D. Molden (eds.) *Water*

Productivity in Agriculture. Limits and Opportunities for Improvement. pp. 145-162. Wallingford, UK: CAB International.

Roose, E. and B. Barthes. (2001). Organic matter management for soil conservation and productivity restoration in Africa: A contribution from francophone research. *Nutrient Cycling in Agroecosystems.* 61: 159-170.

Sahrawat, K.L. (2004). Organic matter accumulation in submerged soils. *Advances in Agronomy.* 81: 169-201.

Sahrawat, K.L., M.K. Abekoe, and S. Diatta. (2001). Application of inorganic phosphorus fertilizer. pp. 225-246. In Tian, G., F. Ishida, and D. Keatinge. (eds.) *Sustaining Soil Fertility in West Africa.* Madison, WI: Soil Science Society of America Special Publication Number 58. Soil Science Society of America and American Society of Agronomy.

Sahrawat, K.L., M.P. Jones, and S. Diatta. (2000). The role of tolerant genotypes and plant nutrients in the management of acid soil infertility in upland rice. In *Management and Conservation of Tropical Acid Soils for Sustainable Crop Production.* pp. 29-43. *Proceedings of a Consultants' Meeting organized by joint FAO/IAEA Division of Nuclear Techniques in Food and Agriculture,* Vienna, Austria, March 1-3, 1999. IAEA-TECDOC 1159. International Atomic Energy Agency (IAEA), Vienna, Austria.

Sanchez, P.A. (1976). *Properties and Management of Soils in the Tropics.* New York, U: John Wiley and Sons.

Sanchez, P.A. (1994). Tropical soil fertility research: Towards the second paradigm. Inaugural and state of the art conferences. *Transactions 15th World Congress of Soil Science.* pp. 65-88.

Sanchez, P.A. and T.L. Logan. (1992). Myths and science about the chemistry and fertility of soils in the tropics. In Lal, R. and P.A. Sanchez (eds.) *Myths and Science of Soils of the Tropics.* pp. 35-46. Madison, WI: SSSA Special Publication No. 29 SSSA and ASA.

Sanchez, P.A., K.D. Shepherd, M.J. Soul, F.M. Place, R.J. Buresh, A.M.N. Izac, A.Z. Mokwunye, F.R. Kwesiga, C.G. Nderitu, and P.L. Woomer. (1997). Soil fertility replenishment in Africa: An investment in natural resource capital. In Buresh, R., P.A. Sanchez, F. Calhoum (eds.) *Replenishing Soil Fertility in Africa.* SSSA Special Publication. 51: 1. Madison, WI: Soil Science Society of America.

Sandford, S.G. (1989). Crop residue/livestock relationships. In Renard, C., R.J. Van Den Beldt, and J.F. Parr (eds.) *Soil, Crop, and Water Management Systems in the Sudano-Sahelian Zone. Proceedings of an International Workshop,* ICRISAT Sahelian Center, Niamey, Niger, January 11-16, 1987, ICRISAT (International Crops Research institute for the Semi-Arid Tropics), Patancheru, Andhra Pradesh, India.

Sedga, Z. (1991). Contribution a la valorization agricole des residus de culture dans le plateau central du Burkina Faso. Invetaire de disponibilites en metiere organique et e'tude des effets de l'inoculum, micro 110 IBF, Memoire d'ingenieur des sciences appliqués, IPDR/Katibougou, 100p.

Sedogo, M.P. (1993). Evolution des sols ferrugineux lessivés sous culture: Influences des modes de gestion sur la fertilité. Thèse de Doctorat Es-Sciences, Abidjan, Université Nationale de Côte d'Ivoire.

Sedogo, D. and S.V.R. Shetty. (1991). Progres de la recherché vers une technologie amelioree de production du sorgho en culture pluviale. Paper presented at the Inter-Network Conference in Food Grain Improvement, March 7-14, Niamey, Niger.

Shepherd, K.D. and M.J. Soule. (1998). Soil fertility management in west Kenya: Dynamic simulation of productivity, profitability and sustainability at different resource endowment levels. *Agriculture, Ecosystems and Environment.* 71(1/3): 131-145.

Shepherd, K., M. Walsh, F. Mugo, C. Ong, S.T. Hansen, B. Swallow, A. Awiti, M. et al. (2000). Likning land and lake, research and extension, catchment and lake basin. Final technical report start-up phase, July 1999 to June 2000. Working Paper 2000-2. ICRAF.

Shetty, H.R. (1984). Crop volume and out-turn table for stands of Acacia nilotica. *Indian Forester.* 110: 586-590.

Singh, V., K.E. Giller, C.A. Palm, J.K. Ladha, and H. Breman. (2001). Synchronizing N Release from Organic Residues: Opportunities for Integrated Management of N. *The Scientific World.* 1(52): 880-886.

Sivakumar, M.V.K., C. Renard, M.C. Klaij, B.R. Ntare, L.K. Fussel, and A. Bationo. (1990). Natural resource management for sustainable agriculture in the Sudano-Sahelian zone. *International Symposium on Natural Resources Management for Sustainable Agriculture,* February 6-19, 1990, New Delhi, India.

Smaling, E.M.A. (1993). *An Agro-ecological Framework of Integrated Nutrient Management with Special Reference to Kenya.* Wageningen, The Netherlands: Wageningen Agricultural University.

Smaling, E.M.A., L.O. Fresco, and A. de Jager. (1996). Classifying, monitoring and improving soil nutrient stocks and flows in African agriculture. *Ambio.* 25: 492-496.

Smaling, E.M.A., S.M. Nandwa, and B.H. Janseen. (1997). Soil fertility in Africa is at stake. In Buresh, R.J., P.A. Sanchez, and F. Calhoun (eds.) *Replenishing Soil Fertility in Africa.* pp. 47-61. Madison, WI: SSSA Special Publ. SI. SSSA.

Smith, J.P. and J.R. Lehr. (1966). An X-ray investigation of carbonate apatites. *Journal Of Agriculture And Food Chemistry.* 14: 342-349.

Somado, E.A., M. Becker, R.F. Kuehne, K.L. Sahrawat, and P.L.G. Vlek. (2003). Combined effects of legumes with rock phosphorus on rice in West Africa. *Agronomy Journal.* 95: 1172-1178.

Spurgeon, W.I. and P.H. Grissom. (1965). Influence of cropping systems on soil properties and crop production. *Mississippi Agriculture Experiment Station Bulletin No. 710.*

Steiner, K.G. (1984). Intercropping in tropical smallholder agriculture with special reference to West Africa. Stein, West Germany, 304p. Report.

Sterk, G., L. Herrmann, and A. Bationo. (1996). Wind-blown nutrient transport and soil productivity changes in southwest Niger. *Land Degradation and Development.* 7: 325-335.

Stoop, W.A. and J.P.V. Staveren. (1981). Effects of cowpea in cereal rotations on subsequent crop yields under semi-arid conditions in Upper-Volta. In Graham, P.C. and S.C. Harris (eds.) *Biological Nitrogen Fixation Technology for Tropical Agriculture.* Cali, Colombia: Centro International de Agricultura Tropical.

Stoorvogel, J. and E.M.A. Smaling. (1990). *Assessment of Soil Nutrient Depletion in Sub-Saharan Africa: 1983-2000.* Vol. 1, Main Report. The Winand Staring Center, Wageningen, The Netherlands.

Stoorvogel, J.J., E.M.A. Smaling, and B.H. Janssen. (1993). Calculating soil nutrient balances in Africa at different scales. I. Supra-national scale. *Fertilizer Research.* 35: 227-235.

Sunnadurai, S. (1973). Crop rotation to control nematodes in tomatoes. *Ghana Journal of Agricultural Science.* 6: 137-139.

Swinton, S.M., G. Numa, and L.A. Samba. (1984). Les cultures associees en milieu paysan dans deux regions du Niger: Filingue et Madarounfa. In *Proceedings of the Regional Workshop on Intercropping in the Sahelian and Sahelo-Sudanian Zones of West Africa.* pp. 183-194. November 7-10, 1984, Niamey, Niger, Bamako, Mali: Institute du Sahel.

Szott, L.T., E.C.M. Fernandez, and P.A. Sanchez. (1991). Soil-plant interactions in agroforestry systems. *Forestry Ecology and Management.* 45: 127-152.

Szott, L.T., C.A. Palm, and P.A. Sanchez. (1991). Agroforestry in acid soils in the humid tropics. *Advances in Agronomy.* 45: 275-301.

Taonda, S.J.B., A. Barro, R. Zougmore, B. Yelemou, and B. Ilboudo. (2003). Review article of the Inter CRSP activities of Burkina Faso. 24p.

Theng, B.K.G. (1991). Soil science in the tropics—The next 75 years. *Soil Science.* 151: 76-90.

Tian, G., F. Ishida, and D. Keatinge, eds. (2001). *Sustaining Soil Fertility in West Africa.* Soil Science Society of America Special Publication Number 58. Madison, WI: Soil Science Society of America and American Society of Agronomy.

Vanlauwe, B. (2004). Integrated soil fertility management research at TSBF: The framework, the principles, and their application. In Bationo, A. (ed.) *Managing Nutrient Cycles to Sustain Soil Fertility in Sub-Saharan Africa.* pp. 25-42. Nairobi, Kenya: Academy Science Publishers.

Vanlauwe, B., C.A. Palm, H.K. Murwira, and R. Merckx. (2002). Organic resource management in sub-Saharan Africa: Validation of a residue quality-driven decision support system. *Agronomie.* 22: 839-846.

van Raay, J.G.T. and P.N. de Leeuw. (1971). The importance of crop residues as fodder: A resource analysis in Katsina Province, Nigeria. *Samaru Research Bulletin 139.* Ahmadu Bello University, Samaru, Zaria, Nigeria. 11p.

Vlaming, J., H. van den Bosch, M.S. van Wijk, A. de Jager, A. Bannink, H. van Keulen (2001). Monitoring nutrient flows and economic performance in tropical farming systems (NUTMON) Alterra, Netherlands.

Walker, B., W. Steffen, J. Canadell, and J. Ingram. (1999). *The Terrestrial Biosphere and Global Change: Implications for Natural and Managed Ecosystems,* Chapters 1 and 9 reproduced with permission for GCTE Focus 3 Conference. Cambridge University Press.

Willet, A. (1998). *Agricultural Education Review: Support for Agricultural Education in the Bank and by Other Donor Executive Summary.* Washington, DC: World Bank.

Williams, T.O., J.M. Powell, and S. Fernandez-Rivera. (1995). Manure utilization, drought cycles and herd dynamics in the Sahel: Implications for crop productivity. In Powell, J.M., S. Fernandez-Rivera, T.O. Williams, and C. Renard. (eds.) *Livestock and Sustainable Nutrient Cycling in Mixed Farming Systems of Sub-Saharan Africa.* Vol. 2. pp. 393-409. Technical Papers. *Proceedings of an International Conference.* November 22-26, 1993. Addis Ababa, Ethiopia: International Livestock Centre for Africa (ILCA).

Windmeijer, P.N. and W. Andriesse. (1993). *Inland Valleys in West Africa: An Agro-Ecological Characterization of Rice Growing Environments.* pp. 9-160.

Yamoah, C.F. and J.R. Burleigh. (1990). Alley cropping *Sesbania sesban* (L.) Merill with crops in the highland region of Rwanda. *Agroforestry Systems.* 10: 169-181.

Young, A. (1997). *Agroforestry for Soil Management.* Wallingford, UK: CAB International.

Zougmore, R.B. (2003). Integrated Water and Nutrient Management for Sorghum Production in Semi-Arid Burkina Faso. Wageningen,The Netherlands. 205p. 35 ref.

Zougmore, R., A. Mando, J. Ringersma, and L. Stroosnidjer. (2003). Effect of combined water and nutrient management on runoff and sorghum yield in semiarid Burkina Faso. *Soil Use and Management.* 19: 257-264.

Zougmore, R. and Z. Zida. (2000). Anti-erosive fight and improvement of the productivity of the ground by the installation of stony cords. Card-index Technique N° 1.

Zougmore, R., Z. Zida, and N.F. Kambou. (2003). Role of nutrient amendments in the success of half-moon soil and water conservation practice in semiarid Burkina Faso. *Soil and Tillage Research.* 71: 143-149.

Chapter 12

Integrated Nutrient Management: Experience and Concepts from the Middle East

Uzi Kafkafi
David J. Bonfil

INTRODUCTION

General Background

The Near East is a dry region. For this reason population is centered largely in the Fertile Crescent around the rivers Tigris, Euphrates, Nile, and on the western mountains slopes of the Mediterranean coast. The rainfed areas have been under traditional agricultural production for thousands of years. Unfortunately, only traditional, old, agricultural practices were in use till the middle of the twentieth century. Since about 1950, rapid development occurred in all the countries in this region. One aspect of this development showed up as a significant increase in NPK consumption by each country (Clawson et al., 1971). The Middle East region in this chapter relates mainly to the countries Cyprus, Israel, Jordan, Lebanon, and Syria. This chapter focuses mainly on changes that occurred in NPK fertilization during the past fifty years. Except Cyprus and Lebanon, most of the area of these countries is desert and the agricultural production is sporadic or nonexistent unless irrigation is available. For generations, the sand dunes along the coast of Israel created a barrier to rainfall floods from the mountain range, resulting in bogged areas that prevented longitude roads along the coast.

This barrier forced the old road "via maris," the land road between Ancient Egypt and Mesopotamia (known now as Iraq), to pass along the foothills of the mountains and reach Egypt through the Sinai desert. Drainage works that started about eighty years ago dried the bogs that for millennium had accumulated floods loaded with clay particles. These soils are today the deep, heavy, clay soils used for irrigated and rainfed field crops. In the irrigated areas the use of fertilizers is usually in excess of the amount required by crops, so deep leaching of nitrate is reaching underground water aquifers that are used for agriculture. In the nonirrigated rainfed areas, the amount of fertilizers used is very restricted. The case of nitrogen application for wheat in rainfed areas is discussed in detail in a special case study at the end of this chapter. Most of the discussion in this chapter uses our own data from Israel and other local sources. The data presented for Syria, Lebanon, Cyprus, and Jordan were derived from IFA and FAO publications.

The fact that water for irrigation was always the limiting factor for crop production was the driving force for increasing irrigation efficiency that led to the development of trickle irrigation systems. Introduction of fertilizers into the trickle irrigation system enabled the quick transformation of nonagricultural soils like the sand dunes (Kafkafi, 1994) and steep mountain slopes into productive agricultural soils for vegetables and tree plantations.

The overall usage of mineral fertilizers and organic manures is usually above the export of minerals in the harvested crops, causing the inevitable accumulation of nutrients in the soil that slowly migrate to the aquifers once they move below the rooting depth.

MAJOR SOILS, CLIMATIC REGIONS, AND MAJOR CROPPING SYSTEMS

Israel lies between the Mediterranean Sea on the west, the Jordan rift on the east, the Sinai desert in the south, and the Lebanon mountains in the north. The main climatic regions used for agriculture are located north of the Be'er Sheva (N31°15' E3448') latitude line with long-term average rainfall of 200 mm. Since the winter rain season lasts from December to April, any intensive agriculture during the warm climate period relies on irrigation. The numbers in parenthesis on Figure 12.1 represent the average number of rainy days in each zone.

The Golan Heights, east of the Jordan rift, contains mainly heavy clay soils of basalt eruption origin. West of the Jordan rift, the soils developed on variable parent materials. The geological regressions and transgressions of the sea are reflected in the current distribution of soils. The region with

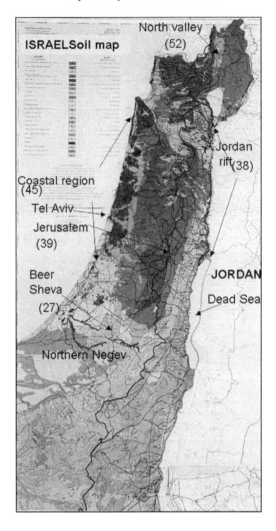

FIGURE 12.1. Israel soil classification map. *Source:* Adapted from Dan et al. (1975).

Jerusalem in the center (see Figure 12.1) represents the mountain range along which lies the rainfall dividing line. About 30 km east of the mountain range lies an area with the world's lowest elevation below sea level. In this region, rainfall drops to less than 300 mm. West of the divide line, rainfall is above

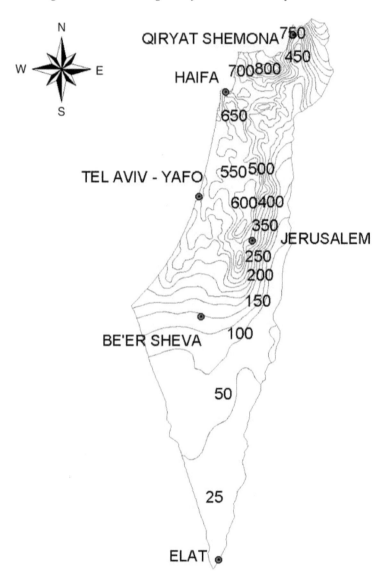

FIGURE 12.2. Israel rainfall contour map. Average annual rainfall (in mm) for the years 1950-1980. *Source:* Prepared by Z. Dorfman from the data of the Israel meteorological service, 1981. Ministry of Agriculture, Soil Conservation and Drainage Division, Israel. Reprinted with permission.

500 mm (Figure 12.2). The soils of the Jordan rift are highly calcareous sediments of the ancient lake that was created when the geological rift initiated. The dark colored soils in the middle of the figure represent the mountain soils that have developed on hard and soft calcareous rocks. The red clay soils (locally termed as Terra Rosa) are the typical clay soils that developed on a hard limestone rock of the Cretaceous geological period.

The soil types found in Syria, Jordan, Lebanon, Turkey, and Greece are similar to those in Israel, being typical soils of the Mediterranean climate. When no irrigation is available these soils grow olives and wheat that depend on stored water from the winter rains. West of the mountain range the flat areas are heavy clay soils created by alluvial flow from the mountains. They are used for irrigated and rainfed field crops such as vegetables, cotton, wheat, and silage crops. Near the coast, sandy and sandy clay soils dominate. These soils originated from sand dunes deposited by the Nile floods. For the past seventy years these soils were the main growing area of the Shamuti orange, known by its trade mark as "Jaffa" orange. All these sandy soils today use trickle irrigation systems or canopy sprinklers. However, due to population pressure, citrus orchards have been replaced by urban construction over the last three decades. The citrus area is moving southward, to the western Negev, a distance of 15-20 km from the sea in the 200 mm rainfall zone. This new growing area is entirely dependent on irrigation. The irrigation water source is recycled sewage water delivered to the processing plant site by pipes from the most populated area of Tel Aviv.

The agriculture pattern in Israel comprises mainly four types of agricultural systems: The traditional Arab farm, the Jewish small farm (Moshav) of up to 4 ha per unit, the cooperative farms of about 2-4 hundred ha, and citrus and other fresh table fruits and grapevine plantations that vary in size per owner.

Land Use

The Negev desert, located in the southern part of Israel, accounts for some 50 percent of Israel's national area and dominates its land use for crop production (Figure 12.1). Two main features of the land use structure (Table 12.1) characterize its impact on agriculture development:

1. very limited area of pastures; and
2. high dependence on water in order to utilize the arable land.

In Israel, Cyprus, and Lebanon, a third of the agricultural area is irrigated, versus only less than 10 percent in Jordan and Syria (Table 12.1). Large areas

TABLE 12.1. Land use in the Middle East (2002)

Land use (hectares 1,000)	Cyprus	Israel	Jordan	Lebanon	Palestine	Syria
Total area	925	2,214	8,921	1,040	621	18,518
Land area	924	2,171	8,893	1,023	602	18,378
Agricultural Area	117	566	1,142	329	381	13,759
Arable and permanent crops	113	424	400	313	231	5,421
Arable land	72	338	295	170	113	4,593
Permanent crops	41	86	105	143	118	828
Permanent pasture	4	142	742	16	150	8,338
Irrigation	40	194	75	104	20	1,333
Nonarable and non permanent	811	1,747	8,493	710	371	12,957

Source: Adapted from FAOSTAT data (2005).

of arable land do not produce a harvest in very low annual rainfall seasons like 1999.

Water

Water scarcity is a main limiting factor in Israeli agriculture. Along the 500 km from north to south, the annual rainfall amount varies from 800 mm to 25 mm (Figures 12.2 and 12.3). The rainy season lasts from October to April. More than half of the area in the region south of Be'er Sheva gets less than 200 mm annual rainfall.

Three main water resources, besides rainfall, supply most of the water demand for agricultural, domestic, and industrial use in Israel. They are the lake of Galilee, from which an average annual amount of 400 million m^3 is pumped to the south part of the country, and the coastal and mountain aquifers. Due to overpumping and frequent droughts, especially during the past decade, the water availability has decreased substantially, making it necessary to reduce the quantity of water allocated to agriculture. In response to these limitations, agriculture not only had to limit cultivated areas but also had to find alternatives to fresh water resources as presented in Table 12.2. Table 12.3 presents the expected demand for fresh water in the next twenty years. Jordan has fewer water sources, while Lebanon and Syria have more water sources (Clawson et al., 1971).

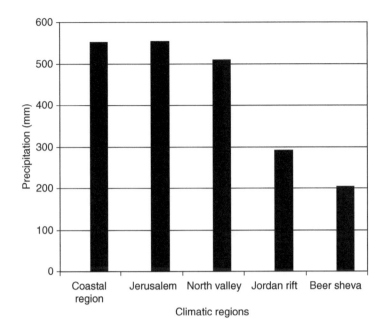

FIGURE 12.3. Rainfall distribution in Israel by climatic regions.

TABLE 12.2. Use of water, 1989-1999

	1989		1999	
	m³ 10⁶	%	m³ 10⁶	%
Total—Israel	1,851	100.0	2,076	100.0
Agricultural water use	1,304	70.4	1,264	60.9
Agricultural use by water quality		100.0		100.0
Fresh water	1,022	78.4	842	66.6
Marginal[a]	226	17.3	224	17.7
Recycled	56	4.3	198	15.7

Source: Rural Planning Authority, Ministry of Agriculture and Rural Development.

[a]Brackish and other low-quality water.

TABLE 12.3. Fresh water demand (m³ 10⁶) in Israel, 2000-2020

Year	Domestic	Industry	Agriculture	PAᵃ and Jordanᵇ	Total
2000	700	90	702	94	1,586
2020	1,120	150	530	200	2,000

Source: Water Commission, Israel.

[a]Palestinian authority.

[b]According to an international agreement.

The agriculture pattern in Israel, until 1956, was dependent on local rainfall or local shallow wells along the coastal region. In 1956, the Israeli central water carrier system was completed and water from the lake of Galilee was pumped throughout the state, bringing water all the way to the northern Negev area. This system allows an increase in the irrigated cropping area in northern Negev, especially for vegetable crops. A similar system that took water from the Jarmuch river down to the Jordan valley was established in Jordan, that increased the irrigated area mostly during the 1980s, while Syria made a significant increase only from 1990 (Figure 12.4) and applied it to most crops (Table 12.4). The price of water, and the increasing demand for fresh water by the growing population, gradually shifted the industrial crops such as cotton and subsequently many plantation crops to the use of recycled city sewage water. The parallel development of trickle irrigation systems enabled the use of recycled water with minimum health risks. The shortage of water and soils, and the high labor costs increased the protected cultivation area within greenhouses that produce high cash value crops and have higher water use efficiency.

Cropping Systems and Distribution

The first documented information on the main crops grown on this land is mentioned in the Bible (Deut. 8:8): "A land of wheat and barley and vines, and fig trees and pomegranates; a land of olive-oil and honey" (meaning dates). Wheat and barley still occupy the largest area of production, together with olive trees (Figure 12.5, Table 12.5). Because of economic constraints, today figs and pomegranates are grown mainly in small gardens and their yield is restricted to a very short season. Dates are grown in the areas where the weather is warm throughout the year, in the Jordan rift south of the Lake

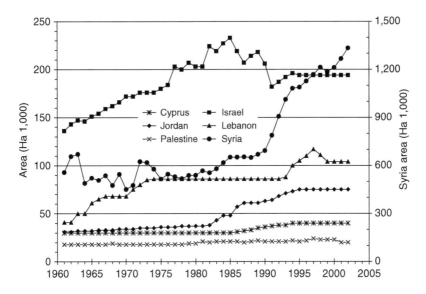

FIGURE 12.4. Agricultural land use with irrigation in the Middle East. *Source:* Adapted from FAOSTAT data (2005).

TABLE 12.4. Average areas of main groups of crops in Syria, 1997-1999 (ha 1,000)

Crop	Total	Irrigated	Rainfed
Summer crops	396	365	31
Summer vegetables	98	59	39
Winter crops	3,577	753	2,824
Winter vegetables	33	28	5
Fruit trees	773	119	654
Grand total	4,877		

Source: Adapted from FAO data.

of Galilee, down to Eilat in the south at the northern tip of the Red Sea. Over the past fifteen years, there has been a steady increase in grape production for wine, mainly at the expense of cotton and citrus. Over the past forty years, there were some significant changes in land use for different crops (Table 12.5). In Israel, the barley and orange area decreased very much, while

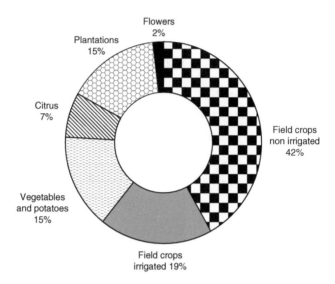

FIGURE 12.5. Crop area percentage distribution of major crops in Israeli agriculture. *Source:* Adapted by Kafkafi from Data Sources of the Ministry of Agriculture, Israel.

major increases were shown for olives, vegetables, and chickpea. In Jordan, a significant reduction was found in land use for all crops, while in Syria land use for all crops increased. Land use for cereals usually presented slow changes (Figure 12.6), but in some cases a rapid decline occurred, such as in Jordan between 1960 and 1975, and a rapid increase occurred in Syria in 1989. In comparison, land use for citrus showed significant increase in Syria (1983-1992) and a parallel decrease in Israel (Figure 12.7).

Rainfall Zones in Syria

Syria lies within the range of a precipitation gradient varying from about 1,500 mm in the west of the country to less than 100 mm in the southeast. The main factors determining the agroecological diversity are climate, landforms, and geological parent materials.

Rainfed winter crops are the main crops grown in Syria (Tables 12.4, 12.5), making rainfall the main factor for crop delineation (Table 12.6). The country can be divided into five agricultural stability zones according to the

TABLE 12.5. Agricultural use of land by crops in the Middle East (ha 1,000; 1961, 2002)

	Cyprus		Israel		Jordan		Lebanon		Palestine	Syria	
	1961	2002	1961	2002	1961	2002	1961	2002	2002	1961	2002
Cereals, total	141.7	59.2	154.9	92.2	377.7	88.7	90.4	56.8	33.7	2,116.3	2,975.7
Barley	60.2	52.9	69.7	10.0	95.0	55.4	12.8	11.5	10.9	727.0	1,234.0
Wheat	78.9	5.9	62.9	75.0	273.2	32.7	68.8	43.5	22.4	1,315.5	1,679.4
Cotton lint and seed	1.6	0.0	54.6	61.4	0.0	0.0	0.0	0.0	0.0	573.9	1,002.0
Chickpea	0.6	0.1	1.9	8.0	5.4	1.7	1.8	2.9	2.1	24.8	102.2
Lentil	1.3	0.0	0.7	0.1	16.6	1.2	1.6	1.6	1.9	57.7	121.2
Vegetables, total	5.9	4.4	23.1	52.6	58.3	31.2	15.1	28.7	18.2	96.5	125.1
Citrus Fruit, total	5.2	5.3	26.5	17.7	2.0	7.8	8.5	15.2	4.1	0.9	28.2
Oranges	3.1	1.4	20.0	5.4	1.6	2.2	5.7	8.8	2.9	0.5	15.3
Olives	20.4	8.6	10.0	18.0	63.5	64.5	27.0	57.6	92.7	80.0	499.0

Source: Adapted from FAOSTAT data (2005).

average annual rainfall. The zones are defined in terms of suitability for rainfed crop production, and to some extent the probability of rainfall:

Zone 1—Annual average rainfall of over 350 mm. The zone is divided into two areas:

    1-A. Annual average rainfall of over 600 mm—rainfed crops are grown successfully.

    1-B. Annual average rainfall of between 350 and 600 mm, the main crops are: wheat, legumes, and summer crops such as melon and watermelon. The area of this zone is 2,698,000 ha, comprising 14.6 percent of the area.

Zone 2—Annual rainfall of 250-350 mm. It is possible to grow two barley crops every three years as well as wheat, legumes, and summer crops. The area of this zone is 2,473,000 ha, comprising 13.4 percent of the area.

Zone 3—Annual rainfall of 250-350 mm with not less than 250 mm during half of the monitored years. It is possible to grow one or two crops every three years. The main crop is barley but legumes may also be grown. The area of this zone is 1,306,000 ha, comprising 7.1 percent of the total area.

Zone 4—A marginal zone between the arable zones and the desert zone, with an annual rainfall between 200 and 250 mm and not less than 200 mm, during half of the monitored years, growing barley or permanent grazing. The area of this zone is 1,823,000 ha, comprising 9.8 percent of the total area.

Zone 5—Desert and steppe zone. This area covers the rest of the country and is not suitable for rainfed cropping. The area of this zone is 10,218,000 ha, comprising 55.1 percent of the total national area. Areas in this zone adjacent to rivers permit irrigated agriculture. As rainfall decreases toward the southeast this zone becomes desert.

AGRICULTURAL PRODUCTION AND NUTRIENT BALANCE

Rainfed crops, wheat and olives, use the strategy of "fertilizing the soil." Very little fertilizer is used in olive production and the wheat fertilization strategy in limited rainfall areas is discussed in the case study later in this chapter. To date, all plantations and vineyards, as well as most of the citrus orchards are using trickle irrigation, or under the canopy sprinkle systems combining application of liquid fertilizer solutions made to order for the farmer.

Fertilizer Consumption in Five Middle East Countries

Fertilizer consumption in the Middle East countries since 1961 is presented in Figures 12.8 through 12.12. The total demand is related to the size

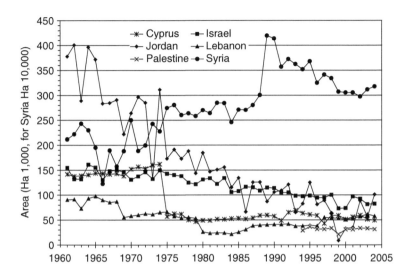

FIGURE 12.6. Agricultural land for cereals in the Middle East. *Source:* Adapted from FAOSTAT data (2005).

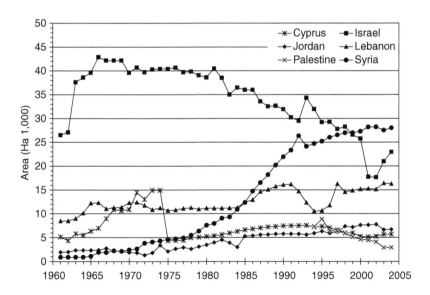

FIGURE 12.7. Agricultural land for citrus in the Middle East. *Source:* Adapted from FAOSTAT data (2005).

TABLE 12.6. Farming systems and commodity groups in Syria

System	Rainfall (mm)	Main commodity group
Desert	<100	Irrigated agriculture/livestock
Steppe/native pastures	<200	Livestock
Barley/livestock	200-350	Barley, livestock, cumin, feed/food legumes, olives
Wheat-based	350-500	Wheat, livestock, feed/food legumes, watermelon, olives
	>500	Wheat, livestock, legumes
	Irrigated	Vegetables, tobacco, fruit trees, wheat, cotton, tomato, sugar beet, maize, faba bean, green forage, potatoes, vegetables, livestock

Source: Adapted from FAO data.

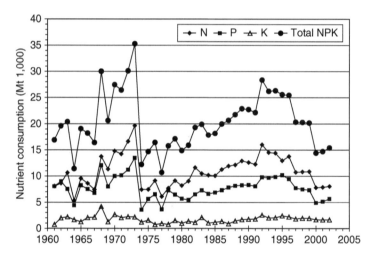

FIGURE 12.8. Major nutrients (N, P_2O_5, and K_2O) consumption in Cyprus. *Source:* Adapted from FAOSTAT data (2005).

of population in each country but the ratio between N, P, and K fertilizers needs some attention. Traditional field research results for wheat usually failed to demonstrate the need for K. These results influenced the total K consumption in Middle East countries for many years. Potassium consumption is very low in Syria and Cyprus.

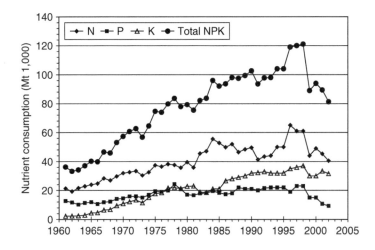

FIGURE 12.9. Major nutrients (N, P_2O_5, and K_2O) consumption in Israel. *Source:* Adapted from FAOSTAT data (2005).

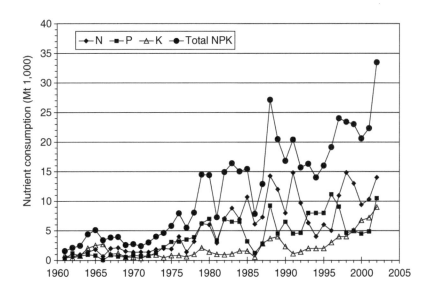

FIGURE 12.10. Major nutrients (N, P_2O_5, and K_2O) consumption in Jordan. *Source:* Adapted from FAOSTAT data (2005).

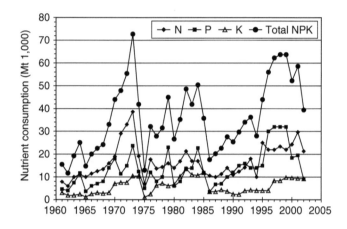

FIGURE 12.11. Major nutrients (N, P_2O_5, and K_2O) consumption in Lebanon. *Source:* Adapted from FAOSTAT data (2005).

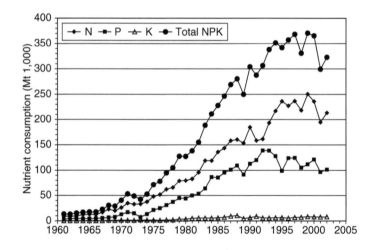

FIGURE 12.12. Major nutrients (N, P_2O_5, and K_2O) consumption in Syria. *Source:* Adapted from FAOSTAT data (2005).

In Cyprus (Figure 12.8) the peak fertilizer use was recorded in 1974 and the total demand has been lower since then, but there is no consistent trend in fertilizer demand. The existing variation is most probably related to rainfall distribution.

In Israel (Figure 12.9), nitrogen consumption rose from 20 to 60,000 tonnes (t) over thirty years and then declined slightly. Nitrogen and potassium consumption has increased continuously due to intensive trickle irrigation of vegetables and plantation crops. The need for N fertilizers is evaluated in different ways: For wheat, in the restricted rainfall area, the biological N test is used (Amir and Ephrat, 1971; Amir et al., 1994; and case study). Phosphate demand reached a plateau of 20,000 t annually and it has remained near this level since 1975. Dry land farmers use soil testing results extensively for P (Na-Bicarbonate extraction based on Olsen's method). Potassium usage rose from almost zero in 1961 to 37,000 t in 1998. From 1971 to 1991 potassium demand tripled. The introduction of intensive fertigation to open fields and greenhouse vegetables and flowers and the supply of combined liquid fertilizers are most probably responsible for this change in usage trend. In fertigation, the decision about N is mainly dictated by specific crop demands. Applications are supplied through the trickle irrigation system and are changed daily according to plant demands, regardless of residual N or K fertilizers.

Only a small increase in Jordan NPK consumption is observed especially during the late 1970s and has continued to increase (Figure 12.10). This could be explained by the reduction in the total land use for agriculture (Figure 12.6, Table 12.5). The annual fluctuations in N and P are still relatively large while K has shown a steady increase since 1996. The ratio between N to P_2O_5 in Jordan was almost 1:1 until 1982. The proportion of N increased from 1985 to 1994, but then the amount of P_2O_5 usage was almost twice that of N for four years. Subsequently, P consumption fell by 50 percent. The steady increase in K_2O usage observed in Jordan since 1997 is most likely related to the increase in fertigation practices.

Fertilizer use in Lebanon (Figure 12.11) demonstrated three peaks of demand of N, P, and K in 1975, 1985, and 2000. The pattern of specific nutrient consumption is also changing. The N to P_2O_5 ratio is close to 1 and from 1996 to 2000 the amount of P fertilizer used was even higher than that of N. Potassium is used in relatively high amounts. The increase of irrigated area in Lebanon from 1993 (Figure 12.4) was accompanied by an increase in NPK usage (Figure 12.11).

Conversely, in spite of the dramatic increase in the Syrian irrigated area since 1990 (Figure 12.4), the increase in NPK usage did not show the same response (Figure 12.12), since the irrigated area in Syria is still negligible in comparison to the rainfed area. In Syria, only the consumption of N and P fertilizers are very high while K fertilizer consumption has been very low over the past forty years. An almost constant N to P_2O_5 ratio was observed until 1993. Since that time, N usage increased from 150 to 250,000 t over three years, and has since varied between 200 and 250,000 t of N. Therefore, it could be that the irrigated area in Syria received mainly nitrogen. There is hardly any usage of K fertilizer, a practice that could be partly explained by the high K content in the soil.

Only in Israel and Jordan in the past few years has the consumption of fertilizer decreased in the order N > P > K. In Syria, Lebanon, and Cyprus the use of K fertilizers is still relatively low.

Fertilizer Use by Crop in Syria

The officially recommended rates of fertilizer use for each crop vary according to the water zone and varieties used (Table 12.7). In fact, very little information is available on farmers' decision making in relation to fertilizer use. Some information is available from farm surveys in northern Syria undertaken by ICARDA since the late 1970s. Most of these surveys were designed to analyze farmers' cultivation practices in general, with some questions directed toward fertilizer use. Two studies concentrated on fertilizer

TABLE 12.7 Recommended rates of fertilizer use by crop (kg ha) in Syria

Crop	Condition	Zone	Nitrogen	Phosphate
Irrigated wheat			150	100
Rainfed wheat	HYV[a]	1	100	80
	HYV	2	80	60
	Local	1	80	60
	Local	2	60	60
	Local	3	30	30
Rainfed barley		1	50	40
		2	40	40
		3	20	20
Cotton	Irrigated		200	150
Maize	Irrigated		120	80
Sugar beet	Autumn		200	120
	Summer		180	120
Potatoes	Autumn		150	120
	Summer		120	120

Source: Adapted from FAO data.

[a]HYV = High-yielding varieties. Local = Local varieties.

use. The first one (Whitaker, 1990) focused on farmers' fertilizer strategies in northern Syria; the second one (Mazid, 1994) concerned fertilizer use on rainfed barley in Syria.

The majority of farmers make their decisions on fertilizer use without consulting the local extension agent. Some policymakers have seriously considered making it compulsory for farmers to apply the recommended rates. Although most wheat farmers in Syria have used fertilizer for fifteen to twenty years, many of them have little information on the recommended rates, or the official fertilizer allocations for each crop. When farmers make decisions about fertilizer use, they consider the allocation of fertilizer between crops, the number of applications per crop, the rates, timing, and method of application. All of these decisions are made in a highly uncertain environment, characterized by wide year-to-year variations in rainfall levels and in seasonal distribution (Whitaker, 1990).

For the rainfed farming system, Whitaker (1990) identified several fertilizer strategies adopted by rainfed wheat farmers in northern Syria. Practically all the farmers surveyed indicated that they applied P_2O_5 only once, at the time of planting (mid-November to early December). Average rates used in the wetter areas (Zone 1) were almost twice the rates applied in the drier regions (Zones 2 and 3). These rates vary very little from year-to-year since P_2O_5 is applied at the beginning of the growing season, when the amount of future rainfall is unknown. Unlike the case with P_2O_5 application, farmers have greater flexibility with N application. This allows them to modify their strategies depending on rainfall levels.

Wheat farmers in Zone 1 generally apply two dressings of nitrogen. The first application is carried out at planting, and the second around tillering (end of February). As with P_2O_5, N rates at planting time show relatively little variation.

The rates of N for the second application depend greatly on rainfall levels during the first half of the growing season (October to February). If prior rainfall is considered to be normal, then farmers usually apply a rate twice as large as that of the first N application. This rate may be cut by one-third if rainfall is below average, or increased by up to 50 percent in a wet year. This depends essentially on the level of previous rainfall and on farmers' expectations about rainfall during the second half of the growing season (early March to early May).

Agriculture Production in Israel

During the decade from 1990 to 2000, the total value of agricultural production in Israel decreased by almost 15 percent at real prices (Table 12.8).

TABLE 12.8. Agricultural production of Israel by division, 1989-1999

	1989		1999	
	Value[a]	%	Value[a]	%
Horticulture and field crops	2,177	56.6	1,813	55.3
Vegetables	524	13.6	585	17.8
Flowers and ornamental plants	230	6.0	215	6.6
Citrus	387	10.1	220	6.7
Other fruit	429	11.2	452	13.8
Field crops[b]	506	13.2	224	6.8
Other crops	101	2.6	117	3.6
Total value of production	*3,844*	*100.0*	*3,279*	*100.0*
Poultry	790	20.6	648	19.8
Dairy and beef	622	16.2	545	16.6
Sheep and goats	101	2.6	97	2.9
Fish	103	2.7	114	3.5
Other livestock	51	1.3	62	1.9
Total Livestock	*1,667*	*43.4*	*1,466*	*44.7*

Source: Israel's Central Bureau of Statistics.

[a]US$ Millions, at 1999 Prices.

[b]A severe rain shortage in this year.

The agriculture production quantity index increased by 30 percent during the 1990-2000 period. However, this development represents the balance between two different trends—declining of agricultural product prices and increasing quantity of production per area. Hence, despite continued reduction in cereal area (Figure 12.6), grain production was almost stable (Figure 12.13), and most of the variation can be related to seasonal rainfall quantity. However, this was not the case for some export crops, in particular citrus and cotton, that suffered a drastic reduction in profits. This reduction in revenues brought reduction in citrus production (Figure 12.14), and led to the decrease in citrus land use (Figure 12.7).The prices received by producers for livestock products were far behind the cost of living index.

The share of horticulture and field crops in the total value of agricultural production during the past ten years remains stable. This part of production is the main export supplier of Israeli agriculture. Most horticulture crops are

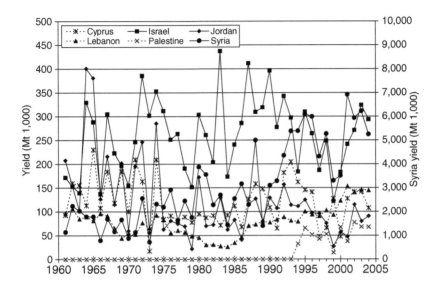

FIGURE 12.13. Cereal production in the Middle East. *Source:* Adapted from FAOSTAT data (2005).

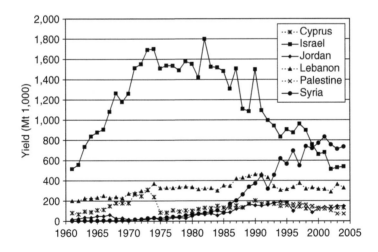

FIGURE 12.14. Citrus production in the Middle East. *Source:* Adapted from FAOSTAT data (2005).

based on subtropical and deciduous fruits, and a large selection of vegetables are grown under plastic cover, or in greenhouses as well as in open areas, but mainly under irrigation (Table 12.9). The quantities of organic manure produced and sold or specific amounts used are not recorded on a country basis. However, due to water shortage, most basic grains for human and livestock consumption are imported. Nonirrigated wheat production can be very low in some years due to rain shortage. The total grain import to the country is shown for the year 2000 in Table 12.10.

Livestock production, depending heavily on imported grains, cannot compete efficiently on foreign markets, and therefore, primarily supplies domestic demand. Production of livestock products is characterized by very intensive use of technology (Table 12.11). As a result, very high yields are obtained in many products. For example, average milk production has increased two and a half times since the 1950s, from 3,900 liters annually, to an average close to 11,000 liters per dairy cow in 1999. Poultry farming, which is the major supplier of meat for domestic demand, has developed under extreme variations of climate.

The total Livestock Inventory by the end of 1999 was (all values in thousands): cattle 388 (including 122 milk cows), laying hens 7,190, broilers 20,150, turkeys 4,900, sheep 350, goats 70, and beehives 72.

Fertilizers and Sustainable Development

Israel imports 93,174 t of N, 31,324 t per annum of P_2O_5, and about 26,000 t of K_2O in animal feed. This amount is equal to about half of the fertilizer P consumption in the country (Table 12.10). The actual load of nitrogen from manures on the soil reserves of N is difficult to calculate due to the lack of data on N losses to the atmosphere, either during animal digestion in the form of ammonia gas, or later by denitrification during irrigation (Bar-Yosef and Kafkafi, 1972). Of the total grain import, 20 percent ends up in the form of organic manures.

After twenty-five years of continuous cropping and fertilization, soil nitrate concentration was measured down to 12 m in selected plots of the Permanent Plots Experiment at Bet Dagan (Figure 12.15). As expected, the minimum nitrate content in the soil was found in the control treatment that had not received any fertilizer for twenty-five years. The nitrate still found in the soil of the control plots is the result of nitrate in the irrigation water and natural organic matter mineralization. The four levels of N tested show a decline in nitrate levels in the first 1 m depth. The excess applied nitrogen that moved below the rooting zone was leached to the deep soil regions by irrigation and rainfall over the twenty-five years of the study. Ongoing long-

TABLE 12.9. Main horticulture and field crops, 1999

Commodity	Value[a] (US$10[6])	% of total	Production[b] (1,000 t)
Vegetables			2,107
Potatoes	122.9	3.7	364
Edible tomatoes	94.4	2.9	242
Peppers	61.1	1.9	102
Cucumbers	33.1	1.0	108
Strawberries	25.8	0.8	16
Flowers and ornamental Plants[b]			1,435
Roses	38.4	1.2	469
Carnations	9.4	0.3	117
Gypsophila	22.7	0.7	155
Fruit			1,299
Citrus			723
Oranges	58.7	1.8	204
Grapefruit	92.5	2.8	372
Easy peelers	43.5	1.3	97
Other fruit			575
Apples	82.1	2.5	128
Apricots	44.9	1.4	46
Table grapes	48.6	1.5	49
Bananas	52.4	1.6	118
Avocado	49.0	1.5	56
Field crops[c]			
Cotton fiber	34.1	1.0	25
Groundnuts	27.3	0.8	23
Spices and medicinal plants	30.2	0.9	

Source: Israel's Central Bureau of Statistics.

[a]At 1999 prices.

[b]Millions of units export only.

[c]No rainfed crops yielded grain due to severe drought.

TABLE 12.10. Total grain imported to Israel in the year 2000 and calculated equivalent of net fertilizer units

Grain type	Quantity ($t \times 10^3$)	Crude protein (%)	N ($kg\ t^{-1}$)	Total N imported (t)	P_2O_5 ($kg\ t^{-1}$)	Total P_2O_5 imported (t)	K_2O ($kg\ t^{-1}$)	Total K_2O imported (t)
Wheat fodder	555	14.2	22.7	12,610	10.1	5,592	4.8	2,664
Soybean	552	36.0	64.0	35,328	11.0	6,072	20.0	11,040
Wheat	947	14.2	22.7	21,516	10.1	9,542	4.8	4,546
Barley	340	13.2	21.1	7,181	8.0	2,725	6.8	2,326
Corn	790	9.8	15.7	12,387	7.3	5,789	5.3	4,171
Sorghum	206	12.6	20.2	4,153	7.8	1,604	5.3	1,088
Total	3,390			93,174		31,324		25,834

Source: Rafi Shternlicht Israel Ministry of Agriculture, Department of Statistics; http://www.ext.nodak.edu/extpubs/ansci/beef/as1238w.htm; http://www.fertilizer.org/ifa/publicat/html/pubman/soybean.htm.

TABLE 12.11. Main livestock products, 1999

Commodity production	Value[a] (US$10⁶)	% of total	Quantity[b] (t × 10³)
Livestock for meat	657.0	20.0	
Cattle	143.7	4.4	88
Sheep and goats	79.5	2.4	22
Poultry	406.3	12.4	391
Broilers	258.9	7.9	260
Turkeys	130.9	4.0	125
Milk	418.8	12.8	
Dairy (liters 10⁶)	401.7	12.3	1,157
Sheep and goats (liters 10⁶)	17.1	0.5	29
Eggs (units 10⁶)	112.1	3.4	1,640
Aquaculture	114.3	3.5	30

Source: Israel's Central Bureau of Statistics.

[a]At December 1999 prices.

[b]Unless otherwise stated.

term permanent plot experiments at Gilat (Bonfil et al., 1999) are the latest sources of information from long-term fertilization applications, and are used to develop fertilization recommendations.

Long-term studies of P application in range areas (Henkin et al., 1996, 1998) have demonstrated the lasting influence of simple superphosphate application on the value and sustainability of forage production during the winter period. The effect of one fertilizer application was apparent seven years later. Continuous goat grazing in the Near East brought about barren mountains with extremely low productivity. It was demonstrated that a shortage of P in the upper 3 cm of mountain soils was the main hindrance to vegetative growth early in the season. It is therefore suggested here that aerial application of superphosphate in the rainfed areas of mountains in the Middle East could bring an increase in forage production for grazing animals. It was demonstrated (Henkin et al., 1998) that the proportion of legume in the natural vegetation was increased to 70 percent of the natural vegetation in the year immediately following P application.

The long term effect of P in manure, on wheat growth, is demonstrated in the infrared aerial picture of a wheat field at the permanent fertilization

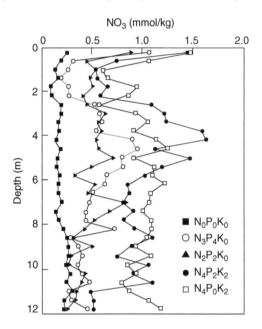

FIGURE 12.15. Nitrate distribution in soil depth to 12 m after 25 year of continuous fertilizer applications in the years 1961-1985 (Total applications in kg elemental nutrients ha^{-1} were: N_0—0, N_2—3120, N_3—4465, N_4—7650, P_0—0, P_2—1210, P_4—2420, K_0—0, K_2—4850). *Source:* Feigin and Halevy (1994), Kafkafi, unpublished. Reprinted with permission.

plots of the Volcani Center in Bet Dagan, Israel (Figure 12.16) (Kafkafi and Halevy, 1974). The horizontal dark rectangle represents zero N levels. The plots marked with P_0 did not get any mineral P fertilizer for seven years before the aerial picture was taken. Plots R_2 received organic manure at a rate of 40 t ha^{-1} three years before wheat. The effect of the residual P from the organic source is clear, but no residual effects of N in the organic plots were detected. The black intensity in the picture is negatively correlated with wheat height (Stanhill et al., 1972). The darker areas occur where the IR reflectance is low since the smaller plants allow greater IR reflectance from the bare soil.

Potato tuber yield is dependant on continuous P uptake from the soil, as the tuber accumulates P from the soil during all stages of growth. Evidence indicates that organic manure benefits potato due to the consistent supply of

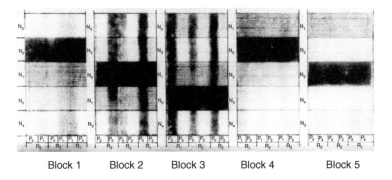

FIGURE 12.16. Aerial photograph of the Permanent Plots Experiment at Bet Dagan, Israel when ten-week-old wheat was grown on it. *Source:* Adapted from Kafkafi U. and Bar-Yosef B. (1971), *Long-term Effects of Cumulative Application of Fertilizers on Irrigated Soil and Crops.* Final report to the Ford Foundation Project A-III/6, Volcani Institute of Agricultural Research, Bet Dagan, Israel. *Notes*: The fertilizer treatments are superimposed on the picture: R_1-only chemical fertilizers, R_2-organic manure once every four years, and R_3-green manure. N level treatments are horizontal and P treatments are vertical. N_0—0, N_1—30, N_2—60, N_3—120, N_4—240 kg N ha^{-1} as ammonium sulfate P_0—0, P_1—90, P_2—180, P_3—360, P_4—720 kg Superphosphate (20% P_2O_5).

P to the tuber over the growing season. Therefore, organic manure is used intensively in potato production. The main crops that use organic manures are potato, strawberry, banana, and carrot. Manures must be incorporated into the soil as soon as possible after spreading or some of the nitrogen can be lost as ammonia gas. However, all other mineral components in the manure will have long-lasting effects on the soil (Figure 12.16). No- and reduce-tillage practices are gaining in popularity, but organic manure application in such systems will normally not be incorporated, inevitably increasing the risk of N losses and application complexity. The main residual value of farmyard manure is due to P, since this is the main component in the manure that remains available in the soil for long periods. No such study is recorded in any report in the Middle East.

Although manures can be an effective source of nutrients for agricultural production, environmental concerns can arise if manure nutrients are not used effectively. The estimated total major nutrient in animal secretions from the cattle and chicken industries in four countries in the Middle East is presented in Table 12.12. The impact of organic manures on the environment depends on the type of the industry. In the dairy and poultry industries

TABLE 12.12. Estimated total nutrient excreted in the year 2002 in farm animals in the Near East

Country	Item	Animals (thousands)	Total annual excretion			Total nutrients excreted		
			N[a] (kg/head)	P (kg/head)	K (kg/head)	N (t)	P (t)	K (t)
Israel	Cattle	390,000	25.0	12.5	25.0	19,500	9,750	19,500
Jordan	Cattle	68,070	25.0	12.5	25.0	3,404	1,702	3,404
Lebanon	Cattle	70,260	25.0	12.5	25.0	3,513	1,757	3,513
Syria	Cattle	771,118	25.0	12.5	25.0	38,556	19,278	38,556
		tonne/year	kg t^{-1}	kg t^{-1}	kg t^{-1}	tonne	tonne	tonne
Israel	Chickens	30,000	630,000	11.3	25.4	9.1	2,381	5,333	1,905
Jordan	Chickens	24,000	504,000	11.3	25.4	9.1	1,905	4,267	1,524
Lebanon	Chickens	20,387	428,127	11.3	25.4	9.1	1,618	3,624	1,294
Syria	Chickens	14,786	310,506	11.3	25.4	9.1	1,173	2,629	939

Source: FAOSTAT data (2005). Adapted from http://www.taa.org.uk/Courses/Week3/FYMnutrientsweb.htm; http://eesc.orst.edu/agcomwebfile/edmat/html/ec1094/ec1094.html

[a]N-including urine, P-P$_2$O$_5$, K-K$_2$O.

in Israel, the cattle and chicken are concentrated in closed yards and air-conditioned houses, respectively. Due to the high animal concentration in buildings and yards in Israel, the load of organic manure produced on a unit area in Israel is the highest in the Middle East, causing an environmental problem. There was a huge transport of urine and nutrients from farm lots down to the underground water aquifers. The case of urine in the milking cow industry is still not completely solved and leaching and aerial ammonia losses are occurring. Only in 2003 were environmental protection efforts introduced forcing all dairy operations to build concrete floors to prevent nutrient leaching. This rule will eliminate smallholdings from the market, as only large units will be economically viable. Most of the solid manure is reprocessed by being dried, formed into compressed pellets, and resold as organic manure for crop production. The excretion from grazing animals is never collected and its destination can only be estimated. The sheep and beef industry is based on range animals that are partly fed by chicken manure. The N from all of their manure is nonrecoverable, and what is not utilized by the pasture is either lost during the hot summer periods by direct volatilization of NH_3 gas to the atmosphere, or alternately moves into the soil where it is used as a plant nutrient in the early growth phases in the next winter or leached to underground water aquifers. In contrast, P and K remain in the upper soil layer and are not washed downward.

Apart from livestock manures, other organic amendments can be effective in recycling nutrients onto agricultural land. Recently, rainfed and irrigated wheat areas have been used as the dumping sites for accumulated sludge from city sewage water purifying plants, with positive results on wheat yield where annual rainfall is more than 300 mm. In drier regions, sludge application encourages excess wheat vegetative growth, resulting in shriveled grain production due to a shortage of water needed to fill the grain. However, caution must be raised since sludge application increased heavy metal content in grain and straw (Han and Banin, 2001; Fine and Mingelgrin, 1996). Therefore, sludge addition to wheat fields must be monitored and the amount allowed is not unlimited.

CASE STUDY OF WHEAT

Integrated Nutrient Management—Wheat Production Under 300 mm Annual Rainfall

Climatic conditions limit wheat yield especially due to water deficiency during the grain filling period. Irrigation with fresh or recycled water ensures horticulture and cash crop production. However, water quantity is limited and most of the area is managed with rainfed field crops. Wheat is the main rainfed crop sown in Israel, and the expected yields are related to rainfall, with the common expectation to harvest 10 kg grain ha^{-1} for each 1 mm of precipitation (only above 100 mm annual rainfall). Wheat is normally grown in rotation, and the rotation chosen is dependent on rainfall. In drier regions, wheat is grown as monoculture continuous wheat, or in rotation with clean fallow. With an increase in annual precipitation wheat is sown in two out of three years and usually rotated with legumes or watermelon. In areas that receive about 500 mm annually, and in irrigated fields, rainfed wheat is grown usually only once in two years and rotated with different crops. Precise fertilization is the main factor that allows the farmer to regulate his rainfed crop growth and production, as well as nutrient use efficiency. Therefore, it is essential that fertilization decisions made for rainfed crops in the Middle East should consider their interaction with precipitation, and/or potential available irrigation, as well as with residual organic and synthetic sources of nutrients from the fallow crop.

Middle East Major Soil/Climatic Regions/Zones, and Major Cropping Systems

Mediterranean type environments of southern Europe, the Middle East, and also of southern Australia are characterized by dry, hot summers alternating with wet, cold winters (Nahal, 1981; Palta et al., 1994; Acevedo et al., 1999). Annual average precipitation varies from about 500 to 600 mm in Israel's central and northern areas, and declines to 200 mm at the southern region, Israel's northern Negev (Figure 12.2). Similar climatic conditions prevail in all Middle East countries. Although the Israeli northern Negev receives less rain, it has become the most important agriculture area for field and horticulture crops in Israel, since urbanization pressure in the past thirty years has modified land use in the northern areas. Mediterranean areas are suitable for the production of high-quality bread-making wheat (*Triticum aestivum* L.). Therefore, wheat is the major field crop grown in Israel. Wheat is sown annually in about 100,000 ha, of which 40-50,000 ha is located in northern Negev.

Rainfall accumulation at Gilat Research Center (Figure 12.17) represents the typical precipitation distribution in the northern Negev area. Gilat is located in the southeastern corner of this agriculture region and receives the lowest winter rainfall. Annual precipitation in this region is between 200 and 500 mm. Gilat receives an average of only 230 mm of precipitation annually. Rainfall accumulation in the rainy season at Gilat represents the average accumulation at the western and northern Negev regions. In addition to

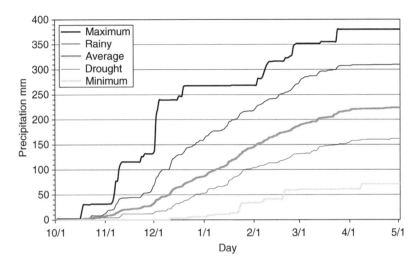

FIGURE 12.17. Annual cumulative rainfall at Gilat Research Center 1975-2003. Minimum and Maximum display the extreme amount for each day. Average is the simple mean including all years. Rainy and Drought represent the mean of all years where their total rainfall was at least 10 percent more or less than the mean of all the years. No rainfall occurs from mid-April to mid-October.

the importance of total rain quantity, variation in rainfall distribution over the growing season plays a major role in rainfall efficiency. The beginning of the rainy season varies from year to year; however, sowing usually takes place around mid-November. The time of wheat emergence is very important, since early germinating plants are susceptible to cold injuries during flowering and grain setting. Also, late germination leads to a shorter grain-filling period and lower grain yields (Borghi et al., 1997; López-Bellido et al., 1998).

The soil type at Gilat and throughout most of northern Negev is loess or sandy loam loess classified as calcic haploxeralf or calcic xerosol (Figure 12.1; Amir et al., 1991; Ben-Dor et al., 2004). Sandy soil covers a small region, especially near the Mediterranean coast, while clay loam and loamy clay soil types dominate northern Negev areas.

Fertilizers and Nutrient Balances

A comprehensive study (Kafkafi and Halevy, 1974), on the NPK accumulation by wheat in a wide range of fertilization levels, is used today in precise fertilization decision-making. A complementary permanent plots experiment, that tests eight combinations of N and P application under different crop managements under drier growing conditions was initiated in 1975 at the Gilat Research Center. This experiment improved nutrient use efficiency in the northern Negev region, and led to the development of a new bio-assay that is used for forecasting precisely available nitrogen and phosphorus content in soils (Amir et al., 1994; Bonfil et al., 2001).

Rainfed wheat fertilization management fifty years ago included ammonium sulphate (30-60 kg N ha^{-1}) and superphosphate (15-20 kg P ha^{-1}), for base fertilization, and ammonium sulfate for topdressing (20-40 kg N ha^{-1}). Phosphorus application was considered for fields if the P-NaHCO$_3$ extraction showed P levels lower than 6-7 mg kg^{-1}, and N was applied considering average rainfall and fallow practices. Twenty years later, some fields kept the same fertilization program, while others started to use newer fertilizers such as aqueous ammonia (50 percent), ammonium nitrate solution (20 percent) 60-100 kg N ha^{-1}, and N-P solid and liquid (20 percent) fertilizer for base application, and solid urea for topdressing (15-50 kg N ha^{-1}). At present, aqueous ammonia and prilled urea, and solid and liquid N-P or triple-phosphate are the most widely used fertilizers. Potassium usually is not applied to rainfed field crops in the Negev area, as local soils are rich in soil K. However, lack of fertilization over long periods could lead to K deficiency. Recently, several experiments examined soil potassium availability, application, and uptake by wheat, and the effects of potassium fertilization on wheat quality parameters (Bonfil, 2001).

Water is the main limiting factor for wheat in the Negev region; therefore only minor effects could be related to fertilization management in grain production. One tonne of wheat grain removes about 25 kg N and 3 kg P. Therefore, highly efficient fertilization would apply the quantity of N and P to balance the expected removal by wheat (adjusted for nutrient use efficiency), which mainly depends on accurate estimation of rainfall predictions.

Fertilizers and Sustainable Development

Many factors can affect nutrient levels in the soil even if the best fertilization decisions are taken. Consequently, nutrient management that applies too much nutrient, or growth conditions which prevent achievement of crop yield potential, would increase residual nutrient levels in soils (or would increase losses to the water and atmosphere). Under favorable growing conditions, high yield production would reduce the residual soil nutrients. The levels of available N and P in the northern Negev fields, which were analyzed by the bio-assay, exhibit the differences in their residual pools (Table 12.13).

TABLE 12.13. Annual rainfall at Gilat and residual nitrogen and phosphorus contents at the northern Negev fields, as determined by the Gilat bio-assay method

Year	Rain (mm)	N (kg ha^{-1})				P (kg ha^{-1})			
		Min.	Mean	Max.	SE	Min.	Mean	Max.	SE
1981	277	13	49	281	1.9	–	–	–	–
1982	219	4	37	161	1.3	–	–	–	–
1983	340	2	30	189	1.5	–	–	–	–
1984	141	3	46	201	1.3	–	–	–	–
1985	207	nd[a]	nd	nd	nd	–	–	–	–
1986	180	14	66	250	1.7	–	–	–	–
1987	268	4	50	201	1.5	–	–	–	–
1988	259	7	54	300	1.0	–	–	–	–
1989	207	5	48	226	1.4	–	–	–	–
1990	244	17	65	274	1.3	–	–	–	–
1991	238	17	72	313	1.5	–	–	–	–
1992	380	14	58	265	1.5	–	–	–	–
1993	257	14	62	269	1.7	–	–	–	–
1994	162	21	83	411	2.0	–	–	–	–
1995	318	16	67	295	1.9	–	–	–	–
1996	217	26	80	605	2.1	–	–	–	–
1997	177	19	81	257	1.8	1.9	9.8	35.0	0.2
1998	222	22	83	440	1.9	1.7	9.9	50.0	0.3
1999	72	33	106	704	2.0	1.6	13.1	84.9	0.4
2000	158	25	102	428	2.0	1.3	14.3	66.6	0.3
2001	245	30	100	530	2.3	2.1	12.1	49.0	0.3
2002	243	30	97	433	1.7	2.0	13.8	85.7	0.3
2003	299	16	94	376	1.9	2.2	13.0	70.7	0.3

Source: Bonfil, personal data.

Note: SE-Standard error.

[a]nd-not detected, P analysis started only in 1997.

Nitrogen content in the soils can be divided into three different periods (1981-1989, 1990-1998, and 1999-2003) with each exhibiting a stepwise increase in soil N content. A new high-yielding wheat variety (Nirit) that is adapted to water shortage during grain filling, and a new high-yielding variety (Dariel) that shows excellence in grain production under favorable conditions were sown from 1990. These high-yielding cultivars require more nutrients, hence farmers applied more nitrogen, and part of this amount remained in the soil. The year 1999 was very dry, and crops died in most commercial fields, leaving the applied nutrients in the soil for the next season. After the highest level of N applied in 1999, we observe a continuous decline in soil N content.

Since 1994, more irrigated wheat fields and fields receiving sewage water, sludge, or city composts have been tested. These practices might explain the increase in residual nitrogen levels in Table 12.13. Soil phosphorus testing by the biological method started routinely only in 1997 and exhibits a very wide range within each year, but no tendency is seen over the years. Therefore, it appears that the soil P content is reflecting the variability in soil management methods as used by specific farmers. The average wheat grain yield in Israel is about 4 t ha^{-1}, and thus requires only 100 kg N and 12 kg P per ha. The highest yielding fields yield about 8 t ha^{-1}. Therefore, the high residual soil N levels (Table 12.13) suggest that excess nutrients are accumulating in the soil especially where sewage, sludge, or city compost applications are used by the farmers. The rising residual nutrient levels in the soil should raise environmental as well as agronomic concerns.

Integrated Nutrient Management (INM)

Insufficient N application results in reduced wheat yield and profit, compared with those obtained from a properly fertilized wheat crop. On the other hand, wheat plants receiving excessive N become susceptible to disease and lodging, resulting in reduced quantity and quality of grain yield. This susceptibility is more pronounced under soil water shortage. The established market adjustments for wheat are based on protein content and test weight, with premiums commonly paid for exceeding the baseline levels, and penalties imposed for falling below them. Efficient use of N fertilizer is important for economical wheat production; it is also important for maintaining the quality of ground and surface waters, since the potential for nitrate enrichment of ground and surface waters also increases with excessive N fertilization. Environmental constraints and economics are forcing farmers to be increasingly precise in determining the rate and timing of nitrogen fertilizer application to crops. Wheat producers in Israel typically use three options for applying N fertilizer: (1) all of the fertilizer applied in the fall, before sowing; (2) some applied in the fall, followed by a mid-winter or early spring topdressing; and (3) the full quota of N applied in the fall, before sowing, followed by topdressing application according to growing conditions in February. Although pre-sowing fertilizer applications decrease the potential for nutrient deficiencies in the early stages of growth, the presence of plant-available residual soil N from the previous season may pose a risk to the environment (leaching), or create growth problems. Excessive pre-sowing applications of N encourage vegetative growth and, therefore, the crop utilizes too much water during the vegetative growth stage (Moore and Tyndale-Biscoe, 1999), leaving insufficient water during the grain filling stage, which results in poor-quality grain (Palta et al., 1994; Bonfil et al., 1999; Bonfil, unpublished data). Late-season N applications, before or immediately following flowering, enhanced the grain protein content by 0.5-2 percent (Bonfil, Mufradi et al., 2004).

Phosphorus application strategy to rainfed wheat differs in two main parameters. First, only a small quantity of P fertilizer is needed annually and second, soil type will affect buffer capacity and phosphate fixation. Considering these factors, farmers in Israel use two options for applying P fertilizer: (1) all applied once every three-four years, usually to the legume crop in the rotation before sowing; (2) small amount applied annually according to bio-assay or Olsen's soil test. Solid and liquid forms are used in the first and second options.

Technical Requirements for INM

In the Negev region, pre-sowing soil tests are usually conducted to determine optimal quantities of fertilizer for wheat. An accurate bioassay was developed to determine soil nitrogen availability (Amir and Ephrat, 1971; Amir et al., 1994; Bonfil et al., 2001), and this bioassay has been commercially used since 1980. Soils are sampled during the summer, by a mechanical auger with 15 cm diameter at 6-10 points per 30-100 ha. The samples are taken to 0-30, 0-45, or 0-60 cm depth according to the average rainfall penetration depth in the fields. From a bore hole sample, 8 kg of air dry soil is used in the summer before wheat seeding to grow maize in the bio-assay tests inside a net house. One month after emergence, the bio-assay plants are harvested and analyzed for total NPK content. Available N, P, and K are calculated (Amir et al., 1994; Bonfil et al., 2001).

Water deficiency as a growth limiting factor is much more critical to grain yield than N stress, especially in the Mediterranean region (Papastyliano and Puckridge, 1981; Borghi et al., 1997). Therefore, it is essential for the management of grain protein in the Mediterranean-type environment that decisions and expenditure on nitrogen fertilizer applications be delayed until later in the season, when climatic conditions and yield potential are clearer. The rain expectation in February is the basis for farmers' decisions on the immediate future of the crop—grain or hay. Once this decision is made, N topdressing at heading becomes a subjective decision, but critical to the potential revenue. A novel decision support system was recently developed (Bonfil, Karnieli et al., 2004) to help wheat producers in the Mediterranean areas make more intelligent and informed decisions about their crop management processes.

Precise fertilization can be achieved only by integrating the data of residual nutrients in soil, rainfall and its distribution, growth condition, and crop management. The permanent long-term experiment at Gilat tests rainfed wheat grown under various crop management systems: different soil tillage and mulching regimes (conventional-CT, minimum-MT, and no tillage-NT), crop rotation (continued wheat-CW, wheat after clean fallow-WF), and fertilization with N (0, 50, 100, or 150 kg ha^{-1}) and phosphorus (0 or 10 kg ha^{-1}). A clear interaction between all these factors is shown in Table 12.14. In 2002, wheat that grew on clean fallow and received P, required only 50 kg ha^{-1} N to reach the highest income, while continuous wheat needed more N. No-tillage management increased yield production (Bonfil et al., 1999), therefore NT shows the highest response to fertilization. In the rainy season (2003, 299 mm), four out of five management systems showed that the highest grain yield value was received by the highest fertilization dose, while in a lower rainfall season (2002, 243 mm), only two management systems showed this pattern. In dry seasons, the negative effects of nutrient application increase and the driest 2000 season showed a dramatic income decrease in parallel to fertilization application. In the 2000 season, the rotation with clean fallow did not improve grain yield as in other seasons, since the previous year (1999) received only 72 mm of rainfall that was not retained in the soil. It is clear from Table 12.14 that only balanced fertilization, which takes into account rotation and tillage, as well as N and P interactions can lead to efficient nutrition application to rainfed wheat in dry land climates.

Actual Implementation and INM

Due to the low response of wheat to additions of K on the loess soils of the Negev region, farmers do not analyze for soil K, and do not apply K fertilizers to field crops. Actual recommendations are based on the bioassay results. Nitrogen (N_{fer}) and phosphorus (P_{fer}) application is done by Equation 12.1,

$$N_{fer} = a(GY \times 2.5\% - N_{bio}), P_{fer} = a(GY \times 0.3\% - P_{bio}) \qquad (12.1)$$

where: GY is the expected grain yield (usually farmers expect a yield of 10 kg per ha grain for each mm of rainfall); N_{bio} and P_{bio} are bioassay results that estimate the available residual N and P; a represents the fertilization efficiency coefficient (2 for rainfed fields, 1.3-1.5 for irrigated fields). Aqueous ammonia and prilled urea, and solid and liquid N-P, or triple-superphosphate are the most commonly used fertilizers for basal application. In the Negev

TABLE 12.14. Grain yield value of wheat grown in the permanent long-term experiment at Gilat as affected by crop management during four seasons

Year	Crop management[a]	N_0P_0	$N_{50}P_0$ ($ ha^{-1})	$N_{100}P_0$	$N_{150}P_0$	N_0P_{10}	$N_{50}P_{10}$	$N_{100}P_{10}$	$N_{150}P_{10}$
1998	WF-CT	347.3	293.5	263.3	270.9	373.7	346.2	357.9	310.9
1998	WF-NT	338.3	241.4	239.8	278.2	398.0	398.2	338.1	298.7
1998	CW-CT	144.2	154.6	109.4	115.3	130.3	255.5	223.0	197.1
1998	CW-MT	135.8	221.1	210.6	166.8	129.4	270.0	231.9	194.4
1998	CW-NT	178.0	239.9	148.8	183.3	138.8	293.9	231.7	208.4
2000	WF-CT	119.4	99.7	106.0	84.8	123.6	110.5	119.6	71.2
2000	WF-NT	183.6	136.3	130.2	169.2	166.5	148.4	149.1	127.6
2000	CW-CT	108.2	98.8	79.5	66.0	115.4	112.6	105.3	111.7
2000	CW-MT	100.5	87.3	92.2	107.9	143.3	132.3	157.2	123.1
2000	CW-NT	166.4	191.0	138.8	160.3	191.5	213.1	190.1	174.4
2002	WF-CT	347.7	327.4	368.0	363.1	377.6	490.7	457.4	484.4
2002	WF-NT	413.2	444.3	439.0	461.1	425.8	548.7	499.0	497.2
2002	CW-CT	87.8	206.6	147.8	159.3	105.0	231.6	299.1	291.7
2002	CW-MT	104.5	169.7	213.1	241.2	113.7	287.4	357.1	386.2
2002	CW-NT	131.7	280.0	318.5	313.6	168.6	301.5	325.0	423.2
2003	WF-CT	236.2	408.3	397.6	393.2	299.1	506.0	437.5	460.6

TABLE 12.14. (continued)

Year	Crop management[a]	N_0P_0	$N_{50}P_0$ ($ ha^{-1})	$N_{100}P_0$	$N_{150}P_0$	N_0P_{10}	$N_{50}P_{10}$	$N_{100}P_{10}$	$N_{150}P_{10}$
2003	WF-NT	372.2	524.6	470.2	490.4	318.5	616.0	549.2	649.9
2003	CW-CT	98.7	180.5	129.0	183.3	148.5	338.6	285.7	381.9
2003	CW-MT	125.4	249.1	202.1	231.6	96.5	368.7	351.0	456.7
2003	CW-NT	109.5	278.6	266.2	241.5	119.8	345.9	415.7	537.8

Note: Annual rainfall was 222, 158, 243, and 299 mm for the 1998, 2000, 2002, and 2003 seasons respectively. In each season, crop management was found to be statistically interacting with nutrition treatment.

[a]Crop management: continued wheat—CW; wheat after clean fallow—WF; conventional tillage—CT; minimum tillage—MT; no tillage—NT.

region, farmers usually apply only the base amount, before seedbed preparation in tilled fields or before sowing in NT fields, without any topdressing application. In the northern areas (Figure 12.1), farmers usually apply topdressing of 30-50 kg N ha^{-1} before the stem elongation stage. The most common fertilizers for that application are solid and liquid urea or urea ammonium nitrate formulation. Some farmers combine this with herbicide application. Since the wheat flag leaf is very sensitive to damage from foliar N application, late fertilization that applies N at heading is done only with solid or liquid urea. Only about 10 percent of the fields received this late application (usually 30-50 kg N ha^{-1}).

Phosphorus is usually applied only once in two-three years. Therefore, in a crop rotation of two years cereals and one year legume (or another crop), producers apply the whole P fertilizer to the legume crop, while the wheat grows on residual P in the soil. Other organic matter is usually not applied to wheat, but wheat can benefit from it if the wheat follows potato or carrot.

Discussion of Cases

A main environmental problem is the question of what to do with the increasing supply of urban waste. An increasing number of farmers have started to test the option of applying sludge or city composts to their fields. In this, they try to build up mulching, which improves crop growth, similar to the natural mulch of the no-tillage management. This applied mulch builds up organic matter on the surface and in the soil, too. Other fields receive recycled and sewage water, which replaces irrigation with fresh water. Both procedures decrease urban waste accumulation. However, it must be considered that with application of these wastes, farmers apply a high level of mineral nutrients. By this application, farmers lose their option to control crop growth by management of the nutrient level. Moreover, high residual nutrient levels (Table 12.13) increase the environmental constraints since the potential for nitrate enrichment of ground and surface waters also increases.

In semiarid regions, with 200-250 mm of annual rainfall, crop management is very important to maintain economically viable rainfed agriculture. No-tillage management is one of the promising practices which enable rainfed agriculture. However, control of weeds and insects in these NT fields is more problematic. On the other hand, clean fallow, and especially NT fallow, stores water in the deep soil from the fallow season for the subsequent growing season (Bonfil et al., 1999). During the fallow year, residual organic matter undergoes microbial degradation and the available N and P levels, as determined by the bioassay, and by nutrient uptake by the wheat, are higher than after a wheat crop. As clean fallow increases soil nutrient level and grain yield, and reduces pest problems, it is considered an attractive management option for rainfed wheat in the marginal cropping area. The combination of all these benefits enabled high income even without any fertilization for thirty years (N_0, P_0 since 1975) or with low balanced application (Table 12.14, WF-NT).

Research Gaps and Future Research Needs

The precise bioassay that was found reliable for estimating the NPK availability to wheat can be improved. The next step would test bioassay calibration for other field crops. At present, the last validation done in bioassay adoption was for estimating available P for pea. Opening the option to use the bioassay results for more field crops will assist fertilization decision making, and increase nutrient application efficiency.

Site-Specific Nitrogen Management

Precision N management is a relatively new practice that combines navigation and variable-rate application technologies to place fertilizers according to spatial patterns in soil productivity within fields. Variable-rate N application to meet plant needs as they vary spatially across the landscape promises to accommodate spatial variability in production potential within fields (Raun and Johnston, 1999). Recent studies have demonstrated that such N management is economically prudent for dryland cropping (Long et al., 2000). At present, research is being done in the Negev region to adopt the bioassay results to site-specific with variable-rate N base application.

Other studies have shown that it is critical to apply N so that it is available to the crop at the time when the crop needs it (Wuest and Cassman, 1992). As the most active period of N uptake in wheat is during the heading growth stage, in-season foliar-applied N effectively ensures adequate N for crop production and reduces the potential for N loss (Solie et al., 1996). A simple decision support system that allows decisions on nitrogen fertilizer applications to be delayed until heading, when climatic conditions and yield potential are clearer, is now at the last validation stage (Bonfil et al., 2004). Using this strategy would improve N application and grain protein quality, and would decrease the amount of shriveled grain produced. All factors together are needed to improve the harvest of quality grains. The next stage will develop the technology to make decisions based on multispectral or hyperspectral data from remote sensing. A combination of spatially variable and in-season foliar N applications will serve as a "keystone" for increasing the consistency of high quality wheat and promoting efficient use of N fertilizer for cereal production.

REFERENCES

Acevedo, E., P. Silva, H. Silva, and B. Solar. (1999). Wheat production in Mediterranean environments. In Satorre, E.H. and G.A. Slafer (eds.) *Wheat: Ecology and Physiology of Yield Determination.* pp. 295-331. Binghamton, NY: The Haworth Press.

Amir, J. and J. Ephrat. (1971). A biological method for evaluating soil nitrogen availability and forecasting nitrogen fertilizer needs of wheat. *Agronomy Journal.* 63: 385-388.

Amir, J., J. Krikun, D. Orion, J. Putter, and S. Klitman. (1991). Wheat production in an arid environment. 1. Water-use efficiency, as affected by management practices. *Field Crops Research.* 27: 351-364.

Amir, J., I. Mufradi, S. Klitman, and S. Asido. (1994). Long-term comparative study of soil nitrate test, Gilat plant indicator method and wheat nitrogen uptake. *Plant and Soil.* 158: 223-231.

Bar-Yosef, B. and U. Kafkafi. (1972). Rates of growth and nutrient uptake of irrigated corn as affected by N and P fertilization. *Soil Science Society of America Journal* 36: 931-936.

Ben-Dor, E., N. Goldshalager, O. Braun, B. Kindel, A.F.H. Goetz, D. Bonfil, N. Margalit, Y. Binaymini, A. Karnieli, and M. Agassi. (2004). Monitoring of infiltration rate in semiarid soils using airborne hyperspectral technology. *International Journal of Remote Sensing.* 25: 2607-2624.

Bonfil, D.J. (2001). Potassium effects on plant growth, development, yield and quality of wheat. *Proc. Potassium in Nutrient Management for Sustainable Crop Production in India.* pp. 275-283. New Delhi, India.

Bonfil, D.J., A. Karnieli, M. Raz, I. Mufradi, S. Asido, H. Egozi, A. Hoffman, and Z. Schmilovitch. (2004). Decision support system for improving wheat grain quality in the Mediterranean area of Israel. *Field Crops Research.* 89: 153-163.

Bonfil, D.J., I. Mufradi, and S. Asido. (2004). Decision support system for improving wheat quality in semi-arid regions. In *7th International Conference on Precision Agriculture and Other Precision Resources Management.* CD-ROM. pp. 933-944. Madison, WI: ASA-CSSA-SSSA.

Bonfil, D.J., I. Mufradi, S. Asido, and B. Dolgin. (2001). Bioassay to improve the determination of phosphorus and potassium availability in soil for fertilization recommendations. In Horst W.J. et al. (eds.) *Plant Nutrition—Food Security and Sustainability of Agro-ecosystems.* pp. 696-697. The Netherlands: Kluwer Academic Press.

Bonfil, D.J., I. Mufradi, S. Klitman, and S. Asido. (1999). Wheat grain yield and soil profile water distribution in a no-till arid environment. *Agronomy Journal.* 91: 368-373.

Borghi, B., M. Corbellini, C. Palumbo, N. DiFonzo, and M. Perenzin. (1997). Effects of Mediterranean climate on wheat bread-making quality. *European Journal of Agronomy.* 6: 145-154.

Clawson, M., H.H. Landsberg, and L.S. Alexander (eds.). (1971). *The Agricultural Potential of the Middle East.* pp. 1-46. New York: Am. Elsevier Pub. Co, Inc.

Dan, Y., Z. Raz, D.H. Yaalon, and H. Koyumdjisky. (1975). *Soil Map of Israel.* Volcani Center BD Israel: Agriculture Research organization, Ministry of Agriculture.

FAOSTAT data. (2005). http://faostat.fao.org/ (last accessed November 2005).

Feigin, A. and J. Halevy. (1994). Effect of crops fertilization regimes on the leaching of solutes in an irrigated soil. In Adriano, D.C., A.K. Iskander, and I.P. Murarka (eds.) *Contamination of Ground Water. Advances in Environmental Science Reviews.* pp. 367-393. UK: Northwood.

Fine, P. and U. Mingelgrin. (1996). Release of phosphorus from waste-activated sludge. *Soil Science Society of America Journal.* 60: 505-511.

Han, F.X. and A. Banin. (2001). Fractional loading isotherm of heavy metals in an arid-zone soil. *Communications in Soil Science and Plant Analysis.* 32: 2691-2708.

Henkin, Z., I. Noy-Meir, U. Kafkafi, M. Gutman, and N. Seligman. (1996). Phosphate fertilization primes production of rangeland on brown rendzina soils in the Galilee, Israel. *Agriculture, Ecosystems and Environment.* 59: 43-53.

Henkin, Z., N. Seligman, I. Noy-Meir, U. Kafkafi, and M. Gutman. (1998). Rehabilitation of Mediterranean dwarf-shrub rangeland with herbicides, fertilizers and fire. *Journal of Range Management.* 51: 193-199.

Kafkafi, U. (1994). Combined irrigation and fertilization in arid zones. *Israel Journal of Plant Sciences.* 42: 301-320.

Kafkafi, U. and B. Bar-Yosef. (1971). *Long term effects of cumulative application of fertilizers on irrigated soil and crops.* Final report to the ford foundation Project A-III/6, Volcani Institute of Agricultural Research, Bet Dagan, Israel.

Kafkafi, U. and J. Halevy. (1974). Rates of growth and nutrients consumption of semi-dwarf wheat. *Trans. 10th Int. Congress Soil Science, Moscow.* 4: 137-143.

Long, D., R. Engel, and G. Carlson. (2000). Method for precision nitrogen management in spring wheat: II. Implementation. *Precision Agriculture.* 2: 25-38.

López-Bellido, L., M. Fuentes, J.E. Castillo, and F.J. López-Garrido. (1998). Effects of tillage, crop rotation and nitrogen fertilization on wheat-grain quality grown under rainfed Mediterranean conditions. *Field Crops Research.* 57: 265-276.

Mazid, A. (1994). *Factors influencing adoption of new agricultural technology in dry areas of Syria,* PhD thesis University of Nottingham, UK.

Moore, G.A. and J.P. Tyndale-Biscoe. (1999). Estimation of the importance of spatially variable nitrogen application and soil moisture holding capacity to wheat production. *Precision Agriculture.* 1: 27-38.

Nahal, I. (1981). The Mediterranean climate from a biological viewpoint. In Di Castri, F., D.W. Goodall, and R.L. Specht (eds.) *Mediterranean-Type Shrublands.* pp. 63-86. Amsterdam, New York: Elsevier Scientific Pub. Co.

Palta, J.A., T. Kobata, N.C. Turner, and I.R. Fillerg. (1994). Remobilization of carbon and nitrogen in wheat as influenced by postanthesis water deficits. *Crop Science.* 34: 118-124.

Papastyliano, I. and D.W. Puckridge. (1981). Nitrogen nutrition of cereals in a short-term rotation. II. Stem nitrate as an indicator of nitrogen availability. *Australian Journal of Agricultural Research.* 32: 713-723.

Raun, W. and G. Johnston. (1999). Improving nitrogen use efficiency for cereal production. *Agronomy Journal*. 91: 357-363.

Solie, J., W. Raun, R. Whitney, M. Stone, and J. Ringer. (1996). Optical sensor based field element size and sensing strategy for nitrogen application. *Transactions of the ASAE*. 39(6): 1983-1992.

Stanhill, G., U. Kafkafi, M. Fuchs, and Y. Kagan. (1972). The effect of fertilizer application on solar reflectance from a wheat crop, Israel. *Australian Journal of Agricultural Research*. 22(2): 109-118.

Whitaker, M. (1990). *Nitrogen fertilizer strategies for rainfed wheat in Northern Syria*. Unpublished PhD thesis. Food Research Institute. Stanford University, USA.

Wuest, S.B. and K.G. Cassman. (1992). Fertilizer nitrogen use efficiency of irrigated wheat: I. Uptake efficiency of preplant versus late-season application. *Agronomy Journal*. 84: 682-688.

INTERNET REFERENCES

ftp://ftp.fao.org/docrep/fao/005/y4732E/y4732E00.pdf (Accessed May 18, 2007).

http://eesc.orst.edu/agcomwebfile/edmat/html/ec/ec1094/ec1094.html (Accessed May 18, 2007).

http://www.ext.nodak.edu/extpubs/ansci/beef/as1238w.htm (Accessed May 18, 2007).

http://www.fertilizer.org/ifa/publicat/html/pubman/soybean.htm (Accessed May 18, 2007).

Index

Page numbers followed by the letter "f" indicate figures and graphs; those followed by the letter "t" indicate tables.

T - #0298 - 071024 - C5 - 229/152/28 - PB - 9780367387730 - Gloss Lamination